Chemistry in Quantitative Language

CHEMISTRY IN QUANTITATIVE LANGUAGE

Fundamentals of General Chemistry Calculations

Christopher O. Oriakhi, CChem, FRSC

OXFORD
UNIVERSITY PRESS

Great Clarendon Street, Oxford, OX2 6DP,
United Kingdom

Oxford University Press is a department of the University of Oxford.
It furthers the University's objective of excellence in research, scholarship,
and education by publishing worldwide. Oxford is a registered trade mark of
Oxford University Press in the UK and in certain other countries

First Edition published in 2009
Second Edition published in 2021
Impression: 1

Published in the United States of America by Oxford University Press
198 Madison Avenue, New York, NY 10016, United States of America

British Library Cataloguing in Publication Data
Data available

Library of Congress Control Number: 2021937954

ISBN 978–0–19–886778–4

DOI: 10.1093/oso/9780198867784.001.0001

Printed and bound by
CPI Group (UK) Ltd, Croydon, CR0 4YY

Preface

This is the second edition of *Chemistry in Quantitative Language*. I am pleased that the first edition, published in 2009, was so well received by both instructors and students. However, as is typical with most books in their first edition, the users graciously provided several constructive suggestions to make the book more useful for the intended audience. In addition, on my part, a decade later I find the need to add a few topics and concepts that were left out of the first edition, as well as the need to reorganize the chapters and update the worked examples and end-of-chapter problems to make the book more useful. All of these have armed me with the motivation and insights necessary to prepare this second edition. Formulating and solving quantitative chemistry problems remains a concern for many students. This edition shares the same objective of making problem-solving in chemistry more pleasurable and exciting, so students can hopefully face their examinations with ease and pass with flying colors.

This second edition has been revised extensively and incorporates many minor changes than can be included here. The most significant changes include:

- The original first two chapters (Essential Mathematics and System of Measurement) have been removed and made available in the Appendix.
- I have added principal new materials broken down into two new chapters (Chemical Bonding 1: Basic Concepts, and Chemical Bonding 2: Modern Theories of Chemical Bonding) as these are foundational college chemistry materials; the latter chapter deals specifically with molecular shapes and structure.
- Throughout the book I have added new problems with varying degree of difficulties, while others have been combined, revised, or replaced.

Throughout the book I have adhered to the philosophy behind the first edition, which was to introduce the concepts of the underlying chemistry followed by a series of worked example to show students how the calculations are done and to hopefully enable them to appreciate what is involved. At the end of each chapter are a series of problems that range from very simple to those requiring more time and energy. Each problem in the book has been solved and checked independently by Ted LaPage. Answers for all the problems are in the Solutions to Problems chapter at the end of the book.

If you have used this book and found any errors or have comments, suggestions, and recommendations for future revisions and additions, please feel free to reach out to me by email at oriakhi@gmail.com or by contacting Oxford University Press.

Acknowledgement

Preparation of the second edition of Chemistry in Quantitative Language benefited greatly from the immense support and encouragement of my family and some amazing friends, colleagues, and students.

I begin by expressing my sincere appreciation to Ted LaPage for his indispensable contributions to this edition. In addition to his great sense of humor, dedication, and friendship, Ted read the entire manuscript, solved the problems, and suggested many improvements in both content and style.

Suggestions and constructive feedback from students and faculty who have used the book are particularly helpful and deeply appreciated. Many of your suggestions have been incorporated into this revision.

I also thank the various persons at Oxford University Press especially Katherine Ward, Sonke Adlung, and Harriet Konishi for their dedication and friendly support in getting this project to the finish line in a record time.

Finally, this acknowledgement would be incomplete without saying how much I appreciate the love, patience, support, and encouragement I received from my family and friends at all times. In particular, I dedicate this edition to the memory of mother, Mrs Ikhiwi Oriakhi, who inspired this work.

Contents

1

Atomic Structure and Isotopes

. . . .

1.1 Atomic Theory

In 1808, based on experimental data and chemical laws known in his day, Dalton proposed his theory of the atom. The theory states that:

1. All chemical elements are made up of tiny indivisible particles called atoms.
2. Atoms cannot be created or destroyed. Chemical reactions only rearrange the *way* that the atoms are combined.
3. Atoms of the same element are identical in all respects and have the same masses and physical and chemical properties. Atoms of different elements have different masses as well as different physical and chemical properties.
4. A combination of elements to form a compound occurs between small, whole-number ratios of atoms.

Dalton's theory resulted in the formulation of the law of conservation of mass and the law of multiple proportions. Along with the law of definite proportion and the law of reciprocal proportion, these form the basis of reaction stoichiometry (see Chapter 7).

1.1.1 The law of conservation of mass

In a chemical reaction, matter is neither created nor destroyed. The total mass of reactants is equal to the total mass of the products. This law was first stated by Lavoisier, based on his findings from combustion experiments.

1.1.2 The law of definite proportion

The law of definite proportions (also known as Proust's law or the law of constant composition) states that a given chemical compound is always composed of the same elements in the same proportions by mass. When elements combine to form a particular compound, they do so in fixed and constant proportions by mass, regardless of the source of the compound. For example, a sample of pure water, regardless of its source, always contains 11.1% hydrogen and 88.9% oxygen.

Chemistry in Quantitative Language: Fundamentals of General Chemistry Calculations. Second Edition. Christopher O. Oriakhi, Oxford University Press. © Christopher O. Oriakhi 2021. DOI: 10.1093/oso/9780198867784.003.0001

1.1.3 The law of multiple proportions

If two elements can form more than one compound between them, the masses of one element, which combine with a fixed mass of the second element, are in a ratio of whole numbers. For example, nitrogen and oxygen combine to form compounds such as NO, NO_2, and N_2O, in which the ratio of nitrogen atoms per oxygen atom is respectively 1, ½, and 2. This is one of the basic laws of stoichiometry, as we discuss in Chapter 7.

1.1.4 The law of reciprocal proportions

According to the Law of Reciprocal Proportions (also known as the law of equivalent proportions or the law of permanent ratios), if two different elements, A and B, chemically combine separately with a fixed mass of a third element, D, the ratio of the masses in which they do so are either the same as or a simple multiple (or simple fraction) of the ratio of the masses in which A and B combine with each other. For example, carbon reacts with sulfur to form carbon disulfide (CS_2) and with oxygen to form carbon dioxide (CO_2). Sulfur and oxygen, meanwhile, combine to form sulfur dioxide (SO_2), in which the ratio of sulfur to oxygen is 1:2, which is ½ the 1:1 ratio of sulfur to oxygen in the carbon compounds.

1.2 The Structure of the Atom

An atom consists of a central nucleus, which contains roughly 99.9% of the total mass of the atom, and a surrounding cloud of electrons. The nucleus is composed of two kinds of particles, protons and neutrons, which are collectively known as nucleons. The proton is positively charged, while the neutron is electrically neutral. The electrons have a negative charge and surround the nucleus in "shells" of definite energy levels. (Note: energy level is discussed in Chapter 8.) In a neutral (unreacted) atom, the number of electrons equals the number of protons, so the atom has a charge of zero. It must be mentioned that the chemistry of a given atom comes from its electrons; all chemical changes take place entirely with regard to the electrons—the nucleus is never affected by chemical reactions. The properties of the three sub-atomic particles are summarized in Table 1.1.

1.2.1 Atomic number (Z)

The atomic number of an element is defined as the number of protons in the nucleus of an atom of the element. It is also equal to the number of electrons in a neutral atom. Atomic number is a characteristic of a given element and determines its chemical properties.

Table 1.1   Properties of subatomic particles

Fundamental particle	Symbol	Location	Mass (g)	Mass (amu)	Charge
Electron	e	Outside the nucleus	9.110×10^{-28}	0.00549	−1
Proton	p	Inside the nucleus	1.673×10^{-24}	1.0073	1
Neutron	n	Inside the nucleus	1.675×10^{-24}	1.0087	0

1.2.2 Mass number (A)

The mass number of an element is the sum of protons and neutrons in the nucleus of the atom.

$$\text{Mass number (A)} = \text{No. of protons (Z)} + \text{No. of neutrons (N)}$$
$$A = Z + N$$

The general symbol for an element, showing its mass number and atomic number, is:

$$^A_Z E$$

A = mass number in atomic mass units (amu)

Z = atomic number

E = symbol of the element as shown on the periodic table

For example, the symbols for C and Mg, showing their mass and atomic numbers, are:

$$^{12}_6 C \quad \text{and} \quad ^{24}_{12} Mg$$

1.2.3 Ions

An ion is an atom or group of atoms that has gained or lost electron(s). A positively charged ion results when an atom loses one or more electrons. Conversely, a negatively charged ion is formed when an atom gains one or more electrons. For example, a sodium atom, Na, loses one electron to form a sodium ion, Na^+; a nitrogen atom, N, gains three electrons to form a nitrogen (or nitride) ion, N^{3-}.

Let us try to determine the number of protons, electrons, and neutrons in the Ca^{2+} ion. A neutral Ca atom has twenty electrons and twenty protons (atomic number = 20). A Ca^{2+} ion however, has twenty protons and eighteen electrons (for a net charge of $+2$). The mass number, which is the sum of protons and neutrons, is forty; since the ion has twenty protons, it must have $40 - 20 = 20$ neutrons.

Example 1.1

Determine the number of protons, neutrons, and electrons in

(a) $^{23}_{11}Na$ (b) $^{16}_8 O^{2-}$ (c) $^{56}_{26}Fe^{3+}$ (d) $^{63}_{29}Cu$

Solution

The solutions are summarized in the following table.

	Symbol	Protons	Neutrons	Electrons
(a)	$^{23}_{11}Na$	11	12	11
(b)	$^{16}_8 O^{2-}$	8	8	10
(c)	$^{56}_{26}Fe^{3+}$	26	30	23
(d)	$^{63}_{29}Cu$	29	34	29

Example 1.2

Complete the following table.

Particle	Atomic number	Mass number	protons	Number of neutrons	electrons	Net charge	Symbol $E^{n\pm}$
A			25	30	25		Mn
B	13			14			Al^{3+}
C	35	80			36		
D	15	31			18		
E			82	125	80		
F		39	19			0	K
G	9			10		−1	
H		56	26		23		

Solution

To complete this table, you need to recall the basic definitions and the general equation: Mass number (A) = No. of protons (Z) + No. of neutrons (N). Use the periodic table to identify the corresponding symbols. Also remember that atomic number is equal to the number of protons in the nucleus of an atom. In a neutral atom (zero charge), the number of electrons is always equal to the number of protons. The difference between the number of protons and the number of electrons is the charge on the atomic particle. If the number of protons is greater than the number of electrons, the particle takes on a net positive charge, and vice-versa. The complete table is

Particle	Atomic number	Mass number	protons	Number of neutrons	electrons	Net charge	Symbol $E^{n\pm}$
A	25	55	25	30	25	0	Mn
B	13	27	13	14	10	3	Al^{3+}
C	35	80	35	45	36	−1	Br^-
D	15	31	15	16	18	−3	P^{3-}
E	82	207	82	125	80	2	Pb^{2+}
F	19	39	19	20	19	0	K
G	9	19	9	10	10	−1	F^-
H	26	56	26	30	23	3	Fe^{3+}

1.3 Isotopes

Isotopes are atoms that have the same atomic number, and hence are the same element, but that have different mass numbers (that is, different numbers of neutrons). The difference in mass is due to the variation in the neutron number. Isotopes have practically the same chemical properties but differ slightly in physical properties. Most elements exhibit isotopy. For example, hydrogen has three isotopes: hydrogen, deuterium, and tritium. The atomic numbers and mass numbers are shown in Table 1.2.

Table 1.2 The hydrogen isotopes

	Hydrogen (H)	Deuterium (D)	Tritium (D)
Atomic number	1	1	1
Mass number	1	2	3

Example 1.3

Which of the following are isotopes?

a) $^{18}_{9}F$ and $^{18}_{10}Ne$

b) $^{2}_{1}H$ and $^{3}_{1}H$

c) $^{40}_{19}K$ and $^{40}_{20}Ca$

d) $^{24}_{12}Mg$ and $^{25}_{12}Mg$

Solution

Isotopes are atoms with the same atomic number but different mass number. Here only (b) and (d) are isotopes.

1.4 Relative Atomic Mass

The relative mass of an atom is called atomic mass (older textbooks call it "atomic weight") and is characteristic of each element. The mass of an individual atom is difficult to measure, but the relative masses of the atoms of different elements can be measured, and this determines the atomic mass.

The masses of individual atoms and molecules are exceedingly small and cannot be conveniently expressed in grams or kilograms. For example, one atom of Ca has a mass of 6.64×10^{-23} g. For this reason, scientists instead use a relative atomic mass scale. Scientists have chosen the carbon-12 isotope as a reference, and its mass is defined as exactly twelve atomic mass units (amu). Therefore one atomic mass unit is equal to $\frac{1}{12}$ of the mass of the carbon-12 atom. One amu has been determined experimentally to be 1.66×10^{-24} g.

The relative atomic mass of an element is defined as the mass of an atom of that element compared with $\frac{1}{12}$ of the mass of the carbon-12 isotope.

The atomic masses of elements given in the periodic table are not the same as mass number. In nature, most elements occur as a mixture of isotopes. Therefore, the atomic mass of an element is an average mass based on the abundance of the isotopes.

1.4.1 Calculating atomic masses

The information needed to calculate the atomic masses of elements includes: the number of isotopes for the element, the relative mass of each isotope on the carbon-12 scale, and the percent abundance of each isotope.

The atomic mass of a given element is then obtained as the sum of the product of the exact mass of each isotope and its percent abundance. For example, naturally occurring carbon consists of a mixture of 98.89% carbon-12 ($^{12}_6C$) with a mass of 12.00000 amu, and 1.11% carbon-13 ($^{13}_6C$) with a mass of 13.00335 amu. The atomic mass is obtained as follows:

$$\text{Atomic mass} = (12.00000 \text{ amu}) \left(\frac{98.89}{100} \right) + (13.00335 \text{ amu}) \left(\frac{1.11}{100} \right) = 12.011 \text{ amu}$$

An instrument called the mass spectrometer measures the exact mass of each isotope and its relative abundance.

Example 1.4

Calculate the average atomic mass of lithium, which is a mixture of 7.5% ^{6}Li and 92.5% ^{7}Li. The exact masses of these isotopes are 6.01 amu and 7.02 amu.
Solution

Multiplying the abundance of each isotope by its atomic mass and then adding these products gives the average atomic mass.

$$\text{Average atomic mass of Li} = (6.01 \text{ amu}) \left(\frac{7.5}{100} \right) + (7.02 \text{ amu}) \left(\frac{92.5}{100} \right) = 6.94 \text{ amu}$$

Example 1.5

Boron, B, with an atomic mass of 10.81, is composed of two isotopes, ^{10}B and ^{11}B, weighing 10.01294 and 11.00931 amu, respectively. What is the fraction and percentage of each isotope in the mixture?

Solution

Let the abundance of ^{10}B be X and that of ^{11}B be $1 - X$ since there are only two isotopes of boron. Therefore,

$$\text{Atomic mass} = 10.81 = (X)(10.01294) + (1 - X)(11.00931)$$

Now solve for X:

$$10.81 = 10.01294 \text{ X} + 11.0093 - 11.0093 \text{ X}$$
$$10.0129 \text{ X} - 11.0093 \text{ X} = 10.81 - 11.0093$$
$$-0.9964 \text{ X} = -0.1993$$
$$X = 0.20 \text{ or } 20\% \text{ and } 1 - X = 0.80 \text{ or } 80\%$$

The fraction and % of ^{10}B is 0.20 or 20%. Also, the fraction of ^{11}B is 0.80 or 80%.

1.5 Problems

1. How many protons, neutrons, and electrons are in each of the following?
   (a) $^{197}_{79}Au$ (b) $^{10}_{5}B$ (c) $^{40}_{20}Ca$ (d) $^{163}_{66}Dy$

2. How many protons, neutrons, and electrons are in each of the following?
   (a) $^{84}_{36}Kr$ (b) $^{24}_{12}Mg$ (c) $^{69}_{31}Ga$ (d) $^{75}_{33}As$

3. Find the number of protons, neutrons, and electrons in the following ions:
   (a) $^{59}_{27}Co^{2+}$ (b) $^{24}_{12}Mg^{2+}$ (c) $^{69}_{31}Ga^{3+}$ (d) $^{118}_{50}Sn^{2+}$

4. Identify the following atoms or ions.

   (a) A halogen (an element in group VII of the periodic table) anion with thirty-six electrons and thirty-five protons.
   (b) An alkali metal (an element in group I of the periodic table) cation with fifty-five protons and fifty-four electrons.
   (c) A transition metal cation with twenty-five protons and twenty-three electrons. (Note: Transition elements are those elements embedded between groups II and III in the periodic table).

5. Complete the following table.

Symbol	$^{40}_{20}C\,a^{2+}$	$^{79}_{34}S\,e^{2-}$	$^{137}_{56}B\,a^{2+}$
Mass number		238	15
Protons	15		
Neutrons	16	146	8
Electrons		88	
Net charge	-3		-3

6. With the aid of a periodic table, identify the following elements:
   (a) $^{56}_{26}X$; (b) $^{131}_{53}X$; (c) $^{202}_{80}X$; (d) $^{19}_{9}X$

7. Fill in the blanks in the following table:

Particle	Atomic number	Mass number	protons	Number of neutrons	electrons	Net charge	Symbol
A	35	80				-1	
B			13	14		3	
C		31	15				P^{3-}
D	24			28	24		
E		35	17			0	
F	54	131				0	
G		204	81		80		
H			40	51		4	Zr^{4+}

8. Find the number of protons, neutrons, and electrons in the following ions:

   (a) $^{32}_{16}S^{2-}$ (b) $^{80}_{35}Br^-$ (c) $^{128}_{52}Te^{2-}$ (d) $^{14}_{7}N^{3-}$

9. Naturally occurring chlorine has two isotopes, which occur in the following abundance: 75.76% $^{35}_{17}Cl$, with an isotopic mass of 34.9689 amu, and 24.24% $^{37}_{17}Cl$, with a isotopic mass 36.9659 amu. Calculate the atomic mass of chlorine.

10. Boron has two naturally occurring isotopes: 80% of $^{11}_{5}B$, and 20% of $^{10}_{5}B$, which has an isotopic mass of 10.02 amu. If the atomic mass of boron is 10.81, what is the isotopic mass of $^{11}_{5}B$?

11. Naturally occurring oxygen gas consist of three isotopes. Use the information in the table to calculate the atomic mass of oxygen gas.

Isotope	Exact mass	Abundance
$^{16}_{8}O$	15.995	99.96%
$^{17}_{8}O$	16.999	0.037%
$^{18}_{8}O$	17.999	0.024%

12. Naturally occurring gallium (Ga) exists in two isotopic forms, ^{69}Ga and ^{71}Ga. The atomic mass of Ga is 69.72. What is the percentage abundance of each isotope? (The exact isotopic masses of ^{69}Ga and ^{71}Ga are 68.9259 and 70.9249, respectively).

2

Formula and Molecular Mass

. . . .

Many chemists use the terms *formula mass* and *molecular mass* interchangeably when dealing with chemical compounds of known formula. This chapter explains the slight difference between the two terms.

2.1 Formula Mass

The formula mass of a compound is the sum of the atomic masses of all the atoms in a formula unit of the compound, whether it is ionic or molecular (covalent).

The formula mass is based on the ratio of different elements in a formula, as opposed to the molecular mass, which depends on the actual number of each kind of atom (compare to Section 4.2). Formula masses are relative since they are derived from relative atomic masses. For example, the formula mass of phosphoric acid, H_3PO_4, is 97.98 atomic mass units (amu) and is obtained by adding the atomic masses (taken from the periodic table) of the elements in one formula unit (i.e. $3 H + 1 P + 4 O$).

$$(3 \times \text{At. wt. of H}) + (1 \times \text{At. wt. of P}) + (4 \times \text{At. wt. of O})$$
$$= (3 \times 1.00) + (1 \times 30.97) + (4 \times 16.00) = 97.97$$

2.2 Molecular Mass

Once the actual formula of a chemical substance is known, the molecular mass can be determined in a manner similar to that for calculating the formula mass. The *molecular mass* of a compound is the sum of the atomic masses of all the atoms in one molecule of the compound. The term applies only to compounds that exist as molecules such as H_2O, SO_2, and glucose, $C_6H_{12}O_6$. For example, the molecular mass of ethanol, C_2H_5OH, is

$$(2 \times C) + (6 \times H) + (1 \times O)$$
$$= (2 \times 12.0) + (6 \times 1.0) + (1 \times 16.0) = 46$$

Chemistry in Quantitative Language: Fundamentals of General Chemistry Calculations. Second Edition. Christopher O. Oriakhi, Oxford University Press. © Christopher O. Oriakhi 2021. DOI: 10.1093/oso/9780198867784.003.0002

When ionic compounds such as NaCl, $Zn(NO_3)_2$, and NH_4Cl are in the crystalline state or in solution, they do not contain physically distinct uncharged molecular entities. Therefore, chemists often use the term formula mass to represent the total composition of such substances.

Example 2.1

Calculate the formula mass (FM) of NaOH.

Solution

$$1 \times \text{AM of Na} = 1 \times 23 \text{ amu} = 23 \text{ amu}$$
$$1 \times \text{AM of O} = 1 \times 16 \text{ amu} = 16 \text{ amu}$$
$$1 \times \text{AM of H} = 1 \times 1 \text{ amu} = \underline{1 \text{ amu}}$$
$$\text{FM} \qquad\qquad\qquad\qquad = 40 \text{ amu}$$

The formula mass of NaOH is 40 g.

Example 2.2

Calculate the formula mass of $MgSO_4 \cdot 7H_2O$ using a table of atomic masses.

Solution

$$1 \times \text{AM of Mg} \quad = \quad 24.3 \text{ amu}$$
$$1 \times \text{AM of S} \quad = \quad 32.1 \text{ amu}$$
$$4 \times \text{AM of O} \quad = \quad 64.0 \text{ amu}$$
$$7 \times 2 \times \text{AM of H} = \quad 14.0 \text{ amu}$$
$$7 \times 1 \times \text{AM of O} = \underline{112.0 \text{ amu}}$$
$$\text{FM} \qquad\qquad = 246.4 \text{ amu}$$

The formula mass of $MgSO_4 \cdot 7H_2O$ is 246.4 amu.

2.3 Molar Mass

The term *molar mass* is now commonly used as a general term for both formula mass and molecular mass. The molar mass of any substance is the mass in grams of one mole of the substance, and it is numerically equal to its formula mass (expressed in amu). NB: the mole will be defined and explained in Chapter 3. For example, the formula mass of glucose, $C_6H_{12}O_6$, is 180.0 amu. So, the molar mass or the mass in grams of 1 mol of glucose is 180.0 g.

Let me mention that, for most purposes, you can round atomic and molecular masses to approximately one decimal place when solving problems.

Example 2.3

Calculate the formula mass and molar mass of $MgSO_4$.

Solution

1. To calculate the formula mass of $MgSO_4$, multiply the atomic mass of each component element by the number of times it appears in the formula, and then add the results.

$$1 \text{ Mg atom} = 1 \times 24.30 \text{ amu} = 24.30 \text{ amu}$$

$$1 \text{ S atom} = 1 \times 23.07 \text{ amu} = 23.07 \text{ amu}$$

$$4 \text{ O atoms} = 4 \times 16.00 \text{ amu} = \underline{64.00 \text{ amu}}$$

$$111.37 \text{ amu}$$

The formula mass of $MgSO_4$ is 111.37 amu.

2. The molar mass of a substance is defined as the mass of 1 mol, or as the formula mass of the substance expressed in grams. Thus, the molar mass of $MgSO_4$ is 111.37 g/mol.

Example 2.4

Mycomycin is an antibiotic produced naturally by the fungus *Nocardia acidophilus*. The molecular formula is $C_{13}H_{10}O_2$. What is the molar mass?

Solution

1. Multiply the atomic mass of each element by the number of times it appears in the molecular formula and then add the results.

$$13 \text{ C atoms} = 13 \times 12.0 \text{ g} = 156.0 \text{ g}$$

$$10 \text{ H atoms} = 10 \times 1.0 \text{ g} = 10.0 \text{ g}$$

$$2 \text{ O atoms} = 2 \times 16.0 \text{ g} = \underline{32.0 \text{ g}}$$

$$198.0 \text{ g}$$

The molar mass of $C_{13}H_{10}O_2$ is 198.0 g/mol.

2.4 Problems

1. Calculate the formula masses of the following:
   (a) $NaHCO_3$ (b) $CaCO_3$ (c) NaH_2PO_4 (d) $Ag_2S_2O_3$ (e) $Al_2(SO_4)_3$

2. Determine the formula or molecular mass of the following substances:
   (a) $C_{12}H_{22}O_{11}$ (b) H_2S (c) Mg_3N_2 (d) $Mg_3(BO_3)_2$ (e) Na_2SO_4

3. When hydrogen sulfide gas is bubbled into an aqueous solution of $SbCl_3$, the compound Sb_2S_3 is precipitated:

$$2\ SbCl_3\,(aq) + 3\ H_2S\,(g) \rightarrow Sb_2S_3\,(s) + 6\ HCl\,(aq)$$

Calculate the formula mass of the precipitated Sb_2S_3.

4. Hydroxyapatite, a chemical compound present in tooth enamel, has the molecular formula $Ca_{10}(PO_4)_6(OH)_2$. Calculate its molar mass in grams.

5. Calculate the formula mass of aspirin, a mild pain reliever with the formula $C_9H_8O_4$.

6. Penicillin V is an important antibiotic. Its chemical formula is $C_{10}H_{14}N_2$. What is its formula mass?

7. The chemical formula of TNT (trinitrotoluene), an industrial explosive, is $C_7H_5O_6N_3$. Calculate its molecular mass.

8. The major chemical component of eucalyptus oil, extracted from eucalyptus leaves, is eucalyptol. Its chemical formula is $C_{10}H_{18}O$. Calculate its formula mass.

9. The formula mass of the compound K_2MCl_6 is 483.14 amu. What is the atomic mass of M? Use the periodic table to identify M.

10. The formula mass of the inorganic acid $H_4X_2O_7$ is 178. What is the atomic mass of X? Use the periodic table to identify X.

3

The Mole and Avogadro's Number

. . . .

3.1 The Mole and Avogadro's Number (N_A)

A mole is defined as the amount of a given substance that contains the same number of atoms, molecules, or formula units as there are atoms in 12 g of carbon-12. For example, one mole of glucose contains the same number of glucose molecules as there are carbon atoms in 12 g of carbon-12. The number of atoms in exactly 12 g of carbon-12 has been determined to be 6.02×10^{23}; this number is called Avogadro's number (N_A). Therefore, a mole is the amount of a substance that contains Avogadro's number of atoms, ions, molecules, or particles. For example:

$$1 \text{ mol He atoms} = 6.02 \times 10^{23} \text{ atoms}$$

$$1 \text{ mol CH}_3\text{OH molecules} = 6.02 \times 10^{23} \text{ molecules}$$

$$1 \text{ mol SO}_4^{2-} \text{ ions} = 6.02 \times 10^{23} \text{ ions}$$

3.2 The Mole and Molar Mass

The term molar mass is now commonly used as a general term for both formula mass and molecular mass. The molar mass of any substance is the mass in grams of one mole of the substance, and it is numerically equal to its formula mass (expressed in amu). For example, the formula mass of glucose, $C_6H_{12}O_6$, is 180 amu. So, the molar mass or the mass in grams of 1 mol of glucose is 180.0 g.

3.3 Calculating the Number of Moles

In terms of chemical arithmetic, the mole is the most important number in chemistry. It provides useful stoichiometric information about reactants and products in any given chemical reaction. The quantities commonly encountered in chemical problems include the number of moles of a substance; the number of atoms, molecules, or formula units of a substance; and the mass in grams. These quantities are related and can be readily inter-converted with the aid of the molar mass and Avogadro's number. Calculations based on the mole can be carried out by using conversion factors, or with simple equations based on those conversion factors.

Chemistry in Quantitative Language: Fundamentals of General Chemistry Calculations. Second Edition.
Christopher O. Oriakhi, Oxford University Press. © Christopher O. Oriakhi 2021.
DOI: 10.1093/oso/9780198867784.003.0003

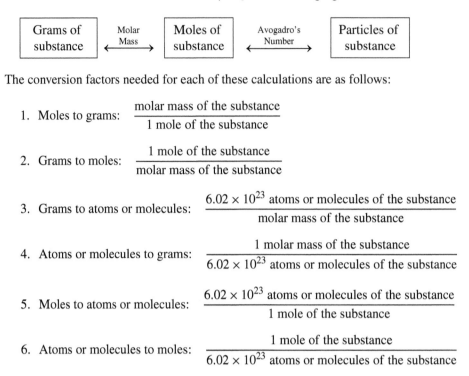

The conversion factors needed for each of these calculations are as follows:

1. Moles to grams: $\dfrac{\text{molar mass of the substance}}{\text{1 mole of the substance}}$

2. Grams to moles: $\dfrac{\text{1 mole of the substance}}{\text{molar mass of the substance}}$

3. Grams to atoms or molecules: $\dfrac{6.02 \times 10^{23} \text{ atoms or molecules of the substance}}{\text{molar mass of the substance}}$

4. Atoms or molecules to grams: $\dfrac{\text{1 molar mass of the substance}}{6.02 \times 10^{23} \text{ atoms or molecules of the substance}}$

5. Moles to atoms or molecules: $\dfrac{6.02 \times 10^{23} \text{ atoms or molecules of the substance}}{\text{1 mole of the substance}}$

6. Atoms or molecules to moles: $\dfrac{\text{1 mole of the substance}}{6.02 \times 10^{23} \text{ atoms or molecules of the substance}}$

Depending on the information provided, the number of moles of a chemical substance can be calculated from the following equations:

1. Number of moles $= \dfrac{\text{mass in grams}}{\text{molar mass}}$

2. Number of moles $= \dfrac{\text{number of atoms or molecules}}{6.02 \times 10^{23} \text{ atoms or molecules/mol}}$

Example 3.1

Calculate the number of moles contained in 5.4 g of each of the following substances:

a) Ag
b) H_2SO_4
c) $Na_2CO_3 \cdot 10\,H_2O$

Solution

First determine the molar mass of each substance. From the information given, decide on the appropriate conversion factor or formula to use. This is a grams-to-moles conversion. Therefore, the following conversion factor can be used:

grams to moles: $\dfrac{\text{1 mole of the substance}}{\text{1 molar mass of the substance}}$

a) $5.4 \text{ g Ag} \times \dfrac{1 \text{mol Ag}}{108 \text{ g Ag}} = 0.050 \text{ mol Ag}$

b) $5.4 \text{ g H}_2\text{SO}_4 \times \dfrac{1 \text{ mol H}_2\text{SO}_4}{98 \text{ g H}_2\text{SO}_4} = 0.055 \text{ mol H}_2\text{SO}_4$

c) $5.4 \text{ g Na}_2\text{CO}_3 \cdot 10\,\text{H}_2\text{O} \times \dfrac{1 \text{ mol Na}_2\text{CO}_3 \cdot 10 \text{ H}_2\text{O}}{286 \text{ g Na}_2\text{CO}_3 \cdot 10 \text{ H}_2\text{O}} = 0.0189 \text{ mol Na}_2\text{CO}_3 \cdot 10 \text{ H}_2\text{O}$

Example 3.2

How many atoms are present in 0.0125 mol of elemental sodium?

Solution

The following conversion factor is applicable:

$$\text{Moles to atoms or molecules:} \quad \dfrac{6.02 \times 10^{23} \text{ atoms or molecules of the substance}}{1 \text{ mol of the substance}}$$

$$? \text{ Na atoms} = 0.0125 \text{ mol Na} \times \dfrac{6.02 \times 10^{23} \text{ atoms Na}}{1 \text{ mol Na}} = 7.5 \times 10^{21} \text{ Na atoms}$$

Example 3.3

Find the number of moles in each of the following samples:
a) 0.027 g Al
b) 1 atom Nitrogen
c) 1.0×10^{15} HPO_4^{2-} ions

Solution

For this problem, we will use the formula method. As demonstrated in the previous examples, we can use the conversion factor method to solve the problem as well.

a) No. of mol Al $= \dfrac{\text{mass of Al}}{\text{molar mass of Al}} = \dfrac{0.027 \text{ g}}{27 \text{ g/mol}} = 0.0010 \text{ mol Al}$

b) No. of mol $N_2 = \dfrac{\text{number } N_2 \text{ atoms}}{6.02 \times 10^{23} \text{ atoms/mol}}$

$\phantom{\text{b) No. of mol } N_2} = \dfrac{1.0 \text{ atom}}{6.02 \times 10^{23} \text{ atoms/mol}} = 1.66 \times 10^{-24} \text{ mol } N_2$

c) No. of mol $HPO_4^{2-} = \dfrac{\text{number } HPO_4^{2-} \text{ ions}}{6.02 \times 10^{23} \text{ ions/mol}} = \dfrac{1.0 \times 10^{15} \text{ ions}}{6.02 \times 10^{23} \text{ ions/mol}}$

$\phantom{\text{c) No. of mol } HPO_4^{2-}} = 1.7 \times 10^{-9} \text{ mol } HPO_4^{2-}$

Example 3.4

Phosphoric acid (H_3PO_4) is used widely in the formulation of detergents, carbonated beverages, fertilizers, and toothpastes. Calculate the number of moles of each type of atom present in 3.5 moles of H_3PO_4.

Solution

The formula given represents 1 mole of H_3PO_4 and it contains 3 moles of H atoms, 1 mole of P atoms, and 4 moles of O atoms. This provides the conversion factors we need to calculate the number of moles of each atom present in 3.5 moles of the acid:

$$\frac{3 \text{ mol H atoms}}{1 \text{ mol } H_3PO_4} ; \quad \frac{1 \text{ mol P atom}}{1 \text{ mol } H_3PO_4} ; \quad \frac{4 \text{ mol O atoms}}{1 \text{ mol } H_3PO_4}$$

$$\text{Moles of H atoms} = 3.5 \text{ mol } H_3PO_4 \times \frac{3 \text{ mol H atoms}}{1 \text{ mol } H_3PO_4} = 10.5 \text{ mol H atoms}$$

$$\text{Moles of P atoms} = 3.5 \text{ mol } H_3PO_4 \times \frac{1 \text{ mol P atom}}{1 \text{ mol } H_3PO_4} = 3.5 \text{ mol P atoms}$$

$$\text{Moles of O atoms} = 3.5 \text{ mol } H_3PO_4 \times \frac{4 \text{ mol O atoms}}{1 \text{ mol } H_3PO_4} = 14 \text{ mol O atoms}$$

Example 3.5

Hemoglobin, which is the oxygen carrier in blood, has the molecular formula $C_{2952}H_{4664}N_{812}S_8Fe_4$.

(a) Calculate its molar mass.

(b) What is the number of moles in 25,000 g of this compound?

(c) Calculate the mass of N, S, and Fe present in 25,000 g of hemoglobin.

Solution

(a) 2,952 atoms of C weigh $2{,}952 \times 12 = 35{,}424$

    4,664 atoms of H weigh $4{,}664 \times 1 = 4{,}664$

      812 atoms of N weigh $812 \times 14 = 11{,}368$

        8 atoms of S weigh $\quad 8 \times 32 = \quad 256$

        4 atoms of Fe weigh $\quad 4 \times 56 = \underline{\quad 224}$

        Formula mass $\qquad\qquad\quad = 51{,}936$

(b) Number of moles

$$C_{2952}H_{4664}N_{812}S_8Fe_4 = 25{,}000 \text{ g} \times \frac{1 \text{ mole } C_{2952}H_{4664}N_{812}S_8Fe_4}{51{,}936 \text{ g } C_{2952}H_{4664}N_{812}S_8Fe_4}$$

$$= 0.4814 \text{ mol.}$$

(c) Number of g of N $= 0.4814$ mole Hemo $\times \dfrac{11{,}368 \text{ g N}}{1 \text{ mole Hemo}}$

$$= 5472 \text{ g N}$$

Number of g of S $= 0.4814$ mole Hemo $\times \dfrac{256 \text{ g S}}{1 \text{ mole Hemo}}$

$$= 123.2 \text{ g S}$$

Number of g of Fe $= 0.4814$ mole Hemo $\times \dfrac{224 \text{ g Fe}}{1 \text{ mole Hemo}}$

$$= 107.8 \text{ g Fe}$$

Example 3.6

White phosphorus burns in excess oxygen to give tetraphosphorus decaoxide, P_4O_{10}. The oxide is highly deliquescent, and absorbs water from the atmosphere to form phosphoric acid, H_3PO_4:

$$P_4O_{10}(s) + 6H_2O(l) \rightarrow 4H_3PO_4(aq)$$

How many kg of phosphoric acid (H_3PO_4) could be obtained from 950 g of white phosphorus (P)?

Solution

In solving this problem, it is not necessary to find the mass of P_4O_{10}. We need to focus only on H_3PO_4. The formula H_3PO_4 indicates that 1 mole of P is contained in 1 mole of H_3PO_4. Using the appropriate conversion factors, we have:

$$\text{Mass of } H_3PO_4 = (950 \text{ g P})\left(\frac{1 \text{ mol P}}{30.97 \text{ g P}}\right)\left(\frac{1 \text{ mol } H_3PO_4}{1 \text{ mol P}}\right)\left(\frac{97.97 \text{g } H_3PO_4}{1 \text{ mol } H_3PO_4}\right)$$

$$= 3{,}005.2 \text{ g} = 3.0 \text{ kg}$$

3.4 Problems

1. Calculate the number of moles present in 100 g of each of the following compounds:
   (a) $CaCO_3$ (b) H_2O (c) $NaOH$ (d) H_2SO_4 (e) N_2

2. Calculate the number of moles present in each of the following:
   (a) 12.5 g $CaSO_4$ (b) 0.25 g H_2O_2 (c) 0.04 g $NaHCO_3$ (d) 9.8 g H_2SO_4

3. Calculate the mass in grams present in 0.0250 mole of each of the following compounds:
   (a) MgO (b) Na_2O (c) NH_3 (d) C_2H_4 (e) $H_4P_2O_7$

4. What is the mass in grams of each of the following?
   (a) 0.50 mole CaO (b) 0.0050 mole Li_2O (c) 2.5 moles P_4O_{10}
   (d) 1.50 moles C_2H_4 (e) 0.010 mole $B_3N_3H_6$

5. Calculate the number of moles present in each of the following:
   (a) 5.0×10^{20} atoms of Mg   (b) 2.5×10^{26} ions of Fe^{3+}
   (c) 1.15×10^{23} molecules of H_2O (d) 6.02×10^0 ions of SO_4^{2-}
   (e) 3.01×10^{-23} molecule of $C_6H_{12}O_6$

6. Calculate the number of atoms of each element contained in 1 mole of
   (a) Al_2O_3 (b) H_2S (c) CO_2 (d) HCl

7. What mass in grams is present in each of the following?
   (a) 5.0×10^{20} atoms of Ca   (b) 2.5×10^{26} atoms of Fe
   (c) 3.01×10^{23} atoms of Ca   (d) 10×10^0 atoms of K
   (e) 6.02×10^{-23} atoms of H

8. Calculate the mass of H_3PO_4 that could be produced from 2,000 kg of phosphorous.

9. Calculate the number of oxygen atoms present in 500 g of the following:
   (a) H_2O_2 (b) $C_{12}H_{22}O_{11}$ (c) $CaCO_3$ (d) $Ca_3(PO_4)_2$

10. Iron (III) sulfate ($Fe_2(SO_4)_3$) ionizes according to the following equation:

$$Fe_2(SO_4)_3 \rightarrow 2\,Fe^{3+} + 3\,SO_4^{2-}$$

Calculate the number of each ion produced from 2.00 moles of $Fe_2(SO_4)_3$.

11. A superconductor is a material that loses all resistance to the flow of electrical current below a characteristic temperature known as the superconducting transition temperature (T_c). The copper-containing oxide $YBa_2Cu_3O_7$, the so-called 1-2-3 compound (1 yttrium, 2 barium, and 3 copper), is an example of a superconductor with a high T_c. Determine:

   (a) the number of moles of Y, Ba, and Cu in 500 g of the oxide, and
   (b) the number of atoms of Y, Ba, and Cu in 500 g of the oxide.

12. Silicon nitride (Si_3N_4) is a high-temperature ceramic that is used in the fabrication of some engine components. How many atoms of Si and N are present in 0.50 mol of Si_3N_4?

4

Formulas of Compounds and Percent Composition

. . . .

4.1 Percent Composition

Percent means 'parts per 100' or 'part divided by whole, multiplied by 100'. If the formula of a compound is known, one can easily calculate the mass percent of each element in the compound. The percent composition of a compound is the mass percent of each element in the compound. The molar mass represents 100% of the compound.

$$\% \text{Element} = \frac{\text{mass of element}}{\text{formula mass}} \times 100$$

The following steps are helpful in calculating the percent composition if the formula of the compound is known:

1. Calculate the formula mass or molar mass of the compound.
2. Calculate the percent composition by dividing the total mass of each element in the formula unit by the molar mass and multiplying by 100.

Example 4.1

Aspirin, which is commonly used as a pain reliever, has the chemical composition $C_9H_8O_4$. Calculate its percent composition.

Solution

First, calculate the molar mass of aspirin.

$$9\,C = 9 \times 12.0 = 108.0 \text{ g}$$
$$8\,H = 8 \times 1.0 = 8.0 \text{ g}$$
$$4\,O = 4 \times 16 = \underline{64.0 \text{ g}}$$
$$\text{Molar mass} = 180.0 \text{ g}$$

Chemistry in Quantitative Language: Fundamentals of General Chemistry Calculations. Second Edition. Christopher O. Oriakhi, Oxford University Press. © Christopher O. Oriakhi 2021. DOI: 10.1093/oso/9780198867784.003.0004

Next, calculate the percent composition.

$$\%C = \frac{9\,C}{C_9H_8O_4} \times 100 = \frac{108}{180} \times 100 = 60.0\%$$

$$\%H = \frac{8\,H}{C_9H_8O_4} \times 100 = \frac{8}{180} \times 100 = 4.4\%$$

$$\%O = \frac{4\,O}{C_9H_8O_4} \times 100 = \frac{64}{180} \times 100 = 35.6\%$$

Now, check to ensure the percentages add to 100%.

$$60.00 + 4.4 + 35.56 = 100.0\%.$$

Thus, aspirin is 60% C, 4.4% H, and 35.5% O by weight.

Example 4.2

Arsenic occurs in several native compounds (or ores): arsenopyrite (FeAsS), realgar (As$_2$S$_2$), and orpiment (As$_2$S$_3$). Which of these ores has the highest sulfur content?

Solution

First calculate the molar mass of each ore:

$$FeAsS = 1 \times Fe + 1 \times As + 1 \times S = 55.85 + 74.92 + 32.06 = 162.83 \text{ g}$$
$$As_2S_2 = 2 \times As + 2 \times S = 149.84 + 64.12 = 213.96 \text{ g}$$
$$As_2S_3 = 2 \times As + 3 \times S = 149.84 + 96.16 = 246.02 \text{ g}$$

Now calculate the percent of S in each ore:

$$\%S \text{ in } FeAsS = \frac{S}{FeAsS} \times 100 = \frac{32.06}{162.83} \times 100 = 19.69\%$$

$$\%S \text{ in } As_2S_2 = \frac{2\,S}{As_2S_2} \times 100 = \frac{64.12}{213.96} \times 100 = 13.72\%$$

$$\%S \text{ in } As_2S_3 = \frac{3\,S}{As_2S_3} \times 100 = \frac{96.16}{246.02} \times 100 = 39.09\%$$

Orpiment (As$_2$S$_3$) has the highest sulfur content.

Example 4.3

Calculate the percentage of water of hydration in the following hydrates:

1. $CuSO_4 \cdot 5H_2O$
2. $Na_2CO_3 \cdot 10\,H_2O$
3. $Zn(NO_3)_2 \cdot 6\,H_2O$

Solution

First calculate the molar mass. Then express water as a mass percent of the compound.

1. $CuSO_4 \cdot 5\ H_2O = 63.55 + 32.06 + 4 \times 16 + 5(2 \times 1 + 16) = 249.61$

$$\%H_2O = \frac{5\ H_2O}{CuSO_4 \cdot 5\ H_2O} \times 100 = \frac{90}{249.61} \times 100 = 36.06\%$$

2. $Na_2CO_3 \cdot 10\ H_2O = 2 \times 23 + 12 + 3 \times 16 + 10\ (2 \times 1 + 16) = 286$

$$\%H_2O = \frac{10\ H_2O}{Na_2CO_3 \cdot 10\ H_2O} \times 100 = \frac{180}{286} \times 100 = 62.94\%$$

3. $Zn(NO_3)_2 \cdot 6\ H_2O = 65.38 + 2\ (14 + 3 \times 16) + 6\ (2 \times 1 + 16) = 297.38$

$$\%H_2O = \frac{6\ H_2O}{Zn(NO_3)_2 \cdot 6\ H_2O} \times 100 = \frac{108}{297.38} \times 100 = 36.32\%$$

Example 4.4

Titanium disulfide is a semiconductor with a layered structure. The lithiated form, $Li_x TiS_2$, is useful as an electrode in rechargeable batteries. Calculate the mass percent of lithium and sulfur present in $Li_x TiS_2$ when $x = 0.25$.

Solution

First, calculate the molar mass of $Li_{0.25} TiS_2$.

$$Li = 6.94 \times 0.25 = 1.735\ g$$
$$Ti = 47.9 \times 1 \quad = 47.900\ g$$
$$S = 32.1 \times 2 \quad = \underline{\ 64.200\ g}$$
$$Molar\ mass \quad = 113.835\ g$$

Now calculate the % of Li and S in $Li_{0.25} TiS_2$.

$$\%Li = \frac{1.735}{113.835} \times 100\% = 1.52\%$$

$$\%S = \frac{64.2}{113.835} \times 100\% = 56.4\%$$

Example 4.5

The boron content of a Pyrex glass was reported as 17% B_2O_3. Determine the percentage of boron in this glass.

Solution

1. Determine the molar mass of B_2O_3.
   $B_2O_3 = 10.81 \times 2 + 16.0 \times 3 = 69.62$

2. Determine the percentage of B in B_2O_3.
   $\%B = \dfrac{21.62}{69.62} \times 100\% = 31.05\%$, or 31.05 g of B in every 100 g of B_2O_3

3. Determine the percentage of B in the glass.
   $(17\% B_2O_3) \left(\dfrac{31.05 \text{ g B}}{100 \text{ g } B_2O_3} \right) = 5.38\% B$ in the glass

Example 4.6

To determine the amount of silver in a metal sample, a chemist dissolved 0.75 g of the metal in dilute nitric acid. He precipitated the silver as AgCl by adding excess NaCl solution. The precipitated AgCl weighed 0.85 g. Calculate the percentage of silver in the dissolved sample.

Solution

1. First, calculate the percent of Ag in AgCl:
   Molar mass of $AgCl = 143.4$

   $$\%\text{Ag in AgCl} = \frac{\text{Mass of Ag}}{\text{Mass of AgCl}} \times 100\%$$

   $$= \frac{107.9}{143.4} \times 100 = 75.2\%$$

2. Calculate the actual mass of Ag in the dissolved sample.
   Mass of Ag in the sample $= 75.2\%$ of the precipitated AgCl

   $$= (0.752)(0.85 \text{ g})$$
   $$= 0.64 \text{ g}$$

3. The percentage of Ag in the original sample is simply the actual amount of silver divided by the total weight:

   $$\%\text{of Ag in the sample} = \frac{0.64}{0.75} \times 100\%$$
   $$= 85.3\%$$

4.2 Types of Chemical Formula

Chemical formulas are often determined from experimental data. Generally, two types of formula are possible, depending on the amount of experimental information available: the empirical formula and the molecular formula. Usually, the empirical formula is determined first, and the molecular formula is calculated from that.

4.2.1 Empirical formula

The empirical formula of a compound represents the smallest whole-number ratio of atoms present in a formula unit of the compound. Empirical formula is found by measuring the mass percent of each element in the compound. The mass percent of each element is converted to the number of moles of atoms of each element per 100 g of the compound. Then the mole ratios between the elements are expressed in terms of small whole numbers.

4.2.2 Steps for determining empirical formula

1. Find the number of grams of each element in a given mass of the compound. Often, we use 100 g for simpler calculation.
2. Find the relative number of moles by dividing the mass of each element by its molar mass.
3. Express the mole ratio between the elements in terms of small whole numbers by dividing through by the smallest number. If this does not result in simple whole numbers for all the elements in the compound, multiply through by the smallest suitable integer to convert all to whole numbers.
4. Obtain the empirical formula by inserting the integers after the symbol of each element.

Example 4.7

Analysis of the percentage composition of a nicotine product yielded 74% C, 8.7% H, and 17.3% N. What is its empirical formula?

Solution

First assume you have 100 g of the sample, so that the given percentage compositions will correspond to masses in grams, i.e. 74 g C, 8.7 g H, and 17.3 g N. Now, convert the relative masses of each element to the relative number of moles of each element. Since this is a mass-to-mole conversion, use the following conversion factor:

$$\text{Grams to mole}: \quad \frac{1 \text{ mole of the substance}}{\text{molar mass of the substance}}$$

$$C: 74 \text{ g C} \times \frac{1 \text{ mole C}}{12.0 \text{ g C}} = 6.17 \text{ mol C}$$

$$H: 8.7 \text{ g H} \times \frac{1 \text{ mole H}}{1.0 \text{ g H}} = 8.70 \text{ mol H}$$

$$N: 17.3 \text{ g N} \times \frac{1 \text{ mole N}}{14.0 \text{ g N}} = 1.24 \text{ mol N}$$

So, 100 g of the compound contains 6.17 mol of C, 8.70 mol H, and 1.24 mol N. Convert these to whole number ratios by dividing by the smallest of the numbers (i.e. 1.24).

$$C: \frac{6.17 \text{ mol}}{1.24 \text{ mol}} = 4.99 \simeq 5.0$$

$$H: \frac{8.7 \text{ mol}}{1.24 \text{ mol}} = 7.02 \simeq 7.0$$

$$N: \frac{1.24 \text{ mol}}{1.24 \text{ mol}} = 1.0$$

The empirical formula of the compound is C_5H_7N.

Example 4.8

A chemist extracted a medicinal compound believed to be effective in treating high blood pressure. Elemental analysis of this compound gave the composition 60.56% C, 11.81% H, and 28.26% N. What is the empirical formula of this compound?

Solution

To solve the problem, we assume that we are starting with 100 g of the material, which will contain 60.56 g C, 11.8 g H, and 28.26 g N.
    Next, convert the gram amount to moles for each element.

$$C: 60.56 \text{ g C} \times \frac{1 \text{ mole C}}{12.0 \text{ g C}} = 5.05 \text{ mol C}$$

$$H: 11.18 \text{ g H} \times \frac{1 \text{ mole H}}{1.0 \text{ g H}} = 11.18 \text{ mol H}$$

$$N: 28.26 \text{ g N} \times \frac{1 \text{ mole N}}{14.0 \text{ g N}} = 2.02 \text{ mol N}$$

Determine the smallest whole-number ratio of moles by dividing each number of moles by the smallest number of moles, 2.02:

$$C: \frac{5.05 \text{ mol}}{2.02 \text{ mol}} = 2.5$$

$$H: \frac{8.7 \text{ mol}}{2.02 \text{ mol}} = 4.3$$

$$N: \frac{2.02 \text{ mol}}{2.02 \text{ mol}} = 1.0$$

We still do not have whole number ratios. Therefore, multiply by the lowest integer that will convert all to whole numbers. The number 10 seems to do the trick.

$$C: 2.5 \times 10 = 25$$

$$H: 4.3 \times 10 = 43$$

$$N: 1.0 \times 10 = 10$$

The empirical formula of the compound is $C_{25}H_{43}N_{10}$.

Example 4.9

A chemical engineer heated 1.8846 g of pure iron powder in an atmosphere of pure oxygen in a quartz tube until all the iron was converted to the oxide, which weighed 2.6946 g. Calculate the empirical formula of the oxide formed.

Solution

First, determine the mass of oxygen gas that has reacted. Recall that in a reaction, mass is conserved. Hence:

$$\text{Mass of Fe} + \text{Mass of O} = \text{Mass of Fe}_x\text{O}_y$$

$$\text{Mass of O} = \text{Mass of Fe}_x\text{O}_y - \text{Mass of Fe}$$

$$= 2.6946 \text{ g} - 1.8846 \text{ g} = 0.81 \text{ g O}$$

Now, convert relative masses to relative numbers of moles.

$$\text{Fe} : 1.8846 \text{ g Fe} \times \frac{1 \text{ mole Fe}}{55.85 \text{ g Fe}} = 0.0337 \text{ mol Fe}$$

$$\text{O} : 0.8100 \text{ g O} \times \frac{1 \text{ mole O}}{16.00 \text{ g O}} = 0.0506 \text{ mol O}$$

Determine the smallest whole-number ratio of atoms:

$$\text{Fe} : \frac{0.0337 \text{ mol}}{0.0337 \text{ mol}} = 1.0$$

$$\text{O} : \frac{0.0506 \text{ mol}}{0.0337 \text{ mol}} = 1.5$$

Multiply the ratios by 2 to clear the decimals. This gives Fe : O = 2:3. The empirical formula of the oxide is thus Fe_2O_3.

4.3 Empirical Formula from Combustion Analysis

Combustion analysis is one of the most common methods used to determine the empirical formula of an unknown compound containing carbon and hydrogen. When a compound containing carbon and hydrogen is burned in a combustion apparatus in the presence of oxygen, all the carbon is converted to CO_2, and the hydrogen to H_2O. The CO_2 formed is trapped by NaOH, and the water is trapped by magnesium perchlorate, $Mg(ClO_4)_2$. The mass of CO_2 is measured as the increase in the mass of the CO_2 trap, and the mass of H_2O produced is measured as the increase in mass of the water trap. From the masses of CO_2 and H_2O, the masses and moles of carbon and hydrogen present in the compound can be determined. (If nitrogen is present in the compound, it must be determined by the Kjeldahl method in a separate analysis, as NH_3.) Oxygen in the compound can then be determined by difference—that is, the percentage remaining when C, H, and N (if present) have been found.

Example 4.10

A 39.0-mg sample of a compound containing C, H, O, and N is burned. The C is recovered as 97.7 mg CO_2, and the H is recovered as 20.81 mg H_2O. A separate Kjeldahl nitrogen analysis gives nitrogen content of 3.8%. Calculate the empirical formula of the compound.

Solution

First convert grams of CO_2 and H_2O to grams of C and H.

$$97.7 \text{ mg CO}_2 \times \frac{1 \text{ mole CO}_2}{44.0 \text{ g CO}_2} \times \frac{1 \text{ mole C}}{1 \text{ mole CO}_2} \times \frac{12 \text{ g C}}{1 \text{ mole C}} \times \frac{1 \text{ g CO}_2}{1,000 \text{ mg CO}_2}$$

$$= 0.0266 \text{ g or } 26.6 \text{ mg C}$$

$$20.81 \text{ mg H}_2O \times \frac{1 \text{ mole H}_2O}{18.0 \text{ g H}_2O} \times \frac{2 \text{ mole H}}{1 \text{ mole H}_2O} \times \frac{1 \text{ g H}}{1 \text{ mole H}} \times \frac{1 \text{ g H}_2O}{1,000 \text{ mg H}_2O}$$

$$= 0.0023 \text{ g or } 2.3 \text{ mg H}$$

For nitrogen, find 3.8% of starting material.

$$\left(\frac{3.8}{100} \times 38.7 \text{ mg} \right) \left(\frac{1 \text{ g N}}{1000 \text{ mg N}} \right) = 0.0015 \text{ g or } 1.5 \text{ mg N}$$

Next, calculate the mass of O as the difference between the mass of the sample and the masses of C, H, and N.

$$\text{Mass of O} = 38.7 \text{ mg}-26.7 \text{ mg}-2.3 \text{ mg}-1.5 \text{ mg} = 8.2 \text{ mg O}$$

Now convert the mass of each element to the corresponding moles.

$$C: 0.0267 \text{ g C} \times \frac{1 \text{ mole C}}{12.0 \text{ g C}} = 0.0022 \text{ mol C}$$

$$H: 0.0023 \text{ g H} \times \frac{1 \text{ mole H}}{1.01 \text{ g H}} = 0.0023 \text{ mol H}$$

$$N: 0.0015 \text{ g N} \times \frac{1 \text{ mole N}}{14.0 \text{ g N}} = 0.00011 \text{ mol N}$$

$$O: 0.0082 \text{ g O} \times \frac{1 \text{ mole O}}{16.0 \text{ g O}} = 0.00051 \text{ mol O}$$

Divide each molar value by the smallest of them.

$$C = \frac{0.0022 \text{ mol}}{0.00011 \text{ mol}} = 20$$

$$H = \frac{0.0023 \text{ mol}}{0.00011 \text{ mol}} = 21$$

$$N = \frac{0.00011 \text{ mol}}{0.00011 \text{ mol}} = 1$$

$$O = \frac{0.00055 \text{ mol}}{0.00011 \text{ mol}} = 5$$

The empirical formula of the compound is $C_{20}H_{21}NO_5$.

4.4 Molecular Formula

The molecular formula of a compound is the true formula, which shows the actual numbers of atoms in a molecule of the compound. The molecular formula may be the same as the empirical formula, or a multiple of it. For example, a compound with the empirical formula CH_2O could have molecular formulas such as $C_2H_4O_2$, $C_3H_6O_3$, or $C_5H_{10}O_5$.

4.4.1 Determination of molecular formula

To determine molecular formula of a molecular compound, the empirical formula (obtained from percent composition) and the molecular mass (obtained experimentally) must be known. Since the molecular formula for a compound is either the same as, or a multiple of, the empirical formula, the following relationship is helpful in determining molecular formula:

$$\text{Molecular weight} = n \text{ (Empirical formula weight) or}$$

$$n = \frac{\text{Molecular weight}}{\text{Empirical formula weight}}$$

Then, multiplying the coefficients in the empirical formula by n gives the actual molecular formula.

Example 4.11

The empirical formula of resorcinol, a common chemical used in the manufacture of plastics, drugs, and paper products, is C_3H_3O. The molecular mass is 110.0. What is the molecular formula?

Solution

We are given the empirical formula and molecular mass of the compound. The unknown is molecular formula.

Begin by calculating the formula mass of C_3H_3O.

$$3 \times C + 3 \times H + 1 \times O = 3 \times 12 + 1 \times 3 + 1 \times 16 = 55 \text{ amu}$$

Next, calculate the multiplication factor, n, from the expression:

$$\text{Molecular weight} = n \text{ (Empirical formula weight) or}$$

$$n = \frac{\text{Molecular weight}}{\text{Empirical formula weight}}$$

$$n = \frac{110}{55} = 2$$

Now, multiply the subscripts in the empirical formula by 2. This yields $C_6H_6O_2$ as the molecular formula.

Example 4.12

The empirical formula of ibuprofen, an active component in the pain remedy Advil, is $C_{13}H_{18}O_2$. Its molecular mass is 206.0. What is its molecular formula?

Solution

The empirical formula, $C_{13}H_{18}O_2$, has a formula weight of 206.0 amu. That is:

$$13 \times 12 + 18 \times 1 + 2 \times 16 = 206 \text{ amu}$$

Since the empirical formula mass and the molecular mass are equal (206.0), n is equal to 1. Hence the molecular formula is the same as the empirical formula, i.e. $C_{13}H_{18}O_2$.

Example 4.13

Epinephrine is an adrenaline hormone secreted into the bloodstream in times of danger or stress. Analysis indicates it has the following composition by mass: 59% C, 7.1% H, 26.2% O, and 7.7% N. The molecular mass is 183.0. Determine the empirical and molecular formula of epinephrine.

Solution

Begin by assuming you have 100 g of sample which contains 59 g C, 7.1 g H, 26.2 g O, and 7.7 g N.
   Next convert each mass of an element into the number of moles of that element:

$$C : 59 \text{ g C} \times \frac{1 \text{ mole C}}{12.0 \text{ g C}} = 4.92 \text{ mol C}$$

$$H : 7.1 \text{ g H} \times \frac{1 \text{ mole H}}{1.0 \text{ g H}} = 7.1 \text{ mol H}$$

$$O : 26.2 \text{ g O} \times \frac{1 \text{ mole O}}{16.0 \text{ g O}} = 1.64 \text{ mol O}$$

$$N : 7.7 \text{ g N} \times \frac{1 \text{ mole N}}{14.0 \text{ g N}} = 0.55 \text{ mol N}$$

Next, determine the relative number of moles of each element by dividing the above numbers by the smallest of the four elements in the compound:

$$C : \frac{4.92 \text{ mol}}{0.55 \text{ mol}} = 8.95 \simeq 9$$

$$H : \frac{7.10 \text{ mol}}{0.55 \text{ mol}} = 13.0$$

$$O : \frac{1.64 \text{ mol}}{0.55 \text{ mol}} = 3.0$$

$$N : \frac{0.55 \text{ mol}}{0.55 \text{ mol}} = 1.00$$

The empirical formula is $C_9H_{13}O_3N$, which has a formula mass of 183 amu. The molecular mass is 183.0. Since the empirical formula mass and the molecular mass are equal, the molecular formula is the same as the empirical formula, i.e. $C_9H_{13}O_3N$.

4.5 Problems

1. Calculate the percentage by mass of sulfur in the following:
   (a) H_2S (b) SO_3 (c) H_2SO_4 (d) $Na_2S_2O_3$

2. Calculate the percent composition of each of the following compounds:
   (a) K_2CO_3 (b) $Ca_3(PO_4)_2$ (c) $Al_2(SO_4)_3$ (d) C_6H_5OH

3. Calculate the percent composition of each of the following organic compounds:
   (a) $C_{20}H_{25}N_3O$ (b) $C_{10}H_{16}N_5P_3O_{13}$ (c) $C_6H_4N_2O_4$ (d) $C_6H_5CONH_2$

4. Calculate the percent by mass of magnesium, chlorine, hydrogen, and oxygen in hydrated magnesium chloride, $MgCl_2 \cdot 6H_2O$.

5. Find the mass of water of crystallization present in 10.85 g of hydrated sodium carbonate, $Na_2CO_3 \cdot 10H_2O$.

6. Caffeine, a stimulant in coffee and tea, has the composition 49.5% C, 5.19% H, 28.9% N, and 16.5% O. It has the molecular formula $C_8H_{10}N_4O_2$. How many grams of carbon and nitrogen are present in 77.25 g of caffeine?

7. Large quantities of hematite, an important ore of iron, are found in Australia, Ukraine, and USA. A 100 g sample of a crude ore contains 46.75 g of Fe_2O_3. What is the percentage of iron (Fe) in the ore? (Assume no other Fe compound.)

8. The element lithium is commonly obtained from the ore spodumene, $LiAlSi_2O_6$, which is found in Brazil, Canada, and the USA. What mass of lithium is present in 17.85 g $LiAlSi_2O_6$?

9. Determine the empirical formulas for compounds having the following compositions:

    (a) 40%S and 60%O

    (b) 14.7%Ca, 67.7%W, and 17.6%O

    (c) 85.63%C and 14.37%H

    (d) 9.90%C, 58.6%Cl and 31.5%F

    (e) 26.52%Cr, 24.52%S, and 48.96%O

10. Given the following percent compositions, determine the empirical formulas:

    (a) 21.85%Mg, 27.83%P, and 50.32%O

    (b) 19.84%C, 2.50%H, 66.08%O, and 11.57%N

    (c) 47.3%C, 2.54%H, and 50.0%Cl

    (d) 23.3%Co, 25.3%Mo, and 51.4%Cl

    (e) 41.87%C, 2.34%H, and 55.78%O

11. Write the empirical formula for each of the following compounds:

    (a) C_3H_6 (b) Fe_2S_3 (c) $C_6H_9O_6$ (d) $C_8H_{10}N_4O_2$ (e) $Al_2(SO_4)_3$

12. A compound that has the empirical formula P_2O_5 is found to have a molecular mass of 283.9. What is its molecular formula?

13. Cholesterol is the compound thought to be responsible for hardening of the arteries. An analysis of the compound gives the following percent composition by mass: 83.99% C, 11.92% H, and 4.15% O. Determine its empirical formula. What is its molecular formula, given that the molar mass is 386?

14. The molar mass of estradiol, a female sex hormone, is 272 g/mole. Is the molecular formula $C_9H_{12}O$ or $C_{18}H_{24}O_2$?

15. A hydrate of sodium carbonate, $Na_2CO_3 \cdot nH_2O$, was found to contain 62.9% by mass of water of crystallization. Determine the empirical formula for the hydrate.

16. A colorless organic liquid A was found on analysis to contain 47.37% C, 10.6% H, and 42% O. The molar mass of the compound is 228 g/mole. Calculate the empirical and molecular formula of A.

17. 25.07 g of a hydrated salt, $FeCl_3 \cdot nH_2O$, on strong heating, gave 15.06 g of the anhydrous salt. Determine the empirical formula of the salt.

18. Combustion analysis of 4.86 g of a sugar yielded 7.92 g of CO_2, and 2.70 g of H_2O. If the compound contains only carbon, hydrogen, and oxygen, what is its empirical formula? If the molar mass is 324 g/mol, what is its molecular formula?

19. 3.2 mg of an unknown organic compound containing C, H, and O gave on combustion analysis 3.48 mg CO_2, and 1.42 mg H_2O. What is the empirical formula of the compound? If the compound has a molecular mass of 244 g/mol, what is its molecular formula?

20. A combustion analysis was carried out on 0.4710 g of an organic compound containing only carbon, hydrogen, nitrogen, and oxygen. The products of combustion were 0.9868 g of CO_2 and 0.2594 g of H_2O. In another experiment, combustion of 0.3090 g of the compound produced 0.0544 g NH_3.

    (a) What is the mass percent composition of the compound?

    (b) Determine the empirical formula of the compound.

5

Chemical Formulas and Nomenclature

· ·

5.1 General Background

5.1.1 Elements

- An element is a pure substance, which cannot be split up into simpler substances. As of 2020, there are 118 known elements. Of these, only ninety-two occur naturally in the earth's crust and atmosphere. Scientists have made the other elements in the laboratory. The elements can be arranged in a *periodic table*, which displays the important relationships between the elements and the structures of their atoms.
- Each element has a name and a *symbol*. The symbol is usually derived from the name of the element by taking the capital form of the first letter of the name. For example, the symbol for nitrogen is N. In most other cases, the first capital letter followed by the next letter in lowercase is used. For example, the symbol for aluminum is Al.
- The symbols for some elements are derived from their Latin names. For example, the symbol for sodium is Na, from its Latin name, natrium. Potassium has the symbol K, taken from its Latin name, kalium.
- Most of the known elements are metals. Only twenty-two are nonmetallic.
- All elements are made up of atoms. Atoms of the same element are identical in all respect (except for isotopic differences; see Chapter 1) but are different from atoms of other elements.

5.1.2 Some basic definitions

- *Atom*: An atom is the smallest particle of an element that can take part in a chemical combination.
- *Molecule*: A molecule is a collection of two or more atoms of the same or different elements held together by covalent bonds.
- *Atomicity of an element*: This is the number of atoms present in one molecule of the element.

  1. *Monatomic* elements exist as single atoms. Examples include sodium (Na), iron (Fe), and the noble gases (He, Ne, Ar, Kr, Xe, and Rn).
  2. *Diatomic* elements contain two atoms per molecule of the element. The atomicity is written as a subscript after the symbol of the element. Examples of some diatomic molecules include H_2, O_2, Cl_2, and N_2.

Chemistry in Quantitative Language: Fundamentals of General Chemistry Calculations. Second Edition.
Christopher O. Oriakhi, Oxford University Press. © Christopher O. Oriakhi 2021.
DOI: 10.1093/oso/9780198867784.003.0005

3. *Triatomic* elements contain three atoms per molecule. An example is ozone (O_3).
4. *Tetra-atomic* elements contain four atoms of the element per molecule. An example is yellow phosphorus (P_4).

- *Ions*: An ion is an electrically charged particle containing one or more atoms. They are formed by gain or loss of electrons.

  1. *Anion*: a negatively charged ion.
  2. *Cation*: a positively charged ion.

- *Polyatomic ions (Radicals)*: Polyatomic ions, also known as radicals, are groups of covalently bonded atoms with an overall charge.
- *Compound*: A pure substance formed when two or more elements combine with each other. For example, water is a compound formed by the combination of two atoms of hydrogen and one atom of oxygen.

  1. The process of forming a compound involves a *chemical reaction*.
  2. The composition of a compound is fixed.

- The *formula* of a compound consists of the symbols of its elements and the number of each kind of atom.
- In every compound, the constituent elements are combined in a fixed proportion. This is known as the *Law of Definite Composition*. The law states that the atoms in a compound are combined in a fixed proportion by mass.
- There are two types of compounds: *molecular compounds* and *ionic compounds*. Molecular compounds are composed of molecules that are not charged. A good example is water, H_2O. Ionic compounds are composed of metallic (positively charged) and nonmetallic (negatively charged) ions held together by attractive forces. An example of an ionic compound is sodium chloride, or table salt (NaCl).

5.2 Chemical Formula

- A chemical formula is used to show the symbols and the combining ratios of all the elements in a compound. For example, ammonia gas is a compound and contains one atom of nitrogen per three atoms of hydrogen. Its formula is NH_3.
- A chemical formula only shows the number and kind of each atom contained in the compound. It usually does not show how the atoms are linked together or the nature of the chemical bonds.

5.3 Oxidation Numbers

- The oxidation number (ON) of an element is a positive or negative number, which expresses the combining capacity of that element in a particular compound or polyatomic ion. The numbers are used to keep track of electron transfer in chemical reactions.

- ONs find applications in writing chemical formulas and equations, balancing oxidation-reduction reactions, and predicting the properties of compounds. Some general rules are used to determine ONs.

5.3.1 Rules for assigning ONs

1. Any atom in an uncombined (or free) element (e.g. N_2, Cl_2, S_8, O_2, O_3, and P_4) has an ON of zero.
2. Hydrogen has an ON of $+1$, except in metal hydrides (e.g. NaH or MgH_2) where it is -1.
3. Oxygen has an ON of -2 in all compounds, except in peroxides (e.g. H_2O_2, Na_2O_2) where it is -1.
4. Metals generally have positive ONs.
5. In simple monoatomic ions. e.g. Na^+, Zn^{2+}, Al^{3+}, Cl^-, and C^{4-}, the ON is equal to the charge on the ion.
6. The algebraic sum of the ONs in a neutral molecule (e.g. $KMnO_4$, $NaClO$, H_2SO_4) is zero.
7. The algebraic sum of the ONs in a polyatomic ion (e.g. SO_4^{2-}, $Cr_2O_7^{2-}$) is equal to the charge on the ion.
8. Generally, in any compound or ion, the more electronegative atom is assigned the negative ON, while the less electronegative atom is assigned the positive ON.

5.3.2 ONs in formulas

The ON of an atom in a compound or ion can be determined using the applicable rules. You will find the following steps helpful in determining the ON of an element within a compound or a polyatomic ion.

1. Write down known ONs below each atom in the formula.
2. Write an algebraic expression summing the product of the number of each atom and its ON and equate it to zero for a compound, or to the net charge on the ion in the case of a polyatomic ion.
3. Solve the equation to obtain the unknown ON.

Example 5.1

What is the ON of silicon (Si) in silicon dioxide (SiO_2)?

Solution

Step 1: ON of $O = -2$
Required, ON of Si
Step 2: Set up the algebraic expression
$1 \times$ ON of Si $+ 2 \times$ ON of O $=$ charge on the compound $= 0$
Step 3: Solve for Si
$Si + 2(-2) = 0$
$Si - 4 = 0$
$Si = +4$
The oxidation number of silicon is $+4$.

Example 5.2

Determine the ON of Cr in $K_2Cr_2O_7$.

Solution

Step 1:   ON of $K = +1$
            ON of $O = -2$
            ON of $Cr = ?$
Step 2: $2 \times$ ON of $K + 2 \times$ ON of $Cr + 7 \times$ ON of $O = 0$
Step 3: $2 \times 1 + 2Cr + 7 \times (-2) = 0$
            $2 + 2Cr - 14 = 0$
            $2Cr - 12 = 0$
            $Cr = +6$
The oxidation number of Cr is $+6$.

Example 5.3

Determine the ON of Mn in MnO_4^-.

Solution

Step 1:   ON of $O = -2$
            Required, ON of Mn
Step 2:   Set up the algebraic expression
            $1 \times$ ON of $Mn + 4 \times$ ON of $O = $ charge on the ion
Step 3:   Solve for Mn
            $Mn + 4 \times (-2) = -1$
            $Mn - 8 = -1$
            $Mn = +7$

5.4 Writing the Formulas of Compounds

In writing the formula of a compound, one must have an accurate knowledge of the ionic charges of anions and cations. Table 5.1 and Table 5.2 show the formulas and charges of some simple and polyatomic ions.

The following rules serve as general guides:

- Where the charge on the cation is not equal to the charge on the anion, use appropriate subscripts to balance the charges.
- The cation (positive ion) is usually written before the anion (negative ion).
- The sum of the ONs for all the atoms in a compound is equal to zero.
- The sum of the ONs for all the atoms in a polyatomic ion is equal to the charge on the ion.

The following examples will illustrate how to write the formula of a compound.

Table 5.1   Some simple cations and anions

Name	Common name	Formula	Charge	Name	Common name	Formula	Charge
Cations				**Cations cont.**			
Cesium		Cs^+	1	Tin (II)	Stannous	Sn^{2+}	2
Copper (I)	Cuprous	Cu^+	1	Aluminum		Al^{3+}	3
Hydrogen		H^+	1	Bismuth (III)		Bi^{3+}	3
Lithium		Li^+	1	Chromium (III)	Chromic	Cr^{3+}	3
Mercury (I)	Mercurous	Hg^+	1	Cobalt (III)	Cobaltic	Co^{3+}	3
Potassium		K^+	1	Iron (III)	Ferric	Fe^{3+}	3
Rubidium		Rb^+	1	Manganese (IV)	Manganic	Mn^{4+}	4
Silver		Ag^+	1	Lead (IV)	Plumbic	Pb^{4+}	4
Sodium		Na^+	1	Tin (IV)	Stannic	Sn^{4+}	4
Barium		Ba^{2+}	2				
Calcium		Ca^{2+}	2	**Anions**			
Chromium (II)		Cr^{2+}	2				
Cobalt (II)		Co^{2+}	2	Bromide		Br^-	-1
Copper (II)	Cupric	Cu^{2+}	2	Chloride		Cl^-	-1
Iron (II)	Ferrous	Fe^{2+}	2	Fluoride		F^-	-1
Lead (II)	Plumbous	Pb^{2+}	2	Hydride		H^-	-1
Magnesium		Mg^{2+}	2	Iodide		I^-	-1
Manganese (II)	Manganous	Mn^{2+}	2	Nitride		N^{3-}	-3
Mercury (II)	Mercuric	Hg^{2+}	2	Oxide		O^{2-}	-2
Nickel (II)		Ni^{2+}	2	Phosphide		P^{3-}	-3
Strontium		Sr^{2+}	2	Sulfide		S^{2-}	-2

Table 5.2   Some common polyatomic ions

Name of polyatomic ions	Formula	Charge	Name of polyatomic ions	Formula	Charge
Cations			**Anions cont.**		
Ammonium	NH_4^+	1	Hydrogen sulfate	HSO_4^-	-1
Hydroxonium	H_3O^+	1	Hydrogen sulfite	HSO_3^-	-1
			Hydroxide	OH^-	-1
Anions			Hypochlorite	ClO^-	-1
Acetate	CH_3COO^-	-1	Hydrosulfide	HS^-	-1
Arsenate	AsO_4^{3-}	-3	Nitrate	NO_3^-	-1
Borate	BO_3^{3-}	-3	Nitrite	NO_2^-	-1
Bromate	BrO_3^-	-1	Oxalate	$C_2O_4^{2-}$	-2
Carbonate	CO_3^{2-}	-2	Perchlorate	ClO_4^-	-1
Chlorate	ClO_3^-	-1	Permaganate	MnO_4^-	-1
Chlorite	ClO_2^-	-1	Peroxide	O_2^{2-}	-2
Chromate	CrO_4^{2-}	-2	Phosphate	PO_4^{3-}	-3
Cyanide	CN^-	-1	Pyrophosphate	$P_2O_7^{4-}$	-4
Dichromate	$Cr_2O_7^{2-}$	-2	Silicate	SiO_3^{2-}	-2
Dihydrogen phosphate	$H_2PO_4^-$	-1	Sulfate	SO_4^{2-}	-2
Hydrogen carbonate	HCO_3^-	-1	Sulfite	SO_3^{2-}	-2
Hydrogen Phosphate	HPO_4^{2-}	-2	Thiosulfate	$S_2O_3^{2-}$	-2

Example 5.4

Write down the correct formula for the following compounds. See Table 5.1 and Table 5.2 for charges and formulas of the various ions.

(a) Magnesium chloride
(b) Aluminum oxide
(c) Iron (III) sulfate
(d) Sodium phosphate
(e) Potassium chromium sulfate

Solutions

(a) Magnesium chloride is composed of magnesium and chloride ions. The formulas of these ions are Mg^{2+} and Cl^-. Since Mg has a charge of +2, it will need two Cl ions to form a compound with a net charge of zero. So we write $Mg^{2+}Cl^-Cl^-$ or $Mg^{2+}(Cl^-)_2$. The formula can be simplified by removing the charges. Thus, the correct formula is $MgCl_2$.
Charge balance:

$$\left[Mg^{2+}\right]+2\left[Cl^-\right]=0$$
$$(+2)+2(-1)=2-2=0$$

(b) The formulas for the cation and anion in aluminum oxide are Al^{3+} and O^{2-}.
Charge balance:

$$2\left[Al^{3+}\right]+3\left[O^{2-}\right]=0$$
$$2(+3)+3(-2)=6-6=0$$

The correct formula is Al_2O_3.

(c) The formulas for the cation and anion in iron (III) sulfate are Fe^{3+} and SO_4^{2-}.
Charge balance:

$$2\left[Fe^{3+}\right]+3\left[SO_4^{2-}\right]=0$$
$$2(+3)+3(-2)=6-6=0$$

The correct formula of Iron (III) sulfate is $Fe_2(SO_4)_3$.

(d) The formulas for the cation and anion in sodium phosphate are Na^+ and PO_3^{3-}.
Charge balance:

$$3\left[Na^+\right]+1\left[PO_4^{3-}\right]=0$$
$$3(+1)+1(-3)=3-3=0$$

The correct formula of sodium phosphate is Na_3PO_4.

(e) The formulas for the cations and anion in potassium chromium sulfate are K^+, Cr^{3+}, and SO_4^{2-}.
Charge balance:

$$\left[K^+\right]+\left[Cr^{3+}\right]+2\left[SO_4^{2-}\right]=0$$
$$(+1)+(+3)+2(-2)=1+3-4=0$$

The correct formula of potassium chromium sulfate is $KCr(SO_4)_2$.

5.5 Nomenclature of Inorganic Compounds

Chemical compounds are named systematically as described by the following rules.

1. All binary compounds are named as ionic compounds, even though some of them may be covalent compounds. The name of the more electropositive (metallic) element is given first, followed by the name of the more electronegative (nonmetallic) element.
2. Monatomic cations retain the name of the parent element while monatomic anions have the ending of the parent element changed to –*ide*. For example, NaCl is named as sodium chlor*ide*.
3. When elements exhibiting variable valency or ONs combine to form more than one compound, differentiate the various compounds by using Greek prefixes such as mono- (one), di- (two), tri- (three), tetra- (four) penta- (five), hexa- (six), hepta- (seven), etc. Table 5.3 gives some examples.

Table 5.3   Compound names using Greek prefixes

Name	Formula
Carbon monoxide	CO
Carbon dioxide	CO_2
Dinitrogen monoxide	N_2O
Nitrogen monoxide	NO
Nitrogen dioxide	NO_2
Dinitrogen trioxide	N_2O_3
Dinitrogen tetroxide	N_2O_4
Dinitrogen pentoxide	N_2O_5
Phosphorus trichloride	PCl_3
Phosphorus pentachloride	PCl_5

- Binary compounds can also be named by using Roman numerals in parentheses to indicate the ON of the more electropositive element, followed by the name of the more electronegative element ending in –*ide*. Examples are shown in Table 5.4. Exception to the –ide ending rule: a few nonbinary compounds are also named according to the -ide rule. These exceptions include compounds of ammonium (NH_4^+) such as NH_4I, ammonium iodide; those of hydroxide, OH^-, such as calcium hydroxide, $Ca(OH)_2$; cyan*ides* such as KCN, potassium cyan*ide*; and hydrosulf*ide* such as NaSH, sodium hydrosulf*ide*.
- *Binary acids:* These acids consist of hydrogen and a nonmetal anion. To name a binary acid, add the prefix *hydro-* to the name of the nonmetallic element, and add the suffix –*ic* after the nonmetal name. Then add the word *acid*. For example, HCl and H_2S are named as *hydro*-chlor-*ic acid* and *hydro*-sulfur-*ic acid*. Table 5.5 gives additional examples.
- *Ternary compounds:* These contain three different elements and are usually made up of a cation and an anion. The anion is normally a polyatomic ion (a radical). They are named like binary compounds. The cationic group is named first, followed by the name of the anion. For example, $Mg(NO_3)_2$ is named as magnesium nitrate. Most polyatomic ions contain oxygen and usually have the suffix –*ate* or –*ite*. The –*ate* form normally indicates more oxygen or higher ON than the –*ite* form. For example, the nitrogen atom in sodium nitrate ($NaNO_3$) is in a +5 oxidation state and has more oxygen than sodium nitrite ($NaNO_2$) in

Table 5.4   Compound name using Roman numerals

Name	Formula
Iron (II) chloride	$FeCl_2$
Iron (III) chloride	$FeCl_3$
Lead (II) oxide	PbO
Lead (IV) oxide	PbO_2
Nitrogen (II) oxide	NO
Nitrogen (IV) oxide	NO_2
Phosphorus (III) chloride	PCl_3
Phosphorus (V) chloride	PCl_5
Sulfur (IV) oxide	SO_2
Sulfur (VI) oxide	SO_3

Table 5.5   Names and formulas of some binary acids

Formula	Acid name	Covalent compound name
HF	Hydrofluoric acid	Hydrogen fluoride
HCl	Hydrochloric acid	Hydrogen chloride
HBr	Hydrobromic acid	Hydrogen bromide
HI	Hydroiodic acid	Hydrogen iodide
H_2S	Hydrosulfuric acid	Hydrogen sulfide
H_2Se	Hydroselenic acid	Hydrogen selenide

which the nitrogen atom is in a $+3$ oxidation state. It is helpful to memorize the names of the various polyatomic ions.

- *Ternary acids:* These contain hydrogen and an oxygen-containing polyatomic ion known as an oxyanion. The name of a ternary acid is formed by adding the suffix *–ic* or *–ous acid* to the root name of the anion. For an anion ending with –ate, the –ate is replaced with –ic *acid*. For example, HNO_3 contains the nitrate (NO_3^-) anion and is named as nitric acid. For an anion ending with –ite, the –ite is replaced with *–ous acid*. For example, HNO_2 contains the nitrite (NO_2^-) anion and is named as nitrous acid. Table 5.6 lists examples of some ternary or oxy-acids.
- *Oxy-halo acids*: Named from the corresponding anion depending on the number of oxygen atoms present. Table 5.6 illustrates the rule for naming the oxy-halogen acids.
- *Ternary bases*: These consist of a metallic or polyatomic cation and a hydroxide ion. Bases have the ending *–ide,* like binary compounds. For example, NaOH is sodium hydroxide, $Al(OH)_3$ is aluminum hydroxide, NH_4OH is ammonium hydroxide, and $Ca(OH)_2$ is calcium hydroxide. See Table 5.7 for further examples.

Example 5.5

Give the systematic names of the following compounds.

(a) BaO
(b) MnO_2
(c) Fe_2O_3

Table 5.6   The oxy-halogen acid series

Formula	No. of oxygen atoms	Prefix used	Suffix used	Name of the oxy-halo acids
$HClO$	1	Hypo-	-ous	Hypochlorous acid
$HClO_2$	2		-ous	Chlorous acid
$HClO_3$	3		-ic	Chloric acid
$HClO_4$	4	Per-	-ic	Perchloric
$HBrO$	1	Hypo-	-ous	Hypobromous acid
$HBrO_2$	2		-ous	Bromous acid
$HBrO_3$	3		-ic	Bromic acid
$HBrO_4$	4	Per-	-ic	Perbromic
HIO	1	Hypo-	-ous	Hypoiodous acid
HIO_2	2		-ous	Iodous acid
HIO_3	3		-ic	Iodic acid
HIO_4	4	Per-	-ic	Periodic acid

Table 5.7   Names of some oxy-acids and bases

Ternary Acids		Bases	
Formula	Acid name	Formula	Name
$HClO$	Hypochlorous acid	$LiOH$	Lithium hydroxide
$HClO_2$	Chlorous acid	$NaOH$	Sodium hydroxide
$HClO_3$	Chloric acid	KOH	Potassium hydroxide
$HClO_4$	Perchloric acid	$Mg(OH)_2$	Magnesium hydroxide
HNO_3	Nitric acid	$Ca(OH)_2$	Calcium hydroxide
HNO_2	Nitrous acid	$Fe(OH)_2$	Iron (II) hydroxide
H_2SO_3	Sulfurous acid	$Fe(OH)_3$	Iron (III) hydroxide
H_2SO_4	Sulfuric acid	$Al(OH)_3$	Aluminum hydroxide
H_2CO_3	Carbonic acid	$Ba(OH)_2$	Barium hydroxide
H_3PO_3	Phosphorous acid	$Pb(OH)_2$	Lead hydroxide
H_3PO_4	Phosphoric acid	$Cu(OH)_2$	Copper (II) hydroxide
$HC_2H_3O_2$	Acetic (ethanoic) acid	$Cd(OH)_2$	Cadmium hydroxide
$H_2C_2O_4$	Oxalic acid	$Sr(OH)_2$	Strontium hydroxide
H_3BO_3	Boric acid	$Zn(OH)_2$	Zinc hydroxide

(d) SO_3
(e) N_2O_5
(f) AlN
(g) Na_3P
(h) V_2O_5

Solution

(a) BaO barium oxide
(b) MnO_2 manganese (IV) oxide
(c) Fe_2O_3 iron (III) oxide
(d) SO_3 sulfur (VI) oxide (or sulfur trioxide)
(e) N_2O_5 nitrogen (V) oxide (or dinitrogen tetroxide)

(f) AlN   aluminum nitride
(g) Na_3P sodium phosphide
(h) V_2O_5 vanadium (V) oxide

Example 5.6

Give the systematic names of the following compounds.

(a) $Cu_3(PO_4)_2$
(b) Cs_2SO_3
(c) $KMnO_4$
(d) $Na_2Cr_2O_7$
(e) CaC_2O_4
(f) $Ca_3(BO_3)_2$
(g) $NaClO_4$
(h) $Hg_3(PO_4)_2$

Solution

(a) $Cu_3(PO_4)_2$ copper (II) phosphate
(b) Cs_2SO_3 cesium sulfite
(c) $KMnO_4$ potassium permanganate
(d) $Na_2Cr_2O_7$ sodium dichromate
(e) CaC_2O_4 calcium oxalate
(f) $Ca_3(BO_3)_2$ calcium borate
(g) $NaClO_4$ sodium perchlorate
(h) $Hg_3(PO_4)_2$ mercury (II) phosphate

5.6 Problems

1. Assign ONs to each element in the following:
   (a) CO_2 (b) SO_3 (c) SF_6 (d) N_2H_4 (e) PbO

2. Assign ONs to nitrogen (N) in the following:
   (a) NO   (b) NO_2 (c) N_2O (d) N_2 (e) NH_3

3. Determine the ONs of the atoms in the following ions:
   (a) N^{3-} (b) Te^{2-} (c) PO_3^{3-} (d) SO_4^{2-} (e) ClO_3^-

4. Determine the ONs of the atoms in the following ions:
   (a) MnO_4^- (b) $Cr_2O_7^{2-}$ (c) UO_2^{2+} (d) $S_2O_3^{2-}$ (e) $S_4O_6^{2-}$*
   (*Hint: The average charge of the sulfur atoms in this ion may not be a whole number. This is uncommon but does happen).

5. What is the oxidation state of the metal present in each species?
   (a) $Fe_2(SO_4)_3$ (b) Cu_2O (c) $NiSO_4$ (d) $Rh_2(CO_3)_3$ (e) Fe_3O_4

6. Assign ONs to the underlined element in each of the following:
   (a) Na$\underline{Au}$Cl$_4$ (b) $H_2\underline{S}O_3$ (c) $Mg(\underline{Cl}O_3)_2$ (d) $K_2\underline{Cr}_2O_7$ (e) $Li_4\underline{P}_2O_7$

7. Assign ONs to the underlined element in each of the following:
   (a) $\underline{Mn}O_4^{2-}$ (b) $\underline{Fe}(CN)_6^{3-}$ (c) $\underline{Cl}O_4^-$ (d) $H\underline{V}_{10}O_{28}^{5-}$ (e) $LiAl\underline{H}_4$

8. Write the formula of the compound that would be formed between the following pairs of elements:
   (a) Mg and N (b) Sn and F (c) H and S (d) In and I (e) Al and Br

9. Write the formula of the compound that would be formed between the following pairs of elements:
   (a) Li and N (b) B and O (c) Ca and O (d) Rb and Cl (e) Cs and S

10. What compounds would form between the following pairs of anions and cations?
    (a) Fe^{3+} and CO_3^{2-} (b) NH_4^+ and PO_4^{3-} (c) Ca^{2+} and NO_3^-
    (d) Li and ClO_4^- (e) K^+ and $Cr_2O_7^{2-}$

11. What are the formulas of the compounds formed between the following pairs of anions and cations?
    (a) Mn^{2+} and CO_3^{2-} (b) Sn^{2+} and AsO_4^{3-} (c) Ca^{2+} and $C_2H_3O_2^-$ (d) Fe^{3+} and ClO_4^-
    (e) Na^+ and BO_3^{3-}

12. Write the formulas of the following binary compounds:
    (a) Sulfur dioxide
    (b) Carbon dioxide
    (c) Nitrogen dioxide
    (d) Dinitrogen pentoxide
    (e) Carbon tetrachloride
    (f) Chlorine dioxide
    (g) Lithium iodide
    (h) Selenium dioxide
    (i) Iron (II) chloride
    (j) Barium phosphide

13. Write the formula of each of the following compounds:
    (a) Vanadium (V) oxide
    (b) Copper (II) sulfide
    (c) Iron (III) sulfide
    (d) Gallium nitride
    (e) Mercury (II) chloride

14. Write the formula for each compound:
    (a) Sodium carbonate
    (b) Ammonium chloride
    (c) Potassium phosphate
    (d) Calcium hydrogen phosphate
    (e) Iron (III) chromate
    (f) Palladium (II) phosphate
    (g) Aluminum hydrogen carbonate
    (h) Potassium dichromate
    (i) Iron (III) hydroxide
    (j) Magnesium borate

15. Write the names of the following binary compounds:
    (a) PCl_3 (b) PCl_5 (c) CO (d) CO_2 (e) SO_2 (f) SO_3 (g) SiO_2
    (h) P_2S_5 (i) N_2O_5

16. Name the following compounds:
    (a) Ni_3N_2 (b) $FeCl_3$ (c) Al_2S_3 (d) GeS_2 (e) TiS_2 (f) CaH_2 (g) $HgCl_2$
    (h) Hg_6P_2 (i) Cu_3N

17. Name the following acids:
    (a) HF (b) HBr (c) H_2Se (d) H_2S (e) HCN

18. Name the following oxygen-containing acids:
    (a) HNO_3 (b) $HBrO_3$ (c) $HClO$ (d) $H_2C_2O_4$ (e) H_3PO_3 (f) HIO_3

19. Write the name of each salt and the name of the acid from which the salt may be derived:
    (a) $Ga(NO_3)_3$ (b) $CoSO_4$ (c) $Fe(C_2H_3O_2)_2$ (d) $Ca_3(BO_3)_2$ (e) PbC_2O_4
    (f) $CrPO_4$ (g) $NiCO_3$ (h) $Fe(CN)_3$ (i) AlI_3 (j) $RbBrO_3$

20. Write the name of each of the following inorganic bases:
    (a) KOH (b) NH_4OH (c) $Co(OH)_2$ (d) $Ba(OH)_2$ (e) $Cr(OH)_3$

6

Chemical Equations

....

6.1 Writing Chemical Equations

A chemical equation is a shorthand way of describing a chemical reaction. It uses symbols of elements and formulas of compounds in place of words to describe a chemical change or reaction.

6.1.1 General rules for writing chemical equations

1. The formulas and symbols of reactants are written on the left side of the equation.
2. The formulas and symbols of products are written on the right side of the equation.
3. A plus sign ($+$) is placed between different reactants and different products.
4. The reactants are separated from the products by an arrow ($\longrightarrow$) pointing in the direction of the reaction. For a reversible reaction, a double arrow ($\rightleftharpoons$) is used.
5. The physical states of substances may be indicated by the symbols (s) for solid, (l) for liquid, (aq) for aqueous solution, and (g) for gases.
6. The equation is then balanced by inserting appropriate coefficients for each product and reactant.

Consider the following word equation:
Gaseous sulfur trioxide reacts with liquid water to form aqueous sulfuric acid.
We can substitute the formulas for reactants and products, following the above rules, and obtain a chemical equation for the reaction as

$$SO_3\,(g) + H_2O\,(l) \longrightarrow H_2SO_4\,(aq)$$

6.2 Balancing Chemical Equations

An equation as translated from words may not be balanced. It only shows the substances present—reactants and products. A chemical equation is balanced when it has the same number of atoms of each element on either side of the equation. Thus, a balanced equation obeys the law of conservation of mass. That is, atoms are not "created" or "destroyed" in writing a chemical equation.

Chemistry in Quantitative Language: Fundamentals of General Chemistry Calculations. Second Edition. Christopher O. Oriakhi, Oxford University Press. © Christopher O. Oriakhi 2021. DOI: 10.1093/oso/9780198867784.003.0006

6.2.1 Guidelines for balancing a chemical equation

Balance the equation to make sure the law of conservation of mass is observed.

1. Inspect both sides of the equation to identify atoms that need to be balanced.
2. Balance one element at a time by placing a suitable coefficient to the left of the formula containing the element. Note that a coefficient placed in front of the formula affects all the atoms in the formula. For example, 2 H_2O implies two molecules of water containing four atoms of hydrogen and two atoms of oxygen.
3. Never attempt to balance an equation by changing subscripts because this will change the formulas of the compounds. For example, to balance the equation

$$H_2\,(g) + Cl_2\,(g) \longrightarrow HCl\,(g)$$

you cannot change the subscripts as below

$$H_2\,(g) + Cl_2\,(g) \longrightarrow H_2Cl_2\,(g)$$

because H_2Cl_2 is *not* the same as the compound HCl.
4. Make sure the elements already balanced are not changed since we are essentially doing this one element at a time.
5. Balance polyatomic ions (e.g. NH_4^+ and CO_3^{2-}) as single entities if they appear unchanged on both sides of the equation.
6. The balanced equation should contain the smallest possible set of whole-number coefficients. For example, a balanced equation showing the decomposition of copper (II) oxide should be expressed as:

$$2\,CuO\,(s) \longrightarrow 2\,Cu\,(s) + O_2\,(g)$$

And not as:

$$4\,CuO\,(s) \longrightarrow 4\,Cu\,(s) + 2\,O_2\,(g)$$

7. Do a final check by counting atoms on each side of the equation.

You may find some equations difficult to balance using these rules, for example, redox equations. These will be treated under Oxidation and Reduction Reactions, in Chapter 22.

Example 6.1

Balance the equation

$$H_2 + O_2 \longrightarrow H_2O \text{ (Unbalanced)}$$

Solution

Hydrogen is balanced since there are two atoms of hydrogen on each side of the equation. Oxygen is not balanced. There are two atoms on the left hand and one atom on the right side. To balance oxygen, place the coefficient 2 in front of H_2O.

$$H_2 + O_2 \longrightarrow 2H_2O \text{ (Unbalanced)}$$

Oxygen is now balanced but hydrogen is no longer balanced. To balance hydrogen, place the coefficient 2 in front of H_2.

$$2\,H_2 + O_2 \longrightarrow 2\,H_2O \quad \text{(Balanced)}$$

There are four atoms of hydrogen and two atoms of oxygen on either side of the equation. The equation is now balanced.

Example 6.2

Sodium metal reacts with water to form sodium hydroxide and hydrogen. Write a balanced chemical equation for this reaction.

Solution

$$\text{Sodium} + \text{Water} \longrightarrow \text{Sodium hydroxide} + \text{hydrogen}$$

$$Na + 2H_2O \longrightarrow 2NaOH + H_2 \quad \text{(Unbalanced)}$$

By inspection, Na and O are balanced on both sides of the equation, but hydrogen is not. Balance hydrogen by placing the coefficient 2 in front of H_2O and NaOH.

$$Na + 2H_2O \longrightarrow 2NaOH + H_2 \quad \text{(Unbalanced)}$$

All the atoms except Na are balanced. Balance Na by placing the coefficient 2 in front of it.

$$2\,Na + 2\,H_2O \longrightarrow 2\,NaOH + H_2 \quad \text{(Balanced)}$$

Final check: There are four atoms of H, two atoms of O, and two atoms of Na on either side of the equation. The equation is balanced.

Example 6.3

Balance the following chemical equations:

(a) $Fe + O_2 \longrightarrow Fe_2O_3$
(b) $Fe_2O_3 + CO \longrightarrow Fe + CO_2$
(c) $H_3PO_4 + Ca(OH)_2 \longrightarrow Ca_3(PO_4)_3 + H_2O$
(d) $C_2H_3Cl + O_2 \longrightarrow CO_2 + H_2O + HCl$

Solution

(a) $4\,Fe + 3\,O_2 \longrightarrow 2\,Fe_2O_3 \quad \text{(Balanced)}$
(b) $Fe_2O_3 + 3\,CO \longrightarrow 2\,Fe + 3\,CO_2 \quad \text{(Balanced)}$
(c) $2\,H_3PO_4 + 3\,Ca(OH)_2 \longrightarrow Ca_3(PO_4)_2 + 6\,H_2O \quad \text{(Balanced)}$
(d) $2\,C_2H_3Cl + 5\,O_2 \longrightarrow 4\,CO_2 + 2\,H_2O + 2HCl \quad \text{(Balanced)}$

6.3 Types of Chemical Reactions

There are numerous types of chemical reactions. Most conform to one of the following:

- Combination or synthesis
- Decomposition
- Displacement
- Double decomposition or metathesis
- Neutralization
- Oxidation-reduction.

The above classification is helpful when writing chemical equations and can assist in predicting the products of other reactions.

6.3.1 Combination or synthesis

A combination reaction is one in which two or more reactants (elements or compounds) combine to form a single product. The general equation for a combination is:

$$A + B \longrightarrow AB$$

Some examples involving elements include:

$$H_2 + Br_2 \longrightarrow 2\,HBr$$
$$S + O_2 \longrightarrow SO_2$$
$$2\,Na + S \longrightarrow Na_2S$$
$$4\,Fe + 3\,O_2 \longrightarrow 2\,Fe_2O_3$$

Some examples involving compounds are:

$$2\,NO_2 + H_2O_2 \longrightarrow 2\,HNO_3$$
$$Na_2O + H_2O \longrightarrow 2\,NaOH$$
$$SO_3 + H_2O \longrightarrow H_2SO_4$$

6.3.2 Decomposition

A decomposition reaction is one in which a single reactant is broken down into two or more simpler substances, usually under the action of heat. The general equation for a decomposition reaction is

$$AB \longrightarrow A + B$$

Representative examples of decomposition reactions include:

$$2 \, NaClO_3 \xrightarrow{Heat} 2 \, NaCl + 3 \, O_2$$

$$CaCO_3 \xrightarrow{Heat} CaO + CO_2$$

$$Mg(OH)_2 \xrightarrow{Heat} MgO + H_2O$$

$$2 \, CuO \xrightarrow{Heat} 2 \, Cu + O_2$$

6.3.3 Displacement

A displacement reaction is one in which a free element replaces another element within a compound. The general equation for a displacement replacement is

$$A + BC \longrightarrow B + AC$$

Representative examples involving metals and non-metals include:

$$Mg + H_2SO_4 \longrightarrow H_2 + MgSO_4$$

$$Cl_2 + 2 \, HBr \longrightarrow Br_2 + 2 \, HCl$$

$$Fe + CuCl_2 \longrightarrow FeCl_2 + Cu$$

6.3.4 Double decomposition or metathesis

A double decomposition (or metathesis) reaction is one in which two compounds exchange ions with each other to form two new compounds. The general equation for a double decomposition reaction is

$$A^+B^- + C^+D^- \longrightarrow A^+D^- + C^+B^-$$

Examples include:

$$AgNO_3 + NaCl \longrightarrow AgCl + NaNO_3$$

$$Ba(NO_3)_2 + Na_2SO_4 \longrightarrow BaSO_4 + 2 \, NaNO_3$$

$$Pb(NO_3)_2 + H_2S \longrightarrow PbS + 2 \, HNO_3$$

6.3.5 Neutralization

A neutralization reaction is one in which an acid reacts with a base to form a salt and water only. The general expression is:

$$Acid + Base \longrightarrow Salt + Water$$

Some examples include:

$$H_2SO_4 + 2 \, KOH \longrightarrow K_2SO_4 + H_2O$$

$$HCl + NaOH \longrightarrow NaCl + H_2O$$

This is a special form of double displacement reaction. Acid-base reactions are discussed in Chapter 14 and Chapter 18.

6.4 Problems

1. Balance the following equations:
   (a) $SO_2 + H_2O \longrightarrow H_2SO_3$
   (b) $H_2O \longrightarrow H_2 + O_2$
   (c) $PCl_3 + Cl_2 \longrightarrow PCl_5$
   (d) $C + O_2 \longrightarrow CO$
   (e) $Mg + O_2 \longrightarrow MgO$

2. Balance the following equations:
   (a) $NH_3 + O_2 \longrightarrow N_2 + H_2O$
   (b) $I_4O_9 \xrightarrow{\Delta} I_2O_5 + I_2 + O_2$
   (c) $K_2S_2O_3 + Cl_2 + H_2O \xrightarrow{\Delta} KHSO_4 + HCl$
   (d) $CS_2 + O_2 \longrightarrow CO_2 + SO_2$

3. Write and balance equations for the following.
   (a) Calcium carbonate $\xrightarrow{\text{Heat}}$ Calcium oxide + Carbon dioxide
   (b) Potassium chlorate $(KClO_3) \xrightarrow{\Delta}$ Potassium chloride + Oxygen
   (c) Lithium nitride $(Li_3N) \xrightarrow{\Delta}$ Lithium + Nitrogen
   (d) Hydrogen peroxide $\xrightarrow{\Delta}$ Water + Oxygen

4. Balance the following equations, which represent the-solid state synthesis of technologically important materials.
   (a) $TiCl_4 + H_2S \longrightarrow TiS_2 + HCl$
   (b) $LaCl_3 \cdot 7H_2O \longrightarrow LaOCl + HCl + H_2O$
   (c) $CsCl + ScCl_3 \longrightarrow Cs_3Sc_2Cl_9$
   (d) $Li_2CO_3 + Fe_2O_3 \longrightarrow LiFe_5O_8 + CO_2$
   (e) $La_2O_3 + B_2O_3 \longrightarrow LaB_6 + O_2$
   (f) $ScCl_3 + SiO_2 \rightleftharpoons Sc_2Si_2O_7 + SiCl_4$

5. Balance the equations for the thermal decomposition of the following compounds:
   (a) $NaN_3 \longrightarrow Na + N_2$
   (b) $(NH_4)_2 Cr_2O_7 \longrightarrow Cr_2O_3 + N_2 + H_2O$
   (c) $Ag_2CO_3 \longrightarrow Ag + CO_2 + O_2$
   (d) $NaHCO_3 \longrightarrow Na_2CO_3 + CO_2 + H_2O$
   (e) $Al_2(CO_3)_3 \longrightarrow Al_2O_3 + CO_2$

6. Complete and balance the following neutralization reactions.
   (a) $H_2SO_4 + KOH \longrightarrow$
   (b) $H_3PO_4 + Ba(OH)_2 \longrightarrow$
   (c) $HBr + Sn(OH)_4 \longrightarrow$
   (d) $HBr + NaOH \longrightarrow$
   (e) $H_4P_2O_7 + NaOH \longrightarrow$

7. Complete and balance the following double decomposition reactions.
   (a) $Ca(NO_3)_2 + Na_2SO_4 \longrightarrow$
   (b) $Mg(NO_3)_2 + Li_2S \longrightarrow$
   (c) $Pb(NO_3)_2 + Cs_2CrO_4 \longrightarrow$

(d) $FeCl_3 + CaS \longrightarrow$

(e) $Na_3PO_4 + Zn(OH)_2 \longrightarrow$

8. The industrial production of phosphorus is represented by the following reaction:

$$Ca_3(PO_4)_2 + SiO_2 + C \longrightarrow P_4 + CaSiO_3 + CO_2$$

Balance the equation for the reaction.

9. Hydrocarbons and their oxygen derivatives react with oxygen to produce carbon dioxide and water. The relative amounts of carbon dioxide and water formed depend on the composition of the parent hydrocarbon. Complete and balance the following equations:

(a) $C_4H_8 + O_2 \longrightarrow$

(b) $C_6H_6 + O_2 \longrightarrow$

(c) $C_6H_{12}O + O_2 \longrightarrow$

(d) $C_{12}H_{22}O_{11} + O_2 \longrightarrow$

(e) $C_4H_{10} + O_2 \longrightarrow$

10. Write balanced chemical equations for the following reactions:

(a) Iron and sulfur react at elevated temperature to form iron (III) sulfide.

(b) Octane, C_8H_{18}, a component of gasoline, burns in oxygen to produce carbon dioxide and water.

(c) Sulfur dioxide burns in oxygen to form sulfur trioxide.

(d) Barium hydroxide reacts with phosphoric acid to produce barium phosphate and water.

(e) Solid aluminum dissolves in hydrochloric acid to form aluminum chloride and hydrogen gas.

11. Write balanced chemical reactions for the following:

(a) Aluminum reacts with chlorine to form aluminum chloride.

(b) Potassium reacts with water to form potassium hydroxide and hydrogen gas.

(c) Boron trichloride gas reacts with steam (gaseous H_2O) to form boron trihydroxide and hydrogen chloride.

(d) Silver nitrate solution reacts with sodium chloride to give aqueous sodium nitrate and precipitate of silver chloride.

(e) Silicon tetrafluoride reacts with sodium hydroxide to form sodium silicate (Na_4SiO_4), sodium fluoride, and water.

(f) Calcium hydroxide reacts with ammonium chloride to form ammonia, calcium chloride, and water.

12. Balance the following chemical reactions.

(a) $Si_4H_{10} + O_2 \longrightarrow SiO_2 + H_2O$

(b) $CH_3NO_2 + Cl_2 \longrightarrow CCl_3NO_2 + HCl$

(c) $NaF + CaO + H_2O \longrightarrow CaF_2 + NaOH$

(d) $Al_4C_3 + H_2O \longrightarrow Al(OH)_3 + CH_4$

(e) $TiO_2 + B_4C + C \longrightarrow TiB_2 + CO$

(f) $C_7H_{16}O_4S_2 + O_2 \longrightarrow CO_2 + H_2O + SO_2$

7

Stoichiometry

. . . .

7.1 Reaction Stoichiometry

Stoichiometry is the study of quantitative relationships among the reactants and products in a chemical reaction. It essentially involves the calculation of the quantities of reactants and products involved in a chemical reaction. The coefficients in a balanced chemical equation, known as stoichiometric coefficients, indicate the number of moles of each reactant taking part in the reaction, and the number of moles of each product formed at the end of the reaction.

7.2 Information From a Balanced Equation

In addition to identifying reactants and products in chemical reaction, a balanced equation gives useful information that is helpful in calculations. Consider the equation for the reaction between ammonia and oxygen to produce nitrogen (II) oxide:

$$4\,NH_3\,(g) + 5\,O_2\,(g) \longrightarrow 4\,NO\,(g) + 6\,H_2O\,(g)$$

The following information can be obtained:

1. Molecules of reactant and products: four molecules of NH_3 react with five molecules of O_2 to form four molecules of NO and six molecules of H_2O.
2. Moles of reactants and products: four moles of NH_3 react with five moles of O_2 to produce four moles of NO and six moles of H_2O.
3. Mass of reactants and products: 68 g of NH_3 react with 160 g of O_2 to produce 120 g of NO and 108 g of H_2O.
4. Volumes of gases: four volumes of NH_3 react with five volumes of O_2 to produce four volumes of NO and six volumes of H_2O at the same temperature and pressure (by Avogadro's law, which is discussed in detail in Section 11.5 of Chapter 11).

7.3 Types of Stoichiometric Problems

There are several types of stoichiometric problems. The common types include:

1. Mole to mole
2. Mass to mass

Chemistry in Quantitative Language: Fundamentals of General Chemistry Calculations. Second Edition. Christopher O. Oriakhi, Oxford University Press. © Christopher O. Oriakhi 2021. DOI: 10.1093/oso/9780198867784.003.0007

3. Mass to mole (or mole to mass)
4. Mole to volume (or volume to mole)
5. Mass to volume (or volume to mass)
6. Volume to volume.

7.3.1 Solving stoichiometric problems

The following general steps can be used to solve many stoichiometric problems:

1. Write the balanced chemical equation for the reaction.
2. Organize your data; determine which quantities you know and which ones you need to find.
3. Write down the mole relationship between the given substance and the required substance.
4. Calculate molar masses and convert masses, molecules, or volumes of the known substance to moles, if necessary.
5. Use stoichiometric coefficients or conversion factors (mole ratios) from the equation to determine the moles of the unknown substance.
6. Convert moles of the unknown substance to the desired mass, molecules, or volume, if necessary.

Figure 7.1 is a summary of the various conversion processes outlined above.

7.3.2 Mole-to-mole stoichiometric problems

In this type of problem, you are given the moles of one component of the reaction and required to find another.

Example 7.1

The balanced equation for the Haber process used for the industrial production of ammonia is:

$$N_2\,(g) + 3\,H_2\,(g) \longrightarrow 2\,NH_3\,(g)$$

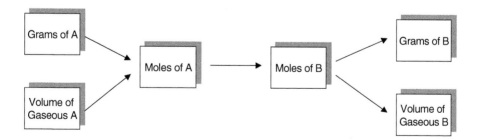

Figure 7.1   A block diagram summarizing the conversion processes used in solving stoichiometric problems.

Calculate the exact moles of nitrogen required to produce 10 moles of ammonia.

Solution

Step 1: Determine whether the equation is balanced—this one is.
Step 2: Organize your information:
        Known: 10 mol of NH_3
        Unknown: moles of N_2
Step 3: Write down the mole relationship between NH_3 and N_2 and formulate the conversion factor.
        1 mol of $N_2 \Rightarrow 2$ moles of NH_3.
        The conversion factors are:

$$\frac{1 \text{ mol } N_2}{2 \text{ mol } NH_3} \quad \text{and} \quad \frac{2 \text{ mol } NH_3}{1 \text{ mol } N_2}$$

Step 4: Select the conversion factor from step 3 that will cancel the units of the known (mol of NH_3) when multiplied by the given amount of the known (i.e. 10 mol NH_3). Use this conversion factor to calculate the moles of N_2:

$$\text{mol } N_2 = (10 \text{ mol } NH_3) \times \left(\frac{1 \text{ mol } N_2}{2 \text{ mol } NH_3} \right) = 5 \text{ mol } N_2$$

Example 7.2

Iron pyrite (FeS_2) burns in air according to the equation:

$$4 \text{ FeS}_2 + 11 \text{ O}_2 \longrightarrow 2 \text{ Fe}_2O_3 + 8 \text{ SO}_2$$

How many moles of iron (III) oxide (Fe_2O_3) are obtained from the combustion of 5 moles of iron pyrite?

Solution

Step 1: We have a balanced equation.
Step 2: Organize your information:
        Known: 5 mol of FeS_2
        Unknown: moles Fe_2O_3
Step 3: Write down the mole relationship between FeS_2 and Fe_2O_3 and formulate the conversion factor. Prepare to calculate molar masses of FeS_2 and Fe_2O_3.
        4 moles of $FeS_2 \Rightarrow 2$ moles of Fe_2O_3
        The conversion factors are:

$$\frac{2 \text{ mol } Fe_2O_3}{4 \text{ mol } FeS_2} \quad \text{and} \quad \frac{4 \text{ mol } FeS_2}{2 \text{ mol } Fe_2O_3}$$

Step 4: Select the conversion factor from step 3 that will cancel the units of the known (mol of FeS_2) when multiplied by the given amount of the known (i.e. 5 mol FeS_2). Use this conversion factor to calculate the moles of Fe_2O_3:

$$\text{mol Fe}_2\text{O}_3 = (5 \text{ mol FeS}_2) \times \left(\frac{2 \text{ mol Fe}_2\text{O}_3}{4 \text{ mol FeS}_2} \right) = 2.5 \text{ mol Fe}_2\text{O}_3$$

7.3.3 Mass-to-mole stoichiometry problems

Here, information about the known quantities is given in mass units, and information about the unknown is required in mole units.

Example 7.3.

The combustion of acetylene gas is represented by the equation:

$$2 \text{ C}_2\text{H}_2 \text{ (g)} + 5 \text{ O}_2 \text{ (g)} \longrightarrow 4 \text{ CO}_2 \text{ (g)} + 2 \text{ H}_2\text{O (g)}$$

Calculate the moles of oxygen (O_2) required for the complete combustion of 25 g of acetylene, (C_2H_2).

Solution

Step 1: We have an equation—in this case, already balanced.
Step 2: Organize your information:
$C_2H_2 = 26$ g/mol
$O_2 = 32$ g/mol
Known: 25 g of C_2H_2
Unknown $=$ moles O_2
Step 3: Write down the mole relationship between C_2H_2 and O_2 and formulate the conversion factor.
2 mole of $C_2 H_2 \Rightarrow 5$ mol of O_2 or
26 g of $C_2 H_2 \Rightarrow 5$ mol of O_2
The conversion factors are:

$$\frac{26 \text{ g C}_2\text{H}_2}{5 \text{ mol O}_2} \text{ and } \frac{5 \text{ mol O}_2}{26 \text{ g C}_2\text{H}_2}$$

Step 4: Select the conversion factor from step 3 that will cancel the units of the known (g of C_2H_2) when multiplied by the given amount of the known (i.e. 25 g C_2H_2). Use this conversion factor to calculate the moles of O_2:

$$\text{moles } O_2 = (25 \text{ g C}_2\text{H}_2) \times \left(\frac{5 \text{ mol O}_2}{52 \text{ g C}_2\text{H}_2} \right) = 2.404 \text{ mol O}_2$$

7.3.4 Mass-to-mass stoichiometry problems

Here, information about the known quantities is given in mass units, and information about the unknown is required in mass units.

Example 7.4

Pure iron can be produced by reacting hematite ore, Fe_2O_3, with carbon monoxide, CO. Write a balanced equation for the reaction. How many grams of carbon monoxide gas are needed to process 500 g of hematite?

Solution

Step 1: Write a balanced equation:

$$Fe_2O_3 + 3\,CO \longrightarrow 2\,Fe + 3\,CO_2$$

The sequence of conversions needed for this calculation is:

$$\text{grams } Fe_2O_3 \longrightarrow \text{mol } Fe_2O_3 \longrightarrow \text{mol CO} \longrightarrow \text{grams CO}$$

Step 2: Organize your information:
$Fe_2O_3 = 159.7$ g/mol
$CO = 26$ g/mol
Known: 500 g of Fe_2O_3
Unknown = grams CO

Step 3: Write down the mole relationship between Fe_2O_3 and CO and formulate the conversion factors.
1 mol of $Fe_2\,O_3$ $\Rightarrow$ 3 mol of CO
159.7 g $Fe_2\,O_3$ $\Rightarrow$ 1 mol $Fe_2\,O_3$
28 g CO $\Rightarrow$ 1 mol CO
The conversion factors are:

$$\frac{1 \text{ mol } Fe_2O_3}{159.7 \text{ g } Fe_2O_3}, \quad \frac{1 \text{ mol } Fe_2O_3}{3 \text{ mol CO}}, \quad \frac{3 \text{ mol CO}}{1 \text{ mol } Fe_2O_3}, \text{ and } \frac{1 \text{ mol CO}}{28 \text{ g CO}}$$

Step 4: Select the conversion factors from step 3 that will cancel the units of the known (g and moles of Fe_2O_3) when multiplied by the given amount of the known (i.e. 500 g Fe_2O_3). Use these conversion factors to find the mass of CO_2. Calculate molar masses and convert mass to moles of Fe_2O_3 using the appropriate conversion factor.

$$\text{g CO} = 500 \text{ g } Fe_2O_3 \times \frac{1 \text{ mol } Fe_2O_3}{159.7 \text{ g } Fe_2O_3} \times \frac{3 \text{ mol CO}}{1 \text{ mol } Fe_2O_3} \times \frac{28 \text{ g CO}}{1 \text{ mol CO}}$$

$$= 262.99 \text{ g CO} \approx 263.0 \text{ g CO}$$

7.3.5 Mass-to-volume stoichiometry problems

Mass-to-volume stoichiometric calculations involve reactions in which one or more of the substances is a gas. Given mass or moles of one component, we are required to find the volume of a gaseous component at a given temperature and pressure. The conversion of moles or grams of a gaseous compound to volume at standard conditions can be accomplished by remembering that 1 mole of a gas at STP occupies 22.4 L. When nonstandard conditions of temperature and pressure are imposed, the required conversion can be accomplished by the Ideal Gas Equation (as discussed in Chapter 11).

Example 7.5

In the laboratory, oxygen can be prepared by heating potassium chlorate. The equation for the reaction is

$$2 \text{ KClO}_3 \text{ (s)} \xrightarrow{\text{heat}} 2 \text{ KCl (s)} + 3 \text{ O}_2 \text{ (g)}$$

Calculate the volume of oxygen produced at STP by heating 20 g of $KClO_3$.

Solution

The sequence of conversions needed for this calculation is:

Step 1: The balanced equation is known.

$$\text{grams KClO}_3 \longrightarrow \text{mol KClO}_3 \longrightarrow \text{mol O}_2 \longrightarrow \text{volume O}_2$$

Step 2: Organize your information:
Calculate molar mass of $KClO_3$
$KClO_3 = 122$ g/mol
Known: 20 g of $KClO_3$
Unknown = volume of O_2 at STP

Step 3: Write down the mole relationship between $KClO_3$ and O_2 and formulate the conversion factors.
2 mole of $KClO_3$ $\Rightarrow$ 3 moles of O_2
The conversion factors are :

$$\frac{\text{I mol KClO}_3}{122 \text{ g KClO}_3}, \quad \frac{2 \text{ mol KClO}_3}{3 \text{ mol O}_2}, \quad \frac{3 \text{ mol O}_2}{2 \text{ mol KClO}_3}, \text{ and} \quad \frac{1 \text{ mol O}_2}{22.4 \text{ dm}^3 \text{ (or L) O}_2}$$

Step 4: Select the conversion factor from step 3 that will cancel the units of the known (g and mol of $KClO_3$) when multiplied by the given amount of the known (i.e. 20 g $KClO_3$). Use this conversion factor to calculate the volume of O_2 as follows:

$$\text{volume O}_2 = 20 \text{ g KClO}_3 \times \frac{1 \text{ mol KClO}_3}{122 \text{ g KClO}_3} \times \frac{3 \text{ mol O}_2}{2 \text{ mol KClO}_3} \times \frac{22.4 \text{ dm}^3 \text{ O}_2}{1 \text{ mol O}_2}$$

$$= 5.51 \text{ dm}^3 \text{ O}_2.$$

Example 7.6

What volume of nitrogen at STP will be released by the thermal decomposition of 40 g of ammonium nitrite?

Solution

Step 1: Write the balanced chemical equation.

$$NH_4NO_2 \text{ (aq)} \xrightarrow{\text{heat}} N_2 \text{ (g)} + 2 \, H_2O \text{ (g)}$$

Step 2: Organize your information:
Calculate molar mass of NH_4NO_2
$NH_4NO_2 = 64 \text{ g/mol}$
Known : 40 g of NH_4NO_2; 1 mole of N_2 gas occupies 22.4 liters at STP
Unknown = vol of N_2 at STP.

Step 3: Write the mole relationship between NH_4NO_2 and N_2 and formulate the conversion factors.

$$1 \text{ mol } NH_4NO_2 \quad \Leftrightarrow \quad 1 \text{ mol } N_2 \quad \Leftrightarrow \quad 22.4 \text{ L } N_2 \text{ at STP}$$

$$\frac{1 \text{ mol } N_2}{1 \text{ mol } NH_4NO_2}, \frac{1 \text{ mol } NH_4NO_2}{64 \text{ g } NH_4NO_2}, \frac{22.4 \text{ L } N_2}{1 \text{ mol } N_2}$$

Step 4: Select the conversion factor from step 3 that will cancel the units of the known (g and mol of NH_4NO_2) when multiplied by the given amount of the known (i.e. 40 g NH_4NO_2). Use this conversion factor to calculate the volume of N_2 as follows:

$$\text{volume } N_2 = 40 \text{ g } NH_4NO_2 \times \frac{1 \text{ mol } NH_4NO_2}{64 \text{ g } NH_4NO_2} \times \frac{1 \text{ mol } N_2}{1 \text{ mol } NH_4NO_2} \times \frac{22.4 \text{ dm}^3 \, N_2}{1 \text{ mol } N_2}$$

$$= 14.00 \text{ L } N_2.$$

7.3.6 Volume-to-volume stoichiometry problems

Volume-volume stoichiometry problems are based on the law of combining volumes, also known as Gay-Lussac's law, which states that at the same temperature and pressure, when gases react, or are formed, they do so in simple whole-number ratios of volume. This volume ratio is directly proportional to the values of the corresponding stoichiometric coefficients in the balanced equation. Hence, we can use the stoichiometric coefficients in the balanced equation to form volume relationships as in other types of stoichiometry problems. For example, consider the following gaseous reaction:

$$N_2 \text{ (g)} + 3 \, H_2 \text{ (g)} \longrightarrow 2 \, NH_3 \text{ (g)}$$

In terms of reacting moles, 1 mole of nitrogen gas reacts with 3 moles of hydrogen gas to form 2 moles of ammonia gas. Now in terms of reacting volumes, the equation indicates that one volume of nitrogen gas reacts with three volumes of hydrogen gas to form two volumes of ammonia gas. For example at STP, 22.4 L of nitrogen (that is, 1 mole) would react with 67.2 L (3×22.4 L) of hydrogen to form 44.8 L (2×22.4 L) of ammonia. The ratio remains the same as in the balanced equation, i.e., $N_2 : H_2 : NH_3$ equals 22.4 L: 67.2 L: 44.8 L (or simply $1 : 3 : 2$).

Example 7.7

Calculate the volume in liters at STP of oxygen gas needed and the volume of nitrogen (II) oxide gas and water vapor produced from the reaction of 1.12 L of ammonia gas. The equation for the reaction is:

$$4\ NH_3\ (g) + 5\ O_2\ (g) \longrightarrow 4\ NO\ (g) + 6\ H_2O\ (g)$$

Solution

All the reactants and products are gases at the same temperature and pressure. Therefore, their volumes are directly proportional to their stoichiometric coefficients in the balanced equation.

Step 1: The balanced equation is known.
Step 2: Organize your information:
  Known: 1.12 L of NH_3
  Unknown $=$ L of O_2, NO, and H_2O
Step 3: Write the mole and volume relationship between NH_3 and each of the unknowns (O_2, NO, and H_2O) and formulate the conversion factors.

$$4\ mol\ NH_3 \quad \Rightarrow \quad 5\ mol\ O_2 \quad \Rightarrow \quad 4\ Mol\ NO \quad \Rightarrow \quad 6\ mol\ H_2O$$

or

$$4\ L\ NH_3 \quad \Rightarrow \quad 5\ L\ O_2 \quad \Rightarrow \quad 4\ L\ NO \quad \Rightarrow \quad 6\ L\ H_2O$$

This corresponds to the following volume conversion factors:

$$\frac{5\ L\ O_2}{4\ L\ NH_3} ; \frac{4\ L\ NO}{4\ L\ NH_3} ; \frac{6\ L\ H_2O}{4\ L\ NH_3}$$

Step 4: Use the conversion factors to calculate the volumes of the unknowns:

$$\text{Vol. of } O_2 : 1.12\ L\ NH_3 \times \frac{5\ L\ O_2}{4\ L\ NH_3} = 1.4\ L\ O_2$$

$$\text{Vol. of } NO : 1.12\ L\ NH_3 \times \frac{4\ L\ NO}{4\ L\ NH_3} = 1.12\ L\ NO$$

$$\text{Vol. of } H_2O : 1.12\ L\ NH_3 \times \frac{6\ L\ H_2O}{4\ L\ NH_3} = 1.68\ L\ H_2O$$

7.4 Limiting Reagents

When chemical reactions are carried out in the laboratory, we do not add reactants in the exact molar ratios indicated by the balanced equation. For various reasons, one or more reactants are usually present in large excess. In most cases, only one reactant is completely consumed at the end of the reaction, and this reactant determines the amount of products formed. This is called the *limiting reagent or reactant*. Once the limiting reagent is entirely consumed, the reaction stops; the other excess reactants will still be present in some amount. *Reactant in excess* is the reactant present in a quantity greater than is needed to completely react with the limiting reagent.

7.4.1 Limiting reagent calculations

For chemical reactions involving two or more reactants, it is necessary to determine which one is the limiting reagent. The following procedure is helpful in determining the limiting reagent:

1. Make sure the chemical equation is balanced.
2. Calculate the number of moles of each of the reactants present from their given amount.
3. Calculate the amount of product that can be formed from the complete reaction of each reactant.
4. Determine which of the reactants would produce the least amount of the product. This is the limiting reagent.

Example 7.8

A mixture of 50 g of CaO and 50 g of H_2O react in an autoclave to form calcium hydroxide according to the equation

$$CaO + H_2O \longrightarrow Ca(OH)_2$$

What is the limiting reagent?

Solution

1. We have a balanced equation.

$$CaO + H_2O \longrightarrow Ca(OH)_2$$

2. Determine the number of moles of each reactant:

$$\text{Moles of CaO} = 50 \text{ g} \times \frac{1 \text{ mol CaO}}{56 \text{ g}} = 0.893 \text{ mol}$$

$$\text{Moles of } H_2O = 50 \text{ g} \times \frac{1 \text{ mol CaO}}{18 \text{ g}} = 2.778 \text{ mol}$$

3. Calculate the moles of $Ca(OH)_2$ formed from each reactant:

Moles of $Ca(OH)_2$ produced from 0.893 mol CaO:

$$\text{Moles Ca(OH)}_2 = 0.893 \text{ mol CaO} \times \frac{1 \text{ mol Ca(OH)}_2}{1 \text{ mol CaO}} = 0.893 \text{ mol}$$

Moles of $Ca(OH)_2$ produced from 2.778 mol H_2O:

$$\text{Moles Ca(OH)}_2 = 2.778 \text{ mol } H_2O \times \frac{1 \text{ mol Ca(OH)}_2}{1 \text{ mol } H_2O} = 2.778 \text{ mol}$$

4. Now determine the limiting reagent. Since CaO produces the least amount of $Ca(OH)_2$, it is the limiting reagent.

7.5 Reaction Yields: Theoretical, Actual, and Percent Yields

When reactions are carried out in the laboratory, the amount of product isolated is always less than the amount predicted from the balanced equation. There are several reasons for this. For example, some of the products may be lost during isolation and purification, some undesired products may be formed due to side reactions, or the reactant may not undergo complete reaction. The *theoretical yield* is the maximum amount of a product that can be formed as calculated from a chemical equation representing the reaction. The *actual yield* of a product is the amount that is actually formed when the experiment is performed. The *percent yield* is the ratio of the actual yield to the theoretical yield multiplied by 100. That is:

$$\text{Percent yield } (\%) = \frac{\text{actual yield}}{\text{theoretical yield}} \times 100$$

Example 7.9

Phosphoric acid can be formed from the combustion of PH_3 in oxygen according to

$$PH_3 \text{ (g)} + 2\,O_2 \text{ (g)} \longrightarrow H_3PO_4 \text{ (s)}$$

If 188 g of phosphoric acid were produced in the reaction of 70 g of PH_3 in excess O_2, calculate the theoretical yield and the percent yield.

Solution

1. We have a balanced equation.
2. Calculate the number of moles of PH_3:

$$\text{Mol } PH_3 = 70 \text{ g} \times \frac{1 \text{ mol } PH_3}{33.97 \text{ g } PH_3} = 2.061 \text{ mol}$$

3. Determine the limiting reagent. Since oxygen is in excess, PH_3 is the limiting reagent.
4. Calculate the theoretical yield of H_3PO_4 using moles of the limiting reactant. Convert moles of H_3PO_4 to grams of H_3PO_4:

$$\text{g } H_3PO_4 = 2.061 \text{ mol } PH_3 \times \frac{1 \text{ mol } H_3PO_4}{1 \text{ mol } PH_3} \times \frac{98 \text{ g } H_3PO_4}{1 \text{ mol } H_3PO_4} = 201.98 \text{ g } H_3PO_4$$

5. Calculate the percent yield:

$$\text{Percent yield } (\%) = \frac{\text{actual yield}}{\text{theoretical yield}} \times 100 = \frac{188 \text{ g}}{201.98 \text{ g}} \times 100 = 93.1\%$$

Example 7.10

In a certain experiment, 70.0 g Al reacted with excess oxygen according to

$$4 \text{ Al (s)} + 3\,O_2 \text{ (g)} \longrightarrow 2 \text{ Al}_2O_3 \text{ (s)}$$

If a yield of 82.5% was obtained, what was the actual yield of Al_2O_3, in grams, from the experiment?

Solution

1. We have a balanced equation.
2. Calculate the number of moles of Al:

$$\text{Mol Al} = 75 \text{ g} \times \frac{1 \text{ mol Al}}{27 \text{ g Al}} = 2.778 \text{ mol}$$

3. Determine the limiting reagent. Since oxygen is in excess, Al is the limiting reagent.
4. Calculate the theoretical yield of Al_2O_3 using moles of the limiting reactant. Convert moles of Al_2O_3 to grams of Al_2O_3:

$$\text{g Al}_2O_3 = 2.778 \text{ mol Al} \times \frac{2 \text{ mol Al}_2O_3}{4 \text{ mol Al}} \times \frac{102 \text{ g Al}_2O_3}{1 \text{ mol Al}_2O_3} = 141.68 \text{ g Al}_2O_3$$

5. Calculate the actual yield:

$$\text{Percent yield } (\%) = \frac{\text{actual yield}}{\text{theoretical yield}} \times 100$$

$$82.5\% = \frac{\text{actual yield}}{141.68} \times 100$$

$$\text{Actual yield} = \frac{141.68 \times 82.5}{100} = 117 \text{ g Al}_2O_3$$

7.6 Problems

1. Consider the formation of ethanol by the fermentation of glucose, given by:

$$C_6H_{12}O_6 \longrightarrow 2 \text{ C}_2H_5OH + 2 \text{ CO}_2$$

Calculate the maximum masses of ethanol and gaseous carbon dioxide that can be produced from the fermentation of 250 g of glucose.

2. What mass of oxygen will react with 10 g of hydrogen to form water?
3. The thermal decomposition of potassium chlorate is represented by:

$$2 \text{ KClO}_3 \longrightarrow 2 \text{ KCl} + 3 \text{ O}_2$$

Calculate:

(a) the mass of $KClO_3$ needed to generate 6.25 moles of oxygen.
(b) the moles of KCl produced from the decomposition of 5.125 g of $KClO_3$.
(c) the mass and number of moles of oxygen produced from the decomposition of 5.125 g of $KClO_3$.

4. Chlorine is prepared in the laboratory by the action of hydrochloric acid on manganese dioxide, MnO_2, according to:

$$4\ HCl + MnO_2 \longrightarrow MnCl_2 + 2\ H_2O + Cl_2$$

What mass of MnO_2 is required to completely react with 10.25 g of HCl?

5. Very pure silicon (used in computer chips) is manufactured by heating silicon tetrachloride, $SiCl_4$, with zinc:

$$SiCl_4 + 2\ Zn \longrightarrow Si + 2\ ZnCl_2$$

If 0.25 mol of Zn is added to excess $SiCl_4$, how many grams of silicon are produced?

6. Tungsten metal, W, is used widely in the industrial production of incandescent bulb filaments. The metal is produced by the action of hydrogen on tungsten (VI) oxide:

$$WO_3 + 3\ H_2 \longrightarrow W + 3\ H_2O$$

If a sample of WO_3 produces 25.825 g of water, how many grams of W are formed?

7. When steam is passed over iron filings at red heat, tri-iron tetroxide (Fe_3O_4) and hydrogen are produced. The balanced equation for the reaction is:

$$3Fe\ (s) + 4H_2O\ (g) \underset{}{\overset{\text{Red heat}}{\rightleftharpoons}} Fe_3O_4\ (\ s) + 4H_2\ (g)$$

Calculate the volume of hydrogen produced at STP by reacting 0.25 mole of Fe with steam.

8. Calculate the volume of carbon dioxide produced at STP when 3.25 g propane burns in a rich supply of oxygen.

$$C_3H_8 + 5O_2 \longrightarrow 3\ CO_2 + 4\ H_2O$$

9. Large quantities of nitrogen gas can be produced from the thermal decomposition of sodium azide (NaN_3):

$$2\ NaN_3 \longrightarrow 2\ Na + 3\ N_2$$

Calculate the volume of nitrogen produced at STP by the thermal decomposition of 85.25 g of NaN_3.

10. Nickel tetracarbonyl, $Ni(CO)_4$, decomposes at elevated temperature according to:

$$Ni(CO)_4\ (g) \rightleftharpoons Ni\ (s) + 4CO\ (g)$$

Calculate the volume of carbon monoxide produced at STP by the decomposition of 0.050 mol of $Ni(CO)_4$.

11. Acetylene gas, which is used in welding, reacts with oxygen according to the equation:

$$C_2H_2 + O_2 \longrightarrow H_2O + CO_2$$

(a) Balance the equation for the reaction.
(b) How many moles of oxygen are needed for the complete combustion of 5.25 g of acetylene?
(c) Calculate the volume at STP of carbon dioxide obtained from 5.25 g of acetylene.

12. Aluminum sulfate has been used as a leather-tanning agent and as an antiperspirant. The compound can be obtained, along with water, by the reaction of aluminum hydroxide with sulfuric acid.

   (a) Write a balanced equation for the reaction.
   (b) Assume that 100 g of aluminum hydroxide reacted with 155 g of sulfuric acid. What is the limiting reagent?
   (c) How many grams of aluminum sulfate are produced?
   (d) How many grams of the excess reactant are left at the end of the reaction?

13. One of the chemical reactions involved in photographic film development is the reaction of the excess AgBr with sodium thiosulfate, $Na_2S_2O_3$.

$$Na_2S_2O_3 + AgBr \longrightarrow Na_3Ag(S_2O_3)_2 + NaBr$$

   (a) Balance the equation for the reaction.
   (b) If 15 g of $Na_2S_2O_2$ and 18 g of AgBr are present in a reaction, which is the limiting reagent for the formation of $Na_3Ag(S_2O_3)_2$?
   (c) How many moles of the excess reagent are left at the end of the reaction?

14. The explosive trinitrotoluene (TNT) can be prepared by reacting toluene with nitric acid:

$$C_7H_8 + HNO_3 \longrightarrow C_7H_5N_3O_6 + H_2O$$
$$(TNT)$$

   (a) Balance the equation.
   (b) Determine the limiting reagent if 65 g C_7H_8 reacts with 50 g of HNO_3.
   (c) What is the theoretical yield of TNT that can be obtained from a reaction mixture that contains 65 g of C_7H_8 and 50 g of HNO_3?
   (d) If the actual yield of TNT for the reaction mixture in part (c) is 55 g, calculate the percent yield of TNT for the reaction.

15. Phosphine, PH_3, can be prepared by the hydrolysis of calcium phosphide, Ca_3P_2:

$$Ca_3P_2 + H_2O \longrightarrow Ca(OH)_2 + PH_3$$

   (a) Balance the equation.
   (b) What is the maximum mass of PH_3 that could be prepared by mixing 44.25 g of Ca_3P_2 and 125.0 mL of water? (Density of water at 25°C is 1.000 g/mL.)

16. Acetylsalicylic acid (the chemical name for aspirin) helps to relieve pain and lowers body temperature. Industrially, it is produced from the reaction of salicylic acid and acetic anhydride:

$$HOC_6H_4COOH + (CH_3CO)_2O \longrightarrow CH_3COOC_6H_4COOH + CH_3COOH$$

   Salicylic acid   Acetic anhydride   Acetysalicylic acid   Acetic acid
   (Aspirin)

Starting with 12.00 g of salicylic acid and excess acetic anhydride, a chemistry student synthesized aspirin in the laboratory and reported a yield of 91%. What was the student's actual yield of aspirin?

17. The main chemical component responsible for artificial banana flavor is isopentyl acetate. An industrial chemist wants to prepare 500 g of isopentyl acetate by the reaction of acetic acid with isopentyl alcohol:

$$CH_3COOH + HO(CH_2)_2CH(CH_3)_2 \longrightarrow CH_3COO(CH_2)_2CH(CH_3)_2 + H_2O$$
Acetic acid   Isopentyl alcohol   Isopentyl acetate

How many grams of acetic acid are needed if the chemist expects a percent yield of only 78%? Assume that isopentyl alcohol is present in excess.

18. An analytical chemist wanted to determine the composition of copper-bearing steel. She placed 5.00 g of the sample in excess aqueous HCl solution. All the iron in the steel dissolved as iron (III) chloride, and 0.15 g of hydrogen was produced. What is the percent composition of the alloy? Assume the alloy is composed only of Cu and Fe.

19. A drug-enforcement agent seized 250.00 g of heroin, $C_{21}H_{23}O_5N$, and burned a 25.00-g sample in 35.25 g of oxygen. The end products were CO_2, NO_2, and H_2O.

   (a) Write a balanced equation for the combustion reaction.
   (b) What is the limiting reagent?
   (c) How many grams of the excess reactant remain unreacted?
   (d) What was the percentage completion of the combustion?
   (e) Determine the total mass of CO_2, NO_2, and H_2O produced in the combustion process.

20. Iron is produced by carbon monoxide reduction of iron ore, usually hematite, Fe_2O_3, in a blast furnace reactor. The following series of three reactions occur in various temperature regions in the reactor, resulting in the formation of molten iron.

$$\begin{aligned} 3\,Fe_2O_3 + CO &\longrightarrow 2\,Fe_3O_4 + CO_2 & (1) \\ Fe_3O_4 + CO &\longrightarrow 3\,FeO + CO_2 & (2) \\ FeO + CO &\longrightarrow Fe + CO_2 & (3) \end{aligned}$$

How many kilograms of iron can be produced from 25.00 kg of the hematite ore? Assume that carbon monoxide is present in excess. (Hint: Manipulate these equations such that FeO and Fe_3O_4 cancel out of the final equation).

8

Structure of the Atom

. . . .

8.1 Electronic Structure of the Atom

The arrangement of electrons around the nucleus of an atom is known as its electronic structure. Since electrons determine all the chemical and most physical properties of an atomic system, it is important to understand the electronic structure. Much of our understanding has come from spectroscopy—the analysis of the light absorbed or emitted by a substance.

8.2 Electromagnetic Radiation

Electronic radiation is a form of energy; light is the most familiar type of electromagnetic radiation. But radio waves, microwaves, X-rays, and many other similar phenomena are also electromagnetic radiation. All these exhibit wave-like properties, and all travel through a vacuum at the speed of light (Figure 8.1). The wave-like propagation of electromagnetic radiation can be described by its frequency (v), wavelength (λ), and speed (c).

Frequency (nu,v): The frequency of a wave is the number of waves (or cycles) that pass a given point in one second. The unit is expressed as the reciprocal of seconds (s^{-1}), or as hertz (Hz). A hertz is one cycle per second ($1\ Hz = 1s^{-1}$).
Wavelength (lambda, λ): The wavelength of a wave is the distance between two successive peaks or troughs.
Speed of light (c): The speed of light in a vacuum is one of the fundamental constants of nature and does not vary with the wavelength. It has a numerical value of 2.9979×10^8 m/s, but for convenience we use 3.0×10^8 m/s.

These measurements are related by the equation: Speed of light = wavelength × frequency, or $c = \lambda v$.

Chemistry in Quantitative Language: Fundamentals of General Chemistry Calculations. Second Edition.
Christopher O. Oriakhi, Oxford University Press. © Christopher O. Oriakhi 2021.
DOI: 10.1093/oso/9780198867784.003.0008

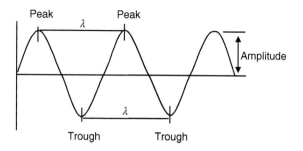

Figure 8.1   The wavelike properties of electromagnetic radiation, illustrating the wavelength and frequency of an electromagnetic wave.

This expression can be rearranged to give:

$$\lambda = \frac{c}{v}, \text{ or } v = \frac{c}{\lambda}$$

Wave Number ($\bar{v}$): The wave number is a characteristic of a wave that is proportional to energy. It is defined as the number of wavelengths per unit of length (usually in centimeters, cm). Wave number may be expressed as:

$$\bar{v} = \frac{1}{\lambda}$$

Example 8.1

Calculate the frequency of red light with a wavelength of 700 nm (note: 1 nm $= 10^{-9}$ m).

Solution

Use the expression

$$v = \frac{c}{\lambda}$$

$c = 3.0 \times 10^8$ m/s

$\lambda = 700$ nm or 7×10^{-7} m

$$v = \frac{3.0 \times 10^8 \text{ m/s}}{7.0 \times 10^{-7} \text{ m}} = 4 \times 10^{14} \text{ s}^{-1} \text{ or } 4 \times 10^{14} \text{ Hz}$$

Example 8.2

What is the wavelength of yellow light that has a frequency of 5.2×10^{14} Hz?

Solution

Use the expression:

$$\lambda = \frac{c}{v}$$

$$c = 3.0 \times 10^8 \text{ m/s}$$

$$v = 5.2 \times 10^{14} \text{ Hz}$$

$$\lambda = \frac{3.0 \times 10^8 \text{ m/s}}{5.2 \times 10^{14} \text{ Hz}} = 5.8 \times 10^{-7} \text{ m (which we can also write as}$$

$$580 \times 10^{-9} \text{ m or 580 nm.)}$$

8.3 The Nature of Matter and Quantum Theory

While electromagnetic radiation behaves like a wave—with characteristic frequency and wavelength—experiment has shown that it also behaves as a continuous stream of particles or energy packets. At the end of the nineteenth century, the general belief among scientists was that matter and energy were unrelated. In 1901, Planck investigated the so-called blackbody radiation—the radiation emitted by hot solids at different temperatures. He observed that the intensity of the blackbody radiation varied with the wavelength. To explain his results, Planck suggested that the emission and absorption of radiant energy are quantized and cannot vary continuously. That is, energy can be gained or lost only in small packets or discrete units called quanta. The smallest amount of radiant energy, E, that can be emitted is called a quantum of energy, and is given by Planck's equation:

$$E = hv$$

where h is Planck's constant with a value of 6.63×10^{-34} J.s and v is the frequency of the radiated light. Planck concluded that radiant energy from a given source must be an exact integral multiple of the simplest quantum hv:

$$E = nhv$$

where $n = 1, 2, 3, \ldots$ etc.

8.3.1 Photoelectric effect

In 1905, Einstein used Planck's results to explain the photoelectric effect. Here, light shining on a metal surface causes electrons to be ejected; but once again, the process occurs only in discrete quantities, rather than continuously. Einstein suggested that electromagnetic radiation is quantized, and that light could be thought of as "particles" called photons. Planck referred to these "particles" as quanta. The energy of a photon is given as:

$$E_{\text{photon}} = hv = \frac{hc}{\lambda}$$

Example 8.3

What is the energy of a photon of infrared radiation that has a frequency of 9.8×10^{13} Hz?

Solution

$$E = hv$$

$$= 6.63 \times 10^{-34} \text{ J.s.photon}^{-1} \times 9.8 \times 10^{13} \text{ s}^{-1}$$

$$= 9.60 \times 10^{-20} \frac{\text{J}}{\text{photon}}$$

8.4 The Hydrogen Atom

When a beam of light passes through a prism it produces a continuous spectrum containing light of all different colors or wavelengths. But when we observe the light given off by a heated gas—whether element or compound—we see only certain colors of light, which are characteristic of the substance.

The emission spectrum of atomic hydrogen consists of several series of isolated lines (four lines are observed in the visible region). The frequencies of the visible emission lines for hydrogen fit the Balmer equation:

$$v = C \left(\frac{1}{2} - \frac{1}{n^2} \right) \quad n = 3, 4, 5, 6$$

C is called the Rydberg constant for hydrogen, whose value is 3.29×10^{15} s^{-1}, and n is an integer greater than 2. For atoms other than hydrogen, the emission-line frequencies follow a more complex relation.

8.4.1 The Bohr model

To account for the line spectrum of the hydrogen atom, Bohr proposed what is commonly referred to as Bohr's model for the atom:

- Electrons move about the central nucleus in circular orbits of fixed radius called energy levels.
- Electrons can occupy only certain (permitted) discrete energy levels in atoms. They do not radiate energy when they are in one of the allowed energy levels.
- The angular momentum of the electron is quantized, and is a whole number multiple of $\frac{h}{2\pi}$:

$$\text{Angular momentum} = mur = n \left(\frac{h}{2\pi} \right)$$

where m is the mass of the electron, u is velocity, r is radius of Bohr's orbit, n is an integer called the *principal quantum number* (discussed later in the chapter), and h is Planck's constant.

- Electrons absorb or emit energy in discrete amounts as they move from one energy level to another. The change in energy for an electron dropping from an excited state E_2 to lower (the ground) state E_1 is proportional to the frequency of the radiation emitted and is given by Planck's relationship:

$$\Delta E = E_2 - E_1 = h\nu.$$

From his model, Bohr derived the following formula for the electron energy levels in the hydrogen atom:

$$E_n = -\left(\frac{R_H}{n^2}\right) \text{ or } -\left(\frac{2.18 \times 10^{-18} \text{ J}}{n^2}\right) \quad n = 1, 2, 3, \ldots \infty$$

where R_H is the Rydberg constant, with a value of 2.18×10^{-18} J, and n is an integer called the *principal quantum number,* which describes the size of the allowed orbits (or shells). The minus sign indicates that E decreases as the electron moves closer to the nucleus. When n is equal to one, the orbit lies closest to the nucleus; an electron in this orbit is said to be in the *ground state.* When n is equal to 2, the electron is at a higher energy level. As n becomes infinitely large, the energy level becomes zero:

$$E_{n=\infty} = -\left(\frac{R_H}{\infty^2}\right) \text{ or } -\left(\frac{2.18 \times 10^{-18} \text{ J}}{\infty^2}\right) = 0$$

This is the reference, or zero-energy, state in which an electron is removed from the nucleus of the atom. Note that the zero-energy state is higher in energy than the states closer to the nucleus (with negative energy).

8.4.2 Emission and absorption spectra

Absorption and emission spectra arise from electrons moving between energy levels. An absorption spectrum is produced when an atom absorbs light of a particular wavelength. The absorbed light energy excites or promotes an electron from a lower energy level to a higher energy level. The wavelength of the absorbed light is obtained from the difference between the two energy levels.

$$\Delta E = \frac{hc}{\lambda} \text{ and } \lambda = \frac{hc}{\Delta E}$$

where $\Delta E = E_2 - E_1$.

Similarly, an emission spectrum arises when an electron previously excited to a higher energy level returns to a lower energy. The wavelength of the emitted light is obtained from the difference between the two energy levels, as before.

Figure 8.2 shows the absorption and emission of photons by an electron in an atom.

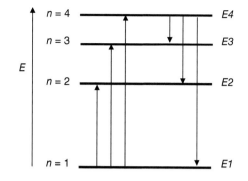

Figure 8.2 Energy levels in hydrogen, from the Bohr model. The arrows indicate the transitions of the electron from one energy level to another.

The upward arrows represent absorption, and the downward arrows represent emission. If an electron jumps from a state with an initial energy level E_i to a state with a final energy level E_f, the change in energy, ΔE, can be obtained from Bohr's equation:

$$\Delta E = \left[\left(\frac{-R_H}{n_f^2} \right) - \left(\frac{-R_H}{n_i^2} \right) \right] = -R_H \left(\frac{1}{n_f^2} - \frac{1}{n_i^2} \right) = R_H \left(\frac{1}{n_i^2} - \frac{1}{n_f^2} \right)$$

Since $\Delta E = E_f - E_i = h\nu$, the frequency of emitted radiation can be obtained from the expression:

$$h\nu = R_H \left(\frac{1}{n_i^2} - \frac{1}{n_f^2} \right) \text{ or } \nu = \frac{R_H}{h} \left(\frac{1}{n_i^2} - \frac{1}{n_f^2} \right)$$

Note:

- For energy absorption, n_f is greater than n_i and ΔE is positive as electrons move from a lower-energy state to a higher energy.
- For energy emission, n_f is less than n_i and ΔE is negative as electrons move from a higher-energy state to a lower energy.

Example 8.4

What is the amount of energy emitted when an electron in an excited hydrogen atom falls from energy level $n = 4$ to energy level $n = 1$?

Solution

This problem describes an electronic transition from high energy ($n = 4$) to lower ($n = 1$). Radiant energy is emitted and ΔE will be negative. Use $R_H = 2.18 \times 10^{-18}$ J.

$$\Delta E = R_H \left(\frac{1}{n_i^2} - \frac{1}{n_f^2} \right) = 2.18 \times 10^{-18} \text{ J} \left(\frac{1}{4^2} - \frac{1}{1^2} \right) = -2.04 \times 10^{-18} \text{ J}$$

Example 8.5

Calculate (a) the frequency and (b) the wavelength of the radiation related to an electron in an excited hydrogen atom falling from $n = 3$ to $n = 1$. Is light emitted or absorbed?

Solution

This is an electronic transition from a higher level to a lower level of the hydrogen atom. Therefore we expect light to be emitted as ΔE will be negative.

(a) To calculate the frequency, use the following expression:

$$v = \frac{R_H}{h}\left(\frac{1}{n_i^2} - \frac{1}{n_f^2}\right) = \left(\frac{2.18 \times 10^{-18} \text{ J}}{6.63 \times 10^{-34} \text{ J.s}}\right)\left(\frac{1}{3^2} - \frac{1}{1^2}\right) = 2.92 \times 10^{15} \text{ s}^{-1}$$

(b) To calculate the wavelength, use the value of the frequency obtained above in the expression:

$$\lambda = \frac{c}{v} = \frac{3.0 \times 10^8 \text{ ms}^{-1}}{2.92 \times 10^{15} \text{ s}^{-1}} = 1.03 \times 10^{-7} \text{ m or } 103 \text{ nm}$$

8.5 The Quantum-Mechanical Description of the Hydrogen Atom

Bohr's model of the hydrogen atom significantly advanced our knowledge of atomic structure, but it could not explain the spectra of heavier atoms. This led to further studies on atomic structure by many scientists.

8.5.1 The wave nature of the electron

The wave-particle dual nature of light led de Broglie in 1924 to propose that any particle with mass m moving with velocity v should have a wavelength (λ) associated with it. The relationship between the velocity and wavelength of the particle can be expressed by the de Broglie equation:

$$\lambda = \frac{h}{mv}$$

where λ is the wavelength, h is Planck's constant, m is the mass of the particle, and v its velocity. This is particularly relevant to particles as small as electrons; the wavelength associated with their motion tends to be significant in comparison to the size of the particle, which is not the case for large objects.

Example 8.6

Calculate the wavelength of an electron moving with a velocity of 6.5×10^6 m/s. The mass of an electron is 9.1×10^{-31} kg.

Solution

$$\lambda = \frac{h}{mv}$$

$$\lambda = \frac{(6.63 \times 10^{-34} \text{Js})}{(9.1 \times 10^{-31} \text{kg}) (6.5 \times 10^{6} \text{m/s})}$$

$$= 1.12 \times 10^{-10} \text{ m} \text{ or } 0.112 \text{ nm}$$

8.5.2 The Heisenberg uncertainty principle

The uncertainty principle states that it is impossible to accurately determine simultaneously both the momentum and position of an individual electron. Mathematically, the law states that the uncertainty in the speed (Δv) times the uncertainty in position (Δx) multiplied by the mass (m) of the particle must be greater than or equal to Planck's constant (h) divided by 4π:

$$(\Delta x)(m \times \Delta v) \geq \frac{h}{4\pi}$$

The more accurately we can determine the position of a moving electron, the less accurately we can measure its momentum. Hence, we can speak only of the probability of finding an electron at any given region in space within the atom.

8.6 Quantum Mechanics and Atomic Orbitals

Schrödinger's work on the wave mechanical model of the atom in the mid-1920s greatly improved our understanding of atomic structure. The model is based on de Broglie's postulate and other fundamental assumptions. Schrödinger formulated what is known as the Schrödinger wave equation, which is the fundamental assumption of quantum theory. This equation successfully predicts every aspect of the hydrogen spectrum, and much of what is seen for heavier elements.

Mathematically, the equation treats the electron as though it were a wave—specifically, a standing wave. Though it is very complex, and often cannot be solved exactly, it gives the explicit mathematical description of the functions that determine the electron's distribution around the nucleus. These take the form of a set of *wave functions* (also known as orbitals) and their energies (which correspond to the energy levels in the Bohr model).

A wave function is represented by the symbol ψ (psi). The square of the wave function, ψ^2, is called the *probability density*; it measures the probability of finding an electron in a region of space. The probability distributions generated by these functions are essentially maps of the electron density around the nucleus, which define the shapes of the various orbitals.

8.6.1 Orbitals and quantum numbers

The solutions to Schrödinger's equation for an electron in a hydrogen atom (which is also applicable to other atoms) give rise to a series of wave functions or atomic orbitals. The equation tells us that the energy of the hydrogen atom is quantized. This means that only certain energies, and hence energy levels that meet certain quantum conditions, are possible. Each orbital is defined by four quantum numbers. Quantum numbers describe the various properties of atomic orbitals such as the number of possible orbitals within a given main energy level and the distribution of the electrons within each orbital.

8.6.1.1 Principal quantum number (n)

The principal quantum, n, indicates the size and energy of the main orbital (or level) an electron occupies in an atom. It can have any positive whole-number value; 1, 2, 3, . . . As n increases, the overall size of the orbital it describes increases, and the orbital can accommodate more electrons. Note that each main orbital, or value of n, is referred to as a *shell*. In the past, energy levels (or values) of n were represented by the letters K, L, M, N, . . . The number of electrons that can be accommodated in each energy level n is given by $2n^2$. Thus the electron capacities of energy levels 1, 2, 3, and 4, are 2, 8, 18, and 32 respectively.

Shell	K	L	M	N
value of n	1	2	3	4
Electron capacity	2	8	18	32

8.6.1.2 Angular-momentum (or azimuthal) quantum number (ℓ)

The angular-momentum quantum number l determines a region—a *subshell*—within the main shell (n) and defines its shape. Only certain values of l are allowed. For each value of n, l can have integral values from 0 to $n - 1$. This implies that there are n different shapes of orbitals within each shell. For example, when n = 4, l = 0, 1, 2, or 3. The first four subshells are generally designated by the letters s, p, d, and f, corresponding to l = 0, 1, 2, and 3. This is summarized below:

Value of l	0	1	2	3
Letter used	s	p	d	f

The number of subshells within a given shell is shown Table 8.1, along with their relative energies.

8.6.1.3 Magnetic (or orbital) quantum number (m_l)

The magnetic quantum number defines the spatial orientation of the subshell. Only certain values of m_l are allowed, and these depend upon l. Values of m_l are whole numbers varying from

Table 8.1   Relationship between shells, subshells, and their relative energies

Shell	Subshells	Relative Energies
$n = 1$	1s	
$n = 2$	2s, 2p	$2p > 2s$
$n = 3$	3s, 3p, 3d	$3d > 3p > 3s$
$n = 4$	4s, 4p, 4d, 4f	$4f > 4d > 4p > 4s$

$-l$ through zero to $+l$, thus the number of allowed values of m_l is given by $2l + 1$. Therefore, there is one s orbital, three p orbitals, five d orbitals, and seven d orbitals.

8.6.1.4 Spin quantum number (m_s)

The spin quantum number m_s specifies the spin of an electron and so the orientation of the magnetic field generated by this spin. The spin can be clockwise or counterclockwise. Therefore for every set of n, l, and m_l values, m_s can take the value $+\frac{1}{2}$ or $-\frac{1}{2}$.

Table 8.2 summarizes the permissible values for each quantum number.

Example 8.7

Calculate the maximum number of electrons that can be placed in

a) A shell with $n = 3$
b) A shell with $n = 4$
c) A shell with $n = 6$

Solution

Use the expression $2n^2$.

a) For $n = 3$, the maximum number of electrons in the shell will be
   $2n^2 = 2(3^2) = 18$
b) For $n = 4$, the maximum number of electrons in the shell will be
   $2n^2 = 2(4^2) = 32$
c) For $n = 6$, the maximum number of electrons in the shell will be
   $2n^2 = 2(6^2) = 72$

Example 8.8

Describe the sublevels in the $n = 4$ energy level.

Solution

For $n = 4$, l can have values of 0, 1, 2, and 3. Therefore the number of sublevels and their designations include:

$$l = 0(4s)$$
$$l = 1(4p)$$
$$l = 2(4d)$$
$$l = 3(4f)$$

Table 8.2 Allowed values of quantum numbers through n = 4

Shell	n	l	m_l	m_s	Designation of Subshells	No. of Orbitals in subshell	No. of electrons in subshell	Total electrons in main shell
K	1	0	0	$+\frac{1}{2}, -\frac{1}{2}$	1s	1	2	2
L	2	0	0	$+\frac{1}{2}, -\frac{1}{2}$	2s	1	2	8
		1	-1,0,+1	$\pm\frac{1}{2}$ for each value of m_l	2p	3	6	
M	3	0	0	$+\frac{1}{2}, -\frac{1}{2}$	3s	1	2	
		1	-1,0,+1	$\pm\frac{1}{2}$ for each value of m_l	3p	3	6	18
		2	-2,-1,0,+1,+2	$\pm\frac{1}{2}$ for each value of m_l	3d	5	10	
N	4	0	0	$+\frac{1}{2}, -\frac{1}{2}$	4s	1	2	
		1	-1,0,+1	$\pm\frac{1}{2}$ for each value of m_l	4p	3	6	32
		2	-2,-1,0,+1,+2	$\pm\frac{1}{2}$ for each value of m_l	4d	5	10	
		3	-3,-2,-1,0,+1,+2,+3	$\pm\frac{1}{2}$ for each value of m_l	4f	7	14	

Example 8.9

What quantum numbers are permissible for a $5p$ orbital?

Solution

For a $5p$ orbital, $n = 5$. The fact that it is a p orbital tells us that $l = 1$. For $l = 1$, m_l can have any of the values -1, 0, or $+1$, hence there are three sets of permissible quantum numbers for a $5p$ orbital. They are:

$$n = 5, l = 1, m_l = -1$$
$$n = 5, l = 1, m_l = 0$$
$$n = 5, l = 1, m_l = +1$$

Note that m_s is not counted here. It describes an electron, not an orbital.

8.7 Electronic Configuration of Multielectron Atoms

An electronic configuration indicates how many electrons an atom has in each of its subshells or energy levels. Knowing the relative energies of the various orbitals allows us to predict for each element or ion the orbitals that are occupied by electrons. The ground-state electronic configurations are obtained by placing electrons in the atomic orbitals of lowest possible energy, provided no single orbital holds more than two electrons. In building up the electronic configuration of an atom, three principles are used. These include the Aufbau Principle, the Pauli Exclusion Principle, and Hund's Rule.

8.7.1 Aufbau principle

Electrons occupy the lowest-energy orbitals available to them, entering the higher-energy orbitals only after the lowest-energy orbitals are filled. The electron building up principle can be summarized as:

- Each added electron will enter the orbitals in the order of increasing energy.
- Each orbital can only take two electrons.

The Aufbau diagram in Figure 8.3 will assist you in writing electron configurations for any element in the periodic table, as long as you know the atomic number.

In a compressed form, electrons fill the subshells available in each main energy level in the following order:

$$1s\ 2s\ 2p\ 3s\ 3p\ 4s\ 3d\ 4p\ 5s\ 4d\ 5p\ 6s\ 4f\ 5d\ 6p\ 7s\ 5f\ 6d$$

For example, the electron configurations of the first ten elements in the periodic table are:

$$_1\text{H} : 1s^1 \qquad _6\text{C} : 1s^2 2s^2 2p^2$$
$$_2\text{He} : 1s^2 \qquad _7\text{N} : 1s^2 2s^2 2p^3$$

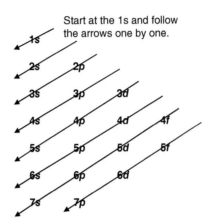

Start at the 1s and follow
the arrows one by one.

Figure 8.3   An aid to remembering the build-up or Aufbau order of atomic orbitals.

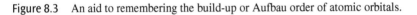

$_3$Li : $1s^2 2s^1$ $_8$O : $1s^2 2s^2 2p^4$

$_4$Be : $1s^2 2s^2$ $_9$F : $1s^2 2s^2 2p^5$

$_5$B : $1s^2 2s^2 2p^1$ $_{10}$Ne : $1s^2 2s^2 2p^6$

Example 8.10

Write the electron configuration for:

(a) Ca (atomic number, Z = 20)
(b) Mn (Z = 25)

Solution

(a) Ca : $1s^2 2s^2 2p^6 3s^2 3p^6 4s^2$
(b) Mn : $1s^2 2s^2 2p^6 3s^2 3p^6 4s^2 3d^5$

Example 8.11

Predict the electron configuration of Ca^{2+}.

Solution

Calcium has atomic number 20, i.e., twenty protons and twenty electrons. But Ca^{2+} has lost two electrons. Therefore, eighteen electrons must be placed in the various subshells in Ca orbitals in order of increasing energy:

$$Ca^{2+} \text{ (18 electrons)} : 1s^2 2s^2 2p^6 3s^2 3p^6$$

8.7.1.1 Shorthand notation for electron configuration

Electron configurations can also be written using a shorthand method, which shows only the outer electrons, i.e. those surrounding the inert noble-gas core. This involves using [He] for $1s^2$, [Ne] for $1s^2 2s^2 2p^6$, [Ar] for $1s^2 2s^2 2p^6 3s^2 3p^6$, etc. Thus, we may represent the electron configurations for Si and K as:

$$\text{Si} : [\text{Ne}] 3s^2 3p^2 \text{ and } \text{K} : [Ar] 4s^1$$

Example 8.12

Using the shorthand notation, write the electron configuration for Mn.

Solution

The atomic number of Mn is 25. Therefore, twenty-five electrons will fill the available orbitals.

$$\text{Mn} : [Ar] 4s^2 3d^5$$

8.7.2 Pauli exclusion principle

The Pauli exclusion principle states that no two electrons in an atom can have the same four quantum numbers. Since there are only two possible values for m_s, $+\frac{1}{2}$ and $-\frac{1}{2}$, any single orbital can accommodate only two electrons, which must have opposite spins.

8.7.3 Hund's principle (or rule)

Hund's Principle states that if two or more orbitals have equal energy (degenerate orbitals), a single electron goes into each until all orbitals are half-filled; only then will a second electron (with an opposite spin quantum number) enter any of the orbitals. In other words, the orbitals in a given subshell will each gain one electron before any of them gain a second one.

8.7.4 Summary of building-up principles

1. The lowest-energy orbitals are filled first.
2. Only two electrons of opposite spin go into any one orbital (Pauli exclusion principle).
3. If two or more orbitals have the same energy (degenerate orbitals), each is half-filled before any one of them is filled completely (Hund's Principle).

8.7.5 Orbital diagrams

Orbital diagrams are used to show the electron occupancy in each orbital. They serve to illustrate both Hund's Principle and Pauli's principle. Each atomic orbital is represented by a box, which contains up to two electrons. An electron is represented by a half-arrow, which points upward (for $m_s = 1/2$) or downward (for $m_s = -1/2$). If two electrons are in the same orbital (box), the two half-arrows representing them will be opposed as illustrated below.

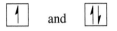 and

An s orbital is represented by one box, a p orbital by three merged boxes, and a d orbital by five merged boxes. That is:

s orbital   p orbital   d orbital

Example 8.13

Draw the orbital diagrams for (a) fluorine, (b) magnesium, and (c) phosphorus.

Solution

(a)

(b)

(c)

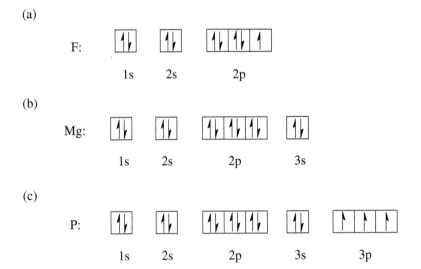

8.8 Problems

1. Express the following in nanometers:
   (a) 2.34×10^{-12} cm
   (b) 1.73×10^{-8} cm
   (c) 5.44×10^3 cm
   (d) 0.575 m

2. Express the following in centimeters:
   (a) 5.34×10^5 nm
   (b) 265 nm

(c) $6.64 \times 10^{12} \text{Å}$

(d) 879Å

3. What is the frequency of yellow light with a wavelength of 589 nm?

4. What is the wavelength of light having a frequency of $9.29 \times 10^{16} \text{ s}^{-1}$?

5. Calculate the wave number, $\bar{v}$, of light having a wavelength of 589 nm.

6. A laser emits light having a frequency of $6.94 \times 10^{14} \text{ s}^{-1}$. Calculate (a) the wavelength, (b) the wave number of this light.

7. What is the energy of photons in kJ that correspond to light with a frequency of $3.70 \times 10^{12} \text{s}^{-1}$?

8. What is the energy of a photon with a wavelength of 408 nm?

9. Blue-green light emitted by a particular atom has a wavelength of 525 nm (in the visible range) of the electromagnetic spectrum. Find (a) frequency, (b) the wave number, and (c) the energy of a photon of this radiation.

10. Nuclear Magnetic Resonance (NMR) spectrometers allow chemists to measure the absorption of energy by some atomic nuclei such as hydrogen. NMR instruments are available that operate at 60 MHz, 200 MHz, 400 MHz, 600 MHz, and more recently, 700 MHz. At what energy do the 60 MHz, 200 MHz, 400 MHz, 600 MHz, and 700 MHz NMR instruments operate, in kJ?

11. What is the energy of an electron in the third energy level of a hydrogen atom?

12. The Lyman series of lines in the hydrogen emission spectrum corresponds to $n = 1$ and is observed in the ultraviolet region. Calculate (a) the wave number of the spectral line in this series for a value of m equal to 5 (note $n = 1$); (b) the energy emitted by an excited hydrogen atom having a spectral line with value of $n = 1$, and $m = 5$.

13. For a hydrogen atom in the ground state $(n = 1)$, calculate the energy necessary to cause ionization $(m = \infty)$.

14. Calculate the energy change and the wavelength that occur when an electron falls from (a) $n = 4$ to $n = 1$, (b) $n = 5$ to $n = 2$ in the hydrogen atom.

15. Calculate the wavelength, in nanometers, of a proton moving at a velocity of $4.56 \times 10^4 \text{ ms}^{-1}$. The mass of a proton is $1.67 \times 10^{-27} \text{ kg}$.

16. If the wavelength of an electron is 25 nm, what is its velocity? The mass of an electron is $9.11 \times 10^{-28} \text{ g}$.

17. In an electron diffraction experiment, a beam of electrons is accelerated by a potential difference of 50 kV. What is the characteristic de Broglie wavelength of the electrons in the beam?

(Hint : $KE = \frac{1}{2} mv^2$; mass of an electron $= 9.11 \times 10^{-31} \text{ kg}$; $1 \text{ J} = 1 \text{ kgm}^2 \text{ s}^{-2}$).

18. What is the maximum number of electrons permissible in the following energy levels?

(a) $n = 1$ energy level

(b) $n = 2$ energy level

(c) $n = 3$ energy level

(d) $n = 4$ energy level

19. For the energy level $n = 3$, what sublevels are possible?

20. What is the shorthand notation describing the orbitals with the following quantum numbers?

(a) $n = 3, l = 1, m_l = 0$
(b) $n = 3, l = 2, m_l = 1$
(c) $n = 4, l = 3, m_l = 3$

21. Give the sets of quantum numbers that describe electrons in the following orbitals:

(a) $2p$ orbital
(b) $3d$ orbital
(b) $4p$ orbital
(c) $4f$ orbital

22. Write the electronic configurations of the following elements:
(a) Na   (b) P   (c) Ca   (d) Rb   (e) Fe

23. Write the electronic configurations of each of the following ions:
(a) S^{2-} (b) Ni^{2+} (c) Cu^{2+} (d) Ti^{4+} (e) Br^-

24. Write the electronic configurations for the following atoms or ions, using the relevant noble-gas inner-core abbreviations:
(a) O   (b) Zn^{2+} (c) Mg   (d) Pb^{2+} (e) V   (f) Te^{2-} (g) Mn^{7+}

25. Identify the elements corresponding to the following electronic configurations:

(a) $[He]2s^2 2p^1$
(b) $[Ar]3d^{10}4s^2 4p^2$
(c) $[Kr]4d^{10}5s^2 4p^5$
(d) $[Xe]4f^{14}5d^{10}6s^2 6p^3$
(e) $[Ar]3d^5 4s^2$
(f) $[Ne]3s^2 3p^3$

26. Identify the elements corresponding to the following electronic configurations:

(a) $1s^2 2s^2 2p^5$
(b) $1s^2 2s^2 2p^6 3s^2 3p^3$
(c) $1s^2 2s^2 2p^6 3s^2 3p^6 4s^2 3d^1$
(d) $1s^2 2s^2 2p^6 3s^2 3p^6 4s^2 3d^8$
(e) $1s^2 2s^2 2p^2$
(f) $1s^2 2s^2 2p^6 3s^2 3p^6 4s^2 3d^2$

27. Draw the orbital diagrams for (a) oxygen (b) sodium (c) titanium (d) zinc.

28. Two elements, A and B, have the following isotopic masses and electronic configurations:

A = 17.99916 amu & $1s^2 2s^2 2p^4$
B = 25.98259 amu & $1s^2 2s^2 2p^6 3s^2$

a) Give the chemical identities of A and B.
b) How many subatomic particles are present in each kind of atom?

9

Chemical Bonding 1: Basic Concepts

. . . .

9.1 Introduction: Types of Chemical Bonds

9.1.1 Types of chemical bonds

The nature and type of a chemical bond are determined primarily by the valence electrons (those in the outermost shell of an atom) shared or transferred between bonded atoms. According to Lewis Theory, a chemical bond is formed when the bonding atoms share or transfer electrons to achieve a stable electronic configuration.

An *ionic bond* is formed if electrons are completely transferred, as in the case of a reaction between a metal and a nonmetal. On the other hand, a *covalent bond* is formed if electrons are shared, such as between two nonmetals with similar electronegativities. There are two other variations of covalent bonding. The first is a *polar covalent bond*, which is a bond between two nonmetal atoms that have significantly different electronegativities and therefore share the bonding pair unequally. The other, a *dative covalent bond* (also called a *coordinate bond*), is a covalent bond in which one of the atoms provides both of the shared electrons. Finally, *metallic bonding* involves the sharing of valence electrons among all the atoms of a metal or metal alloy.

9.2 Lewis Dot Symbols

A Lewis dot symbol (or Lewis structure) is used to describe valence electron configurations of atoms and monatomic ions, and can be helpful in determining how bonds are formed. It consists of the chemical symbol of the element, with dots for each valence electron, placed one at a time around the four sides of the symbol. It is generally written only for main group or representative elements. The number of valence electrons for these elements is equal to the group number. Representative examples are

$$\text{Li}\bullet \qquad \bullet\text{Be}\bullet \qquad \bullet\overset{\bullet}{\text{B}}\bullet \qquad \bullet\overset{\bullet}{\underset{\bullet}{\text{C}}}\bullet \qquad \bullet\overset{\bullet}{\underset{\bullet}{\text{N}}}\bullet \qquad \bullet\overset{\bullet\bullet}{\underset{\bullet\bullet}{\text{O}}}\bullet \qquad \overset{\bullet\bullet}{\underset{\bullet\bullet}{\,:\!\text{F}}}\bullet \qquad \overset{\bullet\bullet}{\underset{\bullet\bullet}{\,:\!\text{Ne}\!:}}$$

Note that elements in the same group have the same number of valence electrons and therefore similar Lewis dot symbols.

Chemistry in Quantitative Language: Fundamentals of General Chemistry Calculations. Second Edition.
Christopher O. Oriakhi, Oxford University Press. © Christopher O. Oriakhi 2021.
DOI: 10.1093/oso/9780198867784.003.0009

Lewis dot symbols can be written for ions just as they are for atoms. The only difference is that cations have fewer electrons than the neutral atoms, while anions have more. Also, the Lewis dot symbol includes square brackets and the charge on the ion. Anions have eight dots (a full octet) surrounding them, but for cations the dots are omitted because they have no valence electrons. Thus, Lewis dot symbols for sulfur and aluminum ions would be written as:

$$\left[:\overset{..}{\underset{..}{S}}: \right]^{2-} \qquad\qquad \left[Al \right]^{3+}$$

Example 9.1

Write the Lewis dot symbols for (a) Al   (b) Ca^{2+} (c) P   (d) O^{2-}

Solution

First write the symbol of the atom or ion. Then determine the valence electrons available and draw them around the atom. Show the ion charge as a superscript outside square brackets. Remember that cations have no valence electrons and so are shown with no dots.

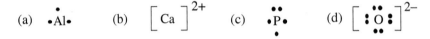

(a) $\cdot \overset{.}{Al} \cdot$ (b) $\left[Ca \right]^{2+}$ (c) $\cdot \overset{..}{\underset{.}{P}} \cdot$ (d) $\left[:\overset{..}{\underset{..}{O}}: \right]^{2-}$

9.2.1 Octet rule

The octet rule states that atoms will gain, lose, or share electrons until they are surrounded by eight valence electrons, with the ns^2np^6 electronic configuration of noble gases. This gives them the remarkable stability associated with the noble gases. For example, in the formation of ionic KF, K loses an electron to achieve the electronic configuration of argon while F gains an electron to attain the electronic configuration of neon.

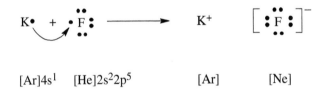

$$K\cdot \quad + \quad \cdot \overset{..}{\underset{..}{F}}: \quad \longrightarrow \quad K^+ \qquad \left[:\overset{..}{\underset{..}{F}}: \right]^-$$

$[Ar]4s^1 \quad [He]2s^22p^5 \qquad\qquad [Ar] \qquad\quad [Ne]$

9.2.2 Duet rule

Atoms with low atomic numbers that have fewer electrons than needed to provide an octet for every atom follow the duet rule. They gain, lose, or share electrons until they are surrounded by two electrons. For example, during bond formation, elements like H, He, and Li gain, lose, or share electrons to form a duet.

9.3 Ionic Bonding: Formation of Ionic Compounds

An ionic compound is formed when a metal reacts with a nonmetal. In the reaction, cations and anions are formed by electron transfer from metals to nonmetals. These oppositely charged ions become bound together as an ionic solid via electrostatic interactions. Consider the formation of sodium chloride (NaCl) as an example. Sodium reacts violently with chlorine gas to form sodium chloride. The reaction is exothermic, releasing a large amount of heat.

$$Na(s) + 1/2Cl_2(g) \longrightarrow NaCl(s) \quad \Delta H_f^\circ = -410.9 \text{ kJ}$$

In this reaction the sodium atom loses an electron to become Na^+, and the chlorine gains the electron to become Cl^-. Consequently both Na^+ and Cl^- acquire an octet of electrons. The ions then aggregate by virtue of electrostatic interactions to form solid NaCl. In this crystalline solid, each Na^+ ion is surrounded by six Cl^- ions and each Cl^- ion is surrounded by six Na^+ ions. This arrangement repeats itself numerous times to yield a regular array or crystal structure of NaCl.

9.3.1 Lattice energies and the strength of the ionic bond

Energy is required to form ionic bonds. This energy and hence the stability of the compound comes largely from the coulomb or electrostatic attraction between oppositely charged ions within the ionic solid. The strength of the electrostatic attraction or the ionic bond is measured by the *lattice energy* of the compound. The lattice energy or lattice enthalpy ($\Delta H_{lattice}$) is defined as the energy released when one mole of a crystalline ionic compound is formed from its free, gaseous elements. For the ionic solid MX, the lattice energy is the enthalpy change for the reaction

$$M^{n+}(g) + X^{n-}(g) \longrightarrow MX(s) \quad \Delta H_{lattice}$$

For example, the formation of crystalline NaCl solid is highly exothermic as shown by the large negative value of the lattice energy.

$$Na^+(g) + Cl^-(g) \longrightarrow NaCl(s) \quad \Delta H_{lattice} = -788 \text{ kJ}$$

For the convention used here, where the free gaseous ions combine to form an ionic solid, the lattice energies will be exothermic (negative values). Note that some textbooks adopt a different convention that defines lattice energy as the energy required (endothermic process) to separate one mole of ionic solid into its component gaseous ions and therefore with positive $\Delta H_{lattice}$ values.

The lattice energy of an ionic solid cannot be measured directly because it is experimentally difficult to isolate gaseous ions. However, the energy value can be estimated using the Born–Haber cycle. This is treated in detail in Chapter 20.

9.3.2 Calculating lattice energies of ionic solids

The lattice energy of an ionic solid can be calculated by using a modified form of Coulomb's law (the point charge model). However, the calculated values are generally higher than the experimentally measured lattice energies.

$$\Delta H_{\text{lattice}} = k\frac{Q_1 Q_2}{d} = 2.31 \times 10^{-19} \text{ J} \cdot \text{nm} \left(\frac{Q_1 Q_2}{d} \right)$$

Here, U is the lattice energy; Q_1 and Q_2 are charges on the ions; d is the interionic distance (the sum of the radii of the positive and negative ions) and k is a constant equal to $1/4\pi\epsilon_0$ or 2.31×10^{-19} J.nm (ϵ_0 is the permittivity of the vacuum).

The lattice energy of an ionic solid depends on:

- The magnitude of the ionic charges
- The radius or size of the ions
- The distance between charges

Large lattice energy is indicative of a tightly bound solid due to a stronger interaction between the constituent ions.

Note that lattice energy increases both as the ionic charge increases and as the ionic size decreases (smaller ions are closer together and interact more strongly).

Example 9.2

Account for the following trend in lattice energy:

Metal oxide	Lattice energy (kJ/mol)
BeO	−4293
MgO	−3795
CaO	−3414
SrO	−3217
BaO	−3029

Solution

Trends in lattice energies of ionic compounds can be explained mainly by considering both the charge on the ions and the ionic radius. All these metal oxides are in group 2A of the periodic table, so the metal ions all have a charge of +2 (and of course the oxygen is always −2). Further, the O^{2-} ion is the same size in all the compounds, so the lattice energy depends on the radius of the positive ion, which increases from Be^{2+} to Ba^{2+}. In other words, the Be^{2+} is closer to the O^{2-} than the Ba^{2+} is, and so can interact more strongly, and hence will have greater lattice energy.

Example 9.3

What is the lattice energy in kJ/mol of the ionic bond between a sodium ion and a chloride ion? The ionic radii of Na^+ and Cl^- are 186 pm and 181 pm, respectively.

Solution

Given:

$$Q_{Cl^-} = -1 \text{(the Cl}^- \text{ion's charge)}$$

$$Q_{Na^+} = +1 \text{ (the charge of the Na}^+ \text{ ion)}$$

$$N_A = 6.022 \times 10^{23} \text{ ion pairs mol}^{-1}$$

1. Compute the distance (d) between the nuclei of Na^+ and Cl^- and express it in nm to use the simplified equation for lattice energy.

$$d = (Na^+ \text{ ionic radius}) + (Cl^- \text{ ionic radius})$$

$$= 186\text{pm} + 181\text{pm} = 367 \text{ pm or } 0.367 \text{ nm}$$

2. Calculate the lattice energy in J using the modified coulombic attraction equation:

$$\Delta H_{\text{lattice}} = 2.31 \times 10^{-19} \text{ J·nm} \left(\frac{Q_1 Q_2}{d} \right)$$

$$= 2.31 \times 10^{-19} \text{ J·nm} \left(\frac{+1 \times -1}{0.367 \text{ nm}} \right)$$

$$= -6.29 \times 10^{-19} \text{J}$$

3. Convert 6.29×10^{-19}J to kJ/mol using N_A:

$$\Delta H_{\text{lattice}} = \left(\frac{6.022 \times 10^{23} \text{ ion pairs}}{1 \text{ mol}} \right) \times \left(\frac{-6.29 \times 10^{-19} \text{ J}}{\text{ion pairs}} \right)$$

$$= -3.77 \times 10^5 \text{ J/mol or} - 377 \text{ kJ/mol}$$

9.3.3 Lewis structure of ionic compounds

Ionic or electrovalent bonds form by the complete transfer of electrons from one atom to another. This results in charged ions which are attracted to each other by electrostatic forces. This attraction is called ionic bonding, and the compounds formed are called ionic compounds. Generally, these bonds form between metals and nonmetals. Examples of ionic compounds include LiBr, Na_2S, CaO, $MgCl_2$, and HBr. Lewis dot symbols can be used to illustrate the formation of cations and anions as shown by the following examples:

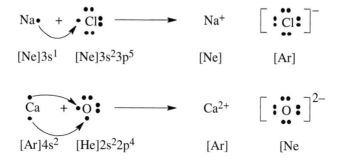

Example 9.4

Use Lewis dot symbols, electronic configurations, and the octet rule to show why aluminum and oxygen can form the ionic compound Al_2S_3.

Solution

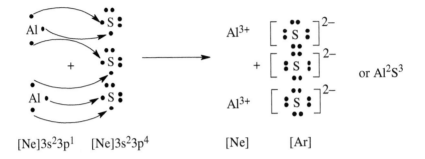

$$[Ne]3s^23p^1 \quad [Ne]3s^23p^4 \qquad\qquad [Ne] \qquad [Ar]$$

9.4 Covalent Bonding: Lewis Structures for Molecules

Covalent bonding involves the donation and sharing of two valence electrons between two atoms in a molecule to attain a stable electronic configuration. This type of bonding is most common between nonmetals. The Lewis structure for a covalent bond uses a line to represent the bond itself (that is, the pair of electrons), with any unshared electrons shown as 'lone pairs' of dots.

Covalent bonds can be:

Single: one pair of electrons is shared between two atoms, as in H_2 and F_2:

H• + •H ⟶ H:H or H—H

:F• + •F: ⟶ :F:F: or :F—F:

Double: two pairs of electrons are shared by two atoms, as in O_2 and CO_2:

:O=O: and :O=C=O:

Triple: three pairs of electrons are shared by two atoms, as in a nitrogen molecule:

:N≡N:

9.5 Covalent Bonding: Writing Lewis Structures

The Lewis structure is an electron-dot representation of the bonding relationship between atoms in a molecule or polyatomic ion. Note that the Lewis structure only describes the bonding, but by itself says nothing about the shape of the molecule.

9.5.1 Rules for writing Lewis structures

1. Determine the total number of valence electrons from all atoms. For anions, add one electron for each net negative charge. For cations, subtract one electron for each net positive charge.
2. Draw the skeletal structure for the molecule or ion and connect bonded atoms with an electron pair bond or a dash.
3. Calculate the number of electrons that remain after forming initial single bonds. Do this by subtracting the number of electrons needed to form single bonds from the total valence electrons.
4. Complete the octets of all the atoms bonded to the central atom (except H, He, Li, which follow the duet rule) by adding lone pairs of electrons.
5. Place any remaining electrons as lone pairs around the central atom(s) even if doing so leads to more than an octet of electrons. Some elements do violate the octet rule (this is discussed later). You have finished when all the valence electrons are used up.
6. If there are not enough electrons to form an octet around the central atom at this point, a multiple bond may be involved. Form a double or triple bond by moving one or more lone pairs from a terminal atom to a region between it and the central atom.

Example 9.5

Write the Lewis structures of (a) CO_3^{2-} (b) NH_3 (c) SiH_4.

Solution

Steps	CO_3^{2-}	NH_3	SiH_4
1. Calculate total valence electrons.	$4 + 3(6) + 2 = 24$	$5 + 3(1) = 8$	$4 + 4(1) = 8$
2. Draw skeletal structure.		H—N—H with H below N	
3. Calculate remaining electrons.	$24 - 6 = 18$	$8 - 6 = 2$	$8 - 8 = 0$
4. Distribute the remaining electrons first to terminal atoms to achieve octet structures.		H—N̈—H with H below N	
5. Place remaining electrons on central atom (if any).	There are not enough electrons to complete the octet on the central atom. Multiple bonding needed.	None	None
6. Make a double bond between O and C by moving a lone pair from one of the oxygen atoms. All the atoms now have complete octets.		N/A	N/A

Example 9.6

Write the Lewis structures of (a) NH_4^+ (b) PO_4^{3-} (c) $HClO_4$.

Solution

Steps	$NH_4{}^+$	$PO_4{}^{3-}$	$HClO_4$
1. Calculate total valence electrons.	$5 + 4(1) - 1 = 8$	$5 + 4(6) + 3 = 32$	$1 + 7 + 4(6) = 32$
2. Draw skeletal structures.	$\begin{array}{c} H \\ \mid \\ H-N^+\!\!-H \\ \mid \\ H \end{array}$	$\begin{array}{c} O \\ \mid \\ O-P-O \\ \mid \\ O \end{array}$	$\begin{array}{c} O \\ \mid \\ H-O-Cl-O \\ \mid \\ O \end{array}$
3. Calculate remaining electrons.	$8 - 8 = 0$	$32 - 8 = 24$	$32 - 10 = 22$
4. Distribute remaining electrons to achieve octet structures.	$\begin{array}{c} H \\ \mid \\ H-N^+\!\!-H \\ \mid \\ H \end{array}$	$\left[\begin{array}{c} \ddot{\text{:}\!O\!:} \\ \mid \\ \ddot{\text{:}\!O}-P-\ddot{O\!:} \\ \mid \\ \ddot{\text{:}\!O\!:} \end{array}\right]^{3-}$	$\begin{array}{c} \ddot{\text{:}\!O\!:} \\ \mid \\ H-\ddot{O}-Cl-\ddot{O\!:} \\ \mid \\ \ddot{\text{:}\!O\!:} \end{array}$
5. Place remaining electrons(if any) on central atom.	None	None	None

9.6 Resonance and Formal Charge

9.6.1 Resonance

A molecule or polyatomic ion is said to show *resonance* when more than one Lewis structure can be drawn to correctly describe its valence electron structure. These possible Lewis structures are called resonance structures. Resonance structures, therefore, are two or more structures with the same arrangement of *atoms*, but a different arrangement of *electrons*. Most often this occurs in structures with double bonds next to single bonds. One resonance form can be converted to another by simply moving lone pairs to bonding positions, and vice-versa. For example, consider SO_3, an intermediate in the production of sulfuric acid. The three Lewis structures below are all valid; they have the same arrangement of atoms and same energy, and differ only in the position of the double bond.

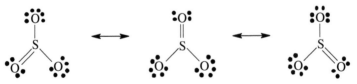

Resonance forms are not real bonding depictions. The actual molecule is a hybrid or an average of the contributing resonance forms. In a resonance hybrid, electrons are said to be "delocalized", and their density is "spread" over a few adjacent atoms.

9.6.2 Rules for writing resonance structures

1. All contributing resonance structures must be correct or valid Lewis structures.
2. Atoms should be left where they are! Only electrons can be moved.
3. Only nonbonding electrons (lone pairs or radicals), or those in double or triple bonds, can readily participate in resonance. Therefore, lone pairs on oxygen, halogens, nitrogen, etc., as well as those on anions, can participate in resonance.

4. The total number of electrons does not change.
5. The number of electron pairs must be the same across all contributing structural forms.
6. A double-headed arrow ($\leftrightarrow$) is placed between resonance structures.
7. The more reasonable resonance form(s) are those that minimize formal charges (as in Section 9.6.3), minimize formal charge separations, and avoid placing like formal charges on adjacent atoms. These will contribute most strongly to the overall description of the molecule.

Example 9.7

Write all the acceptable resonance (Lewis) structures for the nitrate ion, NO_3^-.

Solution

Follow all the steps for writing Lewis structures. Then apply the rules for writing resonance structures.

Steps	NO_3^-
1. Calculate total valence electrons.	$5 + 3(6) + 1 = 24$
2. Draw skeletal structure.	 O | O—N—O
3. Calculate remaining electrons.	$24 - 6 = 18$
4. Distribute remaining electrons to achieve octet structures. In doing so the octet is satisfied on the three oxygen atoms but not on the central N atom.	
5. Place remaining electrons (if any) on central atom.	None
6. Make a double bond between O and N by moving a lone pair from one of the oxygen atoms.	
7. We find three Lewis structures that can satisfy the octet rule, so the nitrate ion shows resonance.	

9.6.3 Formal charge

The formal charge assigned to an atom in a molecule or ion is the charge the atom would have if all electrons were shared equally. This helps determine which of several possible Lewis structures is most correct. It can be calculated using the following formula:

Formal charge (FC) = [# of valence electrons (V)] – [electrons in lone pairs (LP) + 1/2 the number of shared pair electrons (SP)]:

$$FC = V - [LP + 1/2\,(SP)]$$

9.6.4 Rules for assigning formal charge to an atom in a molecule

1. Draw the best Lewis structure for the molecule or ion.
2. Determine the number of valence electrons for the free atom.
3. Determine the number of lone pair electrons on the atom on the structure.
4. Determine the number of bonded or shared electrons.
5. Determine and assign the formal charge, using:

$$FC = V - [LP + 1/2\,(SP)]$$

6. Check that the formal charges for all atoms add up to the total charge on the molecule or ion.
7. The structure with the smallest formal charges is the most preferred structure.

9.6.5 Using formal charge to determine molecular structure

The following guidelines involving formal charge can be helpful in deciding which of the possible Lewis structures is most stable or has the least energy.

1. A molecular structure in which the formal charge on each atom is zero is the best, and is preferable to one in which some formal charges are not zero.
2. If the Lewis structure must have nonzero formal charges, the arrangement with the smallest nonzero formal charges is preferable.
3. The Lewis structure with negative charge on the more electronegative atoms is preferable.
4. Structures with adjacent formal charges that are zero or of opposite sign are preferable.
5. Formal charges other than $+1$, 0, or -1 are uncommon except for metals.

Example 9.8

Calculate the formal charge on each atom in the resonance forms of N_2O. Which of the resonance structures best represents the actual bonding pattern in this molecule?

Solution

Calculating formal charge for the resonance structures of N_2O

Lewis structure	:N̈—N≡O:			N̈=N=Ö			:N≡N—Ö:		
	↓	↓	↓	↓	↓	↓	↓	↓	↓
Number of valence electrons	5	5	6	5	5	6	5	5	6
Number of lone pair electrons	6	0	2	4	0	4	2	0	6
Number of bonded electrons	2	8	6	4	8	4	6	8	2
$FC = V - [LP + \frac{1}{2}(SP)]$	-2	$+1$	$+1$	-1	$+1$	0	0	$+1$	-1
The preferable resonance form							:N≡N—Ö:		

Example 9.9

The following are three possible Lewis structures for carbon disulfide:

$$\ddot{S}=C=\ddot{S} \longleftrightarrow \mathbf{:}S\equiv C-\ddot{\ddot{S}}\mathbf{:} \longleftrightarrow \mathbf{:}\ddot{S}-C\equiv S\mathbf{:}$$

(a) Assign formal charges to the atoms in each structure.
(b) Which resonance form is preferred?

Solution

The Lewis structure is already provided. Use the valence electrons, lone pairs, and shared pair information to determine the FC in the three structures. Finally, use the rules discussed earlier to determine the preferred resonance form.

Calculating Formal Charge for the Resonance Structure of CS$_2$

Lewis structure	$\ddot{S}=C=\ddot{S}$			$\mathbf{:}S\equiv C-\ddot{\ddot{S}}\mathbf{:}$			$\mathbf{:}\ddot{S}-C\equiv S\mathbf{:}$		
	↓	↓	↓	↓	↓	↓	↓	↓	↓
Lewis structure	6	4	6	6	4	6	6	4	6
Number of lone pair electrons	4	0	4	2	0	6	6	0	2
Number of bonded electrons	4	8	4	6	8	2	2	8	6
FC = V − [LP + ½(SP)]	0	0	0	−1	0	−1	−1	0	+1
The preferable resonance form	$\ddot{S}=C=\ddot{S}$								

9.7 Exceptions to the Octet Rule

While most molecules and ions follow the octet rule, some have too few electrons, and others have too many. Thus, we have three general exceptions to the octet rule.

1. Expanded valence shells (hypervalent compounds): molecules such as PCl$_5$, SF$_6$, and XeF$_4$ have *expanded valence shells* in which one or more atoms possess more than eight electrons. Central atoms from periods 3 and later of the periodic table can expand their valence shells by utilizing empty *d* orbitals.
2. Electron-deficient molecules: atoms such as Be and B do not have enough electrons to satisfy the octet rule, regardless of what other atoms they might be combined with.
3. Odd-electron molecules: molecules such as NO (eleven total valence electrons) and NO$_2$ (seventeen total valence electrons), with an odd number of electrons, do not obey the octet

rule. For example, the Lewis structures for NO and NO_2 show only seven electrons around the nitrogen atom.

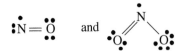

Example 9.10

The inorganic salt lithium hexafluorophosphate ($LiPF_6$) is widely used industrially in the manufacture of electrolyte for rechargeable lithium ion batteries, because of its high solubility in nonaqueous, polar solvents such as ethylene carbonate and dimethyl carbonate. Draw the Lewis structure for the hexafluorophosphate ion, PF_6^-.

Solution

Steps	PF_6^-
1. Calculate total valence electrons.	$5 + 6(7) + 1 = 48$
2. Draw skeletal structure.	
3. Calculate remaining electrons after using twelve to form single bonds.	$48 - 12 = 36$
4. Distribute remaining electrons to achieve octets. P is the central atom in the ion and has 12 valence electrons around it, which is four more than needed for an octet.	
5. Place remaining electrons on central atom (if any).	None left.

Example 9.11

Xenon oxyfluoride, $XeOF_4$, is a colorless volatile liquid. It is generally prepared by the controlled hydrolysis of XeF_6:

$$XeF_6 + H_2O \longrightarrow XeOF_4 + 2HF$$

Write the Lewis structures for the reactant XeF_6 and the product $XeOF_4$, and give the number of electrons in the expanded octet of each molecule.

Solution

Steps	XeF_6	$XeOF_4$
1. Calculate total valence electrons.	$8 + 6(7) = 50$	$8 + 6 + 4(7) = 42$
2. Draw skeletal structure.		
3. Calculate remaining electrons after using twelve to form single bonds.	$50 - 12 = 38$	$42 - 10 = 32$
4. Distribute remaining electrons to achieve octets on terminal atoms.		
5. Place remaining two electrons on Xe in both cases.		

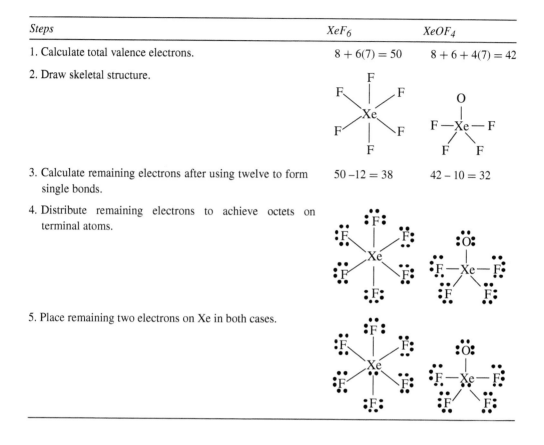

9.8 Polar Covalent Bonds: Bond Polarity and Electronegativity

Electronegativity is a measure of the ability of an atom in a molecule to attract electrons. The difference in the electronegativities of two atoms determines the bond type and, consequently, the physical and chemical properties of materials they are part of. Figure 9.1 gives the electronegativity of elements in the periodic table (Pauling's scale).

Most chemical bonds fall somewhere between ionic (in which one atom has completely donated electrons to another) and covalent (in which both atoms share electrons equally). Bond polarity is a measure of how equally electrons or charge in a bond are shared between two atoms. The polarity of a bond increases with increasing differences in electronegativity between the two atoms. For example, Cl_2 (Cl—Cl) is nonpolar, while HCl (H—Cl) is polar.

Most bonds have at least some polar character, with one end being partially positive ($\delta+$) and the other partially negative ($\delta-$).

9.8.1 Determining polarity of a molecule

Molecules containing covalently bonded atoms may be polar or nonpolar. Two key factors influence whether a molecule will be polar overall:

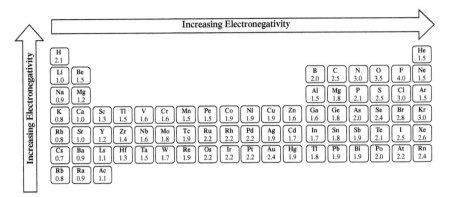

Figure 9.1   Electronegativity of elements (Pauling scale).

Table 9.1   Electronegativity differences and bond types

Electronegativity Difference ($\Delta\chi$)	Bond Type	Example
$\Delta\chi < 0.4$	Nonpolar Covalent	F—F, $\Delta\chi = 0$
$0.4 < \Delta\chi < 2.0$	Polar Covalent	H—Cl, $\Delta\chi = 0.9$
$\Delta\chi > 2.0$	Ionic	Na—F, $\Delta\chi = 3.1$

1. The polarity of individual bonds in the molecule. For a molecule to be polar, it must have at least one polar bond or lone pair.
2. The shape and symmetry of the molecule. If the molecule is symmetrical, it will be nonpolar. A molecule will have overall polarity if the polar bonds are oriented in space such that their dipoles do not offset each other. Diatomic molecules are an exception to this, as the polarity of the molecule is essentially the same as the polarity of the bond.

Use Table 9.1 as a guide. If all the bonds are in the noncovalent range, the molecule will be nonpolar regardless of the shape. On the other hand, if the bond type is in the ionic range, take the bond as polar in judging the polarity of the molecule.

As we discuss in Chapter 10, the VSEPR Theory can predict the shape and bond angle in a molecule and tell if a molecule polar or non-polar overall. The following shapes are symmetrical if all the ligands or terminal atoms bonded to a central atom are the same:

- Linear
- Trigonal planar
- Tetrahedral
- Trigonal bipyramidal
- Octahedral
- Square planar

In summary, a molecule is

1. Nonpolar if it has only nonpolar covalent bonds.
2. Nonpolar if it has polar covalent bonds, but the shape is symmetrical. Polarity is cancelled by symmetry.
3. Polar if it has polar covalent bonds and its shape is not symmetrical.

Example 9.12

In each of the following pairs, identify the more polar bond and indicate the negative and positive poles.

(a) C—F and C—N
(b) Li—Br and Li—F
(c) P—Cl and B—O
(d) Si—O and Ca—Cl

Solution

Use the following steps to answer this question:

1. Calculate the electronegativity difference ($\Delta\chi$) between the bonded atoms using the values for each element in Figure 9.1.
2. Determine the relative polarities of the bonds using the guidelines in Table 9.1. Recall that for

$\Delta\chi \leq 0.4$ Bond is nonpolar covalent

$0.4 < \Delta\chi < 2.0$ Bond is polar covalent

$\Delta\chi > 2.0$ Bond is ionic

4. Indicate the negative and the positive polarity based on your results.

	Bond	$\Delta\chi$	Relative Polarity
(a)	C—F	$4.0 - 2.5 = 1.5$	C—F > C—N
	C—N	$3.0 - 2.5 = 0.5$	
(b)	Li—Br	$2.8 - 1.0 = 1.8$	Li—F > Li—Br
	Li—F	$4.0 - 1.0 = 3.0$	
(c)	P—Cl	$3.0 - 2.1 = 0.9$	B—O > P—Cl
	B—O	$3.5 - 2.0 = 1.5$	
(d)	Si—O	$3.5 - 1.8 = 1.7$	Ca—Cl > Si—O
	Ca—Cl	$3.0 - 1.0 = 2.0$	

Example 9.13

Determine whether the bonds in each of the following compounds are covalent or ionic.

(a) H_2
(b) H_2O
(c) CaO
(d) LiF

Solution

Use Figure 9.1 to obtain the electronegativity difference and Table 9.1 to determine the bond type.

Compound	H_2	HCl	CaO	LiF
Electronegativity difference, $\Delta \chi$	$2.1 - 2.1 = 0$	$3.0 - 2.1 = 0.9$	$3.5 - 1.0 = 2.5$	$4.0 - 1.0 = 3.0$
Bond type	Covalent	Covalent	Ionic	Ionic

9.8.2 Dipole moment and percent ionic character

The *dipole moment* is a measure of the polarity of the molecule and occurs as a result of charge separation due to differences in electronegativity. A larger difference in electronegativity results in a larger dipole moment. The distance between the separated charges is also a deciding factor in the size of the dipole moment.

9.8.3 Dipole moment (μ)

When two atoms with equal but opposite charges, Q, separated by a distance r, share bonding electrons unequally, the result is a bond dipole.

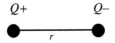

By definition, the dipole moment, μ, is the product of the magnitude of the separated charge and the distance of the separation, i.e.

$$\mu = Q \times r$$

The unit of the dipole moment is the Debye (D). $1D = 3.34 \times 10^{-30}$ coulomb-meters (C.m). For molecules, charge is usually measured in units of the electronic charge e, 1.60×10^{-19} C, and distance in Å. For example, if two charges, $+1$ and -1 (in units of e), are separated by 1.00 Å, the resulting dipole moment is

$$\mu = Q \times r = \left(1.60 \times 10^{-19} \text{C}\right) \left(1.00 \text{Å}\right) \left(\frac{10^{-10}\text{m}}{1\text{Å}}\right) \left(\frac{1\text{D}}{3.34 \times 10^{-30}\text{C.m}}\right) = 4.79\text{D}$$

Example 9.14

The bond length of HF (g) is 0.92 Å.

(a) If the charges on H and F were $+1$ and -1, respectively, what would be the resultant dipole moment, in D?
(b) The experimentally measured dipole moment of gaseous HF is 1.82 D. What magnitude of charge on the H and F atoms, in units of e, would lead to this dipole moment?

Solution

(a) The charge on each atom is the electronic charge, e: 1.60×10^{-19} C. The separation is 0.92 Å. By analogy to the calculation above, the dipole moment is

$$\mu = Q \times r = \left(1.60 \times 10^{-19} \text{C}\right) \left(0.92 \text{Å}\right) \left(\frac{10^{-10}\text{m}}{1\text{Å}}\right) \left(\frac{1\text{D}}{3.34 \times 10^{-30}\text{C.m}}\right) = 4.41\text{D}$$

(b) Given that the dipole moment, μ, is 1.82 D, and that the charge separation distance r is 0.92 Å, we want to calculate the value of Q:

From $\mu = Q \times r$, we get $Q = \dfrac{\mu}{r}$

$$Q = \frac{\mu}{r} = \frac{(1.82D)\left(\dfrac{3.34 \times 10^{-30}C-m}{1.0D}\right)}{(0.92\text{Å})\left(\dfrac{10^{-10}\ m}{1.0\text{Å}}\right)} = 6.61 \times 10^{-20}C$$

Convert 6.607×10^{-20} C to units of e as follows:

$$\text{Charge in } e \text{ units} = (6.61 \times 10^{-20}C)\left(\frac{1e}{1.60 \times 10^{-19}C}\right) = 0.413e$$

Therefore, the separated charge in HF, based on the experimental dipole moment, is:

$$0.413 + 0.413 - \text{ or } \overset{0.413+}{H} - \overset{0.413-}{F}$$

9.8.4 Percent ionic character of a covalent polar bond

Generally, both bond polarity and ionic character increase with an increasing difference in electronegativity. One method for determining the percentage ionic character of chemical bonds is to measure dipole moments. Percent ionic character (I) of a bond is given by

$$I = \frac{\mu_{obs}}{\mu_{ionic}} \times 100$$

where I is the percent ionic character, μ_{obs} is the actual or observed dipole moment, and μ_{ionic} is the calculated dipole moment assuming a completely ionic bond (100%). (Note that no actual bonds have 100% ionic character.)

Example 9.15

Given that the bond length in HI is 1.61 Å and the measured dipole moment is 0.44 D, what is the percent ionic character of the HI bond?

Solution

First calculate the μ_{ionic} assuming HI is 100% ionic:

$$\mu = Q \times r = \left(1.60 \times 10^{-19}C\right)(1.61\text{Å})\left(\frac{10^{-10}\ m}{1\text{Å}}\right)\left(\frac{1D}{3.34 \times 10^{-80}C.m}\right) = 7.71D$$

Next, calculate the percent ionic character knowing that μ_{obs} is 0.44 D.

$$I = \frac{\mu_{obs}}{\mu_{ionic}} \times 100 = \frac{0.44D}{7.71D} \times 100 = 5.71\%$$

9.9 Problems

1. Draw the Lewis symbols for each of the following elements using the periodic table as a guide.
   (a) Rb   (b) Mg   (c) Si   (d) As   (e) Se

2. Write Lewis symbols for the following ions.
   (a) B^{3+} (b) Mg^{2+} (c) N^{3-} (d) Cl^-

3. Considering the Lewis symbols for Be, Mg, Ca, and Sr, what generalization can be made about the Lewis structures for elements within the same group of the periodic table?

4. Use Lewis dot structures, electronic configurations and the octet rule to illustrate the formation of the following ionic compounds.
   (a) MgO   (b) K_2O (c) CaF_2 (d) Li_3N

5. Use Lewis dot structures, electronic configurations and the octet rule to illustrate the formation of the following ionic compounds.
   (a) Li_2S (b) $AlCl_3$ (c) Ca_3N_2 (d) B_2O_3

6. Boron nitride (BN), often called white graphene, has remarkable physical and mechanical properties and finds industrial application as dry lubricant, thermocouple protection sheath, crucible material, and coating for reactor vessels. Show the formation of BN using Lewis symbols.

7. Illustrate the formation of $SiBr_4$ from Si and Br atoms using Lewis symbols and Lewis structures.

8. Which of the following ionic compounds would have the higher lattice energy?
   (a) MgF_2 vs MgI_2 (b) CsI vs CaO   (c) CaS vs CaTe   (d) RbF vs SrO

9. Without any calculation, arrange the following in order of increasing lattice energy: LiF, KF, BeO, MgO, CaO. Explain your rationale.

10. Account for the trend in lattice energies:

Metal oxide	Lattice energy (kJ/mol)
BeH_2	-3205
MgH_2	-2791
CaH_2	-2410
SrH_2	-2250
BaH_2	-2121

11. Calculate the lattice energy in kJ/mol of the ionic bond between a calcium ion and an oxygen ion. The ionic radii of Ca^{2+} and O^{2-} are 100 pm and 140 pm, respectively.

12. Estimate the lattice energy between a pair of beryllium and sulfide ions. The ionic radii of Be^{2+} and S^{2-} are 27 pm and 184 pm, respectively.

13. Give the number of valence electrons in Cs, Mg, Ga, and Cl.

14. How many valence electrons are in each of the following?
    (a) HCN   (b) ClO_4^- (c) IF_6^+ (d) XeF_6

15. Calculate the total number of valence electrons in each of the following.
    (a) CCL_4 (b) PH_4^+ (c) H_3PO_4 (d) SO_4^{2-}

16. On the basis of Lewis theory, determine the compounds formed between:
    (a) Ca and Se (b) Ba and Br (c) Mg and N (d) B and O

17. Write the Lewis structures for each of the following molecules.
    (a) NF_3 (b) H_2O_2 (c) $GeCl_4$ (d) N_2F_2

18. Write the Lewis structures for the following organic compounds.
    (a) Methanol (CH_3OH)
    (b) Methylamine (CH_3NH_2)
    (c) Formic acid (HCOOH)
    (d) Dimethyl ether (CH_3OCH_3)

19. The decomposition of potassium chlorate ($KClO_3$) is commonly used in the laboratory preparation of oxygen as well as in oxygen generators aboard aircraft.

$$2KClO_3 (s) \xrightarrow{\Delta} 2KCl(s) + 3O_2(g)$$

Draw the Lewis structure for ClO_3^-.

20. Draw all possible resonance structures for each molecule or ion.
    (a) O_3 (b) SO_2 (c) SCN^- (d) NO_3^-

21. Draw all possible resonance structures for each molecule or ion.
    (a) N_2O (N is the central atom in all structures) (b) CO_2 (c) N_2O_2 (d) $CH_3SO_3^-$

22. Sodium azide (NaN_3) is used to generate nitrogen gas to inflate airbags upon impact. The impact triggers the following reaction:

$$2NaN_3 (s) \longrightarrow 2Na(s) + 3 N_2(g)$$

Draw the Lewis structure of the azide ion (N_3^-) showing all resonance forms.

23. The sulfate ion, SO_4^{2-}, is observed to have four equal S—O bond lengths that are shorter than a single bond. Write the Lewis structures to account for this observation.

24. The phosphate ion, PO_4^{3-}, contains a central phosphorus atom bonded to four oxygen atoms. The four P—O bond lengths are observed to be equivalent and are at $109.5°$. Write the various resonance structures which contribute to bonding in the phosphate ion.

25. The following structure is one of the four possible resonance structures for the BF_3 molecule:

Draw the Lewis structures for the other three resonance forms.

26. Draw resonance structures that obey the octet rule and calculate the formal charge on each atom in the following molecules or ions.
    (a) NO_2^- (b) SO_2 (c) NO^+ (d) CH_3NCO

27. What is the formal charge on each atom in the following Lewis structures?

(a)   (b)

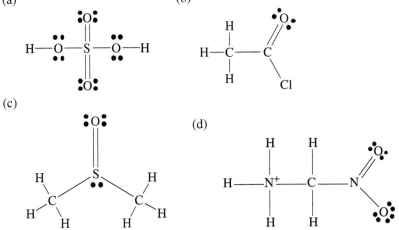

(c)

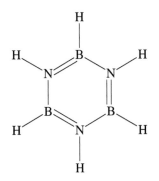

(d)

28. Draw a Lewis structure for nitric acid, HNO_3, and show all possible resonance forms. Use formal charge to determine the most important form contributing to the structure.

29. Carbon dioxide is used in many industrial applications as a supercritical fluid. For example, it is a good solvent, and it is also used for decaffeinating coffee and tea. Draw three Lewis structures for CO_2 and use formal charge to decide the most preferred structure.

30. Borazine, $B_3N_3H_6$, which is often referred to as 'inorganic benzene', is isoelectronic with benzene and has a planar six-member ring structure of alternating B and N atoms.

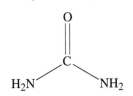

(a) Draw all reasonable resonance forms for the molecule (Hint: Include those in which the octet rule is violated.)
(b) Calculate the formal charges on the atoms in each of the resonance forms.

31. Draw all possible resonance structures for the urea molecule shown below, minimizing formal charges. Which of these forms is preferred in terms of formal charge?

32. Draw Lewis structures for each of the following molecules or ions. Where necessary use expended octets.
   (a) BrF_5 (b) AsF_6^- (c) IF_6^+ (d) XeO_4

33. Which bond of the following pairs is the more polar, based on the electronegativity values given in Figure 9.1?
   (a) H—F or H—Cl   (b) H—H or C—O   (c) B—N or C—F
   (d) H—H or H—F   (d) O—C or O—N

34. Which bond of the following pairs has more ionic character, according to the electronegativity values given in Figure 9.1?
   (a) Li—F or Li—Cl   (b) K—O or K—N   (c) Na—S or Na—P
   (d) Li—N or Na—N   (e) K—Cl or Ca—Cl

35. Classify the following bonds as ionic, non-polar, or polar.
   (a) Boron—Nitrogen   (b) Rubidium—Fluorine   (c) Sulfur—Sulfur
   (d) Hydrogen—Iodine

36. Which of the following molecules has a net dipole moment?
   (a) HF   (b) SO_2 (c) CO_2 (d) Br_2

37. Which of the following molecules has a net dipole moment?
   (a) SF_6 (b) H_2S (c) BF_3 (d) NH_3

38. Determine the percent ionic character of CO if the observed dipole moment is 0.110 D and the C—O interatomic distance is 113 pm. (Note: $q = 1.6 \times 10^{-19}C$; $1 D = 3.34 \times 10^{-30}$ C.m)

39. In the gas phase, the dipole moment of KBr was measured to be 10.41D (3.473×10^{-29} C.m). If the K—Br interatomic distance is 282 pm, calculate:
   (a) The theoretical dipole moment
   (b) The percent ionic character
   (c) The approximate charge distribution on the molecule.

40. The gas phase measured dipole moment and interatomic distance of BrF molecule are 1.29 D and 178 pm, respectively. What are
   (a) The theoretical dipole moment
   (b) The percent ionic character
   (c) Approximate charge distribution on the molecule.

10

Chemical Bonding 2: Modern Theories of Chemical Bonding

....

10.1 VSPER Theory: Molecular Geometry and the Shapes of Molecules

As discussed in Chapter 9, Lewis structures can account for the formulas of covalent compounds by showing the number and types of bonds between atoms. However, Lewis structures alone do not indicate the geometry or shapes of molecules or polyatomic ions.

The Valence Shell Electron Pair Repulsion (VSEPR) theory fills this lack and provides a simple method for predicting the shape or geometry of small molecules and molecular ions.

10.1.1 VSEPR theory

VSEPR theory is a model first proposed by Sidgwick and Powell in 1940 and developed by Gillespie and Nyholm in 1957. The premise of the VSEPR model is that the valence electron groups or pairs surrounding the central atom of a molecule will mutually repel each other. Consequently, the molecule will adopt a spatial arrangement that minimizes this repulsion, thus determining the shape of the molecule or molecular ion.

10.1.2 Assumptions of the VSEPR theory

- The least electronegative atom in a polyatomic molecule (i.e. one consisting of three or more atoms) is identified as the central atom to which all other atoms in the molecule are bonded.
- For polyatomic molecules with more than one central atom (e.g. acetone, CH_3COCH_3, with three central atoms), the VSEPR model assigns an individual geometry to each central atom.
- The geometry or shape of the molecule is determined by the total number of valence shell electron pairs or electron groups around the central atom.
- Electron pairs directly involved in connecting the atoms in a molecule are called bonding pairs. In multiple bonding (double and triple bonds), more than one pair of electrons may bind any two atoms together.
- Some atoms in a molecule may also possess *lone pairs*—nonbonded pairs of electrons not involved in bonding.

Chemistry in Quantitative Language: Fundamentals of General Chemistry Calculations. Second Edition.
Christopher O. Oriakhi, Oxford University Press. © Christopher O. Oriakhi 2021.
DOI: 10.1093/oso/9780198867784.003.0010

- The electron pairs tend to orient themselves in a way that minimizes the repulsion between them, which maximizes the distance between them.
- Lone pairs occupy more space than bonding electron pairs, and double or triple bonds occupy more space than single bonds.
- Repulsive forces decrease rapidly with increasing inter-pair angle. Forces are greatest at $90°$, much weaker at $120°$, and very weak at $180°$.
- Where resonance occurs, the VSEPR theory can be applied to each resonance structure of a molecule.
- After identifying the basic shape of a molecule, adjustments must be made to account for the differences in electrostatic repulsion between bonding regions and nonbonded electron pairs. Repulsions follow the order:
  Lone pair–lone pair > Lone pair–bonding pair > Bonding pair–bonding pair

10.2 VSEPR Theory: Predicting Electron Group Geometry and Molecular Shape with the VSEPR Model

1. Draw the Lewis structure of the molecule or polyatomic ion (for molecules with resonance structures, use any of the structural forms to predict the molecular structure).

Table 10.1   Electron geometries around an atom yielding minimum repulsion

Number of electron groups	Arrangement of electron groups	Electron group geometry	Bond angle
2		Linear	$180°$
3		Trigonal planar	$120°$
4		Tetrahedral	$109.5°$
5		Trigonal bipyramidal	$120°$ Equatorial, $90°$ axial
6		Octahedral	$90°$

2. Count the number of electron groups (lone pairs and bonds) around the central atom.
3. Identify the electron group geometry based on the total number of electron groups as linear, trigonal planar, tetrahedral, trigonal bipyramidal, or octahedral. Use Table 10.1 and Table 10.2 as guides.
4. Use the angular arrangement of the bonded atoms to determine the molecular geometry. If the central atom has no lone pairs, the molecular geometry is the same as the electron group geometry. However, if one or more lone pairs exist on the central atom, the visible molecular geometry will be different than the electron geometry.

10.2.1 Electron group geometries

The central atoms in most molecules have 2, 3, 4, 5, or 6 electron pairs or groups around them. Table 10.1 shows the electron group geometry for each of these cases.

Table 10.2   Using VSEPR to predict shapes of AXE molecules

Molecule class	Electron groups	Bonding groups	Lone pairs	Electron geometry	Molecular geometry	Bond angle	Example
AX_2	2	2	0	Linear	Linear	$180°$	$BeCl_2$, CO_2
AX_3	3	3	0	Trigonal planar	Trigonal planar	$120°$	BF_3, NO_3^-
AX_2E	3	2	1	Trigonal planar	Bent	$< 120°$	O_3, NO_2^-
AX_4	4	4	0	Tetrahedral	Tetrahedral	$109.5°$	CH_4, SO_4^{2-}
AX_3E	4	3	1	Tetrahedral	Trigonal pyramidal	$< 109.5°$	NH_3, $AsCl_3$
AX_2E_2	4	2	2	Tetrahedral	Bent	$< 109.5°$	H_2O, SF_2
AX_5	5	5	0	Trigonal bipyramidal	Trigonal bipyramidal	$120°$ Eq $90°$ Ax	PCl_5, AsF_5
AX_4E	5	4	1	Trigonal bipyramidal	Seesaw	$< 120°$ Eq $< 90°$ Ax	SF_4, SeH_4
AX_3E_2	5	3	2	Trigonal bipyramidal	T-Shaped	$< 90°$	BrF_3, ICl_3
AX_2E_3	5	2	3	Trigonal bipyramidal	Linear	$180°$	XeF_2, I_3^-
AX_6	6	6	0	Octahedral	Octahedral	$90°$	SF_6, IOF_5
AX_5E	6	5	1	Octahedral	Square pyramidal	$< 90°$	BrF_5, $XeOF_4$
AX_4E_2	6	4	2	Octahedral	Square planar	$90°$	XeF_4, BrF_4^-

10.2.2 VSEPR theory: the AXE system

VSEPR theory uses the AX_mE_n notation where $m + n =$ total regions of electron density (sometimes also called charge clouds). Here, A is the central atom; m is the total number of ligands X, and n the number of lone pairs E.

Example 10.1

Predict the electron group geometry, molecular geometry, and bond angles of (a) PF_3 and (b) H_2S.

Solution

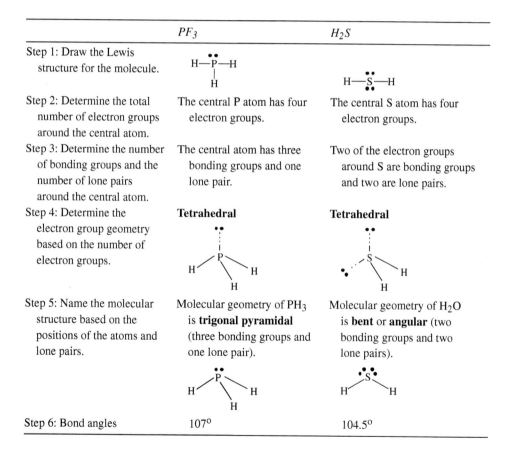

	PF_3	H_2S
Step 1: Draw the Lewis structure for the molecule.	H—P—H with H below	H—S—H
Step 2: Determine the total number of electron groups around the central atom.	The central P atom has four electron groups.	The central S atom has four electron groups.
Step 3: Determine the number of bonding groups and the number of lone pairs around the central atom.	The central atom has three bonding groups and one lone pair.	Two of the electron groups around S are bonding groups and two are lone pairs.
Step 4: Determine the electron group geometry based on the number of electron groups.	**Tetrahedral**	**Tetrahedral**
Step 5: Name the molecular structure based on the positions of the atoms and lone pairs.	Molecular geometry of PH_3 is **trigonal pyramidal** (three bonding groups and one lone pair).	Molecular geometry of H_2O is **bent** or **angular** (two bonding groups and two lone pairs).
Step 6: Bond angles	$107°$	$104.5°$

Example 10.2

Use the VSEPR model to predict the electron group geometry and molecular geometry, including bond angles, of (a) XeF_2 and (b) XeF_4.

Solution

	XeF_2	XeF_4
Step 1: Draw the Lewis structure for the molecule.		
Step 2: Determine the total number of electron groups around the central atom.	The central Xe atom has five electron groups.	The central Xe atom has six electron groups.
Step 3: Determine the number of bonding groups and the number of lone pairs around the central atom.	The Xe atom has two bonding groups and three lone pairs.	Four of the six electron groups around Xe are bonding pairs and two are lone pairs.
Step 4: Determine the electron geometry based on the number of electron groups.	**Trigonal bipyramidal**	**Octahedral**
Step 5: Name the molecular structure based on the positions of the atoms and lone pairs.	Molecular geometry of XeF_2 is **linear** (two bonding groups and three lone pairs).	Molecular geometry of XeF_4 is **square planar** (four bonding groups and two lone pairs).
Step 6: Bond angles	$180°$	$90°$

Example 10.3

Draw the Lewis structures and predict the molecular shapes of (a) CS_3^{2-} and (b) PCl_6^-.

Solution

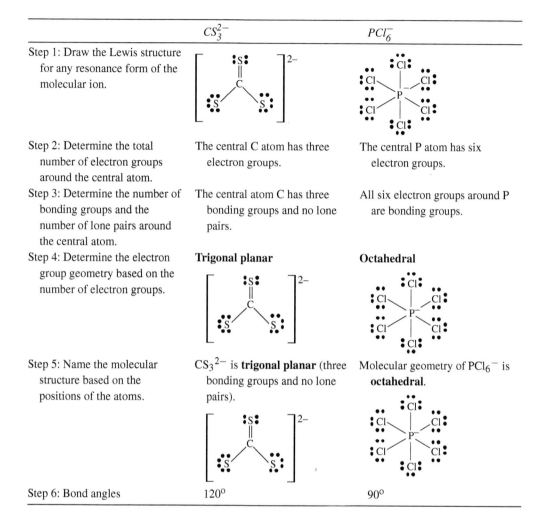

	CS_3^{2-}	PCl_6^-
Step 1: Draw the Lewis structure for any resonance form of the molecular ion.		
Step 2: Determine the total number of electron groups around the central atom.	The central C atom has three electron groups.	The central P atom has six electron groups.
Step 3: Determine the number of bonding groups and the number of lone pairs around the central atom.	The central atom C has three bonding groups and no lone pairs.	All six electron groups around P are bonding groups.
Step 4: Determine the electron group geometry based on the number of electron groups.	**Trigonal planar**	**Octahedral**
Step 5: Name the molecular structure based on the positions of the atoms.	$CS_3{}^{2-}$ is **trigonal planar** (three bonding groups and no lone pairs).	Molecular geometry of PCl_6^- is **octahedral**.
Step 6: Bond angles	120°	90°

10.3 VSEPR Theory: Predicting Molecular Shape and Polarity

A polar molecule is one in which electrons or charges are shared unequally between covalently bonded atoms, causing the molecule to have a net dipole moment. A nonpolar molecule, on the other hand, has zero dipole moment, attributed to equal sharing of bonding electrons and structural symmetry within the molecule.

Two factors determine whether a molecule is polar or nonpolar: the polarity of the bonds and the shape or geometry of the molecule.

A diatomic molecule (such as Cl_2 or HCl) has only one bond between the atoms. Consequently the polarity of that bond determines the polarity of the molecule. If the bond is polar, the molecule will be polar, and if the bond is nonpolar, the molecule is nonpolar by default.

In polyatomic molecules having more than one bond, polarity is determined by both the arrangement and polarity of the bonds. The presence of a polar bond is a necessary condition for a molecule to have inherent polarity. However, a molecule can become nonpolar if the polar bonds present are symmetrically arranged within the molecule or aligned exactly opposite to each other, because the bond polarities will cancel, for example, CO_2. Because oxygen is more electronegative than carbon, the bonds are polar, with electron density concentrated on the two oxygen atoms. However, CO_2 is linear, and the symmetrical arrangement of these polar bonds allows their effects to cancel each other, leaving the molecule nonpolar, as shown.

$$\overset{\delta-}{O} - \overset{\delta+}{C} - \overset{\delta-}{O}$$

On the other hand, the linear molecules HCN and N_2O are polar because both are structurally unsymmetrical, and both contain polar bonds.

10.3.1 Steps for predicting molecular polarity

1. Draw a reasonable Lewis structure and a correct geometrical representation for the molecule.
2. Identify each bond as either polar or nonpolar using the magnitude of the electronegativity difference. Use Table 9.1 as a guide.
3. Describe the polar bonds with arrows pointing toward the more electronegative elements. Use the length of the arrow to show the relative polarities of the different bonds.
4. The molecule is considered nonpolar if:

   • It contains no polar bonds.
   • The central atom has no lone pairs, and all the ligands or terminal atoms bonded to it are the same, as in linear, trigonal planar, tetrahedral, trigonal bipyramidal, octahedral, and square planar geometries.
   • The molecular structure is symmetrical and the vector arrows showing the relative polarities are of equal length.

5. The molecule is polar if:

   • The central atom has at least one polar bond or lone pair.
   • The multiple polar bonds or lone pairs in the molecule are arranged in a way that the dipole moment does not cancel out.
   • It fails to meet all the requirements for being nonpolar.

Example 10.4

Predict whether each of the following molecules is polar or nonpolar.

(a) BF_3
(b) OF_2
(c) SF_6
(d) HCN

Solution

	BF_3	OF_2	SF_6	HCN
Step 1: Draw the Lewis structure for the molecule and determine the molecular geometry.	Trigonal planar	Bent	Octahedral	H—C≡N: Linear
Steps 2 & 3: Determine the polarity of the bonds.	The three B–F bonds are polar ($\Delta\chi = 2.0$).	The two O–F bonds ($\Delta\chi = 0.5$) are polar.	The six S–F bonds are polar ($\Delta\chi = 1.5$).	HCN contains a polar bond H–C ($\Delta\chi = 0.4$) and C–N ($\Delta\chi = 0.5$).
Steps 4 & 5: Determine polarity of the molecule.	All groups on B are the same; no lone pair on B; BF_3 is symmetrical. Vectors cancel out due to symmetry. BF_3 is nonpolar.	OF_2 has a bent shape, so the distribution of the polar bonds is asymmetrical. OF_2 is polar.	All groups on S are the same; no lone pair on S; SF_6 is symmetrical. Vectors cancel out. Therefore SF_6 is nonpolar.	Asymmetric linear geometry. Bond vectors do not cancel out. HCN is polar.

Example 10.5

Draw Lewis structures for the following molecules and predict the polarity of each.
(a) CF_2Cl_2 and CCl_4
(b) SO_2 and SO_3
(c) CS_2 and H_2S
(d) SF_4 and IF_5

Solution

(a) CF_2Cl_2 and CCl_4
  The Lewis structures are:

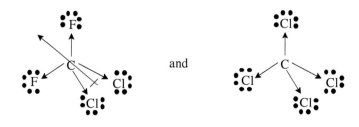

and

Each has a tetrahedral geometry. CF_2Cl_2 has two polar C–F bonds ($\Delta\chi = 1.5$) and two C–Cl bonds ($\Delta\chi = 0.5$). Therefore, the C–F bonds are more polar than the C–Cl bonds. The vector components do not cancel; hence, the molecule is polar. Although CCl_4 contains four polar C–Cl bonds, the molecule is symmetrical and the bond vectors cancel, making CCl_4 nonpolar.

(b) SO_2 and SO_3

The Lewis structures and vector diagrams of SO_2 and SO_3 are

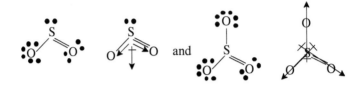

Both contain polar S–O bonds ($\Delta\chi = 1.0$). The molecular geometry of SO_2 is bent and is asymmetrical hence SO_2 is polar. The molecular shape for SO_3 is trigonal planar. All groups attached to the central S atom are the same, and are symmetrical, so the bond vectors cancel, making SO_3 nonpolar.

10.4 Valence Bond Theory

Valence Bond Theory (VBT), originally proposed by Heitler and London, uses quantum mechanics to explain how valence electrons of different atoms combine to form a molecule. The primary focus of this theory is on the formation of individual bonds from the atomic orbitals of the participating atoms composing the molecule.

10.4.1 Postulates of valence bond theory

The basic assumptions of this theory are:

- A covalent bond is formed by the overlap of two atomic orbitals.
- Only atomic orbitals with unpaired electrons are involved in covalent bond formation.
- If the electrons present in the atomic orbitals have parallel spins, no bond will result, and no molecule will be formed.
- If the atomic orbitals have more than one unpaired electron, more than one bond can be formed.
- A covalent bond is formed by the overlap of half-filled valence atomic orbitals. The inner electrons do not play any role in the bond formation.
- During the formation of a covalent bond, only the valence electrons from each bonded atom participate. The other electrons remain unaffected.
- The strength of a covalent bond depends upon the degree of orbital overlap—the greater the overlap, the greater the strength of the resulting covalent bond.
- A covalent bond is directional. The bond will lie along the direction of orbital overlap.
- Based on the nature of the overlap, two types of covalent bonds are possible: sigma (σ) bonds and pi (π) bonds. A σ-bond results from the end-to-end or axial overlap of atomic orbitals, while a π-bond arises from parallel or sideways overlap.

10.5 Valence Bond Theory: Types of Overlap

10.5.1 Sigma (σ) bond

A σ-bond is a molecular orbital formed by the end-to-end overlap of atomic orbitals. It is cylindrically symmetrical and is the strongest type of bond. All single bonds are σ-bonds. A σ-bond can occur between two s-orbitals, s and p orbitals, or two p orbitals. These bonds are then termed σ_{S-S} overlapping, σ_{S-P} overlapping, and σ_{P-P} overlapping, as illustrated below.

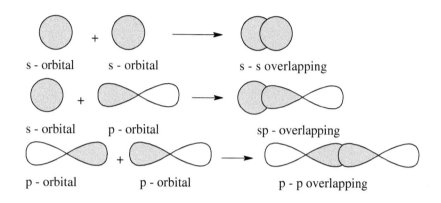

10.5.2 Pi (π) bond

A π-bond results from the sidewise or parallel overlap of p orbitals on adjacent atoms. The electron density in a π-bond is distributed above and below the inter-nuclear axis.

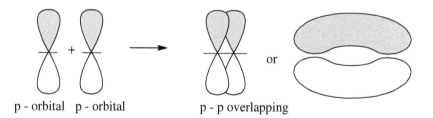

10.5.3 The strength of sigma and pi bonds

A σ-bond is stronger than a π-bond, because the axial or head-on overlap which occurs in a σ-bond is more complete than the sideways overlap in a π-bond.

- A single bond contains one σ-bond.
- A double bond contains one σ-bond and one π-bond.
- A triple bond contains one σ-bond and two π-bonds.

Since a multiple bond contains both σ-bonds and π-bonds, it is stronger than either alone.

10.6 Hybridization

The atomic overlap model proposed in the 1930s worked well to explain covalent bonding and geometry in small molecules like H_2 and HF, but it needed modification as it could not adequately explain the geometrical shapes of polyatomic molecules such as CH_4, NH_3, and H_2O. To resolve the discrepancy between the predicted and observed bond angles in polyatomic molecules, in 1931 Pauling introduced the concept of hybridization. Hybridization is defined as the intermixing of two or more pure atomic orbitals with somewhat different energies (e.g. s and p orbitals) to form new orbitals, known as hybrid orbitals, that have equal energies, identical shapes, and symmetrical orientations in space. These new hybrid orbitals can then overlap with those in other atoms to form σ-bonds.

To form hybrid atomic orbitals, an electron is promoted from a filled ns^2 subshell to an empty np or $(n-1)d$ valence orbital, followed by mixing or hybridization, to give a new set of energetically equivalent orbitals with the proper orientation to form bonds.

10.6.1 Characteristics of hybrid orbitals

- Hybridization takes place only in orbitals of almost similar energies and belonging to the same atom or ion. For example, a $2s$ orbital cannot hybridize with a $3s$ orbital since they have very different energy levels.
- The number of hybrid orbitals formed is equal to the number of pure atomic orbitals used in the hybridization process.
- The hybrid orbitals are degenerate: that is, they are always equivalent in energy and shape.
- Hybridization can take place between filled, half-filled, or empty orbitals.
- The hybrid orbitals are oriented to have the maximum space and minimum repulsion between them, thus defining their geometry.
- The shapes and orientations of hybrid orbitals are very different from the shape of the original atomic orbitals.
- Generally, the d orbitals are unlikely to participate in hybridization because of their large size and high energy compared to the s and p orbitals. However, mixing of d orbitals can take place when highly electronegative atoms are bonded to a central atom causing a contraction in the size of the d orbitals.
- Hybridized orbitals provide more efficient overlap than pure s, p, and d orbitals due to their fixed orientation. Therefore, they form stronger directional bonds than the pure s, p, or d atomic orbitals.
- Hybrid orbitals overlap to form σ-bonds. Unhybridized orbitals overlap to form π-bonds.

10.6.2 Types of hybrid orbitals

Hybridization can be classified as sp^3, sp^2, sp, sp^3d, or sp^3d^2, depending on the orbitals involved in the mixing.

10.6.2.1 sp^3 hybridization—tetrahedral structure

Here, an s orbital and all three of the p orbitals mix to form four new sp^3 orbitals with identical shapes and energies, which are therefore said to be degenerate. Each sp^3 hybrid orbital is used to form a sigma bond and has 25% s and 75% p character. An sp^3 hybridized atom has four regions of electron density, whose mutual electron repulsion favors a tetrahedral geometry with 109.5° bond angles. Note that hybrid orbitals can overlap with any atomic orbital of a different atom to form a bond. For example, in the CH_4 molecule, the $2s$ and $2p$ atomic orbitals of carbon hybridize to form four sp^3 hybrid orbitals with a tetrahedral orientation. These four orbitals on the C overlap with the $1s$ atomic orbitals of the hydrogen to form four CH σ-bonds as illustrated below.

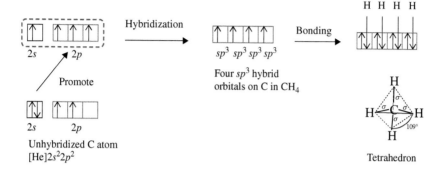

10.6.2.2 sp^2 hybridization—trigonal planar structure

In sp^2 hybridization, one s and two p orbitals combine to form three new hybrid orbitals that can each overlap with another atom's orbitals to form σ-bonds. The new orbitals formed are called sp^2 hybrids, and the unhybridized p-orbital is used for π-bonding. The three hybrid orbitals are directed toward the corners of an equilateral triangle and make an angle of 120° with one another. Each of the hybrid orbitals formed has 33.33% s character and 66.66% p character. Molecules in which the central atom is linked to three atoms through sp^2 hybrid orbitals usually have a trigonal planar shape. We can use BF_3 as an example.

Boron has five electrons and the electronic configuration$1s^2 2s^2 2p^1$.Three of these electrons are in the valence shell. To form bonds with three fluorine atoms, one electron in the $2s$ orbital of B is excited or promoted to its $2p$ orbital. One $2s$ orbital mixes with two $2p$ orbitals to give three new sp^2 hybrid orbitals. One of the $2p$ orbitals of the boron is empty and unhybridized. The three sp^2 hybrid orbitals point to the corners of a triangle. The angle between any two of them is 120°. Each of the three B sp^2 hybrids orbital overlaps with a $2p$ atomic orbital of a fluorine to form three B-F sigma bonds in BF_3.

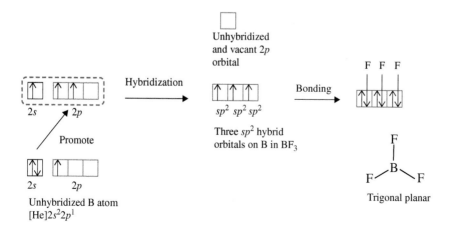

10.6.2.3 *sp* hybridization

In a system involving *sp* hybridization, the central atom's *s* orbital and one of its *p* orbitals have mixed to form two new *sp* hybrids, leaving two unhybridized *p* orbitals, which are used for π-bond formation as needed. The two *sp* hybrid orbitals arranged linearly with a bond angle of 180°. The unhybridized *p* orbitals are oriented in a plane perpendicular to the hybridized *sp* orbitals. To illustrate *sp* hybridization, consider BeF_2. In the ground state, Be cannot form any bonds. With appropriate energy it promotes an electron from the $2s$ to the $2p$ orbital. One *s* orbital and one *p* orbital combine to form two *sp* hybrid orbitals. The two valence electrons distribute themselves among the two *sp* hybrid orbitals, making it possible to form two identical Be–F bonds. The hybridization process is illustrated below.

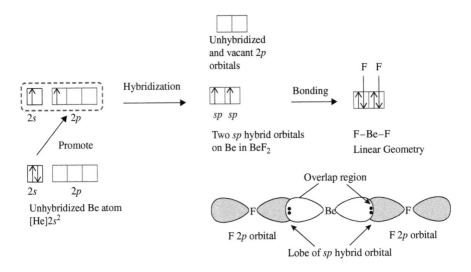

10.6.3 Hybridization and multiple bonds

Hybridization may be applied to describe covalent bonding in molecules containing double and triple bonds. When there is more than one central atom, the bonding around each atom is considered independently. Let us consider ethene (ethylene) and ethyne (acetylene).

10.6.3.1 Ethene, C_2H_4

Ethylene, which is a planar molecule, has the following Lewis structure:

Each carbon is surrounded by three electron groups. Using the VSEPR model, we would predict a trigonal planar geometry around each carbon atom in the structure, with an H–C–H bond angle of 120°. This geometry calls for a planar set of orbitals at angles of 120°, which the $2s$ and $2p$ valence orbitals of carbon do not have. Thus hybrid orbitals are needed.

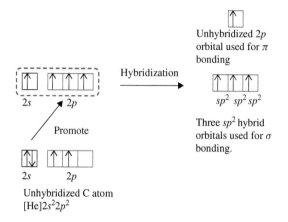

After hybridization, two of the three new identical sp^2 orbitals on each carbon atom overlap with the $1s$ orbitals of the hydrogen to form σ-bonds. The remaining sp^2 hybrid orbital on each carbon overlaps with the sp^2 hybrid orbital of the adjacent C atom to form a σ-bond. The two unhybridized $2p$ orbitals of the C atoms lie perpendicular to the plane of the trigonal sp^2 hybrids and overlap sideways to form a π-bond (Figure 10.1).

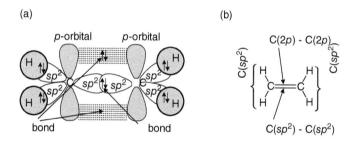

Figure 10.1 Hybridization of carbon orbitals in ethylene. (a) Orbitals involved in bond formation (b) Bonding framework

10.6.3.2 Ethyne (acetylene), C_2H_2

The Lewis structure of ethyne is

$$H—C\equiv C—H$$

Each carbon atom is surrounded by two electron groups. This requires a linear arrangement with a bond angle of $180°$, which means that both carbons are sp hybridized. Two of the carbon's four valence orbitals hybridize to sp orbitals, which form sigma bonds.

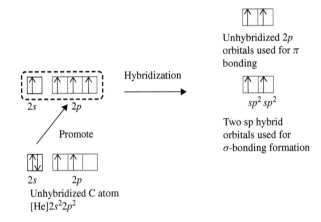

The remaining unhybridized p orbitals each contain one electron, and can overlap to form π-bonds, as seen in Figure 10.2.

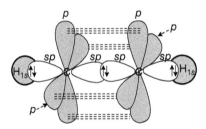

Figure 10.2   Formation of two π bonds in ethyne from the overlap of two sets of unhybridized carbon $2p$ orbitals.

Example 10.6

Consider the structure of cumulene:

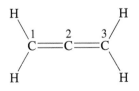

a.   What are the geometry and hybridization of carbon atoms 1, 2, and 3?
b.   Determine the total number of σ-bonds and π-bonds in the molecule.

Table 10.3 Common hybridization schemes

Effective electron pairs	Hybrid orbitals	Atomic orbitals used	Geometrical arrangement and bond angle	Examples
2	sp	s,p	Linear (180°)	BeH_2, CO_2, C_2H_2, $HgCl_2$
3	sp^2	s,p,p	Trigonal planar (120°)	BF_3, C_2H_4, NO_3^-, CO_3^{2-}
4	sp^3	s,p,p,p	Tetrahedral (109.5°)	CH_4, NH_4^+, $SnCl_4$, BF_4^-
5	sp^3d	s,p,p,p,d	Trigonal bipyramidal (120°, 90°)	SF_4, PF_5, XeF_2, I_3^-, BrF_3
6	sp^3d^2	s,p,p,p,d,d	Octahedral (90°)	SF_6, $[CrF_6]^{3-}$, $[Co(NH_3)_6]^{3+}$
7	sp^3d^3	s,p,p,p,d,d,d	Pentagonal bipyramidal	IF_7

Solution

Use the VSEPR model to determine the electron groups and molecular geometry around each carbon atom. The geometry will suggest the hybrid orbitals used.

a. C1 and C3 are identical, having three electron groups around the central atom; geometry is trigonal planar, sp^2 hybridized. Geometry about C2 is linear and so sp hybridized.
b. There are six σ-bonds and two π-bonds:

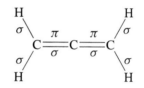

10.6.4 Hybridization of elements involving *d*-orbitals

Elements in the third period and beyond, with more than four electron pairs, require the involvement of the *d* orbitals in covalent bond formation to accommodate the expanded octet. This is possible because the energy of the $3d$ orbitals is comparable to the energy of the $3s$, $3p$, $4s$, and $4p$ orbitals. They can form hybrid orbitals by mixing with the appropriate *s* and *p* orbitals. The most important hybridized orbitals in this category are described in Table 10.3.

10.6.4.1 sp^3d hybridization

In sp^3d hybridization, one *s* orbital, three *p* and one *d* combine to form five sp^3d hybrid orbitals that point to the corners of a trigonal bipyramid. PCl_5 offers an example of this hybridization.

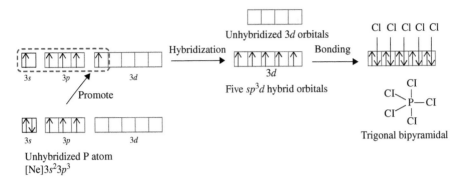

The covalent bond between phosphorous and each of the five chlorine atoms is formed by the overlap between an sp^3d hybrid orbital on phosphorous and the p orbital on chlorine.

10.6.4.2 sp^3d^2 hybridization

In sp^3d^2 hybridization, one s, three p and two d orbitals combine to form six sp^3d^2 hybrid orbitals, which are directed toward the corners of a regular octahedron.

The bonding in SF_6 is a good example. The single $3s$, three $3p$, and two $3d$ orbitals of the sulfur combine to give six sp^3d^2 hybrid orbitals with an octahedral structure in which all bond angles are $90°$. In the process, sulfur promotes one s and one p electron to its $3d$ orbitals resulting in six orbitals each containing one electron available for bonding. The resulting orbitals are shown below:

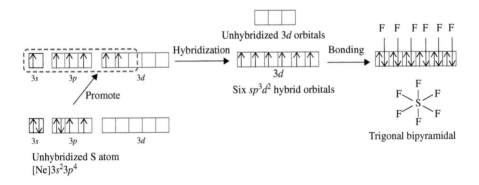

10.6.5 Predicting the hybrid orbitals used by an atom in bonding

To determine the hybrid orbitals used by an atom in a molecule, follow the same procedure used in determining electron group geometry and molecular structure:

1. Draw the Lewis structure for the molecule or ion.
2. Determine the electron group geometry using the VSEPR model.
3. Specify the hybrid orbitals corresponding to the geometry using Table 10.3 as a guide.

Example 10.7

Indicate the hybridization of orbitals used by the central atom in (a) NH_4^+ (b) PF_6^-.

Solution

	NH_4^+	PF_6^-
Step 1: Draw the Lewis structure for the molecule or ion.	H |+ H—N—H | H	
Step 2: Determine the total number of electron groups around the central atom.	The central N atom has four electron groups.	The central P atom has six electron groups.
Step 3: Determine the number of bonding groups and the number of lone pairs around the central atom.	All four electron groups around the central N atom are bonding groups.	All six electron groups around P are bonding groups.
Step 4: Determine the electron group geometry due to bonding groups and lone pairs.	Tetrahedral	Octahedral
Step 5: Determine the correct hybridization type of the central atom based on the electron geometry	sp^3	sp^3d^2

10.7 Limitations of Valence Bond Theory

The Valence Bond Theory has its own set of limitations. For example, it:

- Fails to explain the tetravalency of carbon.
- Offers no explanation of the colors observed for complex ions.
- Does not consider the splitting of *d*-orbital energy levels.
- Assumes electrons are localized to specific locations (and thus understates the importance of resonance).
- Fails to account for various magnetic, electronic, and spectroscopic properties of metal complexes.
- Leads to erroneous ionic and covalent structure assignments based on magnetic data.
- Is not able to account for or predict even the relative energies of different structures.

10.8 Molecular Orbital Theory

To overcome the limitations of Valence Bond Theory, in 1932, Hund and Mulliken derived the more powerful Molecular Orbital (MO) Theory, which explains the bonding in molecules as well as the magnetic and spectral properties of molecular compounds. According to MO theory, linear combinations of the atomic orbitals of different atoms form new orbitals called molecular orbitals. Electrons in these molecular orbitals are distributed over the entire molecule.

10.8.1 Linear combination of atomic orbitals (LCAO)

The wave function for the molecular orbitals can be approximated by taking linear combinations of atomic orbitals of comparable energies and of suitable symmetry. Essentially, when two atoms come together, their two atomic orbitals interact either constructively to form bonding molecular orbitals having a lower energy, or destructively to form antibonding molecular orbitals having a higher energy than the original atomic orbital. Electrons in the bonding MO are responsible for the stability of the molecule, while electrons in the antibonding orbital destabilize it.

Various rules and restrictions govern the formation of molecular orbitals:

1. When a set of atomic orbitals combines to form molecular orbitals (MOs), the number of MOs formed is equal to the number of atomic orbitals combined.
2. The bonding MO is at a lower energy relative to the original atomic orbitals and the antibonding MO is at a higher energy.
3. Atomic orbitals combine most effectively with other atomic orbitals of comparable energy and with proper symmetry to form molecular orbitals.
4. The degree of overlap between two atomic orbitals is dependent upon the effectiveness with which they combine. Orbitals with zero overlap cannot successfully combine at all.
5. In the ground-state configuration, electrons enter the lowest energy MO first.
6. The maximum number of electrons in each MO is two, in accordance with the Pauli Exclusion Principle.
7. In the ground-state electronic configurations, electrons enter MOs of identical energies singly, pairing only when necessary, in accordance with Hund's rule of maximum multiplicity.
8. Bond order (BO) is given by

$$\text{Bond Order} = \frac{(\text{\# of electrons in bonding MOs} - \text{\# of electrons in anti-bonding MOs})}{2}$$

A bond order of 1 represents a single bond, 0 means no bond is formed, and 2 or 3 indicate double or triple bonds, respectively.

10.8.2 Molecular orbitals for simple diatomic molecules (H_2 and He_2)

Figure 10.3 shows an MO energy diagram for H_2, where the two $1s$ atomic orbitals combine to form two MOs, one a bonding MO (σ_{1s}) and the other an antibonding MO (σ^*_{1s}). For a diatomic molecule such as H_2, the atomic orbitals of one atom are shown on the left, and those of the other on the right. Each box or horizontal line (as shown in later examples) represents one orbital that can hold two electrons. The molecular orbitals formed by the combination of the atomic orbitals

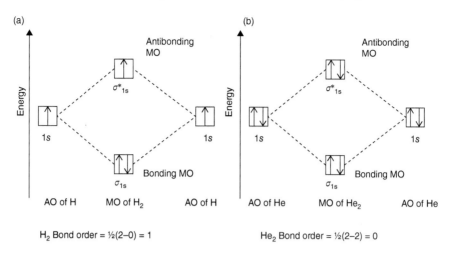

Figure 10.3   Molecular orbital energy diagram for (a) H_2 and (b) He_2.

are shown in the center. The dashed lines show which of the atomic orbitals combine to form the molecular orbitals.

The electronic configuration for molecules is written in the same way as for individual atoms. Therefore, the MO configuration of H_2 is $(\sigma_{1s})^2$. This gives a bond order of 1, which represents a single bond. Similarly, the MO configuration of He_2 is written as $(\sigma_{1s})^2(\sigma^*_{1s})^2$, resulting in a bond order of 0, so the molecule will not exist (unstable).

10.8.3   Bonding in Li_2 molecules

The formation of Li_2 from two Li atoms involves combining the various $1s$ and $2s$ orbitals to give appropriate MOs. Keep in mind the rule that filled subshells do not contribute to bond formation, so the $1s$ orbitals here have no effect on bonding. Also, the $1s$ orbitals will not interact with the $2s$ orbitals due to the large energy difference between them, so all the bonding in Li_2 is due to the $2s$ electrons (Figure 10.4).

10.8.4   Molecular orbital energy-level diagram for homonuclear diatomic molecules

The second-period atoms, notably B, C, N, O, and F, can form molecular bonds using both the $2s$ and $2p$ orbitals. A molecular orbital diagram is handy in showing the relative energy levels of atomic and molecular orbitals, as illustrated by Figure 10.5. Characteristics that always hold true include:

- MOs obtained from $2p$ AOs have a higher energy than those from the $2s$ AOs.
- There is always one σ_{2p} orbital, two π_{2p} orbitals, two π^*_{2p} orbitals, and one σ^*_{2p} orbital.
- The π_{2p} orbitals are degenerate, i.e., they have exactly the same energy and are drawn side by side in the MO diagram. The same is true for the π^*_{2p} orbitals.
- The σ^*_{2p} orbital has the highest energy and is placed at the top of the MO diagram. This is followed by the degenerate π^*_{2p} orbitals.

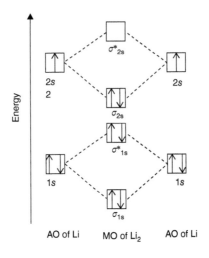

Figure 10.4   Molecular orbital energy diagram for Li_2.

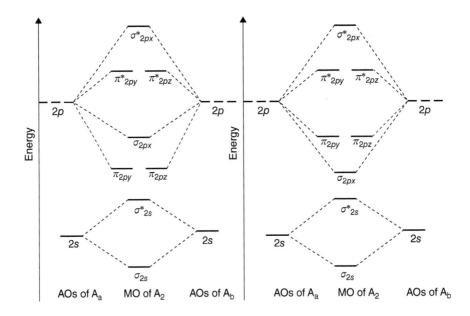

Figure 10.5   MO energy level diagram for second-row p-block diatomic orbitals of (a) B_2, C_2, & N_2 (b) O_2, F_2, and Ne_2.

- The MO diagram of homonuclear diatomic molecules depends on the amount of *sp* mixing. Mixing of the 2s and $2p_x$ orbitals is greater for B_2, C_2, and N_2 than for O_2, F_2, and Ne_2. This is because the energy levels of the atomic orbitals in B, C, and N are more closely spaced than in O, F, and Ne. The result is a change in energy ordering for the π_{2p} and σ_{2p} MOs as shown in Figure 10.5 (a) and (b).

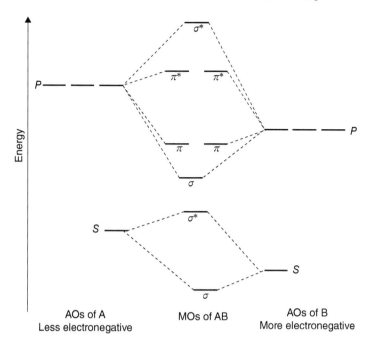

Figure 10.6 Molecular orbital energy-level diagram for a heteronuclear diatomic molecule AB, where B is more electronegative than A.

10.8.5 Molecular orbitals for heteronuclear diatomic molecules

A diatomic molecule with two different atoms, such as CO, NO, or HF, is referred to as **heteronuclear**. Because the two interacting atoms are nonidentical, the interacting atomic orbitals will not have the same energy. As a result, the molecular energy-level diagram will no longer be symmetrical, but skewed toward the more electronegative atom (Figure 10.6). This is how molecular orbital theory can describe a polar covalent bond.

10.8.6 MO electronic configuration and properties of the molecule

The following steps will assist you in writing orbital configurations predicting the properties of molecules.

1. Determine the number of valence electrons in the molecule or ion. These valence electrons are what you will use to populate the placed atomic orbitals for that atom. For example, a nitrogen molecule (atomic number $= 7$ and electronic configuration $= [He]2s^2 2p^3$) has ten valence electrons (five for each atom).
2. Determine whether the molecule or ion is homonuclear or heteronuclear and draw the MO energy-level diagram accordingly.
3. Fill the molecular orbitals with the valence electrons in order of increasing energy, in accordance with the Pauli Principle and Hund's rule.

4. For homonuclear diatomic molecules (and ions) of the second row p-block elements like B_2, C_2, and N_2, the energy of the σ_{2pz} MO is higher than π_{2px} and π_{2py} MOs. For these atoms, the electron filling order is:

$$\sigma_{1s} < \sigma_{1s}^* < \sigma_{2s} < \sigma_{2s}^* < \left(\pi_{2px}, \pi_{2py}\right) < \sigma_{2pz} < \left(\pi_{2px}^*, \pi_{2py}^*\right) < \sigma_{2pz}^*$$

For O_2 and higher molecules, the order of filling is

$$\sigma_{1s} < \sigma_{1s}^* < \sigma_{2s} < \sigma_{2s}^* < \sigma_{2pz} < \left(\pi_{2px}, \pi_{2py}\right) < \left(\pi_{2px}^*, \pi_{2py}^*\right) < \sigma_{2pz}^*$$

5. Describe the bonding and the magnetic properties (diamagnetic, with no unpaired electrons vs paramagnetic, with at least one unpaired electron). Calculate the bond order.

Example 10.8

Draw the MO energy-level diagram for O_2^+ and O_2 and calculate the bond order of each.

Solution

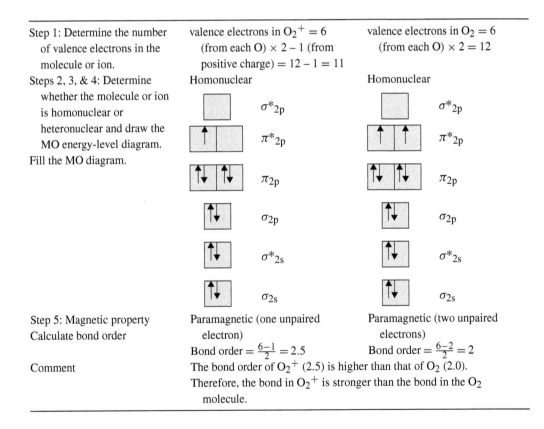

Step 1: Determine the number of valence electrons in the molecule or ion.	valence electrons in $O_2^+ = 6$ (from each O) $\times$ 2 – 1 (from positive charge) $= 12 - 1 = 11$	valence electrons in $O_2 = 6$ (from each O) $\times$ 2 $= 12$
Steps 2, 3, & 4: Determine whether the molecule or ion is homonuclear or heteronuclear and draw the MO energy-level diagram. Fill the MO diagram.	Homonuclear	Homonuclear
Step 5: Magnetic property Calculate bond order	Paramagnetic (one unpaired electron) Bond order $= \frac{6-1}{2} = 2.5$	Paramagnetic (two unpaired electrons) Bond order $= \frac{6-2}{2} = 2$
Comment		The bond order of O_2^+ (2.5) is higher than that of O_2 (2.0). Therefore, the bond in O_2^+ is stronger than the bond in the O_2 molecule.

Example 10.9

Calcium carbide, used in the laboratory preparation of acetylene gas, contains the acetylide ion, with the structure C_2^{2-}. Draw the MO energy-level diagram for acetylide by using C^- ions on both sides and calculate the bond order. Is C_2^{2-} diamagnetic or paramagnetic?

Solution

Step 1: Determine the number of valence electrons in the molecule or ion.	valence electrons in $C_2^{2-} = 4$ (from each C) $\times 2 + 2$ (from negative charge) $= 8 + 2 = 10$
Steps 2, 3, & 4: Determine if the molecule or ion is homonuclear or heteronuclear and draw the MO energy-level diagram. Fill the MO diagram.	Homonuclear

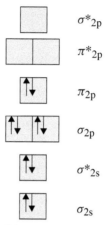

Step 5: Magnetic property Calculate bond order	Diamagnetic (All electrons are paired) Bond order $= \frac{6-0}{2} = 3$; triple bond

The MO energy diagram shows that C_2^{2-} has all its electrons paired and is therefore diamagnetic. A bond order of 3 indicates a C–C triple bond.

10.9 Problems

1. Predict the electron geometry and molecular geometry for each of the following molecules using the VSEPR model.
   (a) BrF_3 (b) SO_2 (c) XeF_2 (d) XeF_4

2. Predict the electron geometry and molecular geometry for each of the following using the VSEPR model.
   (a) F_3 (b) OF_2 (c) CIF_2^+ (d) BF_4^-

3. Draw the Lewis structures for each of the following molecules or ions. For each give the electron geometry and molecular geometry.
   (a) CHF_3 (b) XeO_3 (c) IF_4^- (d) S_3^{2-}

4. Draw the Lewis structure for each of the following molecules or ions. For each give the electron geometry and molecular geometry.
   (a) PH_4^+ (b) $Si(OH)_4$ (c) SF_5^+ (d) OSF_4

5. Hydroxylamine, H_2NOH, is a weak base in water at 25°C. It can remove a proton from water to form the conjugate acid H_3NOH^+ according to

$$H_2NOH(aq) + H_2O(l) \rightleftharpoons H_3NOH^+(aq) + OH^-(aq)$$

Draw the Lewis structures and predict the molecular geometry around the N atom for hydroxylamine and its conjugate acid, H_3NOH^+.

6. Aluminum hydroxide, $Al(OH)_3$, is an amphoteric compound with numerous industrial applications. In the pharmaceutical industry it finds use as an antacid, for alleviating constipation, for controlling hyperphosphatemia (elevated phosphate or phosphorus in the blood), and as an adjuvant in some vaccine formulations. In a strong base, $Al(OH)_3(s)$ dissolves to form the complex ion $Al(OH)_4^-$:

$$Al(OH)_3(s) + OH^-(aq) \rightleftharpoons Al(OH)_4^-(aq)$$

Draw the Lewis structures and predict the molecular geometry for $Al(OH)_3$ and its complex ion, $Al(OH)_4^-$.

7. Phosphorus tribromide, PBr_3, can be prepared by the reaction of liquid bromine with red phosphorus. With excess bromine, phosphorus pentabromide, PBr_5, is formed. However, PBr_5 is unstable and dissociates readily in the gaseous state:

$$PBr_5(g) \rightleftharpoons PBr_3(g) + Br_2(g)$$

Draw the Lewis structures and predict the molecular geometry for PBr_3 and PBr_5.

8. Predict the polarity of the following molecules:

   (a) BF_3 (trigonal planar)
   (b) PBr_3 (trigonal pyramid)
   (c) ClF_3 (T-shape)
   (d) BrF_5 (square pyramidal)

9. Predict the polarity of the following molecules:

   (a) CS_2 (linear)
   (b) NO_2 (bent)
   (c) CH_3OH (tetrahedral)
   (d) BHF_2 (trigonal planar)

10. Predict the molecular geometry for each of the following molecules and determine whether each is polar or nonpolar.
    (a) BeF_2 (b) H_2S (c) $SeBr_2$ (d) CHI_3

11. Determine the molecular geometry of the following molecules and predict each molecule's polarity based on its shape.
    (a) $COBr_2$ (b) $HCOOH$ (c) SO_3 (d) CBr_4

12. Determine the molecular geometry of the following molecules and predict each molecule's polarity based on its shape.
    (a) CF_4 (b) SeF_4 (c) KrF_4 (d) SiF_4

13. What is the electron geometry of the central atom in each of the following ions, and what hybrid orbital set does it use?
    (a) BH_4^- (b) ICl_4^- (c) F_4ClO^- (d) $F_2BrO_2^+$

14. Determine the electron geometry and the hybridization of the orbitals used by the central atom in the following molecules.
    (a) BeF_2 (b) NH_3 (c) SO_3 (d) SiH_4

15. Predict the electron group geometry and the hybridization of the central atom in the following molecules or ions.
    (a) NH_2^- (b) BrF_3 (c) SO_3^{2-} (d) IF_5

16. Describe the change in hybridization (if any) of the Al atom in the following reaction.

$$Al(OH)_3(s) + OH^-(aq) \rightleftharpoons Al(OH)_4^{-(aq)}$$

17. The interhalogen compound bromine trifluoride, BrF_3, undergoes autoionization through F^- transfer according to the equation:

$$2BrF_3 \longrightarrow BrF_2^+ + BrF_4^-$$

What is the hybridization state of Br in BrF_3, BrF_2^+ and BrF_4^-?

18. Vitamin C (also known as L-ascorbic acid) occurs naturally in foods such as guavas, kiwifruit, papayas, persimmons, strawberries, citrus fruits, and leafy vegetables. The chemical structure is as follows:

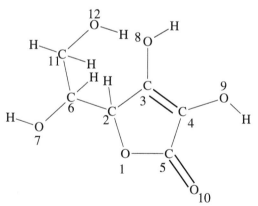

State the hybridization of carbon atoms 2, 3, 5, and 11, and oxygen atoms 1, 7, and 10.

19. Para-aminobenzoic acid, or PABA, is a naturally occurring compound used by plants, bacteria, and fungi to synthesize folic acid. It also used in most sunscreen formulations. The Lewis structure is

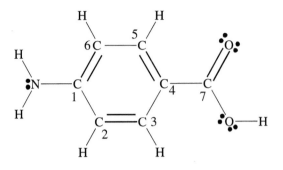

(a) How many π-bonds and σ-bonds are present in the PABA molecule?

(b) Predict the hybridization around the nitrogen atom and carbon atoms 1, 3, and 7.

20. Predict the type of hybridization of Se that would explain why all the Cl–Se–Cl bond angles in $SeCl_6$ are 90°.

21. Determine the electronic configuration, bond order, and stability for
(a) H_2^+ (b) H_2^- (c) He_2 (d) He_2^+

22. Use MO theory to determine which of the following molecules or ions is expected to exist.
(a) F_2^{2-} (b) O_2^{4-} (c) N_2^{2-} (d) C_2^{2+}

23. Determine the number of unpaired electrons in Be_2^{2+}. Is it diamagnetic or paramagnetic?

24. Using molecular orbital notation, write the electronic configuration for B_2^{2-} and calculate its bond order.

25. Write the MO electronic configuration for N_2^{2+}. Would you expect this ion to be paramagnetic or diamagnetic, and why?

26. Write the MO electronic configuration for CO^+. What is the bond order for CO? Would you expect CO to be paramagnetic or diamagnetic, and why?

27. Which of the following ions have electrons in the π antibonding orbital?
(a) F_2^{2-} (b) O_2^{2-} (c) N_2^{2+} (d) C_2^{2+}

28. Using MO theory, determine the bond order and magnetism of the nitrogen oxide species NO, NO^+, and NO^-.

29. Use MO theory to predict the bond order and magnetism of CN^- and CO.

30. Draw the MO diagram of the HF molecule.

11

Gas Laws

. . . .

11.1 Standard Temperature and Pressure

Volumes and densities of gases vary significantly with changes in pressure and temperature. This means that experiments which involve measuring the volumes of gases will likely vary from one laboratory to another, or from one country to another. To correct for this, scientists have adopted a set of standard conditions of temperature and pressure (STP) as a reference point in reporting all measurements involving gases. These are a temperature of 0°C or 273 K and a pressure of 760 mm Hg or 1 atm (or 1.013×10^5 Nm^{-2} in SI units). Therefore *standard temperature and pressure, as used in calculations involving gases, are defined as $0°C$ (or 273 K) and 1 atmosphere (or 760 torr).*

11.2 Boyle's Law: Volume vs Pressure

Boyle's law states that *the volume of a given mass of gas at constant temperature is inversely proportional to the pressure.* In mathematical terms,

$$V \propto \frac{1}{P} \text{ or } PV = k \text{ at constant } n \text{ and } T$$

Since $P \times V = $ constant, we can write

$$P_1 V_1 = P_2 V_2$$

Here P_1, V_1 represent one set of conditions and P_2, V_2 represent another set of conditions for a given mass of gas.

Example 11.1

A gas occupies 5.0 L at 760 mmHg. Calculate the volume of the gas if the pressure is reduced to 650 mmHg without changing the temperature.

Chemistry in Quantitative Language: Fundamentals of General Chemistry Calculations. Second Edition.
Christopher O. Oriakhi, Oxford University Press. © Christopher O. Oriakhi 2021.
DOI: 10.1093/oso/9780198867784.003.0011

Solution

First, list the known quantities and the unknown quantity:

$$\text{Known: } V_1 = 5.0 \text{ L}$$
$$P_1 = 760 \text{ mmHg}$$
$$P_2 = 650 \text{ mmHg}$$
$$\text{Unknown: } V_2 = ?$$

Then solve for the unknown quantity:

$$P_1 V_1 = P_2 V_2$$

$$V_2 = V_1 \times \frac{P_1}{P_2} = 5.0 \text{ L} \times \frac{760 \text{ mm}}{650 \text{ mm}} = 5.85 \text{ L}$$

Example 11.2

A given mass of gas occupies 300 mL at an atmospheric pressure of 10^5 Nm^{-2}. What will be the pressure if the volume is reduced to 50 mL, assuming no change in temperature?

Solution

First, list the known quantities and the unknown quantity:

$$\text{Known: } V_1 = 300.0 \text{ mL}$$
$$P_1 = 10^5 \text{ Nm}^{-2}$$
$$V_2 = 50 \text{ mL}$$
$$\text{Unknown: } P_2 = ?$$

Then solve for the unknown quantity:

$$P_1 V_1 = P_2 V_2$$

$$P_2 = P_1 \times \frac{V_1}{V_2} = 10^5 \text{ Nm}^{-2} \times \frac{300 \text{ mL}}{50 \text{ mL}} = 6 \times 10^5 \text{ Nm}^{-2}$$

The final pressure of the gas is 6.0×10^5 Nm^{-2}.

11.3 Charles's Law: Volume vs Temperature

Charles' law states that *the volume of a given mass of gas is directly proportional to its absolute temperature.* So, if the absolute temperature is doubled, say from 300 K to 600 K, the volume of the gas will also double. A plot of the volume of a gas versus its temperature (°C) gives a straight line. A notable feature of such plot for numerous gases is that the volume of all gases extrapolates to zero at the same temperature, $-273.2°C$.

This point is defined as 0 K on the Kelvin temperature scale and is called *absolute zero*. Thus, the relationship between the Kelvin and Celsius temperature scales is given as: $K = °C + 273$. Scientists believe that absolute zero temperature, 0 K, cannot be attained, although some laboratories have reported producing below 0.0001 K.

In mathematical terms, Charles's Law can be stated as:

$$V \propto T \text{ or } \frac{V}{T} = k, \text{ at constant } n \text{ and } P.$$

Since $\frac{V}{T} =$ constant, we can write

$$\frac{V_1}{T_1} = \frac{V_2}{T_2}$$

Here V_1, T_1 represent one set of conditions and V_2, T_2 represent another set of conditions for a given mass of gas.

Example 11.3

6.0 L of a gas was heated from $-5°C$ to $80°C$ at constant pressure. Calculate the volume of the gas at $80°C$.

Solution

First, identify the known quantities and the unknown quantity:

$$\text{Known: } V_1 = 6.0 \text{ L}$$
$$T_1 = -5 + 273 = 268 \text{ K}$$
$$T_2 = 80 + 273 = 353 \text{ K}$$
$$\text{Unknown: } V_2 = ?$$

Then solve for the unknown quantity using the expression:

$$\frac{V_1}{T_1} = \frac{V_2}{T_2}$$

$$V_2 = V_1 \times \frac{T_2}{T_1} = 6.0 \text{ L} \times \frac{353 \text{ K}}{268 \text{ K}} = 7.90 \text{ L}$$

The final volume of the gas is 7.9 L.

Example 11.4

A given mass of gas occupies 500 mL at 15°C in a closed cylinder. If the maximum capacity of the cylinder is 950 mL, what is the highest temperature to which the cylinder can be heated at constant pressure?

Solution

First, identify the known quantities and the unknown quantity:

$$\text{Known: } V_1 = 500 \text{ mL}$$
$$T_1 = 15 + 273 = 288 \text{ K}$$
$$V_2 = 950 \text{ mL}$$
$$\text{Unknown: } T_2 = ?$$

Then solve for the unknown quantity using the expression:

$$\frac{V_1}{T_1} = \frac{V_2}{T_2}$$

$$T_2 = T_1 \times \frac{V_2}{V_1} = 288 \text{ K} \times \frac{950 \text{ mL}}{500 \text{ mL}} = 547.2 \text{ K or } 274.2°\text{C}$$

The maximum temperature to which the cylinder can be heated is 547 K or 274°C.

11.4 The Combined Gas Law

For a fixed mass of gas we can combine Boyle's law and Charles' law to express the relationships among pressure, volume, and temperature. The combined gas law may be stated as follows: *For a given mass of gas, the volume is directly proportional to the absolute temperature and inversely proportional to the pressure.* Mathematically, we have:

$$\frac{P_1 V_1}{T_1} = \frac{P_2 V_2}{T_2}$$

Example 11.5

A 250-mL sample of nitrogen gas was collected over mercury at −24°C and 735 torr. Calculate the volume at STP.

Solution

First, identify the known quantities and the unknown quantity:

Known: $V_1 = 250$ mL

$$T_1 = -24 + 273 = 249 \text{ K}; \quad T_2 = 273 \text{ K}$$

$$P_1 = 735 \text{ torr}; \quad P_2 = 760 \text{ torr}$$

Unknown: $V_2 = ?$

Then solve for the unknown quantity:

$$\frac{P_1 V_1}{T_1} = \frac{P_2 V_2}{T_2}$$

$$V_2 = (V_1) \left(\frac{T_2}{T_1} \right) \left(\frac{P_1}{P_2} \right) = 250 \text{ mL} \times \left(\frac{273 \text{ K}}{249 \text{ K}} \right) \left(\frac{735 \text{ torr}}{760 \text{ torr}} \right) = 262.17 \text{ mL}$$

The final volume of the gas is 262.2 mL.

Example 11.6

If 20.0 L of nitrogen at STP are heated to 450 K, what will be the new pressure of gas if its volume expands to 25.0 L?

Solution

First, identify the known quantities and the unknown quantity:

Known: $V_1 = 20$ L; $V_2 = 25$ L

$$T_1 = 273 \text{ K}; \quad T_2 = 450 \text{ K}$$

$$P_1 = 760 \text{ torr}$$

Unknown: $P_2 = ?$

Then solve for the unknown quantity:

$$\frac{P_1 V_1}{T_1} = \frac{P_2 V_2}{T_2}$$

$$P_2 = (P_1) \left(\frac{T_2}{T_1} \right) \left(\frac{V_1}{V_2} \right) = 760 \text{ torr} \times \left(\frac{450 \text{ K}}{273 \text{ K}} \right) \left(\frac{20 \text{ L}}{25 \text{ L}} \right) = 1,002.2 \text{ torr}.$$

The final pressure of the gas is 1,002 torr.

Example 11.7

At 300 K and 2.25 atm, a sample of helium occupies 3.5 L. Determine the temperature at which the gas will occupy 4.50 L at 1.50 atm.

Solution

Identify the known quantities and the unknown quantity:

$$\text{Known: } V_1 = 3.5 \text{ L}; V_2 = 4.5 \text{ L}$$

$$T_1 = 300 \text{ K}$$

$$P_1 = 2.25 \text{ atm}; P_2 = 1.5 \text{ atm}$$

$$\text{Unknown: } T_2 = ?$$

Then solve for the unknown quantity:

$$\frac{P_1 V_1}{T_1} = \frac{P_2 V_2}{T_2}$$

$$T_2 = (T_1)\left(\frac{P_2}{P_1}\right)\left(\frac{V_2}{V_1}\right) = 300 \text{ K} \times \left(\frac{1.5 \text{ atm}}{2.25 \text{ atm}}\right)\left(\frac{4.5 \text{ L}}{3.5 \text{ L}}\right) = 257.1 \text{ K}.$$

The final temperature of the gas is 257.1 K.

11.5 Gay-Lussac's Law and Reactions Involving Gases

Chemical reactions involving gases occur in definite and simple proportions by volume. In 1808, after an extensive study of the volumes of gases that react chemically together, Gay-Lussac published his observations, which are now known as Gay-Lussac's law of combining volumes. The law states that:

> *When gases react, they do so in volumes which are in simple ratios to one another and to the volume of the gaseous products, provided the volumes are measured at the same temperature and pressure.*

For example, it has been determined experimentally that one volume of hydrogen and one volume of chlorine react to give two volumes of hydrogen chloride:

$$1 \text{ volume of } H_2 + 1 \text{ volume of } Cl_2 \rightarrow 2 \text{ volumes of HCl}$$

Ratio   1   1   2

Similarly, one volume of nitrogen will react with three volumes of hydrogen to form two volumes of ammonia:

$$1 \text{ volume of } N_2 + 3 \text{ volumes of } H_2 \rightarrow 2 \text{ volumes of } NH_3$$

Ratio   1   3   2

Example 11.8

50 mL of CH_4 are sparked with 100 mL of oxygen at 100°C and 760 torr, according to the following equation:

$$CH_4 \text{ (g)} + 2 O_2 \text{ (g)} \rightarrow CO_2 \text{ (g)} + 2 H_2O \text{ (g)}$$

At the end of the reaction, 50 mL of CO_2 and 100 mL of H_2O are produced. Show that the data illustrate Gay-Lussac's law.

Solution

Check the volume relationships between reactants and products to see if they are in simple whole-number ratios.

$$\frac{\text{Volume of CH}_4}{\text{Volume of O}_2} = \frac{50}{100} = \frac{1}{2}$$

$$\frac{\text{Volume of CH}_4}{\text{Volume of CO}_2} = \frac{50}{50} = \frac{1}{1}$$

$$\frac{\text{Volume of CH}_4}{\text{Volume of H}_2O} = \frac{50}{100} = \frac{1}{2}$$

$$\frac{\text{Volume of O}_2}{\text{Volume of CO}_2} = \frac{100}{50} = \frac{2}{1}$$

$$\frac{\text{Volume of O}_2}{\text{Volume of H}_2O} = \frac{100}{100} = \frac{2}{2} = \frac{1}{1}$$

These ratios fit Gay-Lussac's law.

Example 11.9

$100\ cm^3$ of acetylene are mixed and exploded with $500\ cm^3$ of oxygen. If all the volumes are measured at $110°C$ and 1 atm, calculate the volume of the resulting gases. What is the percentage of steam in the resulting gas mixture?

Solution

First, write a balanced equation for the reaction.

$$2\ C_2H_2\ (g) + 5\ O_2\ (g) \rightarrow 4\ CO_2\ (g) + 2\ H_2O\ (g)$$

Then, determine the volume of O_2 required to react completely with the acetylene. Applying Gay-Lussac's law, C_2H_2 and O_2 react in the ratio of 2:5 by volume. The volume of O_2 is calculated as:

$$\text{Vol of O}_2 = 100\ cm^3\ C_2H_2 \times \frac{5\ cm^3\ O_2}{2\ cm^3\ C_2H_2} = 250\ cm^3\ O_2$$

Next, determine the excess oxygen. Since only $250\ cm^3$ was used out of the initial $500\ cm^3$, the excess O_2 is

$$500\text{--}250 = 250\ cm^3$$

From here, determine the volume of the products. Since two volumes of C_2H_2 produced four volumes of CO_2 and two volumes of H_2O:

$$\text{Vol of } CO_2 = 100 \text{ cm}^3 \text{ } C_2H_2 \times \frac{4 \text{ cm}^3 \text{ } CO_2}{2 \text{ cm}^3 \text{ } C_2H_2} = 200 \text{ cm}^3 CO_2$$

$$\text{Vol of } H_2O = 100 \text{ cm}^3 \text{ } C_2H_2 \times \frac{2 \text{ cm}^3 \text{ } H_2O}{2 \text{ cm}^3 \text{ } C_2H_2} = 100 \text{ cm}^3 H_2O$$

Then, determine the total resulting volume.

$$\text{Total volume of resulting gases } = \text{vol of } CO_2 + \text{vol of } H_2O + \text{excess } O_2$$

$$= 200 \text{ cm}^3 + 100 \text{ cm}^3 + 250 \text{ cm}^3 = 550 \text{ cm}^3$$

Finally, calculate the percentage of steam in the gas mixture.

$$\% H_2O = \frac{\text{vol of } H_2O}{\text{total vol of resulting gases}} \times 100 = \frac{100}{550} \times 100 = 18.2\%$$

11.6 Avogadro's Law

Initially, scientists, including Gay-Lussac himself, could not explain the law of combining volumes of gases. In 1811, Avogadro, an Italian professor of physics, put forward a hypothesis that:

Equal volumes of all gases at the same temperature and pressure contain the same number of molecules.

This became known as Avogadro's law after it was experimentally proved to be correct with $\pm 2\%$ margin of error. This law nicely explains Gay-Lussac's law of combining volumes. For example, it can be shown experimentally that nitrogen reacts with hydrogen to form ammonia in a volume ratio of one to three:

$$1 \text{ volume of } N_2 + 3 \text{ volumes of } H_2 \rightarrow 2 \text{ volumes of } NH_3$$

Furthermore, a similar experiment indicates that they react in a ratio of 1 mol to 3 mol (as shown by measuring masses which react). Thus, the reacting volumes are proportional to the number of moles involved—which we can take as support for Avogadro's Law and Gay-Lussac's law.

Mathematically, the law can be represented as:

$$V \propto n, \text{ or } V = kn$$

where V = volume, n = number of moles, and k = a proportionality constant. Essentially, this equation means that the volume of a sample of gas at constant temperature and pressure is directly proportional to the number of moles of molecules in the gas. Hence:

$$\frac{V_1}{n_1} = \frac{V_2}{n_2}$$

Example 11.10

A balloon filled with 5.0 moles of He has a volume of 2.5 L at a given temperature and pressure. If an additional 1.0 mole of He is introduced into the balloon without altering the temperature and pressure, what will be its new volume?

Solution

First, identify the known quantities and the unknown quantity:

$$\text{Known: } V_1 = 2.5 \text{ L}$$

$$n_1 = 5.0 \text{ mol}$$

$$n_2 = n_1 + 1.0 = 5.0 + 1.0 = 6 \text{ mol}$$

$$\text{Unknown: } V_2 = ?$$

Then solve for the unknown quantity:

$$\frac{V_1}{n_1} = \frac{V_2}{n_2}$$

$$V_2 = V_1 \times \frac{n_2}{n_1} = 2.5 \text{ L} \times \frac{6.0 \text{ mol}}{5.0 \text{ mol}} = 3.0 \text{ L}$$

The new volume of the balloon is 3.0 L.

11.7 The Ideal Gas Law

Boyle's law, Charles' law, and Avogadro's law show that the volume of a given mass of gas depends on pressure, temperature, and number of moles of the gas.

$$\text{Boyle's law} : V \propto \frac{1}{P} \quad \text{(constant } n, T)$$

$$\text{Charles's law} : V \propto T \quad \text{(constant } n, P)$$

$$\text{Avogadro's law} : V \propto n \quad \text{(constant } P, T)$$

Mathematically, if we combine these laws, we have:

$$V \propto \frac{nT}{P} \text{ or } V = R\left(\frac{nT}{P}\right)$$

R is a proportionality constant known as the ideal gas constant. Depending on the units of volume and pressure, R can have a value of 0.082 atm L K^{-1} mol^{-1} or 8.31 J $K^{-1}mol^{-1}$.

This equation can be rearranged to give what is generally known as the ideal gas law:

$$PV = nRT$$

We can also write this as

$$\frac{PV}{T} = nR = \text{constant}$$

Since 1 mole of any gas occupies 22.4138 L under standard conditions, we can compute the value of R as follows:

$$R = \frac{PV}{nT} = \frac{(1 \text{ atm})(22.4138 \text{ L})}{(1 \text{ mol})(273.15 \text{ K})} = 0.08206 \text{ L atm K}^{-1} \text{ mol}^{-1}$$

Example 11.11

Calculate the number of moles of an ideal gas that occupies a volume of 15 L at 25°C and a pressure of 3 atm.

Solution

First, identify the known quantities and the unknown quantity:

$$\text{Known: } V = 15 \text{ L}$$
$$T = 25 + 273 = 298 \text{ K}$$
$$P = 3 \text{ atm}$$
$$\text{Unknown: } n = ?$$

Then solve for the unknown quantity:

$$PV = nRT$$

$$n = \frac{PV}{RT} = \frac{(3 \text{ atm})(15 \text{ L})}{(0.08206 \text{ L atm K}^{-1} \text{ mol}^{-1})(298 \text{ K})} = 1.84 \text{ mol mL}$$

Example 11.12

A 0.25-mol sample of an ideal gas occupies 56 L at 0.3 atm. Calculate the temperature of this gas.

Solution

First, identify the known quantities and the unknown quantity:

$$\text{Known: } V = 56 \text{ L}$$
$$n = 0.25 \text{ mol}$$
$$P = 0.3 \text{ atm}$$
$$R = 0.08206 \text{ L atm K}^{-1} \text{ mol}^{-1}$$
$$\text{Unknown: } T = ?$$

Then solve for the unknown quantity:

$$PV = nRT$$

$$T = \frac{PV}{nR} = \frac{(0.3 \ \text{atm})(56 \ \text{L})}{(0.25 \ \text{mol})(0.08206 \ \text{L atm K}^{-1} \text{mol}^{-1})} = 819 \ \text{K}$$

11.8 Density and Molecular Mass of a Gas

Recall that density is defined as mass per unit volume and can be used to convert volume into mass. By applying the ideal gas equation, the molecular mass of a gas can be calculated from its density, ρ. The relationship between the density and molecular weight of a gas is derived as follows:

From the ideal gas equation

$$V = \frac{nRT}{P}$$

$$n = \frac{\text{mass (g)}}{\text{mol.wt. (g/mol.)}} = \frac{m}{M_w}$$

Substituting for n in the ideal gas equation gives:

$$P = \frac{nRT}{V} = \frac{mRT}{VM_w}$$

This equation is the basis of the Dumas method for finding the molar mass of volatile compounds.

But density, ρ, is equal to $\frac{m}{V}$, hence,

$$P = \rho \frac{RT}{M_w}$$

Thus

$$M_w = \rho \frac{RT}{P}$$

So if you know the density of a gas, its pressure, and its temperature, you can calculate its molecular weight.

Example 11.13

What is the density of nitrogen gas at STP?

Solution

First, identify the known quantities and the unknown quantity:

$$\text{Known: } P = 1.0 \text{ atm}$$

$$T = 273 \text{ K}$$

$$M_w = 28 \text{ g/mol}$$

$$R = 0.08206 \text{ L atm K}^{-1} \text{ mol}^{-1}$$

$$\text{Unknown: } \rho = ?$$

Then solve for the unknown quantity:

$$\rho = \frac{PM_w}{RT} = \frac{(1.0 \text{ atm})(28 \text{ g/mol})}{(0.08206 \text{ L atm K}^{-1} \text{ mol}^{-1})(273 \text{ K})} = 1.25 \text{ g/L}$$

Example 11.14

An organic vapor suspected to be CCl_4 has a density of 4.43 g/L at 714 mmHg and 125°C. What is its molecular mass?

Solution

First, identify the known quantities and the unknown quantity:

$$\text{Known: } P = 714 \text{ mmHg}$$

$$T = 125 + 273 = 398 \text{ K}$$

$$\rho = 4.43 \text{ g/L}$$

$$R = 0.08206 \text{ L atm K}^{-1} \text{ mol}^{-1}$$

$$\text{Unknown: } M_w = ?$$

Then solve for the unknown quantity:

$$M_{CCl_4} = \frac{\rho RT}{P} = \frac{(4.43 \text{ g/L})(0.08206 \text{ L atm K}^{-1} \text{ mol}^{-1})(398 \text{ K})}{(714 \text{ mm Hg})\left(\dfrac{1 \text{ atm}}{760 \text{ mm Hg/atm}}\right)} = 154.0 \text{ g/mol}$$

The molar mass is 154.0 g/mol, and this corresponds to the molar mass of CCl_4.

11.9 Molar Volume of an Ideal Gas

Molar volume is defined as the volume occupied by one mole of any gas at STP. Using the ideal gas law, the molar volume of a gas has been determined experimentally to be 22.4 L. In other words, 1 mol of hydrogen gas will occupy 22.4 L, 1 mol of nitrogen will occupy 22.4 L, and

1 mol of oxygen will occupy 22.4 L at STP. However, the molar volume of many real gases does show deviations from the ideal value, for reasons to be discussed later.

We can rearrange the ideal gas law:

$$PV = nRT$$

to calculate volume:

$$V = \frac{nRT}{P}$$

At STP, $T = 273.2$ K and $P = 1$ atm. R has the usual value of 0.08206 L atm/K mol. Since we want to find the volume of 1 mol of gas, $n = 1.0$ mol. Thus:

$$V = \frac{(1.000 \text{ mol})(0.08206 \text{L·atm}/K \cdot \text{mol})(273.2 \ K)}{1.000 \text{ atm}} = 22.42 \text{ L}$$

Example 11.15

Quicklime (CaO) is produced by the thermal decomposition of limestone ($CaCO_3$) in the manufacture of cement. Calculate the volume of carbon dioxide, CO_2, at 30°C and 700 mm Hg, produced from the decomposition of 300 g of $CaCO_3$.

Solution

First, write a balanced equation for the reaction:

$$CaCO_3 \text{ (s)} \xrightarrow{\text{heat}} CaO \text{ (s)} + CO_2 \text{ (g)}$$

Next, calculate molar masses of $CaCO_3$ and CO_2:

$$CaCO_3 = 100 \text{ g/mol}$$
$$CO_2 = 44 \text{ g/mol}$$
$$\text{Known}: 300 \text{ g of } CaCO_3$$
$$\text{Unknown} = \text{vol of } CO_2$$

Then, formulate appropriate conversion factors and calculate the volume of CO_2 produced at STP:

$$\text{vol of } CO_2 = 300 \text{ g } CaCO_3 \times \frac{1 \text{ mol } CaCO_3}{100 \text{ g } CaCO_3} \times \frac{1 \text{ mol } CO_2}{1 \text{ mol } CaCO_3} \times \frac{22.4 \text{ L } CO_2}{1 \text{ mol } CO_2} = 67.2 \text{ L}$$

Finally, convert 67.2 L of CO_2 at STP to the volume at 30°C and 700 mm Hg using the general gas equation:

$$\frac{P_1 V_1}{T_1} = \frac{P_2 V_2}{T_2}$$

$P_1 = 760$ mm Hg, $V_1 = 67.2$ L, $T_1 = 273$ K, $P_2 = 700$ mm Hg

$T_2 = 30 + 273 = 303$ K

$$V_2 = (V_1)\left(\frac{T_2}{T_1}\right)\left(\frac{P_1}{P_2}\right) = 67.2 \text{ L} \times \left(\frac{273 \text{ K}}{303 \text{ K}}\right)\left(\frac{700 \text{ mm}}{760 \text{ mm}}\right) = 55.77 \text{ L}$$

Thus, the decomposition of 300 g of $CaCO_3$ will liberate 55.8 L of CO_2 at 30°C and 700 mm Hg.

11.10 Dalton's Law of Partial Pressure

Dalton experimented with the pressure exerted by mixtures of gases, and, in 1803, he summarized his observations in terms of partial pressures of gases. The *partial pressure* of a gas is the pressure that a gas would exert on the container if it were present alone. Dalton's law of partial pressure can be stated as:

In a mixture of gases, which do not react chemically with one another, the total pressure exerted is the sum of the pressures that each gas would exert if it were alone.

In mathematical terms, the law can be expressed as

$$P_{total} = P_1 + P_2 + P_3 + ...,$$

where P_{total} is the total pressure of the mixture and P_1, P_2, P_2, . . . , are the partial pressures exerted by the individual gases 1, 2, 3, . . . , that constitute the mixture.

If each gas obeys the ideal gas equation, then,

$$P_1 = n_1 \frac{RT}{V}, P_2 = n_2 \frac{RT}{V}, P_3 = n_3 \frac{RT}{V}, \text{etc.}$$

Substituting for P_1, P_2, P_3, etc., in Dalton's law expression and factorizing gives:

$$P_{total} = \frac{RT}{V}(n_1 + n_2 + n_3 + ...) = n_t \left(\frac{RT}{V}\right)$$

where $n_t = n_1 + n_2 + n_3 + ...$

In other words, the total gas pressure depends on how many moles of gas are present, but not on their individual identities.

Example 11.16

A tank containing a mixture of He, Ne, Ar, and N_2 has a total pressure of 6 atm. The pressure exerted by He is 0.5 atm, by Ne is 1.0 atm, and by Ar is 2.5 atm. Calculate the partial pressure of N_2 in the tank.

Solution

Use the expression for total pressure and solve for the unknown (P_{N_2}) as:

$$P_t = P_{He} + P_{Ne} + P_{Ar} + P_{N_2}$$

$$P_{N_2} = P_t - (P_{He} + P_{Ne} + P_{Ar}) = 6.0 - 4.0 = 2.0 \text{ atm}$$

11.10.1 Collecting gases over water

In the laboratory, many gases, for example, oxygen and hydrogen, are collected by allowing them to displace water from a container (downward displacement of water). The gas collected will be mixed with water vapor. The total pressure of the sample is the sum of the pressure of the gas collected and the water vapor pressure present in the sample:

$$P_t = P_{gas} + P_{H_2O}$$

Example 11.17

What is the partial pressure of hydrogen gas collected over water at 25°C at 745 torr? The water vapor pressure at 25°C is 23.8 torr.

Solution

Applying Dalton's law, we have:

$$P_t = P_{gas} + P_{H_2O}$$

$$P_{gas} = P_t - P_{H_2O} = 745 \text{ torr} - 23.8 \text{ torr} = 721.2 \text{ torr}.$$

11.11 Partial Pressure and Mole Fraction

The mole fraction of a substance in a mixture is defined as the ratio of the number of moles of that substance in the mixture to the total number of moles in the mixture. In a mixture containing three components 1, 2, and 3, the mole fractions $X_1, X_2,$ and X_3 of the components 1, 2, and 3 are expressed as:

$$X_1 = \frac{n_1}{n_1 + n_2 + n_3} = \frac{n_1}{n_{total}}$$

where $n_t = n_1 + n_2 + n_3$ and similarly for X_2 and X_3. From our discussion of Dalton's law, we can express P_t and P_1 as follows:

$$P_1 = n_1 \frac{RT}{V}, \quad P_{total} = n_{total} \frac{RT}{V}$$

$$\frac{P_1}{P_{total}} = \frac{n_1 (RT/V)}{n_{total} (RT/V)} = \frac{n_1}{n_{total}}$$

From our definition of mole fraction, we can rewrite the above equation as:

$$\frac{P_1}{P_{total}} = \frac{n_1}{n_{total}} = X_1$$

Hence: $P_1 = X_1 P_{total}$

In other words, the partial pressure of a gas in a mixture is directly related to its mole fraction.

Example 11.18

A flask contains 2.5 mol He, 1.5 mol Ar, and 3.0 mol O_2. If the total pressure is 700 mmHg, what is the mole fraction and partial pressure of each gas?

Solution

First, we calculate the mole fraction of each gas using the definition of mole fraction:

$$X_{He} = \frac{n_{He}}{n_{He} + n_{Ar} + n_{O_2}} = \frac{2.5 \text{ mol}}{2.5 \text{ mol} + 1.5 \text{ mol} + 3.0 \text{ mol}} = 0.357$$

$$X_{Ar} = \frac{n_{Ar}}{n_{He} + n_{Ar} + n_{O_2}} = \frac{1.5 \text{ mol}}{2.5 \text{ mol} + 1.5 \text{ mol} + 3.0 \text{ mol}} = 0.214$$

$$X_{O_2} = \frac{n_{O_2}}{n_{He} + n_{Ar} + n_{O_2}} = \frac{3.0 \text{ mol}}{2.5 \text{ mol} + 1.5 \text{ mol} + 3.0 \text{ mol}} = 0.429$$

We know the total pressure. Therefore we can calculate the partial pressure of each gas from the mole fraction:

$$P_{He} = X_{He} P_{total}$$

$$P_{He} = 0.357 \times 700 \text{ mm Hg} = 250 \text{ mm Hg}$$

$$P_{Ar} = X_{Ar} P_{total}$$

$$P_{Ar} = 0.214 \times 700 \text{ mm Hg} = 150 \text{ mm Hg}$$

$$P_{O_2} = X_{O_2} P_{total}$$

$$P_{He} = 0.429 \times 700 \text{ mm Hg} = 300 \text{ mm Hg}$$

11.12 Real Gases and Deviation from the Gas Laws

Real gases obey gas laws (i.e. behave like ideal gases) only at high temperatures and low pressures. At high pressures or low temperatures, real gases show marked deviation from the behavior described by the gas laws. There are two main reasons for this deviation:

1. Real gas molecules have a small but finite volume, in contrast to the postulates of the kinetic theory for ideal gases.

2. The force of attraction and repulsion between real gas molecules (at high concentration) is not negligible. Gas molecules do attract each other, contrary to the assumption of the kinetic theory.

In 1873, van der Waals, a professor of physics at the University of Amsterdam, developed an equation for real gases. He modified the ideal gas law by adding correction factors a and b to account for the effects of intermolecular attraction and repulsion due to the finite volume of the gas molecules. His equation, known as the van der Waals equation, is:

$$P_{obs} = \frac{nRT}{V - nb} - a\left(\frac{n}{V}\right)^2$$

For easy comparison with the ideal gas, this equation can be rearranged to give

$$\left[P_{obs} + a\left(\frac{n}{V}\right)^2\right](V - nb) = nRT$$

The values $a(n/V)^2$ and nb correct for the attractive forces and volume of the gas molecules. The values for the van der Waal constants, a and b, are different for each gas, and are determined experimentally. The values of a and b for some gases are represented in Table 11.1. Professor van der Waals was awarded the Nobel Prize in 1910 for his work.

Table 11.1   van der Waals constants for some gases

Gas	a (atm-L^2/mol^2)	b (L/mol)
He	0.034	0.0237
Ne	0.211	0.0171
Ar	1.35	0.0322
Kr	2.32	0.0398
Xe	4.19	0.0511
H_2	0.244	0.0266
N_2	1.39	0.0391
O_2	1.36	0.0318
Cl_2	6.49	0.0562
H_2O	5.46	0.0305
CH_4	2.25	0.0428
CO_2	3.59	0.0427
NH_3	4.17	0.0371

Example 11.19

Using the van der Waals equation, calculate the pressure for a 2.5-mol sample of ammonia gas confined to 1.12 L at 25°C. (For ammonia, $a = 4.17$ atm.L^2/mol^2 and $b = 0.0371$ L/mol). How does the result compare with the prediction from the ideal gas law?

Solution

The following data are given:

$$V = 1.12 \text{ L}$$

$$n = 2.5 \text{ mol}$$

$$T = 25 + 273 = 298 \text{ K}$$

$$R = 0.08206 \frac{\text{L atm}}{\text{mol} - \text{K}}$$

$$a = 4.17 \frac{\text{atm} - \text{L}^2}{\text{mol}^2}, \text{ and } b = 0.0371 \frac{\text{L}}{\text{mol}}$$

Substitute the values given directly into the van der Waal equation:

$$P_{obs} = \frac{nRT}{V - nb} - a\left(\frac{n}{V}\right)^2$$

$$P_{obs} = \frac{(2.5 \text{ mol})(0.08206 \text{ L} - \text{atm/mol} - \text{K})(298 \text{ K})}{1.12 \text{ L} - (2.5 \text{ mol})(0.0371 \text{ L/mol})} - \left(4.17 \frac{\text{atm} - \text{L}^2}{\text{mol}^2}\right)\left(\frac{2.5 \text{ mol}}{1.12 \text{ L}}\right)^2$$

$$= 59.5101 \text{ atm} - 20.7769 \text{ atm} = 38.7332 \text{ atm} \simeq 38.7 \text{ atm.}$$

The prediction of the ideal gas law would be:

$$P = \frac{nRT}{V}$$

$$P = \frac{(2.5 \text{ mol})(0.08206 \text{ L} - \text{atm/K} - \text{mol})(298K)}{1.12 \text{ L}} = 54.5846 \text{ atm} \simeq 54.6 \text{ atm}$$

The pressure predicted by the ideal gas law is 41% higher than the van der Waals equation indicates.

11.13 Graham's Law of Diffusion

Graham's law of diffusion states that, at constant pressure and temperature, gases diffuse at rates which are inversely proportional to the square roots of their molecular weights (or vapor densities). In mathematical terms:

$$\text{Rate} \propto \frac{1}{\sqrt{d}}, \text{ or Rate} = \frac{k}{\sqrt{d}}$$

For two gases A and B, diffusing at rates R_A and R_B, at the same temperature and pressure,

$$\frac{R_A}{R_B} = \frac{\sqrt{d_B}}{\sqrt{d_A}} = \sqrt{\frac{d_B}{d_A}}$$

where d_A and d_B are the densities of gases A and B.

But density is proportional to the relative molecular mass M of a gas. Hence,

$$\frac{R_A}{R_B} = \sqrt{\frac{M_B}{M_A}}$$

Furthermore, rate is also inversely proportional to time of diffusion, t, i.e.:

$$R \propto \frac{1}{t}$$

Therefore,

$$\frac{R_A}{R_B} = \sqrt{\frac{M_B}{M_A}} = \frac{t_B}{t_A}$$

Example 11.20

An unknown gas X diffuses one-third as fast as ammonia, NH_3. What is the molecular mass of X?

Solution

Use the following expression:

$$\frac{R_X}{R_{NH_3}} = \sqrt{\frac{M_{NH_3}}{M_X}}$$

$$R_X = 1; R_{NH_3} = 3; M_{NH_3} = 17 \text{ g/mol}; M_X = ?$$

$$\left(\frac{1}{3}\right)^2 = \left(\sqrt{\frac{17}{M_X}}\right)^2 = \frac{17}{M_X}$$

$$\frac{1}{9} = \frac{17}{M_x}$$

$$M_X = 9 \times 17 \text{ g/mol} = 153 \text{ g/mol}$$

Example 11.21

In forty-five seconds, 100 mL of oxygen diffused through a small hole. 150 mL of an unknown gas A diffused in sixty seconds. Calculate: (a) the rates of diffusion of the two gases, (b) the molecular mass of gas A.

Solution

(a) Rate of diffusion $= \dfrac{\text{Volume}}{\text{Time}}$

Rate of diffusion of $O_2 = \dfrac{100 \text{ mL}}{45 \text{ s}} = 2.22 \text{ mL/s}$

Rate of diffusion of gas A $= \dfrac{250 \text{ mL}}{60 \text{ s}} = 4.17 \text{ mL/s}$

(b) Use Graham's law:

$$\frac{R_A}{R_{O_2}} = \sqrt{\frac{M_{O_2}}{M_A}}$$

$$\frac{4.17 \text{ mL/s}}{2.22 \text{ mL/s}} = \sqrt{\frac{32}{M_A}}$$

$$1.878 = \sqrt{\frac{32}{M_A}}$$

$$3.52 = \frac{32}{M_A}$$

$$M_A = 9.1 \text{ g/mol}$$

11.14 Problems

1. A certain mass of gas occupies a volume of 500 mL at a pressure of 700 mm Hg. If the temperature remains constant, what would be the pressure when the volume is increased to 1500 mmHg?

2. A sample of nitrogen occupies 1.5 L at 25°C and 0.79 atm. Calculate the volume at 25°C and 5.5 atm.

3. A 25-g sample of argon occupies 650 cm^3 at a pressure of 1.0×10^5 Nm^{-2}. What will be its pressure if the volume is increased to 1,950 cm^3, assuming no change in temperature?

4. A fixed mass of gas occupies a volume of 300 cm^3 at 25°C and at 101.3 kPa pressure. What is the volume of the gas at 5°C and at a pressure of 101.3 kPa?

5. A fixed mass of gas initially occupying a volume of 430 cm^3 is heated from 0°C to 120°C at a constant pressure of 740 mmHg. What is the final volume, assuming no change in pressure?

6. A sample of gas occupies a volume of 15.0 L at 1.15 atm and 30°C. Find the temperature in °C at which its volume becomes 7.5 L, assuming that its pressure remains constant.

7. A 2.50-L container is filled with helium gas to a pressure of 700 mmHg at 225°C. Determine the volume of a new container that can be used to store the same gas at STP.

8. Find the temperature to which 20.0 L of CO_2 at 20°C and 650 mmHg must be brought to change the volume to 12 L at a pressure of 760 mmHg.

9. The volume of SO_2 gas initially at 100°C is to be decreased by 25%. Calculate the final temperature to which the gas must be heated if the pressure of the gas is to be tripled.

10. A 3.0-L steel tank contains nitrogen gas at 23.5°C and a pressure of 7.5 atm. What is the internal gas pressure when the temperature of the tank and its contents increases to 95°C?

11. A 350-cm^3 sample of hydrogen gas was collected over water at 55°C and 745 mmHg. What would be the volume of the dry gas at STP? The vapor pressure of water at 55°C is 118.04 mmHg.

12. Butane (C_4H_{10}), which is used as fuel in some cigarette lighters, reacts with oxygen in air to produce steam and carbon dioxide.

    (a) Write a balanced chemical equation for this reaction.

    (b) If 1.50 cm^3 of C_4H_{10} initially present at 760 mmHg and 27°C was used in the combustion process in a carefully controlled experiment, how many cm^3 of steam measured at 110°C and 800 mmHg are formed?

13. An ideal gas originally at 1.25 atm and 45°C was allowed to expand to a volume of 25 L at 0.75 atm and 95°C. What was the initial volume of this gas?

14. The production of chlorine trifluoride is given by the equation

$$Cl_2(g) + 3 F_2(g) \longrightarrow 2ClF_3(g)$$

What volume of ClF_3 is produced if 10 L of F_2 are allowed to react completely with Cl_2, with all gases at the same temperature and pressure?

15. A 1.25-mole sample of NO_2 gas at 1.25 atm and 50°C occupies a volume of 2.50 L. What volume would 1.25 mole of N_2O_5 occupy at the same temperature and pressure?

16. Using the standard-condition values for P (atm), T (K), and V (L), evaluate the constant R in the ideal gas equation for one mole of an ideal gas.

17. What is the volume occupied by 10.25 g of CO_2 at STP?

18. What volume will 3.125 g of NH_3 gas occupy at 33°C and 0.85 atm?

19. A 50.0-L cylinder contains fluorine gas at 24°C and 2,650 mmHg. Find the number of moles of fluorine in the cylinder.

20. At STP, 0.50 L of a gas weighs 2.0 g. What is the molecular weight of the gas?

21. At 725 mmHg and 30°C, 0.725 g of a gas occupies 0.85 L. What is the molecular weight of the gas?

22. What is the density of sulfur trioxide (SO_3) in grams per liter at 700 mmHg and 27°C?

23. The molecular weight of an unknown gas was determined to be 88 g/mol at 25°C and 1 atm using the Dumas method. What is the density of the unknown gas under these conditions?

24. One of the reactions involved in the production of elemental sulfur from the hydrogen sulfide content of sour natural gas is burning the gas in oxygen in the front-end furnace. The equation for the reaction is:

$$2 H_2 S(g) + 3 O_2(g) \longrightarrow 2 H_2O(g) + 2 SO_2(g)$$

If all gases are present at the same temperature and pressure, calculate:

    (a) The volume of oxygen needed to burn 1,000 L of H_2S.

    (b) The volume of SO_2 formed.

25. A 19.51-g mixture of K_2CO_3 and $CaHCO_3$ was treated with excess hydrochloric acid (HCl). The CO_2 gas liberated occupied 4.0 L at 28°C and 0.95 atm. Calculate the percentage by mass of $CaHCO_3$ in the mixture.

26. 10 cm^3 of a gaseous hydrocarbon required 50 cm^3 of oxygen for complete combustion. If 30 cm^3 of CO_2 gas was evolved, what is the molecular formula of the hydrocarbon?

27. An organic compound containing only carbon, hydrogen, and oxygen was burned in a rich supply of oxygen. A 51-cm^3 sample of the organic compound needed 204 cm^3 of oxygen for complete combustion. If 153 cm^3 of CO_2 and 153 cm^3 of steam (H_2O) were collected,

(a) Determine the molecular formula of the compound, and
(b) Write a balanced equation to show the complete combustion in oxygen.

All measurements are at the same temperature and pressure.
(Hint: the combustion process may be represented by the equation:

$$C_xH_yO_z + \left(x + \frac{y}{4} - \frac{z}{2}\right)O_2 \rightarrow x\,CO_2 + \frac{y}{2}H_2O\right)$$

28. Oxygen gas may be produced in the laboratory by decomposing solid $KClO_3$ at elevated temperature (products are KCl and O_2).

(a) Write a balanced equation for the reaction.
(b) Calculate the mass of $KClO_3$ needed to produce 10.5 L of oxygen at STP.

29. Acetylene gas may be prepared by the hydrolysis of calcium carbide according to:

$$CaC_2 + 2H_2O \rightarrow C_2H_2 + Ca(OH)_2$$

Calculate the volume of acetylene gas collected at 25°C and 1.3 atm if 55 g of water were added to solid calcium carbide.

30. Dry air is composed of 78.08% N_2, 20.95% O_2, 0.93% Ar, and 0.036% CO_2 by volume. What is the partial pressure of each gas (N_2, O_2, Ar, and CO_2) if the total pressure is 1900 mm Hg?

31. A mixture of gases contains 1.5 moles of helium, 2.0 moles of neon, and 2.5 moles of argon.

(a) What is the mole fraction of each gas in the mixture?
(b) Calculate the partial pressure of each gas if the total pressure is 2.5 atm.

32. At STP, a 25.0-L flask contains 8.25 g of oxygen, 12.50 g of nitrogen, and 15.75 g of carbon dioxide. Calculate:

(a) the mole percent of each gas in the mixture.
(b) the volume percent of nitrogen in the mixture.

33. What are the relative rates of diffusion of the following pairs of gases?

(a) N_2 and CO_2
(b) H_2 and O_2
(c) Ne and Ar

34. At STP, the density of helium is 0.18 g/L, and of oxygen is 1.43 g/L. Under the same conditions of temperature and pressure, determine which gas diffuses faster through a small orifice in a cylinder and by how much.

35. In twenty-five minutes, 300 mL of sulfur dioxide diffused through a porous partition. Under the same conditions of temperature and pressure, 250 mL of an unknown gas Y diffused through the same partition in 9.11 minutes. Calculate:

(a) the rates of diffusion of the two gases.
(b) the molecular mass of gas Y.

36. The van der Waals constants for steam, H_2O, are $a = 5.464 \text{ L}^2 \text{ atm/mol}^2$, $b = 0.03049 \text{ L/mol}$. Calculate the pressure for 2.50 moles of steam contained in a volume of 5.0 L at 100°C:

(a) assuming H_2O obeys the ideal gas law, and
(b) using the van der Waals equation.

37. A 4.5-L pressurized gas cylinder was filled with 12.25 moles of hydrogen gas at 25°C. The van der Waals constants for hydrogen, H_2, are $a = 0.244 \text{ L}^2 \text{ atm/mol}^2$, $b = 0.0266 \text{ L/mol}$. Find the pressure of the gas:

(a) using the ideal gas law, and
(b) using the van der Waals equation.

38. A 1.0-mol sample of dry air contains 0.7808 mol N_2, 0.2095 mol O_2, 0.0093 mol Ar, and 0.00036 mol CO_2 confined to 0.750 L at −25°C.

(a) Using the van der Waals equation of state, calculate the partial pressures of the individual component gases.
(b) Use these partial pressures to predict the pressure of the sample under these conditions. Select the appropriate van der Waals constants from Table 11.1.

12

Liquids and Solids

. . . .

12.1 The Liquid State

The atoms or molecules in a liquid have enough kinetic energy to partially overcome the forces of attraction between them. Therefore, they are in constant random motion (as in a gas) but remain relatively close together, with much less free space than gases have. However, they are not as tightly packed, or as well ordered, as in a solid. The atoms or molecules may aggregate together to form chains or rings that readily move relative to one another; this gives a liquid its fluid (flow) properties.

Liquids generally occur as compounds. For example, water, ethanol, and carbon tetrachloride are liquids at room temperature. However, a few elements are also liquids at room temperature: bromine, cesium, gallium, mercury, and rubidium.

12.1.1 Properties of liquids

A liquid is characterized by several physical properties: boiling point and freezing point, density, compressibility, surface tension, and viscosity. These properties are greatly influenced by the strength of a liquid's intermolecular forces.

In summary:

- Liquids have definite volume but no definite shape. They take on the shapes of their containers.
- Liquids are characterized by low compressibility, low rigidity, and high density relative to gases.
- Liquids diffuse through other liquids.
- Liquids can vaporize into the space above them and produce a vapor pressure.

12.2 Vapor Pressure and the Clausius–Clapeyron Equation

The vapor pressures of all substances increase with increasing temperature. The dependence of the vapor pressure of a liquid on the absolute temperature is described by the Clausius–Clapeyron equation:

Chemistry in Quantitative Language: Fundamentals of General Chemistry Calculations. Second Edition.
Christopher O. Oriakhi, Oxford University Press. © Christopher O. Oriakhi 2021.
DOI: 10.1093/oso/9780198867784.003.0012

$$\ln P = \frac{-\Delta H_{vap}}{RT} + C$$

where P is the vapor pressure, ΔH_{vap} is the molar heat of vaporization, R is the gas constant (8.314 J/K.mol), T is the absolute temperature, and C is a constant. The equation can be expressed in a linear form as:

$$\ln P = \left(\frac{-\Delta H_{vap}}{R}\right)\left(\frac{1}{T}\right) + C$$

$$\updownarrow \qquad \updownarrow \qquad \updownarrow \quad \updownarrow$$

$$y = \quad (m) \quad \cdot \quad (x) + b$$

The above equation predicts that a graph of $\ln P$ against $1/T$ should give a straight line with a slope equal to $-\Delta H_{vap}/R$. From this slope, the heat of vaporization (assumed to be independent of temperature) can be determined.

$$\Delta H_{vap} = -\text{slope} \times R$$

The Clausius–Clapeyron equation can also be used to calculate the vapor pressure of a liquid at different temperatures if ΔH_{vap} and the vapor pressure at one temperature of a liquid are known. Suppose the vapor pressure of a liquid at temperature T_1 is P_1 and at temperature T_2 is P_2; then we can write the corresponding Clausius–Clapeyron equations as:

$$\ln P_1 = \frac{-\Delta H_{vap}}{RT_1} + C$$

$$\ln P_2 = \frac{-\Delta H_{vap}}{RT_2} + C$$

Subtracting the first equation from the second,

$$\ln P_2 - \ln P_1 = \frac{\Delta H_{vap}}{R}\left(\frac{1}{T_1} - \frac{1}{T_2}\right)$$

On rearrangement, the equation becomes:

$$\ln \frac{P_2}{P_1} = \frac{\Delta H_{vap}}{R}\left(\frac{1}{T_1} - \frac{1}{T_2}\right) \text{ or } \frac{\Delta H_{vap}}{R}\left(\frac{T_2 - T_1}{T_1 T_2}\right)$$

Converting this equation to ordinary logarithms gives:

$$\log \frac{P_2}{P_1} = \frac{\Delta H_{vap}}{2.303R}\left(\frac{1}{T_1} - \frac{1}{T_2}\right) \text{ or } \frac{\Delta H_{vap}}{2.303R}\left(\frac{T_2 - T_1}{T_1 T_2}\right)$$

Example 12.1

The vapor pressure of water at 50°C is 92.4 mmHg and its molar heat of vaporization is 42.21 kJ/mol. Calculate the vapor pressure at the boiling point of water (100°C). (Note that the vapor pressure at the boiling point is 1 atm = 760 mmHg by definition).

Solution

To begin solving this problem, organize the data, convert ΔH_{vap} from kJ/mol to J/mol, and change temperatures in degrees Celsius to Kelvin.

$$\Delta H_{vap} = 42.21 \text{ kJ/mol} = 42210 \text{ J/mol}$$

$$T_1 = 50^{\circ}C = 323 \text{ K}; T_2 = 100^{\circ}C = 373 \text{ K}$$

$$P_1 = 92.4 \text{ mmHg}; P_2 = ?$$

Next, write the modified Clausius–Clapeyron equation and solve for P_2 using the given data.

$$\ln \frac{P_2}{P_1} = \frac{\Delta H_{vap}}{R} \left(\frac{1}{T_1} - \frac{1}{T_2} \right) \text{ or } \frac{\Delta H_{vap}}{R} \left(\frac{T_2 - T_1}{T_1 T_2} \right)$$

$$\ln \frac{P_2}{92.4} = \frac{42210 \text{ J/mol}}{8.314 \text{ J/K}} \left(\frac{373K - 323K}{373K \times 323K} \right)$$

$$\ln \frac{P_2}{92.4} = 2.107$$

$$\frac{P_2}{92.4} = e^{2.107} = 8.22$$

$$\frac{P_2}{92.4} = 8.22, \Rightarrow P_2 = 760 \text{ mmHg}$$

Example 12.2

The vapor pressure of ethanol at 30°C is 98.5 mmHg and the molar heat of vaporization is 39.3 kJ/mol. Determine the approximate normal boiling point of ethanol from these data.

Solution

First, we are required to solve for the normal boiling point, T_2. At this temperature, the vapor pressure is equal to atmospheric pressure, 760 mmHg. To begin, we organize the data, and change temperatures in degrees Celsius to Kelvins.

$$T_1 = 30^{\circ}C = 303 \text{ K}; T_2 = ?$$

$$P_1 = 98.5 \text{ mmHg}; P_2 = 760 \text{ mm Hg}$$

$$\Delta H_{vap} = 39.3 \text{ kJ/mol}$$

Next, write the modified Clausius–Clapeyron equation and solve for T_2.

$$\ln\frac{P_2}{P_1} = \frac{\Delta H_{vap}}{R}\left(\frac{1}{T_1} - \frac{1}{T_2}\right)$$

$$\ln\frac{760}{98.5} = \frac{39,300}{8.314}\left(\frac{1}{303} - \frac{1}{T_2}\right)$$

$$4.322 \times 10^{-4} = \left(\frac{1}{303} - \frac{1}{T_2}\right)$$

$$\frac{1}{T_2} = \frac{1}{303} - 4.322 \times 10^{-4} = 0.0029$$

$$T_2 = 344.7 \text{ K or } 71.7^\circ\text{C}$$

Example 12.3

The vapor pressure of diethyl ether at 19°C is 400 mmHg and its normal boiling point is 34.6°C. Calculate the molar enthalpy of vaporization of diethyl ether. (The pressure at the boiling point is 1 atm = 760 mmHg).

Solution

First, organize the data, and change temperature in degrees Celsius to Kelvin.

$$T_1 = 19^\circ\text{C} = 292 \text{ K}; T_2 = 34.6^\circ\text{C} = 307.6 \text{ K}$$

$$P_1 = 400 \text{ mm Hg}; P_2 = 760 \text{ mm Hg}$$

$$\Delta H_{vap} = ?$$

Next, write the modified Clausius–Clapeyron equation and solve for ΔH_{vap} using the given data.

$$\ln\frac{P_2}{P_1} = \frac{\Delta H_{vap}}{R}\left(\frac{T_2 - T_1}{T_1 T_2}\right)$$

$$\ln\frac{760}{400} = \frac{\Delta H_{vap}}{8.314 \text{ J/K.mol}}\left(\frac{307.6 \text{ K} - 292 \text{ K}}{307.6 \text{ K} \times 292 \text{ K}}\right)$$

$$0.642 = 2.08 \times 10^{-5}\Delta H_{vap}$$

$$\Delta H_{vap} = 30,732 \text{ J/mol} = 30.7 \text{ kJ/mol}$$

12.3 The Solid State

Unlike liquids and gases, solids have a definite volume and shape. This is because the structural units (atoms, ions, or molecules) that make up the solid are held in close proximity and (usually) rigid order by chemical bonds, electrostatic attraction, or intermolecular forces.

12.3.1 Types of solids

Solids may be crystalline or amorphous. A *crystalline solid* is composed of one or more crystals characterized by a well-defined three-dimensional ordering of the basic structural units. An *amorphous solid* has a random and disordered three-dimensional arrangement of basic structural units. An amorphous solid lacks long-range molecular order.

There are four different types of crystalline solids, distinguished by the forces holding the structural units together: covalent, ionic, metallic, and molecular solids.

Covalent Solids: A covalent solid is made up of atoms joined together by covalent bonds. Examples include diamond, graphite, quartz (SiO_2), and silicon carbide (SiC).

Ionic Solids: An ionic solid consists of positive and negative ions held together in an ordered three-dimensional structure by electrostatic attraction between oppositely charged nearest neighbors, i.e. ionic bonds. Examples include NaCl, CsI, and $CaSO_4$.

Metallic Solids: A metallic solid consists of metal cations held together by a "sea" of delocalized electrons. Examples include K, Fe, Zn, and Cu.

Molecular solids: A molecular solid is composed of atoms and molecules held together by relatively weak intermolecular forces such as dispersion forces, dipole–dipole forces, and hydrogen bonds. Examples include solid water (ice), solid carbon dioxide (dry ice), solid argon, and sucrose.

12.3.2 Crystal lattices

A *crystal* is a solid with a regular, three-dimensional arrangement of atoms, molecules, or ions. Crystals often have plane surfaces, sharp edges, and regular polyhedral shapes, but the defining characteristic is the regular internal arrangement.

The constituents of a crystal are placed according to a geometrical pattern known as the *crystal lattice* or *space lattice*, which is a system of points representing the repeating pattern. The *crystal structure* of a solid is generated by an indefinite (or infinite) three-dimensional repetition of identical structural units called the unit cell. Crystal lattices may be *primitive*, with one lattice point per unit cell, or *centered*, having two or four points per cell.

12.3.3 Unit cells

A *unit cell* is the fundamental repeating structural unit of a crystalline solid. It is the smallest fraction of the crystal lattice that contains a representative portion of the crystal structure. Thus, when a unit cell is stacked repeatedly in three dimensions without gaps, it reproduces the entire crystal structure. In general, the unit cell is described by the lengths of its edges (a, b, c) and the angles between the edges (α, β, γ).

12.4 The Crystal System

There are seven basic shapes of unit cells, commonly referred to as the seven crystal systems (Table 12.1). From symmetry considerations, variations of the seven types of unit cells are possible. If we combine the crystal systems (primitive translational symmetry) and the centering translations, this will give rise to the fourteen possible unique crystal lattices, also known as

Table 12.1   The unit cell relationships for the seven crystal systems

Crystal system	Bravais lattices (Symbol)	Unit cell		Example
		Lengths	Angles	
Cubic	P, I, F	$a = b = c$	$\alpha = \beta = \gamma = 90°$	NaCl (rock salt)
Tetragonal	P, I	$a = b \neq c$	$\alpha = \beta = \gamma = 90°$	TiO_2 (rutile)
Orthorhombic	P, C, I, F	$a \neq b \neq c$	$\alpha = \beta = \gamma = 90°$	$MgSO_4{\cdot}7H_2O$ (epsomite)
Monoclinic	P, C	$a \neq b \neq c$	$\alpha = \gamma = 90°; \beta \neq 90°$	$KClO_4$ (potassium chlorate)
Triclinic	P	$a \neq b \neq c$	$\alpha \neq \beta \neq \gamma \neq 90°$	$K_2Cr_2O_7$ (potassium dichromate)
Hexagonal	P	$a = b \neq c$	$\alpha = \beta = 90°; \gamma = 120°$	SiO_2 (silica)
Rhombohedral	R	$a = b = c$	$\alpha = \beta = \gamma \neq 90°$	$CaCO_3$ (calcite)

the Bravais lattices (Table 12.1). Hence, fourteen different unit cell geometries can occur in a crystalline solid.

In this book, however, we consider only unit cells with cubic symmetry.

12.4.1 Close-packed structure

Most metals and several molecular substances have close-packed structures. These can be thought of as consisting of identical objects packed together as closely as possible. There are two possible close-packed arrangements:

1. Cubic close-packed (CCP)
2. Hexagonal close-packed (HCP)

These two arrangements represent the most efficient way of packing spheres. In this book, we focus only on CCP structures.

12.4.2 Cubic unit cells

There are three types of cubic unit cells. They are primitive or simple cubic (SC), body-centered cubic (BCC), and face-centered cubic (FCC) (Figure 12.1).

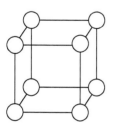
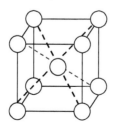
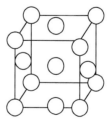

Simple or Primitive cubic   Body-centered cubic   Face-centered cubic

Figure 12.1   Types of cubic cells.

12.4.2.1 Guidelines for determining the number of atoms in a unit cell

1. An atom occupying a corner is shared equally among eight adjacent unit cells. Therefore it contributes only 1/8 of an atom to each unit cell.
2. An atom occupying an edge is shared equally among four adjacent unit cells. Therefore it contributes only 1/4 of an atom to each unit cell.
3. An atom occupying a face is shared equally between two adjacent unit cells. Therefore it contributes only 1/2 of an atom to each unit cell.
4. An atom completely inside the unit cell, i.e. a body-centered atom, is not shared with neighboring unit cells. Therefore, a body-centered atom contributes one atom to the unit cell.

12.4.2.2 Simple or primitive cubic unit cell

An SC unit cell has one lattice point (or atom) at each of its eight corners; each corner atom is shared with neighboring cells such that only 1/8 of the atom is in each unit cell. Therefore the number of atoms per unit cell is:

$$\left(\frac{\frac{1}{8} \text{ atom}}{\text{corner}} \right) \times \frac{8 \text{ corners}}{\text{unit cell}} = 1 \frac{\text{atom}}{\text{unit cell}}$$

In this book, the number of atoms in a unit cell is denoted by the letter "Z," so that $Z = 1$ for the primitive unit cell.

12.4.2.3 Body-centered cubic unit cell

A BCC unit cell has one lattice point or atom in the center of each unit cell and one atom at each of its eight corners. Each corner atom is shared with neighboring cells such that only 1/8 of the atom is in each unit cell. Therefore the number of atoms per unit cell is:

$$\left(\frac{\frac{1}{8} \text{ atom}}{\text{corner}} \right) \times \left(\frac{8 \text{ corners}}{\text{unit cell}} \right) + \left(\frac{1 \text{ body center}}{\text{unit cell}} \right) \left(\frac{1 \text{ atom}}{\text{body center}} \right) = 2 \frac{\text{atoms}}{\text{unit cell}}$$

Thus $Z = 2$ for the BCC unit cell.

12.4.2.4 Face-centered cubic unit cell

An FCC unit cell has one atom in each face of the unit cell and one atom at each corner. There are eight corners and six faces per unit cell. Each corner atom is shared with neighboring cells such that only 1/8 of the atom is in each unit cell; each face contributes 1/2 face atom to each unit cell. Therefore, the number of atoms per unit cell is:

$$\left(\frac{\frac{1}{8} \text{ atom}}{\text{corner}} \right) \times \left(\frac{8 \text{ corners}}{\text{unit cell}} \right) + \left(\frac{6 \text{ faces}}{\text{unit cell}} \right) \left(\frac{\frac{1}{2} \text{ atom}}{\text{face}} \right) = 4 \frac{\text{atoms}}{\text{unit cell}}$$

Thus $Z = 4$ for the FCC unit cell.

Example 12.4

Tungsten (W) crystallizes in a BCC lattice. What is the number of W atoms per unit cell?

Solution

The BCC system contains one atom in the center of each unit cell and one atom at each corner. The number of atoms per unit cell is:

$$\left(\frac{\frac{1}{8}\,\text{W atom}}{\text{corner}}\right) \times \left(\frac{8\,\text{corners}}{\text{unit cell}}\right) + \left(\frac{1\,\text{body center}}{\text{unit cell}}\right)\left(\frac{1\,\text{W atom}}{\text{body center}}\right) = 2\frac{\text{W atoms}}{\text{unit cell}}$$

Example 12.5

Copper crystallizes in an FCC system with a unit cell edge of 3.62Å at 20°C. Calculate the number of copper atoms in one of the unit cells.

Solution

The FCC system contains one atom in each face of each unit cell and one atom at each corner. The number of atoms per unit cell is:

$$\left(\frac{\frac{1}{8}\text{Cu atom}}{\text{corner}}\right) \times \left(\frac{8\,\text{corners}}{\text{unit cell}}\right) + \left(\frac{6\,\text{faces}}{\text{unit cell}}\right)\left(\frac{\frac{1}{2}\text{Cu atom}}{\text{face}}\right) = 4\frac{\text{Cu atoms}}{\text{unit cell}}$$

12.4.3 Coordination number

Coordination number, as used here, is defined as the number of nearest neighbors of an atom or an ion in a crystal structure.

12.5 Calculations Involving Unit Cell Dimensions

The structure and dimensions of a unit cell can be determined by X-ray diffraction studies (described in Section 12.9). Once these are known, we can calculate several physical parameters such as volume, density, atomic mass, and Avogadro's number.
  The theoretical density of a crystalline solid can be computed from the following equation:

$$\rho = \frac{ZM}{N_A V}$$

where ρ is the density, Z is the total number of atoms in the unit cell, M is the molecular mass, N_A is Avogadro's number, and V is the volume of the unit cell. For a cubic unit cell, $V = a^3$, where a is the unit cell edge.

Example 12.6

Calcium crystallizes in a cubic lattice system with a unit cell edge of 5.6 Å. Its density is 1.55 g/cm³.

(a) Calculate the number of Ca atoms per unit cell.
(b) Use the answer in part (a) to determine what cubic crystal system Ca belongs to.

Solution

First, determine the mass of Ca in one unit cell from the density and unit cell dimensions:

$$\text{Density} = \frac{\text{Mass of unit cell}}{\text{Volume of unit cell}}$$

$$\text{Mass of unit cell} = \text{Density} \times \text{Volume of unit cell}$$

For a cubic unit cell,

$$V = a^3 = \left(5.6\text{Å} \times \frac{1\text{ cm}}{10^8\text{Å}}\right)^3 = 1.756 \times 10^{-22} \text{ cm}^3$$

$$\text{Mass of unit cell} = \left(1.55\frac{\text{g Ca}}{\text{cm}^3}\right)\left(1.756 \times 10^{-22}\frac{\text{cm}^3}{1 \text{ unit cell}}\right)$$

$$= 2.722 \times 10^{-22}\frac{\text{g Ca}}{\text{unit cell}}$$

Next, calculate the mass of a Ca atom from its atomic mass and N_A (i.e. 6.02×10^{23} atoms/mol):

$$\text{Mass of 1 Ca atom} = \left(\frac{40.08\text{ g}}{1\text{ mol Ca}}\right)\left(\frac{1\text{ mol Ca}}{6.02 \times 10^{23}\text{ atom Ca}}\right)$$

$$= \frac{6.66 \times 10^{-23}\text{ g Ca}}{\text{atom Ca}}$$

Then, determine the number of Ca atoms per unit cell from the mass of Ca per unit cell and the mass of Ca per atom:

$$\text{Number of Ca atoms per unit cell} = \left(2.722 \times 10^{-22}\frac{\text{g Ca}}{\text{unit cell}}\right)\left(\frac{1\text{ atom Ca}}{6.66 \times 10^{-23}\text{ g Ca}}\right)$$

$$= 4\frac{\text{atoms Ca}}{\text{unit cell}}$$

Thus, the number of atoms per unit cell corresponds to an FCC system. That is,

$$\left(\frac{\frac{1}{8}\text{Cu atom}}{\text{corner}}\right) \times \left(\frac{8\text{ corners}}{\text{unit cell}}\right) + \left(\frac{6\text{ faces}}{\text{unit cell}}\right)\left(\frac{\frac{1}{2}\text{Cu atom}}{\text{face}}\right) = 4\frac{\text{Cu atoms}}{\text{unit cell}}$$

Example 12.7

Elemental silver, Ag, crystallizes in an FCC lattice with a unit cell edge of 4.10 Å. The density of Ag is 10.51 g/cm³. Determine the number of atoms in 108.0 g of Ag.

Solution

First, calculate the volume of the unit cell:

$$V = a^3 = \left(4.10 \text{ Å} \times \frac{1 \text{ cm}}{10^8 \text{ Å}}\right)^3 = 6.89 \times 10^{-23} \frac{\text{cm}^3}{\text{unit cell}}$$

Next, calculate the volume of 108 g of Ag:

$$\text{Volume} = \frac{\text{Mass}}{\text{Density}} = \left(\frac{108 \text{ g}}{10.51 \frac{\text{g}}{\text{cm}^3}}\right) = 10.28 \text{ cm}^3$$

Then, calculate the number of unit cells in this volume and hence the total number of atoms:

$$\text{Number of unit cells} = \left(10.28 \text{ cm}^3\right)\left(\frac{1 \text{ unit cell}}{6.89 \times 10^{-23} \text{ cm}^3}\right)$$

$$= 1.50 \times 10^{23} \text{ unit cells}$$

There are four atoms per unit cell in a an FCC system

$$\text{Thus the total number of atoms} = \left(4\frac{\text{atoms}}{\text{unit cell}}\right)\left(1.50 \times 10^{23} \text{ unit cells}\right)$$

$$= 6.0 \times 10^{23} \text{ atoms}$$

Example 12.8

Nickel crystallizes in an FCC structure with a unit-cell edge length of 352.5 pm. Calculate:

(a) the radius of a Ni atom in pm, and
(b) the density of Ni in g/cm³.

Solution

(a) In an FCC unit cell, the face atoms touch the corner atoms along the diagonal of each face. However, the corner atoms do not touch each other along the edges. Therefore, the length of the diagonal is equal to four times the radius ($4r$). Note that the diagonal and the edges of the cube form a right-angled triangle (Figure 12.2). Hence, we can solve for r using the Pythagorean theorem.

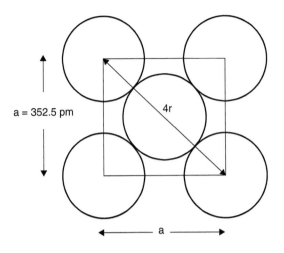

Figure 12.2   A representation of the unit cell of nickel, indicating the relative sizes and positions of the atoms.

$$a^2 + a^2 = (4r)^2$$

$$2a^2 = 16r^2$$

$$r = \sqrt{\frac{2a^2}{16}} = \sqrt{\frac{a^2}{8}} = \sqrt{\frac{352.5^2}{8}} = 124.6 \text{ pm}$$

(b) To calculate the density of Ni:

$$\text{Density (d)} = \frac{\text{mass of a single unit cell (m)}}{\text{volume of a single unit cell (v)}}$$

$$\text{Volume} = a^3 = (352.5 \text{ pm})^3 = \left(\frac{4.38 \times 10^7 \text{pm}^3}{\text{unit cell}}\right)\left(\frac{10^{-10} \text{ cm}}{1 \text{pm}}\right)^3$$

$$= 4.38 \times 10^{-23} \text{ cm}^3$$

To calculate the mass of a unit cell, we will need to know the number of atoms per cell and then convert the number of atoms to mass in grams using the atomic mass and Avogadro's number, N_A.

In an FCC unit cell, there are 4 atoms, i.e.:

$$\left(\frac{1}{8} \times 8\right) + \left(\frac{1}{2} \times 6\right) = 4$$

The atomic mass of Ni is 58.7 g/mol, and $N_A = 6.022 \times 10^{23}$ atoms/mol.

$$\text{Mass of a Ni unit cell} = (4 \text{ atoms})\left(\frac{58.7 \text{ g/mol}}{6.022 \times 10^{23} \text{ atom/mol}}\right)$$

$$= 3.90 \times 10^{-22} \text{ g}$$

$$\text{Density} = \frac{3.90 \times 10^{-22} \text{ g}}{4.38 \times 10^{-23} \text{ cm}^3} = 8.90 \text{ g/cm}^3$$

Example 12.9

An unknown metal M crystallizes in a BCC structure with a unit cell dimension of 315 pm.

(a) Calculate the number of M atoms per unit cell.
(b) If the measured density of M is 10.15 g/cm³, determine the relative atomic mass and use the periodic table to identify the unknown metal.

Solution

In a BCC unit cell, there are eight corner atoms and one center atom. Therefore, the number of M atoms will be:

$$\left(\frac{\frac{1}{8}\text{ atom}}{\text{corner}}\right) \times \left(\frac{8\text{ corners}}{\text{unit cell}}\right) + \left(\frac{1\text{ body center}}{\text{unit cell}}\right)\left(\frac{1\text{ atom}}{\text{body center}}\right) = 2\frac{\text{M atoms}}{\text{unit cell}}$$

To calculate the atomic mass of M, we will need to know the volume of the unit cell. This can be calculated from the cell dimensions:

$$\text{Volume} = a^3 = (315\text{ pm})^3 = \left(\frac{315\text{ pm}}{\text{unit cell}}\right)^3\left(\frac{10^{-10}\text{ cm}}{1\text{pm}}\right)^3$$

$$= 3.126 \times 10^{-23}\text{ cm}^3$$

We can now calculate the atomic mass of M from the mass of the unit cell.

$$\text{Density} = \frac{\text{mass}}{\text{volume}}$$

$$10.15\frac{\text{g}}{\text{cm}^3} = \frac{\text{mass}}{3.126 \times 10^{-23}\text{ cm}^3}$$

Mass of an M unit cell $= 3.172 \times 10^{-22}$ g

$$\text{Mass of unit cell (g)} = \frac{(\text{number of atoms per cell})(\text{atomic mass in g/mol})}{N_A\left(\frac{\text{atom}}{\text{mol}}\right)}$$

$$3.172 \times 10^{-22}\text{ g} = \frac{(2\text{ atoms})(\text{relative atomic mass})}{6.022 \times 10^{23}\frac{\text{atom}}{\text{mol}}}$$

$$\text{Relative atomic mass} = 95.5\frac{\text{g}}{\text{mol}}$$

From the periodic table M is most likely Mo, which has a relative atomic mass of 95.9.

Example 12.10

Metallic gold (Au) crystallizes in a face-centered cubic system with a unit cell dimension of 4.07Å. What is

(a) the distance between centers of nearest neighbors
(b) the radius of a gold atom
(c) the volume of a gold atom in cm³, given that for a spherical atom, $V = \frac{4}{3}\pi r^3$
(d) the % volume of a unit cell occupied by Au atoms, and
(e) the density of Au.

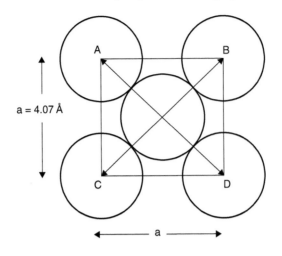

Figure 12.3   One face of the face-centered cubic (FCC) cell of gold.

Solution

(a) Figure 12.3 represents the face of an FCC unit cell. Each sphere is a gold atom. Therefore, the distance between nearest neighbors is one-half the distance of the diagonal AD (for example) of the inscribed square. Using the Pythagorean theorem, we can calculate the distance AD, and AD/2 is the distance between centers of neighbors.

$$(AD)^2 = a^2 + a^2 = 2a^2$$

$$AD = \sqrt{2a^2} = a\sqrt{2} = 1.414a$$

$$AD = 1.414 \times 4.07 \text{ Å} = 5.76 \text{ Å}$$

$$\text{Distance between centers of nearest neighbors} = \frac{AD}{2} = \frac{5.76}{2} = 2.88 \text{ Å}$$

(b) Take a close look at the diagram in Figure 12.3. Notice that the atoms touch along the diagonals of each face but not on the edges of the cube. The length of the diagonal is $4r$ (or $r + 2r + r$). Once again, we can use the Pythagorean theorem to show the relationship between the unit cell dimension and the radius of the atom.

$$a^2 + a^2 = (4r)^2$$

$$2a^2 = (4r)^2 = 16r^2$$

$$a^2 = 8r^2; a = r\sqrt{8}$$

$$r = \frac{a}{\sqrt{8}} = \frac{4.07 \text{ Å}}{\sqrt{8}} = 1.44 \text{ Å}$$

(c) To calculate the volume of a gold atom, substitute its atomic radius in the volume equation:

$$V_{\text{Au atom}} = \frac{4}{3}\pi r^3 = \left(\frac{4}{3}\right)(\pi)\left(1.44\text{Å} \times \frac{10^{-8}}{\text{Å}}\right)^3 = 1.25 \times 10^{-23} \text{ cm}^3$$

(d) There are four atoms in an FCC crystal lattice unit cell. Each Au atom occupies a volume of 1.25×10^{-23} cm^3. Therefore, four Au atoms will occupy

$$V_{\text{Au atoms}} = (4)\left(1.25 \times 10^{-23} \text{ cm}^3\right) = 5.00 \times 10^{-23} \text{ cm}^3.$$

This is the volume of the Au atoms themselves.

Now calculate the volume of the unit cell using $V = a^3$ (since the unit cell is cubic):

$$V_{\text{unit cell}} = a^3 = \left(4.07\text{Å} \times \frac{10^{-8} \text{ cm}}{\text{Å}}\right)^3 = 6.74 \times 10^{-23} \text{ cm}^3$$

$$\%\text{Volume of unit cell occupied by Au atoms} = \frac{V_{\text{Au atoms}}}{V_{\text{unit cell}}} \times 100$$

$$= \frac{5.00 \times 10^{-23} \text{ cm}^3}{6.74 \times 10^{-23} \text{ cm}^3} \times 100$$

$$= 74.2\%$$

(e) To calculate the density of Au, we need to determine the mass of the unit cell:

$$\text{Mass of a Au unit cell} = (4 \text{ atoms})\left(\frac{197 \text{ g/mol}}{6.022 \times 10^{23} \text{ atom/mol}}\right)$$

$$= 1.31 \times 10^{-21} \text{ g}$$

$$\text{Density} = \frac{\text{Mass of unit cell}}{\text{Volume of unit cell}} = \frac{1.31 \times 10^{-21} \text{ g}}{6.74 \times 10^{-23} \text{ cm}^3} = 19.42 \text{ g/cm}^3$$

12.6 Ionic Crystal Structure

Most salts crystallize as ionic solids, with anions and cations (rather than atoms) occupying the unit cell. The sizes of cations are usually different from those of anions. Hence, ionic crystal structures are different from those of metals. The structure, properties, and stability of ionic compounds can be understood from the relative radii of the constituent ions. Ionic compounds of the general formula $M^{n+}X^{x-}$ usually crystallize in the NaCl structure (FCC lattice), CsCl structure (SC lattice), or ZnS structure (FCC lattice).

12.6.1 The sodium chloride (NaCl), or "rock-salt" structure

The NaCl structure is based on an FCC lattice (Figure 12.4). It consists of cubic-close-packed chloride anions with sodium cations occupying all the available octahedral holes to achieve the required 1:1 stoichiometry. Each sodium ion is surrounded by six chloride ions located at the corners of a regular octahedron. Similarly, each chloride ion is surrounded octahedrally by six sodium ions, resulting in a coordination number of 6:6. The NaCl structure is generally adopted when the cation-to-anion radius ratio is in the range 0.41–0.73. Examples of compounds with the NaCl structure include halides of Li, Na, K, and Rb, as well as divalent metal oxides and sulfides such as CaO, MgO, MnO, and CaS.

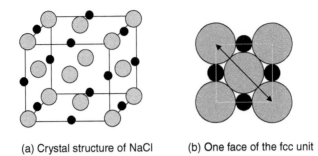

(a) Crystal structure of NaCl   (b) One face of the fcc unit

Figure 12.4   The unit cell of NaCl showing (a) skeletal view and (b) space-filling view. The larger chloride anions adopt an FCC unit cell with the smaller sodium cations occupying the holes between adjacent anions.

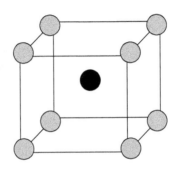

Figure 12.5   The structure of CsCl.

12.6.2 The cesium chloride (CsCl) structure

In the CsCl structure, a Cs cation is located at the center of the cube with eight chloride ions occupying the eight corners (Figure 12.5). Similarly, each chloride ion is at the center of a cube, surrounded by eight nearest-neighbor cesium ions. The coordination number of Cs and Cl is 8:8, with one Cs ion and one Cl ion per unit cell. The CsCl structure is preferred when the cation/anion radius ratio is 0.73 or greater. Examples of compounds with the CsCl structure include the halides of monovalent Cs, Tl, and NH_4^+.

12.6.3 The zinc blende (ZnS) structure

Zinc sulfide occurs in two crystalline forms, zincblende and wurtzite. The zincblende structure is based on the FCC lattice, and is adopted when the cation to anion radius ratio is between 0.22 and 0.41. The structure can be viewed as a cubic close-packed (FCC) array of S atoms with Zn ions occupying half of the tetrahedral sites, resulting in a ZnS stoichiometry with eight Zn and sixteen S ions in the unit cell (Figure 12.6). Examples of compounds with the zincblende structure include BeS, CuCl, CdS, HgS, and SiC.

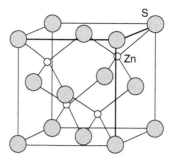

Figure 12.6   The structure of ZnS.

Table 12.2   Radius ratio and crystal types of selected ionic compounds

Formula	Crystal structure type	Approximate limiting radius ratio (r^+/r^-)	Coordination number of M^-	Coordination number of X^-	Examples with radius ratio in parenthesis
MX	Cesium chloride	> 0.73	8	8	CsCl (0.93), & TlCl (0.83)
MX$_2$	Fluorite	> 0.73	8	4	CaF$_2$ (0.73), & PbF$_2$ (0.88)
MX	Sodium chloride	0.41 - 0.73	6	6	NaCl (0.52), & AgCl (0.70)
MX$_2$	Rutile	0.41 - 0.73	6	3	MgF$_2$ (0.48), & ZnF$_2$ (0.54)
MX	Zincblende	< 0.41	4	4	ZnS (0.40), & BeO (0.22)

12.7 The Radius Ratio Rule for Ionic Compounds

The structure of a crystalline ionic compound depends on the coordination number of each ion and the relative sizes of the constituent cations and anions. Each ion in an ionic crystal always tries to surround itself with as many oppositely charged ions as possible. This results in a stable and neutral structure. The radius ratio is defined as the ratio of cation radius to anion radius; its value is useful in determining the coordination number of the smaller ion, which in most cases is the cation. Table 12.2 shows the coordination number, crystal structure, and the limiting values of the radius ratio for several common cases.

Example 12.11

The unit cell of potassium chloride, KBr, is an FCC with edge length of 6.60 Å. Assuming anion–anion contact and anion–cation contact, calculate:

(a) the ionic radius of the bromide ion
(b) the ionic radius of the potassium ion, and
(c) the radius ratio.

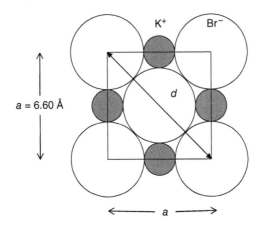

Figure 12.7   One face of the FCC cell of KBr. Notice that the bromide ions touch along the diagonal d.

Solution

(a) Figure 12.7 shows one face of the face-centered cubic unit cell of KBr. Notice that the bromide ions touch along the diagonal d, which is also the hypotenuse of the two right isosceles triangles and is equal to four times the radius of the bromide ion. That is, $d = r_{Br^-} + 2r_{Br^-} + r_{Br^-} = 4r_{Br^-}$. We can calculate d using the Pythagorean theorem since we know the sides of the right-angled triangle ($a = 6.60$ Å):

$$d^2 = a^2 + a^2 = 2a^2$$

$$d = \sqrt{2a^2} = \sqrt{(2)(6.60)^2} = 9.33 \text{ Å}$$

$$\text{But } d = 4r_{Br^-} = 9.33 \text{ Å}$$

$$r_{Br^-} = \frac{9.33 \text{ Å}}{4} = 2.33 \text{ Å}$$

(b) The unit cell edge length, a, is two times the radius of the bromide ion plus two times the radius of the potassium ion.

$$a = 2r_{K^+} + 2r_{Br^-} = 6.60\text{Å}$$

$$\text{From part (a)}, r_{Br^-} = 2.33\text{Å}$$

$$2r_{K^+} = 6.60\text{Å} - 2r_{Br^-} = 6.60 - 2 \times 2.33$$

$$2r_{K^+} = 1.93\text{Å}$$

$$r_{K^+} = 0.97\text{Å}$$

(c) Radius ratio $= \dfrac{\text{radius of cation}}{\text{radius of anion}} = \dfrac{r_{K^+}}{r_{Br^-}} = \dfrac{0.97}{2.33} = 0.41$

Example 12.12

CsBr crystallizes in the CsCl structure with a bromide ion at each corner and a cesium ion at the center of the unit cell. The unit cell edge length is 4.29Å. Calculate:

(a) the number of ions per unit cell
(b) the distance between the center of a cesium ion and the center of the nearest bromide ion
(c) the radius of the bromide ion, if the radius of the cesium ion is 1.75Å, and
(d) the density of CsBr in g/cm³.

Solution

(a) One Cs^+ ion is at the center of the cell and is in contact with eight corner Br^- ions. Therefore the number of Br^- ions in the unit cell is

$$8 \text{ corners} \times \frac{1}{8}\left(\frac{Br^-}{corner}\right) = 1 \ Br^-$$

Thus the unit cell of CsBr contains the equivalent of 1 Cs^+ ion and 1 Br^- ion. This is consistent with one formula unit per unit cell.

(b) In a body-centered cubic unit cell, all the atoms or ions touch along a diagonal passing through the center of the cube, as illustrated in Figure 12.8b. So the shortest distance between the center of a cesium ion and the center of the nearest bromide ion is along the diagonal of the unit cell and is $(AG)/2$ from Figure 12.8a. Therefore we need to calculate the length of the diagonal AG from Figure 12.8a.
By the Pythagorean theorem:

$$(AG)^2 = (AE)^2 + (EG)^2$$

$$\text{But } (EG)^2 = (EF)^2 + (FG)^2$$

$$(AG)^2 = (AE)^2 + \left(EF\right)^2 + (FG)^2 \ (\text{or } 3a^2)$$

$$(AG)^2 = 4.29^2 + 4.29^2 + 4.29^2 = 55.2\text{Å}^2$$

$$AG = 7.43\text{Å}$$

The distance between the centers of Cs^+ and Br^- is $\frac{7.43}{2}$ or 3.72Å.

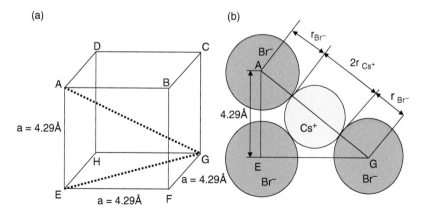

Figure 12.8 CsBr unit cell geometry.

(c) Since we know the Cs–Br distance, if the radius of the cesium ion is 1.75 Å, we can easily find the radius of Br^-.

$$2r_{Cs^+} + 2r_{Br^-} = AG$$

$$r_{Cs^+} + r_{Br^-} = \frac{AG}{2} = 3.72\text{Å}$$

$$1.75\text{Å} + r_{Br^-} = 3.72\text{Å}$$

$$r_{Br^-} = 1.97\text{Å}$$

(d) To find the density of CsBr, first we need to calculate the volume of the unit cell.

$$V = a^3 = \left(4.29\text{Å} \times \frac{10^{-8}\ \text{cm}}{1\text{Å}}\right)^3 = 7.90 \times 10^{-23}\ \text{cm}^3$$

Next, we calculate density:

There are 2 ions per unit cell, 1 Cs^+ and 1 Br^-

Mw of CsBr $= 213$ g/mol

$$\text{Mass of per unit cell} = (2\ \text{ions})\left(213\frac{\text{g CsCl}}{2\ \text{ions}}\right)\left(\frac{1\ \text{mol}}{6.02 \times 10^{23}\ \text{ions}}\right)$$

$$= 3.54 \times 10^{-22}\ \text{g}$$

$$\text{Density} = \frac{\text{Mass}}{\text{Volume}} = \frac{3.54 \times 10^{-22}\ \text{g}}{7.90 \times 10^{-23}\ \text{cm}^3} = 4.48\ \text{g/cm}^3$$

Example 12.13

Predict the unit cell structure of CaS, given that the ionic radii of Ca^{2+} and S^{2-} are 0.99 Å and 1.84 Å, respectively.

Solution

$$\text{Radius ratio} = \frac{r_{Ca^{2+}}}{r_{S^{2-}}} = \frac{0.99}{1.84} = 0.54$$

CaS should crystallize in the NaCl unit cell structure.

Example 12.14

The cubic unit cell of TlCl is similar to that of CsCl and has a radius ratio of 0.83. Estimate the ionic radius of Tl^+ given that the ionic radius of Cl^- is 1.81 Å.

Solution

$$\text{Radius ratio} = \frac{r_{Tl^+}}{r_{Cl^-}}$$

The radius ratio for TlCl is 0.83.

$$0.83 = \frac{r_{Tl^+}}{1.84} \Rightarrow r_{Tl^+} = 1.53\text{Å}$$

The ionic radius of Tl^+ is 1.53Å.

12.8 Determination of Crystal Structure by X-Ray Diffraction

X-rays are electromagnetic radiation of very short wavelength, comparable to the distance between atoms or (lattice points) in a crystal. When passed through crystals, X-rays are scattered—diffracted—in ways that depend on the arrangement of atoms, molecules, or ions within the crystals. Thus, X-rays can be used to determine the important parameters of crystal structure.

In 1913, Bragg and Bragg (father and son) formulated a fundamental geometrical equation to explain X-ray diffraction. To understand how a diffraction pattern may be produced, consider the reflection of monochromatic X-rays of wavelength λ incident on atoms in two different layers of a crystalline solid (Figure 12.9). Let d be the lattice-point spacing between the two layers and let θ be the grazing angle. The lower wave travels an extra distance equal to the sum of BC + CD. The two waves will be in phase after reflection and reinforce each other as they exit the crystal if the path difference (i.e. BC + CD) is an integral number of wavelengths ($n = 1, 2, 3 \ldots$).

In mathematical terms,

$$BC + CD = n\lambda$$

Applying trigonometry, we can show that

$$BC + CD = 2d\sin\theta$$

Combining these two equations yields:

$$2d\sin\theta = n\lambda \ (\text{where } n = 1, 2, 3 \ldots)$$

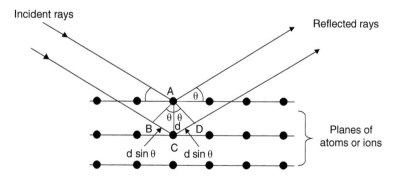

Figure 12.9   Reflection of X-rays by atoms in two different planes of a crystal.

This is referred to as Bragg's equation or Bragg's law. For a first-order or principal reflection, $n = 1$. So the equation just above becomes:

$$2d\sin\theta = \lambda$$

Examples 12.15

The principal diffraction angle of a crystal is $28°$ for X-rays of wavelength 1.62 Å. What is the inter-planar spacing?

Solution

To determine the inter-planar distance d, we use Bragg's equation.

$$n\lambda = 2\,d\,\sin\theta$$

Given

$$n = 1 \text{ for principal diffraction angle}$$

$$\lambda = 1.62\text{Å}$$

$$\theta = 28°$$

Solve for d by substituting in Bragg's equation

$$d = \frac{n\lambda}{2\sin\theta} = \frac{1 \times 1.62\text{Å}}{2 \times \sin 28°} = \frac{1.62\text{Å}}{0.9389} = 1.73\text{Å}$$

Example 12.16

What is the principal diffraction angle of a crystal if X-rays of wavelength 1.54 Å are used? The sets of parallel crystal planes are 1.90 Å apart.

Solution

To determine the diffraction angle, we use Bragg's equation.

$$n\lambda = 2d\,\sin\theta$$

Given

$$n = 1 \text{ for principal diffraction angle}$$

$$\lambda = 1.54 \text{ Å}$$

$$d = 1.90 \text{ Å}$$

Solve for θ by substituting in Bragg's equation

$$\sin \theta = \frac{n\lambda}{2d} = \frac{1 \times 1.54 \text{ Å}}{2 \times 1.90 \text{ Å}} = 0.4053$$

$$\theta = \sin^{-1}(0.4053) = 23.9^\circ$$

Example 12.17

The so-called Miller indexes are very helpful in describing the separation between atomic planes. For example, for a cubic lattice, the separation of the (hkl) planes is given by the equation

$$d_{hkl} = \frac{a}{\left(h^2 + k^2 + l^2\right)^{\frac{1}{2}}}$$

where a is the unit cell length. Sodium chloride crystallizes in an FCC lattice with a unit cell edge length of 5.64 Å. Calculate the perpendicular distances (d_{hkl}) between the planes having the following Miller indices (100), (111), (210), and (221).

Solution

Simply substitute the given information in the given equation.

$$d_{hkl} = \frac{a}{\left(h^2 + k^2 + l^2\right)^{\frac{1}{2}}}$$

$$d_{100} = \frac{5.64 \text{ Å}}{\left(1^2 + 0^2 + 0^2\right)^{\frac{1}{2}}} = 5.64 \text{ Å}$$

$$d_{111} = \frac{5.64 \text{ Å}}{\left(1^2 + 1^2 + 1^2\right)^{\frac{1}{2}}} = 3.26 \text{ Å}$$

$$d_{210} = \frac{5.64 \text{ Å}}{\left(2^2 + 1^2 + 0^2\right)^{\frac{1}{2}}} = 2.52 \text{ Å}$$

$$d_{221} = \frac{5.64 \text{ Å}}{\left(2^2 + 2^2 + 1^2\right)^{\frac{1}{2}}} = 1.88 \text{ Å}$$

Example 12.18

Cesium metal crystallizes in a BCC lattice with a unit cell length of 6.10 Å. Calculate:

(a) the glancing angle at which the reflection from the (222) planes of this crystal lattice will be observed when an X-ray beam of wavelength 1.54 Å is used, and
(b) the density of cesium.

Solution

(a) Use the equation given below to solve for d_{222}. Then substitute this value in the Bragg equation and solve for the angle θ.

$$d_{hkl} = \frac{a}{\left(h^2 + k^2 + l^2\right)^{\frac{1}{2}}}$$

$$d_{222} = \frac{6.10 \text{ Å}}{\left(2^2 + 2^2 + 2^2\right)^{\frac{1}{2}}} = 1.76 \text{ Å}$$

$$d_{222} = \frac{\lambda}{2\sin\theta} \quad \Rightarrow \quad \sin\theta = \frac{\lambda}{2d_{222}} = \frac{1.54 \text{ Å}}{2 \times 1.76 \text{ Å}} = 0.4375$$

$$\theta = 25.9^{\circ}$$

12.9 Problems

1. In a laboratory experiment, a student determined the vapor pressure of ethanol to be 350 mmHg at 60°C. If the normal boiling point of ethanol is 78.3°C, calculate the molar enthalpy of vaporization of ethanol.

2. The normal boiling point of mercury is 357°C and its molar heat of vaporization is 59 kJ/mol. At what pressure will mercury have its boiling point reduced by 37°C?

3. The molar heat of vaporization of chloroform ($CHCl_3$) is 29.4 kJ/mole. If the vapor pressure at 80°C is 1,455 mmHg, what is the normal boiling point of $CHCl_3$?

4. At 30°C and 100°C, the vapor pressure of carbon tetrachloride (CCl_4) is 142.5 mmHg and 1,463 mmHg, respectively. Calculate the heat of vaporization of CCl_4.

5. The heat of vaporization of water is 40.8 kJ/mole and its vapor pressure at 50°C is 92.7 mmHg. What will be the vapor pressure at 110°C?

6. The vapor pressure of hexafluorobenzene, C_6F_6, is 92.5 mmHg at 27°C and 483 mmHg at 67°C. Calculate:

   (a) the molar heat of vaporization
   (b) the vapor pressure at 0°C, and
   (c) the normal boiling point of C_6F_6.

7. The vapor pressure of methyl iodide (CH_3I) at −45.5°C and 25.5°C are 9.5 mmHg and 400 mmHg, respectively. Calculate:

   (a) the molar heat of vaporization
   (b) the vapor pressure at 100°C, and
   (c) the normal boiling point of CH_3I.

8. Aluminum (Al) crystallizes in an FCC cell. Determine the number of Al atoms in each unit cell.

9. Vanadium (V) crystallizes in a BCC cell. Calculate the number of V atoms in each unit cell.

10. The element iridium (Ir) crystallizes in an FCC cell, and the length of the unit cell is 3.83 Å. Calculate the density of Ir in g/cm³.

11. Polonium (Po) crystallizes in an SC structure with a unit cell edge of 3.34 Å. Calculate:

    (a) the number of Po atoms in each unit cell
    (b) the volume of the unit cell, and
    (c) the density of Po in g/cm³.

12. Sodium chloride (NaCl) crystallizes in an FCC lattice.

    (a) Calculate the number of Na^+ ions and Cl^- ions in each NaCl unit cell.
    (b) If the density of NaCl is 2.15 g/cm³, what is the unit cell length in Å?

13. An unknown metal M crystallizes in an FCC structure with a unit cell length of 4.08 Å. The density was determined to be 19.3 g/cm³. What is the atomic mass of the metal?

14. Nickel metal (Ni) has an FCC unit cell with an edge length of 3.52 Å. The density of Ni is 8.90 g/cm³. Use these data to calculate the value of Avogadro's number. (The atomic mass of Ni is 58.7 g/mol.)

15. Gold (Au) crystallizes in an FCC lattice with a unit cell length of 4.08 Å. What is the radius of a gold atom in Å?

16. Europium crystallizes in a BCC lattice. The unit cell edge length is 4.60 Å. Calculate the radius of a europium atom.

17. Aluminum crystallizes in a cubic lattice with a unit cell edge of 4.04 Å. The density is 2.70 g/cm³. How many Al atoms are in each unit cell?

18. Zinc oxide crystallizes with a cubic crystal structure. The ionic radii are 0.74 Å for Zn^{2+} and 1.40 Å for O^{2-}. Calculate the radius ratio and suggest the possible crystal structure of ZnO.

19. Indium phosphide (InP) crystallizes with a cubic structure. The ionic radii are 0.81 Å for In^{3+} and 2.12 Å for P^{3-}. Calculate the radius ratio and describe the crystal structure of InP.

20. X-rays of wavelength 1.54 Å are diffracted from a gallium crystal at an angle of 12.7°. What is the spacing between the planes of gallium atoms responsible for this diffraction, assuming $n = 1$?

21. The distance between layers of atoms parallel to the cube face of a sodium chloride crystal is 2.80 Å. What X-ray wavelength will be diffracted at an angle of 6.25° from this crystal, assuming $n = 1$?

22. Bragg's law states that $n\lambda = 2d\sin\theta$ where $n = 1, 2, 3 \ldots$ are called the first-order reflection, second-order reflection, third-order reflection, etc. If X-rays of wavelength 1.54 Å are diffracted from a NaCl crystal, calculate the glancing angles at which the first-, second-, and third-order reflections are obtained from atomic planes separated by 2.80 Å.

23. Gold metal (Au) crystallizes in an FCC lattice with a unit cell length of 4.08 Å. Calculate the separations of the planes (100), (111), and (311).

24. Cesium bromide (CsBr) crystallizes in a BCC structure and has a density of 4.44 g/cm³. Determine:

    (a) the unit cell edge length, and
    (b) the separations of the planes (002), (211), and (222).

25. Diamond crystallizes in an FCC lattice. There are four unshared atoms in the unit cell, along with eight atoms shared by eight unit cells at the corners and six atoms shared by two unit cells on the faces. The reflection from the (222) planes was observed at a glancing angle of 20.2° using X-rays of wavelength 0.712 Å. Calculate:

(a) the unit cell edge length, and
(b) the density of diamond in g/cm^3.

13

Solution Chemistry

. . . .

13.1 Solution and Solubility

13.1.1 Some definitions

A *solution* is a homogeneous mixture of two or more substances. It is usually made up of a solute and a solvent. Generally,

$$\text{Solute} + \text{Solvent} = \text{Solution}$$

A *solute* is any substance that is dissolved in a solvent. For example, when sugar dissolves in water to give a clear sugar solution, the sugar is the solute, while water is the solvent. Relative to the solvent, a solute is usually present in smaller amount.

A *solvent* is any substance in which a solute dissolves. It is usually the part of the solution that is present in the largest amount.

Two liquids are said to be *miscible* if they form a single phase (homogeneous solution) or dissolve in each other in all proportion. For example, ethanol and water are miscible. If two liquids do not form a single phase (or do not dissolve in each other) in any appreciable amount, they are said to be *immiscible*. For example, water and toluene are immiscible. When mixed, they phase separate into two distinct layers.

Substances that are only slightly soluble in a given solvent are said to be *insoluble*.

An *aqueous* solution is a solution in which water is the solvent.

A *dilute* solution is one that contains a small amount of solute in solution compared to maximum amount that can dissolve at that temperature.

A *concentrated* solution is one that contains a large amount of solute compared to the maximum amount that can dissolve at that temperature.

A *saturated solution* is a solution that is in equilibrium with undissolved solute at a given temperature and pressure:

$$\text{Solute}_{(\text{solid})} \rightleftharpoons \text{Solute}_{(\text{dissolved})}$$

In other words, it is one that contains the *maximum amount of solute that can be dissolved* at that particular temperature.

Chemistry in Quantitative Language: Fundamentals of General Chemistry Calculations. Second Edition.
Christopher O. Oriakhi, Oxford University Press. © Christopher O. Oriakhi 2021.
DOI: 10.1093/oso/9780198867784.003.0013

An *unsaturated* solution is a solution that contains less solute than the maximum amount (saturated solution) possible at the same temperature.

A *supersaturated* solution is a solution that contains more solute than the saturated solution at the same temperature. This type of solution is very unstable. When agitated, or a speck of the solute is added to it, the excess solute will begin to crystallize out rapidly from the solution until the concentration becomes equal to that of the saturated solution.

The *solubility* of a solute in a solvent, at a given temperature, is the maximum amount of solute in grams that can dissolve in 100 g of solvent at that temperature.

13.2 Concentration of Solutions

13.2.1 Percent by mass

The *percent concentration of a solution* may be expressed as *the mass of the solute per 100 mass units of solution*:

$$\%\text{Solute} = \frac{\text{mass of solute}}{\text{mass of solution}} \times 100\%$$

$$= \frac{\text{mass of solute}}{\text{mass of solute} + \text{mass of solvent}} \times 100\%$$

For example, a 10% solution of NaCl contains 10 g of NaCl per 100 g of solution.

Example 13.1

Calculate the percentage of Na_2SO_4 in a solution prepared by dissolving 20 g of Na_2SO_4 in 80 g of H_2O.

Solution

$$\%Na_2SO_4 = \frac{\text{mass of } Na_2SO_4}{\text{mass of } Na_2SO_4 + \text{mass of } H_2O} \times 100$$

$$= \frac{20\,g}{20\,g + 80\,g} \times 100 = 20\%$$

Example 13.2

How many grams of ethanol are dissolved in 60 g of water to make a 40% solution?

Solution

$$\%\text{Ethanol} = \frac{\text{mass of Ethanol}}{\text{mass of Ethanol} + \text{mass of } H_2O} \times 100$$

Let the mass of ethanol be y g.

$$40\% = \frac{y}{y + 60 \text{ g}} \times 100$$

$$0.40 = \frac{y}{y + 60 \text{ g}}$$

$$0.40y + 24 = y$$

$$24 = 0.60y$$

$$y = 40 \text{ g}$$

13.2.2 Parts per million (ppm) and parts per billion (ppb)

For very dilute solutions the concentration may be expressed in parts per thousand (ppt), parts per million (ppm) and parts per billion (ppb) by multiplying with a factor of 10^3, 10^6 and 10^9 respectively.

$$\text{ppt} = \frac{\text{mass of solute (or analyte)}}{\text{mass of solution (or sample)}} \times 10^3$$

$$\text{ppm} = \frac{\text{mass of solute (or analyte)}}{\text{mass of solution (or sample)}} \times 10^6$$

$$\text{ppb} = \frac{\text{mass of solute (or analyte)}}{\text{mass of solution (or sample)}} \times 10^9$$

For example, a 20 ppm glucose solution contains 20 g of glucose per 10^6 g of solution. Also ppm and ppb can simply be expressed as mg of solute per kg (or mg/L) of solution and μg of solute per kg (or μg/L) of solution. Note that mg/kg and mg/L are equivalent only for water where a liter of water weighs 1 kg. Otherwise you will have to factor in the density.

Example 13.3

A 750.0-g sample of drinking water is found to contain 150.0 mg Pb. Calculate the concentration of Pb in parts per million.

Solution

It is important to note that when the mass of the solute is very small, the mass of the solution is essentially the same as the mass of solvent.
   Mass of solute = 150 mg or 0.150 g
   Mass of solution = 750 g

Substitute these values in the expression for ppm:

$$\text{ppm of Pb} = \frac{\text{g of Pb}}{\text{g of solution}} \times 10^6$$

$$\text{ppm of Pb} = \frac{0.150\,\text{g}}{750\,\text{g}} \times 10^6$$

$$= 200\,\text{ppm}$$

Example 13.4

The concentration of Br^- in a water sample taken from a given lake was measured to be 79.5 ppm. Determine the mass of Br^- present in 600.0 mL of the water, which has a density of 1.00 g/mL.

Solution

First you need to determine the mass of the sample using the density of water. Then, using the definition of ppm, you can calculate the mass of solute (Br^-).

$$\text{Mass of sample} = 600.0\,\text{mL} \times \frac{1.0\,\text{g}}{1.0\,\text{mL}} = 600.0\,\text{g}$$

$$\text{ppm Br}^- = \frac{\text{mass of Br}^-}{\text{mass of solution}} \times 10^6$$

$$79.5\,\text{ppm Br}^- = \frac{\text{mass of Br}^-}{600.0\,\text{g}} \times 10^6$$

Rearrange the equation and solve for the mass of Br^-

$$\text{mass of Br}^- = \frac{(79.5\,\text{ppm Br}^-)(600.0\,\text{g})}{1,000,000} = 0.0477\,\text{g} = 47.7\,\text{mg}$$

Example 13.5

Chemical spillage near a farm land resulted in contamination by 1,4-dioxane. What is the concentration of the dioxane in ppb if 80 mg of it was found in 2,500 kg of the soil sample analyzed?

Solution

We have the mass of the solute 1,4-dioxane as 80 mg or 0.080 g. We also know the mass of the mixture is 2,500 kg or 2,500,000 g. To determine the concentration of the 1,4-dioxane in ppb, use the ppb formula.

$$\text{ppb} = \frac{\text{mass of analyte}}{\text{mass of sample}} \times 10^9$$

$$\text{ppb of } 1,4\text{-dioxane} = \frac{0.080 \text{ g}}{2,500,000 \text{ g}} \times 10^9 = 32 \text{ ppb}$$

13.2.3 Percent by volume

Sometimes the concentration of solutions may be expressed as parts by volume. This is achieved by taking the ratio of the volumes multiplied by a multiplication factor.

$$\text{Parts by volume} = \frac{\text{volume of solute}}{\text{volume of solution}} \times \text{multiplication factor}$$

For example, volume percent is expressed as

$$\text{Percent by volume} = \frac{\text{Volume of solute}}{\text{Volume of solution}} \times 100$$

A 10% ethanoic acid solution by volume contains 10 mL of ethanoic acid per 100 mL of solution.

13.2.4 Molarity

Molarity (M), or molar concentration, is defined as *moles of solute per liter of solution*. (Note: some textbooks may use dm^3 in place of liter.)

$$\text{Molarity} = \frac{\text{number of moles of solute}}{\text{number of liters of solution}}$$

This implies that:

$$\text{Number of moles of solute} = \text{Molarity} \times \text{Volume in liters}$$

If the concentration is expressed in grams per liters or dm^3, then molarity may be calculated as follows:

$$\text{Molarity} = \frac{\text{Concentration in grams per liter}}{\text{Formula mass}}$$

$$= \frac{\text{Grams of solute}}{\text{Formula mass of solute} \times \text{volume in liters of solution}}$$

Example 13.6

Calculate the molarity of a solution which contains 4.9 g of H_2SO_4 in 3 liters of solution.

Solution

Method 1
   Use the expression for mass concentration, i.e.,

$$\text{Molarity} = \frac{\text{Mass in grams of solute}}{\text{MW of solute} \times \text{Volume in liters of solution}}$$

MW of $H_2SO_4 = 98$ g/mole

Mass of $H_2SO_4 = 4.9$ g

Volume of solution $= 3$ Liters

$$\text{Molarity} = \frac{4.9 \text{ g}}{98 \text{ g/mole} \times 3 \text{ liters}} = 0.0167 \frac{\text{mol}}{\text{L}} \text{ (or M)}$$

Method 2

Step 1
    Calculate the number of moles of H_2SO_4

$$\text{No. of moles} = \frac{\text{mass of } H_2SO_4}{\text{MW of } H_2SO_4} = \frac{4.9 \text{ g}}{98 \text{ g/mole}} = 0.050 \text{ mole}$$

Step 2
    Calculate the molarity of H_2SO_4 from the definition:
    Molarity $= 0.05$ mole/3 liters $= 0.0167$ M

Example 13.7

Calculate the molarity of each of the following solutions:

(a) 4.90 g of H_2SO_4 per liter of solution
(b) 0.40 g of NaOH in enough water to make 0.50 L of solution
(c) 6.00 g HNO_3 in enough water to make 250 cm^3 of solution

Solution

1. Determine the number of moles of H_2SO_4:

$$\text{Molar mass of } H_2SO_4 = 2 \times 1 + 32 + 4 \times 16 = 98$$

$$(4.90 \text{ g } H_2SO_4)\left(\frac{1 \text{ mole of } H_2SO_4}{98 \text{ g of } H_2SO_4}\right) = 0.0500 \text{ mole}$$

Find the molarity using the definition:

$$\text{Molarity} = \frac{\text{number of moles of solute}}{\text{number of liters of solution}}$$

$$= \frac{0.0500 \text{ mole}}{1.00 \text{ L}} = 0.0500 \text{ M}$$

2. Determine the number of moles of NaOH:

$$(0.40 \text{ g NaOH})\left(\frac{1 \text{ mole of NaOH}}{40 \text{ g of NaOH}}\right) = 0.010 \text{ mol}$$

Find the molarity from the definition:

$$\text{Molarity} = \frac{\text{number of moles of solute}}{\text{number of liters of solution}}$$

$$= \frac{0.010\,\text{mol}}{0.50\,\text{L}} = 0.020\,\text{M}$$

3. Determine the number of moles of HNO_3:

$$(6.0\,\text{g } HNO_3)\left(\frac{1\,\text{mol of } HNO_3}{63.0\,\text{g of } HNO_3}\right) = 0.095\,\text{mol}$$

Find the molarity from the definition:

$$\text{Molarity} = \frac{\text{number of moles of solute}}{\text{number of liters of solution}}$$

$$= \frac{0.095\,\text{mol}}{0.250\,\text{L}} = 0.38\,\text{M}$$

13.2.5 Normality

At times, it may be desirable to express the concentration of a solution as "Normality." The normality (N) of a solution is defined as *the number of equivalent weights (or simply equivalents) of solute per liter of solution.*

$$\text{Normality (N)} = \frac{\text{number of equivalent weights of solute}}{\text{number of liters of solution}} \quad \text{or} \quad \left(\frac{\text{eq}}{\text{L}}\right)$$

Normality may also be expressed as:

$$\text{Normality (N)} = \frac{\text{mass of solute in grams}}{\text{equivalent weight of solute} \times \text{Liters of solution}}$$

The *equivalent weight of an element* is defined as *the atomic weight divided by its valency* (or oxidation number). In acid-base reactions, this definition differs. The equivalent weight of an acid is defined as *the mass of the acid that will yield 1 mol of hydrogen ions* or react with 1 mol of hydroxide ions. It is obtained by dividing the formula mass of the acid by the acidity (i.e., the number of ionizable hydrogen ions). Similarly, the equivalent weight of a base is defined as *the mass of the base that will produce 1 mole of hydroxide ions*, or react with 1 mole of hydrogen ions. It can be obtained by dividing the formula mass by the number of ionizable hydroxide ions per molecule. Table 13.1 shows the equivalent weights of some common acids and bases.

Example 13.8

What is the normality of a solution containing 9.8 g of H_2SO_4 in 2 L of solution? (MW of $H_2SO_4 = 98$ g/mole, equivalent weight (EW) $= 49$ g/equiv.)

Table 13.1   Molar mass and equivalent mass of some acids and bases

Substance	Formula	Molar mass	Equivalent mass	No. of Replaceable H^+ or OH^- ions
Acids				
Nitric acid	HNO_3	63	63	1
Acetic acid	CH_3COOH	60	60	1
Alanine	$CH_3CH(NH_2)COOH$	89	89	1
Oxalic acid	H_2OCCO_2H	90	45	2
Sulfuric acid	H_2SO_4	98	49	2
Phosphoric acid	H_3PO_4	98	32.7	3
Bases				
Sodium hydroxide	NaOH	40	40	1
Ammonia	NH_3	17	17	1
Barium hydroxide	$Ba(OH)_2$	171.4	85.7	2
Aluminum hydroxide	$Al(OH)_3$	78	26	2

Solution

First, calculate the number of equivalents:

$$\text{No. of equivalents} = \frac{\text{mass in grams}}{\text{EW}} = \frac{9.8\ g}{49\ g/equiv} = 0.2\ \text{equiv}$$

Then, calculate the normality:

$$\text{Normality (N)} = \frac{\text{number of equivalent weights of solute}}{\text{number of liters of solution}} \quad \text{or} \quad \left(\frac{eq}{L}\right)$$

$$= \frac{0.2 eq}{2L} = 0.1\ N$$

Example 13.9

Calculate the normality of a solution of phosphoric acid (H_3PO_4) prepared by dissolving 49 g of H_3PO_4 in enough water to make 750 cm^3 of solution.

Solution

First, change the grams of H_3PO_4 to equivalents:

$$\text{Equivalent mass of } H_3PO_4 = \frac{\text{molar mass of } H_3PO_4}{\text{number of replaceable } H^+ \text{ ions}}$$

$$= \frac{98\ g}{3\ eq} = 32.7\ g/eq$$

Next, determine the equivalents of H_3PO_4:

$$\text{Equivalents of } H_3PO_4 = (49.0 \text{ g}) \left(\frac{1 \text{ equivalent}}{32.7 \text{ g}} \right) = 1.50 \text{ equivalents of } H_3PO_4$$

$$\text{Normality} = \frac{\text{number of equivalents}}{\text{liters of solution}}$$

$$= \frac{1.50 \text{ equivalents}}{0.75 \text{L}} = 2.0 \text{ N}$$

13.2.6 Mole fraction

Mole fraction is defined as *the ratio of the number of moles of a given component in a mixture to the total number of moles of all the components* in the mixture. For example, the mole fraction (X) of a component A in a mixture containing components A, B, C, and D is given by

$$X_A = \frac{n_A}{n_{Total}} = \frac{n_A}{n_A + n_B + n_C + n_D}$$

Where n_A, n_B, n_C, and n_D are the number of moles of components A, B, C, and D.

Example 13.10

Calculate the mole fraction of NaOH in a solution made by dissolving 1.0 mole of solute in 1,000 g of water.

Solution

$$X_{NaOH} = \frac{n_{NaOH}}{n_{NaOH} + n_{H_2O}}$$

$$n_{NaOH} = 1.0 \text{ mol}$$

$$n_{H_2O} = \frac{\text{Grams } H_2O}{\text{MW of } H_2O} = \frac{1000 \text{ g}}{18.0 \text{ g/mol}} = 55.5 \text{ mol}$$

$$X_{NaOH} = \frac{1}{1 + 55.5} = 0.018$$

13.2.7 Molality

Molality (m) is defined as *the number of moles of solute per kilogram of solvent*. Thus, a 1-molal solution contains 1 mol of solute in 1 kg of solvent.

$$\text{Molality} = \frac{\text{moles of solute}}{\text{kg of solvent}}$$

But

$$\text{Moles of solute} = \frac{\text{grams of solute}}{\text{MW of solute}}$$

Hence, molality may also be expressed as:

$$\text{Molality} = \frac{\text{mass of solute}}{\text{MW of solute} \times \text{kg of solvent}}$$

Example 13.11

A solution of sodium carbonate (Na_2CO_3) was prepared by dissolving 1.06 g of Na_2CO_3 in 100 g of deionized water. What is the molality of Na_2CO_3 in this solution? (Density of water $= 1.00$ g/mL.)

Solution

$$\text{Molality} = \frac{\text{moles of solute}}{\text{kg of solvent}}$$

$$\text{Moles of } Na_2CO_3 = \frac{1.06 \text{ g}}{106 \text{ g/mol}} = 0.010$$

$$\text{Molality} = \frac{0.010 \text{ mol}}{0.100 \text{ kg}} = 0.10 \ m$$

Example 13.12

What is the molality of a 20% aqueous solution of ethanol (C_2H_5OH)?

Solution

For convenience, assume we have 100 g of solution. Of that, 20%, or 20 g, will be ethanol.

$$\text{Mass of } C_2H_5OH = 20 \text{ g}$$

$$\text{Moles of } C_2H_5OH = \frac{\text{moles of ethanol}}{\text{MW of ethanol}} = \frac{20 \text{ g}}{46 \text{ g/mole}} = 0.435$$

$$\text{Mass of water} = 100\text{g} - 20\text{g} = 80 \text{ g} = 0.080 \text{ kg}$$

$$\text{Molality} = \frac{\text{moles of solute}}{\text{kg of solvent}}$$

$$\text{Molality} = \frac{0.435 \text{ mol}}{0.080 \text{ kg}} = 5.435 \ m$$

13.2.8 Dilute solutions

A dilute solution is prepared by adding solvent to a concentrated solution. When a concentrated solution is diluted, the number of moles of solute remains unchanged. Only the net volume is changed, having added more solvent. From the molarity equation, we can write an expression for the number of moles of solute, i.e.:

$$\text{Molarity (M)} = \frac{\text{number of moles of solute}}{\text{number of liters of solution}} \quad \text{or} \quad \left(\frac{\text{mol}}{\text{L}}\right), \text{and}$$

$$\text{Moles of solute} = \text{Molarity (M)} \times \text{Liters of solution (L)} = \text{constant}$$

Therefore:

$$M_1\left(\frac{\text{mol}}{\text{L}}\right)V_1\,(\text{L}) = M_2\left(\frac{\text{mol}}{\text{L}}\right)V_2\,(\text{L})$$

or simply

$$M_1 \times V_1 = M_2 \times V_2$$

where M_1 and V_1 are the initial molarity and volume (before dilution), and M_2 and V_2 are the final molarity and volume (after dilution).

Example 13.13

Hydrochloric acid is normally purchased at a 12.0 M concentration. Calculate the volume of this stock solution needed to prepare 500 mL of 0.500 M aqueous solution.

Solution

$$M_1 \times V_1 = M_2 \times V_2$$
$$M_1 = 12.0 \text{ M}$$
$$V_1 = ?$$
$$M_2 = 0.500 \text{ M}$$
$$V_2 = 500 \text{ mL}$$
$$V_1 = \frac{M_2 V_2}{M_1}$$
$$V_1 = \frac{0.5 \text{ M} \times 500 \text{ mL}}{12 \text{ M}} = 20.8 \text{ mL}$$

For problems involving mixing, always remember that no solute is lost or gained. The total amount of solute present before mixing is the same as the total amount of solute after mixing. That is:

$$\text{Moles of solute \#1} + \text{moles of solute \#2} = \text{Final moles of solute}$$
$$M_1 V_1 + M_2 V_2 = M_3 V_3$$

Here, M_1, M_2, V_1, and V_2 represent the molar concentration and volume of solutions 1 and 2 before mixing. M_3 and V_3 represent the molar concentration and volume obtained after mixing. Note that $V_3 = V_1 + V_2$.

Example 13.14

50 cm^3 of 0.25 M NaOH solution was mixed with 250 cm^3 of 0.75 M NaOH solution. What is the molarity of the resulting solution?

Solution

To solve this problem, remember that the number of moles of solute ($=$ molarity $\times$ volume) after mixing is the sum of the amounts that were mixed together. Now, substitute the values for V_1, V_2, and V_3 into the equation:

$$M_1 V_1 + M_2 V_2 = M_3 V_3$$

$$M_1 = 0.25 \text{ M}, V_1 = 50 \text{ cm}^3, M_2 = 0.75 \text{ M}, V_2 = 250 \text{ cm}^3, V_3 = 300 \text{ cm}^3$$

$$M_3 = ?$$

$$0.25 \text{ M} \times 50 \text{ cm}^3 + 0.75 \text{ M} \times 250 \text{ cm}^3 = M_3 \times 300 \text{ cm}^3$$

$$M_3 = 0.67 \text{ M}$$

13.3 Solving Solubility Problems

13.3.1 Solubility in grams per 100 g of solvent

1. Determine the amount of solute in the solution.
2. Write an expression for concentration (C) in terms of mass ratio, i.e.,

$$\text{Concentration (C)} = \frac{\text{mass of solute (g)}}{\text{mass of solvent (g)}}$$

3. Using the expression for C as a conversion factor, determine the amount of solute in 100 g of H_2O.
4. Make sure solubility is in proper units.

Example 13.15

A laboratory technician added 35 g of KCl to 50 g of water and heated to 80°C until all the solids were completely dissolved. When the solution was cooled to 55°C, large crystals of KCl began to form. What is the solubility of KCl at 55°C?

Solution

1. Mass of solute = 35 g; mass of solvent = 50 g.

2. Concentration (C) = $\dfrac{\text{mass of solute (g)}}{\text{mass of solvent (g)}} = \dfrac{35 \text{ g KCl}}{50 \text{ g H}_2\text{O}}$

3. Amount of KCl in dissolved in 100g of water at 55°C

$$= 100\text{g of H}_2\text{O} \times \dfrac{35 \text{ g KCl}}{50 \text{ g H}_2\text{O}} = 70 \text{ g KCl}$$

4. The solubility of KCl at 55°C = 70 g per 100 g of H_2O.

Example 13.16

To prepare a saturated solution of sodium carbonate, an undergraduate student added 106 g of the salt to 1,000 g of water at 24°C under constant stirring until no more solute dissolved. Upon filtration, the student recovered 26 g of undissolved Na_2CO_3. Calculate the solubility of Na_2CO_3 in water at 24°C.

Solution

1. Mass of dissolved Na_2CO_3 = Initial mass – Mass of undissolved Na_2CO_3 = 106 g – 26 g = 80 g

2. Mass of solvent = 1,000 g

3. Concentration (C) = $\dfrac{\text{mass of solute (g)}}{\text{mass of solvent (g)}} = \dfrac{80 \text{ g Na}_2\text{CO}_3}{1,000 \text{ g H}_2\text{O}}$

4. Amount of Na_2CO_3 in dissolved in 100 g of water at 24°C

$$= 100 \text{ g of H}_2\text{O} \times \dfrac{80 \text{ g Na}_2\text{CO}_3}{1,000 \text{ g H}_2\text{O}} = 8 \text{ g Na}_2\text{CO}_3$$

5. The solubility of Na_2CO_3 at 24°C = 8 g per 100 g of H_2O.

13.3.2 Solubility in moles per liter of solvent

Solubility can also be expressed in moles per liter of the solvent (or molarity):

$$\text{Solubility in moles per liter} = \dfrac{\text{mass concentration (g/liter)}}{\text{molar mass of solute (g/mole)}}$$

Example 13.17

A saturated solution was made by dissolving 25 g of sodium nitrate, $NaNO_3$, in 75 g of deionized water. Calculate the solubility of $NaNO_3$ in moles/L at 25°C.

Solution

1. Molar mass of $NaNO_3$ = 85 g

2. Calculate solubility in grams per 100 g of solvent

$$\text{Mass of solute} = 25\text{g; mass of solvent} = 75\text{g}$$

$$\text{Solubility in grams per 100 g of solvent} = \frac{25\text{ g}}{75\text{ g}} \times 100\text{ g H}_2\text{O} = 33.33\text{ g}$$

3. Convert solubility in grams per 100 g of solvent to mass concentration in grams per 1,000 g of solvent or g/L:

$$33.3\text{ g per }100\text{ g of H}_2\text{O} = 333\text{ g per }1,000\text{ g of H}_2\text{O or }333\text{ g/L}(1,000\text{ g} \cong 1\text{ L})$$

4. Calculate the solubility in moles per liter (molarity):

$$\text{Solubility in moles per liter} = \frac{\text{mass concentration (g/liter)}}{\text{molar mass of solute (g/mole)}}$$

$$\text{Solubility in moles per liter} = \frac{333.0\text{ g/liter}}{85.0\text{ g/mole}} = 3.92\text{ moles/liter (or M)}$$

Example 13.18

The solubility of lead (II) nitrate, $Pb(NO_3)_2$, is 2.5 M. Calculate the mass of solute in 1,500 g of saturated solution.

Solution

$$\text{Molarity of saturated Pb(NO}_3)_2 = 2.5$$

$$\text{Molar mass of Pb(NO}_3)_2 = 331\text{g/mole}$$

$$\text{Solubility in moles per liter} = \frac{\text{mass concentration (g/L)}}{\text{molar mass of solute (g/mol)}}$$

$$\text{Mass concentration}\left(\frac{\text{g}}{\text{L}}\right) = \text{solubility}\left(\frac{\text{mol}}{\text{L}}\right) \times \text{molar mass}\left(\frac{\text{g}}{\text{mol}}\right)$$

$$= 2.5\left(\frac{\text{mol}}{\text{L}}\right) \times 331\left(\frac{\text{g}}{\text{mol}}\right) = 827.4\text{ g/L}$$

$$= 2.5\left(\frac{\text{moles}}{\text{Liter}}\right) \times 331\left(\frac{\text{g}}{\text{moles}}\right) = 827.4\text{ g/Liter}$$

Thus, 1,000 g of water dissolves 827.5 g of $Pb(NO_3)_2$.
   Recall that:

$$\text{Mass of solute} + \text{Mass of solvent} = \text{Mass of solution}$$

$$\text{Mass of solution} = 1,000\text{ g} + 827.5\text{ g} = 1,827.5\text{ g}$$

1,827.5 g of solution contain 827.5 g of $Pb(NO_3)_2$, so 1,500 g of solution will contain

$$1,500 \text{ g} \times \frac{827.5 \text{ g}}{1,827.5 \text{ g}} = 679.2 \text{ g of } Pb(NO_3)_2$$

Example 13.19

The solubility of potassium chlorate, $KClO_3$, is 25.5 g per 100 g H_2O at 70°C, and 7.15 g per 100 g H_2O at 20°C. What mass of $KClO_3$ will crystallize out of solution if 95 g of the saturated solution at 70°C is cooled to 20°C?

Solution

Mass of solute + Mass of solvent = Mass of solution

At 70°C, we have   25.5 g + 100 g = 125.5 g

At 20°C, we have   7.15 g + 100 g = 107.5 g

Mass of solute precipitated on cooling from 70°C to 20°C = 25.5 − 7.15 = 18.35 g
When cooled from 70°C to 20°C, 125.5 g of saturated solution precipitates 18.35 g.

$$95\text{g of saturated solution will precipitate } 95 \text{ g} \times \frac{18.35 \text{ g}}{125.5\text{g}} = 13.89 \text{ g of } KClO_3.$$

13.4 Effect of Temperature on Solubility

In general, the solubility of a solute in a given solvent varies with temperature. However, it is very difficult to predict the temperature dependence of solubility. For most ionic and molecular solids, e.g. potassium nitrate, calcium chloride, potassium chlorate, and sugar, solubility increases with temperature. For a few substances, e.g. NaCl, temperature has little effect on solubility; but others, like Na_2SO_4 and $CeSO_4$, become less soluble with increasing temperature. The solubility of gases is more predictable. In general, the solubility of a gas in water decreases with increasing temperature.

If the solubility of a substance is plotted as a function of temperature, a graph known as a solubility curve is obtained.

13.5 Solubility Curves

A solubility curve is a graph that represents the change in concentration of a saturated solution with change in temperature. The solubility is plotted on the vertical axis (y-axis) while the temperature is plotted on the horizontal axis (x-axis). From the graph, the solubility of a solute at a particular temperature can be determined. Points on the curve represent saturated solution; points above the curve represent supersaturated solution, while those below represent unsaturated solution.

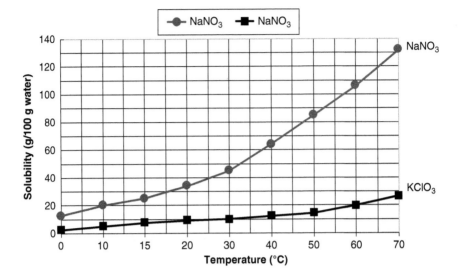

Figure 13.1   The solubility of sodium nitrate and potassium chlorate at various temperatures.

Example 13.20

Figure 13.1 shows the solubility curves of sodium nitrate and potassium chlorate at various temperatures. From the graph answer the following questions:

1. What is the solubility of sodium nitrate and potassium chlorate at 65°C?
2. What mass of sodium nitrate crystals will be deposited when the saturated solution is cooled from 60°C to 20°C? Note: solubility is in unit of g/100g water.
3. At what temperature will a solution containing 38 g of sodium nitrate in 100 g of water become saturated?

Solution

1. A vertical line from 65°C meets the $NaNO_3$ solubility curve at a point corresponding to about 119 g on the solubility axis. Therefore, the solubility of sodium nitrate at 65°C is 119 g/100 g water. Similarly, a vertical line from 65°C meets the $KClO_3$ solubility curve at a point corresponding to about 24 g. Therefore, the solubility of $KClO_3$ at 65°C is 24 g per 100 g water.
2. Draw vertical lines from 60°C and 20°C on the x-axis to the solubility curve of sodium nitrate. Read off the corresponding solubility values from the solubility (y) axis. These are approximately 19.8 g at 60°C and 9.2 g at 20°C. So, when the solution is cooled from 60°C to 20°C, the mass of sodium nitrate deposited will be 19.8 g − 9.2 g = 10.6 g per 100 g of water.
3. A horizontal line from 38 g meets the solubility curve of sodium nitrate at a point which corresponds to about 23°C. Therefore the required temperature is 23°C.

Example 13.21

The solubility of copper (II) sulfate at various temperatures is given in the following table:

Temperature (°C)		0	10	20	30	40	50	60	70	80
Solubility (g/100 g H_2O)	23.1	27.7	31.7	37.9	44.4	53.5	62.1	73.1	83.5

Plot the solubility curve for copper (II) sulfate, and determine from it:

a) The solubility of copper (II) sulfate at 55°C.
b) The temperature at which a solution containing 65 g of salt in 100 g of water becomes saturated.
c) The mass of the salt that will be deposited when the solution is cooled from 60°C to 10°C.
d) Whether a solution containing 50 g of copper (II) sulfate in 100 g water at 20°C is saturated, unsaturated, or supersaturated.

Solution

Figure 13.2 shows the solubility of copper (II) sulfate at various temperatures. The solutions are determined directly from the graph.

a) 58 g/100 g water
b) 63°C
c) 34.4 g
d) Supersaturated

13.6 Effect of Pressure on Solubility

Generally, pressure change has little or no effect on the solubility of solids and liquids. However, the solubility of gases in a liquid is greatly affected by changes in pressure. Henry's Law describes the effect of pressure on the solubility of a gas in quantitative terms. According to Henry's Law,

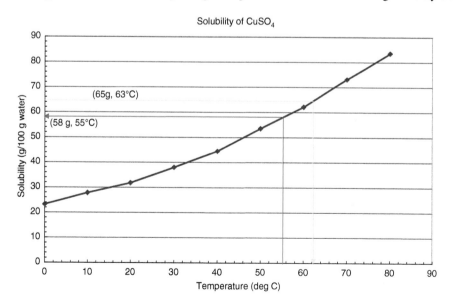

Figure 13.2 The solubility of copper (II) sulfate at various temperatures.

the solubility of a gas in a liquid at a particular temperature is directly proportional to the partial pressure of that gas above the solution:

$$\text{Solubility}(S) = k_H P$$

where k_H is the Henry's Law constant characteristic for each gas at a given temperature (in mol/L-atm), and P is the partial pressure (in atm).

Example 13.22

What is the solubility of CO_2 in a soda drink pressurized to 1.5 atm at 25°C? The Henry's Law constant for CO_2 is 3.2×10^{-2} mol/L-atm at 25°C.

Solution

Substitute the given data directly into the gas solubility formula:

$$\text{Solubility}(S) = k_H P = \left(3.2 \times 10^{-2} \tfrac{\text{mol}}{\text{L-atm}} \right)(1.5\,\text{atm})$$
$$= 4.8 \times 10^{-2}\,\text{M}$$

Example 13.23

The solubility of argon in water at 20°C is 1.5×10^{-3} M when the pressure over the solution is 1.0 atm. What is the solubility at this temperature if the argon pressure is increased to 5 atm?

Solution

Using the initial data, calculate the Henry's Law constant. Since the temperature is kept constant, use the k_H so obtained to calculate the solubility at a partial pressure of 5 atm.

$$S = k_H P$$

$$k_H = \frac{S}{P} = \frac{(1.5 \times 10^{-3}\,\text{mol/L})}{(1\,\text{atm})} = 1.5 \times 10^{-3}\,\text{mol/L} - \text{atm}$$

When $P = 5$ atm, the solubility, S, will be

$$S = k_H P = \left(1.5 \times 10^{-3} \frac{\text{mol}}{\text{L} - \text{atm}} \right)(5\text{atm}) = 7.5 \times 10^{-3}\,\text{M}$$

13.7 Problems

1. A 2.5569-g sample of sodium hydrosulfide (NaSH) is dissolved in 18.75 mL of water. Calculate the percent by mass of NaSH in this solution. Assume the density of water to be 1.00 g/mL.

2. Calculate the mass in grams of a 20.00% by mass $LiNO_3$ solution that contains 6.125 g $LiNO_3$.

3. How many grams of NaCl would be required to make 500 g of 2% NaCl solution?

4. The density of concentrated hydrochloric acid (HCl) is 1.19 g/cm^3. Calculate the mass of anhydrous HCl in 2.5 cm^3 of the concentrated acid containing 37% HCl by mass.

5. A 170-g tube of a popular brand of toothpaste sample contains 225 mg of fluoride as an active ingredient, enough to prevent or reduce the severity of dental carries for all ages. Determine the concentration of the fluoride ion in (a) ppm and (b) ppt.

6. Analysis of a 100.0-μL blood serum sample for glucose content yielded 75.0 μg. What is the concentration of glucose in ppm?

7. Arsenic is toxic at low levels and according to the World Health Organization, long-term exposure to it can cause skin cancer as well as cancer of the bladder and the lungs. Consequently, in 2001 the US Environmental Protection Agency lowered the maximum contamination level (MCL) for arsenic in public-water supplies to 10 μg/L from 50 μg/L. Express the concentration corresponding to the new MCL in (a) ppb and (b) ppm.

8. What is the molarity of a 500.00 ppm Na_2CO_3 solution?

9. What is the molarity of a solution containing:

   (a) 10.4 g of HNO_3 per liter (L) of solution?
   (b) 6.3 g of KOH per 500 cm^3 of solution?
   (c) 0.25 mole of Na_2CO_3 per 250 cm^3 of solution?
   (d) 5.00 moles of glucose ($C_6H_{12}O_6$) per 20 L of solution?

10. Calculate the molarity of a calcium hydrogen carbonate solution containing 25 g of $Ca(HCO_3)_2$ in 850 cm^3 of solution.

11. The density of concentrated sulfuric acid (H_2SO_4) containing 98% by mass of H_2SO_4 is 1.84 g/cm^3. Calculate the volume of the liquid needed to prepare 500 cm^3 of 0.125 M sulfuric acid.

12. Calculate the volume in cm^3 of sulfur trioxide gas at STP that will dissolve in 500 cm^3 of water to produce 0.10 M solution of sulfuric acid.

13. Calculate the normality of a solution of 6.25 g of carbonic acid (H_2CO_3) in 500 cm^3 of solution. (Note: carbonic acid can supply two hydrogen ions per molecule.)

14. What volume of solution is required to provide 22.40 g barium hydroxide ($Ba(OH)_2$) from a 2.5 N $Ba(OH)_2$ solution in reactions that replace the two hydroxide ions?

15. How many grams of solute are required to prepare 2.00 L of 1.00 N solution of phosphoric acid (H_3PO_4)?

16. Calculate the mole fractions of solute and solvent in a solution prepared by dissolving 50 g of sodium nitrate ($NaNO_3$) in 750 g of water (H_2O).

17. Calculate the mole fractions of all the components in a solution containing 0.375 moles of methanol (CH_3OH), 1.275 moles of acetone (CH_3COCH_3), and 1.750 moles of carbon tetrachloride (CCl_4).

18. What is the mole fraction of table salt (sodium chloride, NaCl) in an aqueous solution containing 40.0 g of the salt and 100 g of water?

19. Calculate the molality of a sulfuric acid solution containing 35 g of sulfuric acid in 250 g of water.

20. A chemist wants to prepare a 0.05 m solution of iron (II) sulfate ($FeSO_4$). Calculate the number of grams of $FeSO_4$ that she must add to 300 g of water to achieve this concentration.

21. The density of a 1.50-M aqueous solution of ethyl alcohol C_2H_5OH is $0.857 g/cm^3$ at 25°C. Calculate the molality of the solution.

22. The density of a 6.50-molal acetic acid (CH_3COOH) solution is $1.045 g/cm^3$. What is the molarity of the solution?

23. The density of an aqueous solution of sulfuric acid containing 20% H_2SO_4 by mass is $1.175 g/cm^3$. Find the molar concentration and molality of the solution.

24. A 1.00-g sample of codeine, $C_{18}H_{21}NO_3$, a commonly prescribed painkiller, is dissolved in 0.500 L of water at 25°C. Determine the:

    (a) molarity
    (b) molality, and
    (c) mole fraction of codeine in the solution.

25. What is the molarity of nitric acid solution prepared by mixing 150 cm^3 of 7.50 M HNO_3 with 2 L of water?

26. Calculate the concentration of the resulting solution if 10 mL of a 12.0 M stock solution of hydrochloric acid is diluted to 500 mL.

27. What is the molarity of the solution prepared by mixing 125 cm^3 of 2.50 M KOH solution with 250 cm^3 of 0.75 M KOH solution?

28. How would you prepare 5.00 L of 1.00 M HCl solution from a stock solution that contains 18% HCl by mass and has a density of $1.05 g/cm^3$?

29. A solution is prepared by dissolving 25 g of anhydrous sodium carbonate, Na_2CO_3, in sufficient double-distilled water to produce 500.00 cm^3 of the solution. A 10.00-cm^3 portion of this solution was further diluted to 500.00 cm^3. Calculate the concentration of the final Na_2CO_3 solution in

    (a) moles/L, and
    (b) g/L.

30. A solution was prepared by diluting 100 cm^3 of 0.75 M Li_3PO_4 with water to a final volume of 1.0 L. Determine:

    (a) the molar concentrations of Li^+ and PO_4^{3-} in the original solution, and
    (b) the molar concentrations of Li_3PO_4, Li^+, and PO_4^{3-} in the final solution after dilution.

31. The following data were obtained in the determination of the solubility of copper (II) sulfate at 60°C:

    • Mass of dish = 30.25 g
    • Mass of dish + saturated solution of $CuSO_4$ at 60°C = 62.15 g
    • Mass of dish + solid $CuSO_4$ after heating to dryness = 46.70 g

    Calculate the solubility of $CuSO_4$:

    (a) in grams per 100 g of water at 60°C, and
    (b) in moles per liter. Assume the density of water at this temperature to be 1.00 g/mL.

32. The solubility of NH_4Cl is 70 g per 100 g H_2O at 90°C, and 41.75 g per 100 g H_2O at 30°C. What mass of NH_4Cl will crystallize out if 200 g of the saturated solution is cooled from 90°C to 30°C?

33. A 20-mL saturated solution of $AgNO_3$ was prepared by adding 125.25 g of $AgNO_3$ to deionized water at 25°C. If the solubility of $AgNO_3$ is 9.75 moles/L, determine the mass of the salt remaining undissolved under these conditions.

34. At a particular temperature, the solubility of anhydrous Na_2SO_4 was found to be 4.25 M. Calculate the amount of this salt that dissolves in 20 mL of water. (Density of water = 1.00 g/mL.)

35. The solubility of an organic salt at 22°C is determined to be 45 g per 100 g of water. Calculate the mass of the substance present in 100 g of its saturated solution at this temperature.

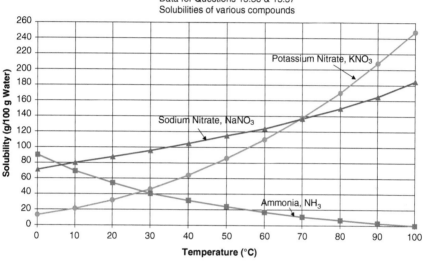

Data for Questions 13.36 & 13.37
Solubilities of various compounds

36. Use the graph above to answer the following questions:

   (a) How many grams of KNO_3 will dissolve in 100 g of water at 64°C?
   (b) How many grams of KNO_3 will dissolve in 50 g of water at 60°C?
   (c) How many grams of $NaNO_3$ will dissolve in 400 g of water at 73°C?
   (d) Describe how you would prepare a saturated solution of $NaNO_3$ at 40°C in 25 g of water.
   (e) Estimate the solubility of ammonia, NH_3, at 15°C.
   (f) Starting with a saturated solution of ammonia in 100 g of water at 15°C, estimate the mass of ammonia that will be liberated as the temperature of the solution is raised to 55°C.

37. Use the solubility curve above to answer the following questions:

   (a) At what temperatures will the solutions containing

      1) 100 g of KNO_3
      2) 70 g of NH_3

      in 100 g of water become saturated?

(b) At what temperature do $NaNO_3$ and KNO_3 have the same solubility?

(c) Calculate the mass of KNO_3 that will be deposited by cooling 300 g of the saturated solution from 80°C to 50°C.

(d) What mass of ammonia will be evolved if 200 g of the saturated solution at 10°C is gently warmed to 30°C?

38. The Henry's law constant for N_2 is 6.8×10^{-4} mol/L.atm at 25°C. What is the solubility of N_2 in water at 25°C if the partial pressure of nitrogen is 0.78 atm?

39. If the solubility of pure oxygen gas in water at 25°C and 1.00 atm is 2.56×10^{-5} M, what is the concentration of oxygen in water at 0.21 atm and 25°C?

40. The Henry's law constant for CO_2 is 3.2×10^{-2} mol/L.atm at 25°C, and the partial pressure of CO_2 in air is 0.0004 atm. What is the concentration of CO_2 in:

(a) A 2.0-L bottle of soda open to the atmosphere at 25°C?

(b) A 2.0-L bottle of soda covered under a CO_2 pressure of 4.5 atm and at 25°C?

14

Volumetric Analysis

. . . .

14.1 Introduction

Volumetric analysis is a chemical analytical procedure based on measurement of volumes of reacting solutions. It uses *titration* to determine the concentration of a solution by carefully measuring the volume of one solution needed to react with another. In this process, a measured volume of a standard solution, the *titrant*, is added from a *burette* to the solution of unknown concentration. When the two substances are present in exact stoichiometric ratio, the reaction is said to have reached the *equivalence* or *stoichiometric point*.

To determine when this occurs, another substance, the *indicator*, is also added to the reaction mixture. This is an organic dye, which changes color when the reaction is complete. This color change is known as the *end point*; ideally, it will coincide with the equivalence point. For various reasons, there is usually some difference between the two, though if the indicator is carefully chosen, the difference will be negligible. Some indicators include Litmus, Methyl Orange, Methyl Red, Phenolphthalein, and Thymol Blue.

A typical titration is based on a reaction of the general type:

$$aA + bB \longrightarrow products$$

where A is the titrant, B the substance titrated, and a:b is the stoichiometric ratio between the two.

14.2 Applications of Titration

Titration can be applied to any of the following chemical reactions:

- Acid-base
- Complexation
- Oxidation-reduction
- Precipitation

Only acid-base titration will be treated here, though the basic principles are the same in all cases.

Chemistry in Quantitative Language: Fundamentals of General Chemistry Calculations. Second Edition. Christopher O. Oriakhi, Oxford University Press. © Christopher O. Oriakhi 2021. DOI: 10.1093/oso/9780198867784.003.0014

14.2.1 Acid-base titrations

Acid-base titration involves measuring the volume of a solution of the acid (or base) that is required to completely react with a known volume of a standardized solution of a base (or acid). The relative amounts of acid and base required to reach the equivalence point depend on their stoichiometric coefficients. Therefore, it is critical to have a balanced equation before attempting calculations based on these reactions.

Next, we define some of the common terms associated with acid-base reactions.

14.2.2 A molar solution

A molar solution contains one mole of the substance per liter of solution. For example, a molar solution of sodium hydroxide contains 40 g ($NaOH = 40$ g/mole) of the solute per liter of solution. As described previously, the concentration of a solution expressed in moles per liter of solution is known as the *molarity* of the solution.

14.2.3 Standard solutions

A standard solution is one of known concentration. For example, titration of an unknown acid might use a carefully prepared 0.1250 M solution of sodium carbonate.

14.2.4 Standardization

Standardization involves determining the concentration of a solution by accurately measuring the volume of that solution required to react with an exactly known amount of a primary standard. The solution thus standardized is called a secondary standard and is used to analyze unknown samples. An example of a primary standard is sodium carbonate solution. It can readily be prepared by dissolving pure anhydrous sodium carbonate in water. A standardized base can be used to determine the concentration of an unknown acid solution, and vice versa.

14.3 Calculations Involving Acid-Base Titration

Generally, calculations based on acid-base reactions involve the concepts of reaction stoichiometry, number of moles, mole ratio, and molarity. The key requirements here are: the volume and concentration (molarity, or normality) of titrant; a balanced chemical equation between the reacting substances; and the mole ratio of the titrant to the substance being titrated.

14.3.1 Calculation involving mass and percentage of substance titrated

Consider the reaction between titrant A and titrated substance B. The balanced equation is

$$aA + bB \longrightarrow cC + dD$$

The mole ratio of B to A is given by

$$R = \frac{b}{a}.$$

Note: R as used here is just a symbol, and not the gas constant. Any other letter such as K, U, or B could have been used.

To calculate the mass and percentage of the substance being titrated, follow these steps:

1. Write a balanced equation for the reaction between A and B.
2. Calculate the moles of titrant A:

$$\text{Moles of } A = \frac{mL_A M_A}{1000}$$

3. Calculate the moles of titrated substance B:

$$\text{Moles of } B = \text{moles of } A \times \frac{b}{a} = \frac{mL_A M_A}{1000} \times \frac{b}{a}$$

4. Calculate the mass of substance (B) titrated:

$$\text{Mass of } B \text{ in grams} = \text{moles of } B \times \text{MW of } B = \frac{mL_A M_A}{1000} \times \frac{b}{a} \times MW_B$$

5. Calculate the percentage of substance B in the sample:

$$\% B = \frac{\text{mass of } B \text{ in grams}}{\text{mass of sample in grams}} \times 100 = \left(\frac{g_B}{g_{\text{sample}}} \right) \times 100$$

Example 14.1

A chemistry student was given a mini-project to analyze for ethanoic acid in a 100-g sample of a marine plant extract. The student conducted a titration and found that it took 20 mL of 0.625 M NaOH solution to neutralize 25 mL of a solution of the marine plant extract. Calculate:

(a) The mass of ethanoic acid in the sample, and
(b) The percentage of ethanoic acid in the sample.

Solution

1. $NaOH + CH_3COOH \longrightarrow CH_3COONa + H_2O$
2. Moles of NaOH $= \dfrac{mL_{NaOH} \times M_{NaOH}}{1000} = \dfrac{20 \text{ mL} \times 0.625 \text{ M}}{1000} = 0.0125$
3. Moles of ethanoic acid $=$ moles of NaOH $\times R$

$$R = 1$$

$$\text{Moles of ethanoic acid} = \frac{mL_{NaOH} \times M_{NaOH}}{1000} = \frac{20 \text{ mL} \times 0.625 \text{ M}}{1000}$$

$$= 0.0125$$

4. Mass of ethanoic acid = moles of ethanoic acid × MW of ethanoic acid

$$= 0.0125 \text{ mole} \times 60.03 \text{ g/mole}$$

$$= 0.7504 \text{ g}$$

5. % Ethanoic acid $= \left(\dfrac{g_{\text{ethanoic acid}}}{g_{\text{sample}}} \right) \times 100 = \dfrac{0.7504}{100} \times 100 = 0.75 \%$

Example 14.2

A research chemist analyzed a 25-g sample of polluted water taken from a lake suspected to contain trichloroethanoic acid (Cl_3CCOOH, with one acidic hydrogen per molecule). Titration of this sample with 0.25 M NaOH required 55 mL to reach the equivalence point. Calculate the mass percent of trichloroethanoic acid in the water sample.

Solution

$$Cl_3CCOOH + NaOH \longrightarrow Cl_3CCOONa + H_2O$$

$$R = 1$$

$$\text{Moles of NaOH} = \frac{mL_{NaOH} \times M_{NaOH}}{1000} = \frac{55 \text{ mL} \times 0.25M}{1000} = 0.0138$$

Moles of Cl_3CCOOH = Moles of NaOH = 0.0138 since $R = 1$

Mass of Cl_3CCOOH = moles of Cl_3CCOOH × MW of Cl_3CCOOH

$$= 0.0138 \text{ mole} \times 163.5 \text{ g/mole}$$

$$= 2.256 \text{ g}$$

$$\text{Mass percent of } Cl_3CCOOH = \frac{2.256 \text{ g}}{25 \text{ g}} \times 100 = 9.03 \%$$

14.3.2 Calculations involving molarity, mass concentration, solubility, and percentage purity from a standardizing titration

In an acid-base titration, the following relationship holds between reacting solutions at the equivalence point:

$$\frac{\text{Volume of Acid} \times \text{Molarity of Acid}}{\text{Volume of Base} \times \text{Molarity of Base}} = \frac{\text{Moles of Acid}}{\text{Moles of Base}}$$

$$\left(\text{Note that mole ratio or stoichiometric ratio} = \frac{\text{Moles of Acid}}{\text{Moles of Base}} \right)$$

This can be shortened to:

$$\frac{M_A V_A}{M_B V_B} = \frac{n_A}{n_B} = \text{mole ratio}$$

where

M_A = molarity of acid
M_B = molarity of base
V_A = volume of acid
V_B = volume of base
n_A = moles of acid
n_B = moles of base

For example, the balanced equation for the reaction between HNO_3 and Na_2CO_3 is:

$$2HNO_3 + Na_2CO_3 \longrightarrow 2NaNO_3 + H_2O + CO_2$$

The equation indicates that 2 mol of the acid react with 1 mol of the base. The relationship at the equivalence point is given by:

$$\frac{M_A V_A}{M_B V_B} = \frac{n_A}{n_B} = \frac{2}{1}$$

Other formulas of interest include:

$$\text{Molarity} = \frac{\text{Concentration in g/liter or dm}^3}{\text{Molar mass}}$$

and

$$\text{Molarity} = \frac{\text{Moles of solute}}{\text{Liters of solution}}$$

Example 14.3

A sample of solid $Mg(OH)_2$ was stirred in water at 50°C for several hours until the solution became saturated. Undissolved solid $Mg(OH)_2$ was filtered out and 150 mL of the resulting solution was titrated with 0.100 M HNO_3. The titration required 75.50 mL of the acid solution for complete neutralization. Calculate:

a) the molarity of $Mg(OH)_2$
b) the mass of $Mg(OH)_2$ contained in 150 mL of the saturated solution, and
c) the solubility of $Mg(OH)_2$ in water at 50°C, in g per 100 mL of water.

Solution

This is an example of an acid-base reaction. First we write the balanced equation for the reaction; then, using the formula method (or the conversion factor method) we can determine the molarity of the base, $Mg(OH)_2$.

$$Mg(OH)_2 (aq) + 2 HNO_3 (aq) \longrightarrow Mg(NO_3)_2 (aq) + 2 H_2O (l)$$

(a) Calculate the molarity of $Mg(OH)_2$:

$$\frac{M_A V_A}{M_B V_B} = \frac{n_A}{n_B}$$

where

$$M_A = \text{molarity of the acid} = 0.100 \text{ M}$$

$$M_B = \text{molarity of the base} = ?$$

$$V_A = \text{volume of acid} = 75.5 \text{ mL}$$

$$V_B = \text{volume of base} = 150 \text{ mL}$$

$$n_A = \text{mol of acid} = 1$$

$$n_B = \text{mol of base} = 2$$

$$M_B = \frac{M_A V_A n_B}{V_B n_A} = \frac{0.100 \text{ M} \times 75.5 \text{ mL} \times 1 \text{ mol Mg(OH)}_2}{150 \text{ mL} \times 2 \text{ mol HNO}_3} = 0.0252 \text{ M}$$

(b) Calculate the moles and hence the mass of $Mg(OH)_2$ contained in 150 mL of the saturated solution:

$$\text{Mol of Mg(OH)}_2 = \left(0.0252 \frac{\text{mol}}{\text{L}}\right)\left(\frac{1 \text{ L}}{1000 \text{ mL}}\right)(150 \text{ mL}) = 3.8 \times 10^{-3} \text{ mol}$$

$$\text{Mass of Mg(OH)}_2 = \left(3.8 \times 10^{-3} \text{ mol}\right)\left(58.3 \frac{\text{g}}{\text{mol}}\right) = 0.22 \text{ g}$$

(c) Now calculate the solubility of $Mg(OH)_2$ in water, in g per 100 mL of the solution at 50°C:

$$\text{Solubility in} \frac{\text{mol}}{\text{L}} = \text{molarity in} \frac{\text{mol}}{\text{L}}$$

Using the molarity of $Mg(OH)_2$ from part (a):

$$\text{Solubility} = \left(\frac{0.0252 \text{ mol Mg(OH)}_2}{\text{L}}\right)\left(\frac{58.3 \text{ g}}{\text{mol Mg(OH)}_2}\right)\left(\frac{1 \text{ L}}{1000 \text{ mL}}\right)$$

$$= 1.5 \times 10^{-3} \text{ g/mL or } 0.15 \text{ g/100 mL}$$

Example 14.4

14.30 g of hydrated sodium carbonate ($Na_2CO_3 \cdot yH_2O$) were dissolved in water and the solution made up to 500 mL. 25 mL of this solution required 20.0 mL of 0.25 M HNO_3 for complete neutralization using methyl orange indicator. Calculate the number of moles of water of crystallization, y, in the $Na_2CO_3 \cdot yH_2O$ sample.

Solution

We will start by determining the molarity of the sodium carbonate solution, as in previous titration problems. From this we can calculate the actual moles of sodium carbonate present, which will tell us the mass. Subtracting this from the mass of the original sample will tell us the mass of water present, which we can use to find the moles of water; then the ratio between these two numbers of moles will give us y.

1. Write a balanced equation for the reaction and calculate the molarity of Na_2CO_3:

$$Na_2CO_3 + 2\,HNO_3 \longrightarrow NaNO_3 + 2\,H_2O + CO_2$$

$$\frac{M_A V_A}{n_A} = \frac{M_B V_B}{n_B} \Rightarrow M_B = \frac{M_A V_A n_B}{V_B n_A} = \frac{0.25\ \text{M} \times 20\ \text{mL} \times 1\ \text{mol}\ Na_2CO_3}{25\ \text{mL} \times 2\ \text{mol}\ HNO_3} = 0.100\ \text{M}$$

2. Calculate the number of moles and hence the mass of Na_2CO_3:

$$\text{Mol}\ Na_2CO_3\ \text{in 500 mL} = \left(0.100\ \frac{\text{mol}}{\text{L}}\right)\left(\frac{1\ \text{L}}{1000\ \text{mL}}\right)(500\ \text{mL}) = 0.050\ \text{mol}$$

$$\text{Mass of anhydrous}\ Na_2CO_3 = (0.050\ \text{mol})\left(106\ \frac{\text{g}}{\text{mol}}\right) = 5.30\ \text{g}$$

$$\text{Mass of water of crystallization} = 14.3 - 5.3 = 9\ \text{g}$$

3. Calculate the moles of water of crystallization. To do this, calculate the ratio of moles H_2O to moles Na_2CO_3:

$$\text{Moles}\ Na_2CO_3 = 5.30\ \text{g} \times \frac{1\ \text{mol}\ Na_2CO_3}{106\ \text{g}} = 0.050\ \text{mol}\ Na_2CO_3$$

$$\text{Moles}\ H_2O = 9.0\ \text{g} \times \frac{1\ \text{mol}\ H_2O}{18\ \text{g}} = 0.500\ \text{mol}\ H_2O$$

$$\text{Then}\ y = \frac{0.500\ \text{mol}\ H_2O}{0.050\ \text{mol}\ Na_2CO_3} = 10,\ \text{so the formula for the hydrate is}$$

$$Na_2CO_3 \cdot 10\ H_2O.$$

Alternatively, we can use the formula method to calculate y.

$$\frac{\text{Mass of anhydrous}\ Na_2CO_3}{\text{Mass of water of crystallization}} = \frac{\text{Molar mass of anhydrous}\ Na_2CO_3}{\text{Molar mass of}\ yH_2O}$$

$$\frac{5.3}{9.0} = \frac{106}{18y}$$

$$y = \frac{106 \times 9}{18 \times 5.3} = 10$$

The formula is $Na_2CO_3 \cdot 10H_2O$.

Example 14.5

8.5 g of potassium hydroxide were dissolved in water and the solution made up to 1 liter. With methyl orange as indicator, 25 mL of this solution neutralized 30.25 mL of a solution containing 5.75 g/L of impure HCl. Calculate:

(a) The concentration of the pure HCl in mol/L
(b) The concentration of the pure acid in g/L
(c) The percentage by mass of pure HCl, and
(d) The percentage of impurity in the HCl.

Solution

(a) First calculate the molarity of the KOH:

$$\text{Molar mass of KOH} = 56.1 \text{ g/mol}$$

$$\text{Molarity of KOH} = \left(8.5 \, \frac{\text{g KOH}}{\text{L}}\right)\left(\frac{1 \text{ mol}}{56.1 \text{ g}}\right) = 0.152 \text{ mol/L} = 0.15 \text{ M}$$

(b) Write a balanced equation for the reaction and calculate the molarity (concentration in mol/L) of the acid:

$$KOH(aq) + HCl(aq) \longrightarrow KCl(aq) + H_2O(l)$$

$$\frac{M_A V_A}{n_A} = \frac{M_B V_B}{n_B} \Rightarrow M_A = \frac{M_B V_B n_A}{V_A n_B} = \frac{0.152 \text{ M} \times 25 \text{ mL} \times 1 \text{ mol HCl}}{30.25 \text{ mL} \times 1 \text{ mol KOH}} = 0.126 \text{ M}$$

(c) Calculate the concentration of the pure acid in grams per liter:

$$\text{Concentration of HCl in g/L} = \left(0.125 \, \frac{\text{mol}}{\text{L}}\right)\left(36.5 \, \frac{\text{g}}{\text{mol}}\right) = 4.57 \text{ g/L}$$

(d) Calculate the percentage of pure HCl:

$$\% \text{ by mass of pure HCl} = \frac{\text{mass of pure HCl}}{\text{mass of impure HCl}} \times 100$$

$$= \frac{4.57 \text{ g/L}}{5.75 \text{ g/L}} \times 100 = 79.5 \, \%$$

(e) Calculate the percentage of impurity in the HCl:

$$\% \text{ impurity} = \frac{\text{mass of impurity}}{\text{mass of impure HCl}} = \frac{(5.75 - 4.57)}{5.75} \times 100 = 20.51 \, \% = 20.5 \, \%$$

Alternatively, we can calculate the % of impurity by subtracting the % of pure HCl from 100, i.e. % impurity = $(100.00 - 79.49) = 20.51 \, \% = 20.5 \, \%$.

14.4 Back Titrations

The direct titration method is only applicable to substances that are soluble in water. Some compounds such as metal carbonates or hydroxides that are insoluble in water can be determined by using a procedure known as *back titration*. Here, an excess of a standard solution of a titrant (say acid A) is deliberately added to a given amount of the substance (say base B) to be determined. Part of acid A is consumed in reacting with base B, but since A is present in excess, some remains unreacted. The excess of acid A is then determined by titrating against a second standard titrant (say base C).

$$\text{Acid } A + \text{Base } B \longrightarrow \text{products} + \text{excess Acid } A$$

$$\text{excess Acid } A + \text{Base } C \longrightarrow \text{products}$$

Thus two standard titrant solutions are employed, both of which are used in the calculations. We know the initial amount of A; the second titration lets us calculate the unreacted excess, which in turn lets us calculate the amount of A consumed in the reaction with B.

Example 14.6

50 mL of 0.500 M HCl were added to 1.125 g of limestone (impure $CaCO_3$). The excess acid required 25 mL of 0.500 M KOH for neutralization. Calculate the percentage purity of the $CaCO_3$.

Solution

1. Write a balanced equation for the neutralization of excess HCl and determine the number of moles of excess HCl.

$$HCl(aq) + KOH(aq) \longrightarrow KCl(aq) + H_2O(l)$$

$$\text{Moles of KOH used to neutralize excess HCl} = \left(\frac{0.500 \text{ mol}}{L}\right)\left(\frac{1 \text{ L}}{1000 \text{ mL}}\right)$$

$$\times (25 \text{ mL}) = 0.0125 \text{ mol.}$$

From the balanced equation, 1 mol KOH = 1 mol HCl

$$\text{mol of excess HCl} = 0.0125 \text{ mol.}$$

2. Determine the moles of HCl that actually reacted with the limestone. This is obtained as the difference between the initial amount of HCl added to the limestone, and the excess HCl.

$$\text{Mol of HCl initially added} = (50 \text{ mL})\left(\frac{0.500 \text{ mol}}{L}\right)\left(\frac{1 \text{ L}}{1000 \text{ mL}}\right)$$

$$= 0.0250 \text{ mol}$$

$$\text{HCl used} = \text{initial HCl added} - \text{excess HCl}$$

$$= 0.0250 \text{ mol} - 0.0125 \text{ mol} = 0.0125 \text{ mol}$$

3. Write a balanced equation for the initial reaction between HCl and $CaCO_3$. From the equation determine the moles of $CaCO_3$ reacted.

$$2 \text{ HCl(aq)} + CaCO_3(s) \longrightarrow CaCl_2(aq) + H_2O(l) + CO_2(g)$$

2 mol of HCl react with 1 mol of $CaCO_3$

$$0.0125 \text{ mol HCl will react with } (0.0125 \text{ mol HCl})\left(\frac{1 \text{ mol } CaCO_3}{2 \text{ mol HCl}}\right) = 0.0063 \text{ mol } CaCO_3$$

Thus mol of $CaCO_3 = 0.0063$ mol.

4. Calculate the percentage purity.

$$\% \text{ purity of CaCO}_3 = \frac{\text{mass of pure CaCO}_3 \text{ in the limestone}}{\text{mass of limestone (impure CaCO}_3)} \times 100$$

$$\text{Mass of pure CaCO}_3 = (0.0063 \text{ mol}) \times \left(100 \frac{g}{\text{mol}}\right) = 0.63 \text{ g}$$

$$\text{Mass of limestone} = 1.125 \text{ g}$$

$$\% \text{ purity of CaCO}_3 = \frac{0.6300}{1.125} \times 100 \% = 56 \%$$

5. Thus the percentage purity of CaCO$_3$ is 56%.

Example 14.7

2.214 g of a divalent metal carbonate (MCO$_3$) were dissolved in 50 mL of 1.0 M HNO$_3$. The resulting solution was made up to 250 mL. Using a suitable indicator, 25 mL of the acid solution required 20 mL of 0.10 M LiOH solution for complete neutralization. Calculate the molecular mass of the metal carbonate and the atomic mass of M.

Solution

1. Write a balanced equation for the neutralization of HNO$_3$ and determine the number of moles of excess HNO$_3$.

$$\text{HNO}_3(\text{aq}) + \text{LiOH}(\text{aq}) \longrightarrow \text{LiOH}(\text{aq}) + \text{H}_2\text{O}(\text{l})$$

$$\text{Concentration of HNO}_3 \text{ solution that reacted with LiOH} = \left(0.100 \frac{\text{mol}}{\text{L}}\right)\left(\frac{20 \text{ mL}}{25 \text{ mL}}\right)$$

$$= 0.080 \frac{\text{mol}}{\text{L}}.$$

The actual volume of HNO$_3$ solution used is 250 mL, so

$$\text{Total moles of excess HNO}_3 = (250 \text{ mL HNO}_3)\left(\frac{0.080 \text{ mol}}{1000 \text{ mL}}\right) = 0.020 \text{ mol}.$$

2. Determine the moles of HNO$_3$ that actually reacted with the metal carbonate. This is obtained as the difference between the initial amount of HNO$_3$ added to the carbonate, and the excess HNO$_3$.

$$\text{mol of HNO}_3 \text{ initially added} = \left(1.00 \frac{\text{mol}}{\text{L}}\right)\left(\frac{1 \text{ L}}{1000 \text{ mL}}\right)(50 \text{ mL})$$

$$= 0.050 \text{ mol}$$

$$\text{HNO}_3 \text{ used} = \text{initial HNO}_3 \text{ added} - \text{excess HNO}_3$$

$$= 0.0500 \text{ mol} - 0.020 \text{ mol} = 0.030 \text{ mol}$$

3. Write a balanced equation for the initial reaction between HNO_3 and MCO_3. From this determine the moles of MCO_3 reacted.

$$2\ HNO_3(aq) + MCO_3(s) \longrightarrow M(NO_3)_2\ (aq) + H_2O(l) + CO_2(g)$$

2 mol of HNO_3 reacted with 1 mol of MCO_3

$$\text{Mol of } MCO_3 = (0.030 \text{ mol } HNO_3) \left(\frac{1 \text{ mol } MCO_3}{2 \text{ mol } HNO_3} \right) = 0.015 \text{ mol}$$

4. Calculate the relative molecular mass of MCO_3 and the relative atomic mass of M.

$$\text{Molar mass of } MCO_3 = \frac{2.214 \text{ g}}{0.0150 \text{ mol}} = 147.6 \text{ g/mol}$$

The relative atomic mass is calculated as follows:

$$M + C + 3 \times O = 147.6$$
$$M + 12 + 3 \times 16 = 147.6$$
$$M = 87.6$$

A quick look at the periodic table suggests $M = Sr$. Thus the metal carbonate is $SrCO3$.

Example 14.8

The active ingredient in a "Milk of Magnesia" tablet is $Mg(OH)_2$. In a titration experiment, a student dissolved one tablet weighing 500.0 mg in 45 mL of 1.0 M HCl solution. The resulting solution was made up to 500 mL with water, and 20.0 mL of this diluted solution needed 15.5 mL of 0.100 M NaOH solution to neutralize the excess acid. Determine the mass of $Mg(OH)_2$ in each tablet. What is the mass percent of $Mg(OH)_2$ in the tablet?

Solution

1. Write a balanced equation for the neutralization of excess HCl and determine the number of moles of excess HCl.

$$HCl(aq) + NaOH(aq) \longrightarrow NaCl(aq) + H_2O(l)$$

$$\text{Concentration of HCl remaining in solution} = \left(0.100 \frac{\text{mol}}{L} \right) \left(\frac{15.5 \text{ mL}}{20 \text{ mL}} \right) = 0.0775 \frac{\text{mol}}{L}.$$

Since the total volume of HCl is 500 mL, the total number of mole of HCl is

$$\text{Mol HCl} = (500 \text{ mL}) \left(\frac{0.0775 \text{ mol}}{1000 \text{ mL}} \right) = 0.0388 \text{ mol}$$

From the balanced equation, 1 mol NaOH = 1 mol HCl

Thus mol of excess HCl = 0.0388 mol.

2. Determine the moles of HCl that actually reacted with the $Mg(OH)_2$. This is obtained as the difference between the initial amount of HCl added to the "Milk of Magnesia", and the excess HCl.

$$\text{Mol of HCl initially added} = (45 \text{ mL}) \left(\frac{1.00 \text{ mol}}{\text{L}} \right) \left(\frac{1 \text{ L}}{1000 \text{ mL}} \right) = 0.0450 \text{ mol}$$

$$\text{HCl used} = \text{initial HCl added} - \text{excess HCl}$$

$$= 0.0450 \text{ mol} - 0.0388 \text{ mol} = 0.00630 \text{ mol}$$

3. Write a balanced equation for the initial reaction between HCl and $Mg(OH)_2$. From the equation determine the moles of $Mg(OH)_2$ reacted.

$$2 \text{ HCl}(aq) + Mg(OH)_2 (s) \longrightarrow MgCl_2 (aq) + 2 \text{ H}_2O(l)$$

2 mol of HCl reacted with 1 mol of $Mg(OH)_2$

$$\text{Mol of } Mg(OH)_2 = (0.0063 \text{ mol HCl}) \left(\frac{1 \text{ mol of } Mg(OH)_2}{2 \text{ mol HCl}} \right) = 0.0031 \text{ mol } Mg(OH)_2$$

4. The mass of $Mg(OH)_2$ per tablet is obtained from the number of moles and the molar mass.

$$\text{Mass of } Mg(OH)_2 \text{ in 1 tablet} = (0.0031 \text{ mol}) \left(58.30 \frac{\text{g}}{\text{moL}} \right) = 0.1822 \text{ g or } 182.2 \text{ mg}$$

$$\text{Mass percent of } Mg(OH)_2 = \frac{\text{mass of } Mg(OH) 2}{\text{mass of 1 tablet}} \times 100 \%$$

$$= \frac{182.2 \text{ mg}}{500 \text{ mg}} \times 100 \%$$

$$= 36.4 \%$$

The active $Mg(OH)_2$ represents 36.4% of the tablet.

14.5 Kjeldahl Nitrogen Determination

The Kjeldahl titration is a volumetric analysis method used for the determination of nitrogen and/or protein in organic samples. The analysis, which employs the principles of back-titration, consists of three parts:

- The sample is digested in boiling H_2SO_4, which contains a catalyst to convert all the organic nitrogen into NH_4^+ ions.
- The solution is made basic to convert the NH_4^+ ions into NH_3, which is distilled into a known excess of an acid such as HCl or HNO_3.
- Excess acid (not consumed by reacting with NH_3) is back-titrated with an alkaline solution such as NaOH to determine the amount of NH_3 in the sample.

The following chemical conversions are involved:

1. Digestion

$$N \text{ (in protein or organic compound)} \longrightarrow NH_4^+$$

2. Distillation of NH_3

$$NH_4^+(aq) + OH^-(aq) \longrightarrow NH_3(g) + H_2O(l)$$

3. Collection of NH_3 in HCl or HNO_3

$$NH_3(g) + H^+(aq) \longrightarrow NH_4^+(aq)$$

4. Back Titration of Unreacted Acid

$$H^+(aq) + OH^-(aq) \longrightarrow H_2O(l)$$

Example 14.9

1.2505 g of an unknown protein was analyzed by the Kjeldahl method. The sample was digested in H_2SO_4 to convert N to $NH_4{}^+$ ions. A concentrated solution of NaOH was added to convert all of the nitrogen in the protein to NH_3. The liberated ammonia was collected into a flask containing 100 mL of 0.125 M HCl. The excess acid required 9.50 mL of 0.0625 M NaOH for complete titration. Calculate the percent of nitrogen in the sample.

Solution

Step 1: Calculate the total moles of HCl in the distillation flask which received the NH_3.

$$\left(0.100L \times 0.125 \ \frac{mol}{L} \right) = 0.0125 \ mol \ HCl$$

Step 2: Calculate the total moles of NaOH required for complete titration of the excess acid.

$$\left(0.0095L \times 0.0625 \ \frac{mol}{L} \right) = 0.000594 \ mol \ NaOH$$

Step 3: Calculate the moles of ammonia produced. This is the difference between the moles of HCl and the moles of NaOH needed to completely neutralize the excess acid.

$$(0.0125 \ mol \ HCl) - (0.000594 \ mol \ NaOH) = 0.01191 \ mol \ NH_3$$

Step 4: Calculate the mass of nitrogen in the sample. From the reaction stoichiometry, one mole of nitrogen in the protein sample produces one mole of ammonia. Therefore, the mass of nitrogen in the sample is equal to the mass of nitrogen produced as ammonia.

$$\text{mass of N} = (0.01191 \ mol \ NH_3) \left(\frac{1 \ mol \ N}{1 \ mol \ NH_3} \right) \left(\frac{14.01 \ g \ N}{1 \ mol \ N} \right) = 0.1668 \ g \ N$$

Step 5: Calculate the percentage by mass of N in the sample:

$$\% \, N = \frac{\text{mass of nitrogen in sample}}{\text{mass of sample analyzed}} \times 100$$

$$= \frac{0.1668}{1.2505} \times 100 = 13.3\%$$

14.6 Problems

1. A 35.25-cm^3 sample of HNO_3 required 27.50 cm^3 of 0.250 M NaOH solution for complete neutralization. What is the molarity of the acid?

2. A 50.00-cm^3 sample of $H_2C_2O_4$ required 45.00 cm^3 of 0.45 M NaOH solution for complete neutralization. What is the molarity of the acid?

3. Calculate the mass of potassium hydroxide required to completely neutralize 250 cm^3 of a solution containing 24.5 g of H_2SO_4 per liter of solution.

4. In a certain titration experiment, 7.50 cm^3 of 0.0075 M benzoic acid, C_6H_5COOH, completely neutralized 12.75 cm^3 of NaOH. What is the concentration of NaOH in g/L?

5. In a standardization experiment, exactly 0.7525 g of primary standard potassium hydrogen phthalate, KHP (molar mass = 204.23), was dissolved in water. If titrating this solution required 22.30 cm^3 of a sodium hydroxide solution, what is the concentration of NaOH

    (a) in moles per liter, and
    (b) in grams per liter?

6. A solution of nitric acid contains 5.85 g of HNO_3 per liter of solution. 22.50 cm^3 of this solution neutralized 25.00 cm^3 of an unknown alkali, MOH, whose concentration is 8.57 g/L. Find:

    (a) the molarity of MOH
    (b) the atomic mass of M, and
    (c) the chemical identity of M, with the aid of the periodic table.

7. 7.125 g of soda ash (impure Na_2CO_3) is dissolved in water and the solution made up to 250 cm^3. Using methyl orange indicator, 25 cm^3 of this solution requires 23.50 cm^3 of 0.50 M HNO_3 for neutralization. Find the percentage purity of the Na_2CO_3.

8. A 7.575-g sample of impure oxalic acid, $H_2C_2O_4$, was dissolved in water and the solution was made up to 500 cm^3. 25.00 cm^3 of this solution required 20.25 cm^3 of 0.400 M KOH for complete reaction. Find the percentage by mass of impurities in the sample.

9. Exactly 10.1225 g of washing soda crystals, $(Na_2CO_3 \cdot xH_2O)$, were dissolved in water and made up to 250 cm^3 of solution. Using methyl orange indicator, 25.00 cm^3 of this required 14.10 cm^3 of 0.50 M HCl solution for complete neutralization. Calculate:

    (a) the percentage of water in the crystal
    (b) the number of molecules of water of crystallization, x, in the sample of Na_2CO_3.

10. 2.075 g of a diprotic organic acid (H_2A) was dissolved in water and made up to 250 cm^3 of solution. By titration, it was found that 20 cm^3 of this solution required 25 cm^3 of 0.100 M NaOH for complete neutralization. Calculate the molar mass of the acid.

11. The nitrogen content in an egg sample was determined by the Kjeldahl procedure. The ammonia produced from 0.4557g of the sample was absorbed in a flask containing 50.00 mL of 0.1087 M HCl. If exactly 15 mL of 0.1125 M NaOH solution is required for the complete neutralization of the excess acid, what is the percentage of nitrogen in the sample?

12. A bioactive product was known to contain 17.25% (wt/wt) of nitrogen. 1.975 mL of a solution of this product was analyzed by the Kjeldahl method. The liberated ammonia was absorbed in 10.00 mL of 0.0525 M HNO_3. What is the concentration of the natural product (in mg/mL) in the original solution if the excess acid required 5.75 mL of 0.0215 M NaOH for complete neutralization?

13. A 5.25-g sample of an ammonium salt was heated to boiling in 70.00 mL of 1.20 M NaOH solution until all ammonia was liberated. What is the mass percent of ammonia in the salt if the excess NaOH required 22.75 mL of 0.500 M HNO_3 for complete neutralization?

14. A 25.0-g sample of impure LiOH was dissolved in water and the solution made up to 1 L in a volumetric flask. 25.0 mL of this solution required 23.15 mL of 0.918 M HNO_3 to reach end-point. What is the percentage purity of LiOH sample used in this experiment?

15

Ideal Solutions and Colligative Properties

15.1 Colligative Properties

Colligative properties of solutions are those that depend only on the number of solute particles (molecules or ions) in the solution, rather than on their chemical or physical properties. The colligative properties that can be measured experimentally include:

- Vapor pressure depression
- Boiling point elevation
- Freezing point depression
- Osmotic pressure.

On the other hand, noncolligative properties depend on the identity of the dissolved species and the solvent. Examples include solubility, surface tension, and viscosity.

15.2 Vapor Pressure and Raoult's Law

15.2.1 Vapor pressure

The addition of a solute to a solvent typically causes the vapor pressure of the solvent (above the resulting solution) to be lower than the vapor pressure above the pure solvent. As the concentration of the solute in the solution changes so does the vapor pressure of the solvent above a solution.

The vapor pressure of a solution of a nonvolatile solute is always lower than that of the pure solvent. For example, an aqueous solution of NaCl has a lower vapor pressure than pure water at the same temperature.

15.2.2 Raoult's law

The addition of solute to a pure solvent depresses the vapor pressure of the solvent. This observation, first made by Raoult, is now commonly known as Raoult's law, which states that *the lowering of vapor pressure of a solution containing nonvolatile solute is proportional to the mole*

Chemistry in Quantitative Language: Fundamentals of General Chemistry Calculations. Second Edition.
Christopher O. Oriakhi, Oxford University Press. © Christopher O. Oriakhi 2021.
DOI: 10.1093/oso/9780198867784.003.0015

fraction of the solute. If P_A^0 represents the vapor pressure of pure solvent, P_A the vapor pressure of solution, and X_B the mole fraction of the solute, then according to Raoult's law

$$\frac{P_A^0 - P_A}{P_A^0} = X_B$$

This equation may be rearranged as follows:

$$\frac{P_A^0}{P_A^0} - \frac{P_A}{P_A^0} = 1 - \frac{P_A}{P_A^0} = X_B$$

$$\frac{P_A}{P_A^0} = 1 - X_B$$

or

$$P_A = P_A^0(1 - X_B)$$

Recall that for a solute B dissolved in a solvent A,

$$X_A + X_B = 1$$

so

$$X_A = 1 - X_B$$

Therefore,

$$P_A = P_A^0 X_A$$

In other words, the law may be stated as *the vapor pressure of a solution is proportional to the mole fraction of the solvent.*

The amount of vapor pressure lowering (ΔP) can be calculated by multiplying the vapor pressure of the pure solvent by the mole fraction of the solute. If we are to calculate the vapor pressure lowering, ΔP, rather than the actual vapor pressure, we can do so using the original form of Raoult's law. Here is a quick derivation:

Since

$$\Delta P_A = P_A^0 - P_A$$

we have

$$\frac{P_A^0 - P_A}{P_A^0} = X_B$$

$$\frac{\Delta P_A}{P_A^0} = X_B$$

$$\Delta P_A = P_A^0 . X_B$$

so that *the change in vapor pressure depends on the mole fraction of the solute.*

Example 15.1

What is the vapor pressure of a solution prepared by dissolving 100 g of glucose ($C_6H_{12}O_6$) in 150 g of water at 25°C? The vapor pressure of pure water at 25°C is 23.8 mmHg.

Solution

First, we need to determine the mole fraction of water. To do this, we need the number of moles of water and glucose.

$$n_{H_2O} = 150 \text{ g H}_2\text{O} \times \frac{1 \text{ mol H}_2\text{O}}{18.0 \text{ g H}_2\text{O}} = 8.33$$

$$n_{C_6H_{12}O_6} = 100 \text{ g C}_6\text{H}_{12}\text{O}_6 \times \frac{1 \text{ mol C}_6\text{H}_{12}\text{O}_6}{180 \text{ g C}_6\text{H}_{12}\text{O}_6} = 0.56$$

The mole fraction of water is

$$X_{H_2O} = \frac{n_{H_2O}}{n_{H_2O} + n_{C_6H_{12}O_6}} = \frac{8.33}{8.33 + 0.55} = 0.938$$

The vapor pressure of the solution is

$$P_{soln.} = P^0_{H_2O} X_{H_2O} = 23.8 \text{ mmHg} \times 0.938 = 22.3 \text{ mmHg}$$

Example 15.2

The vapor pressure of water at 25°C is 23.8 torr. If 2.25 g of a nonvolatile solute dissolved in 25 g of water lowers the vapor pressure by 0.22 torr, what is the molecular weight of the solute?

Solution

Calculate the mole fraction of the solute using the formula:

$$\Delta P_{soln} = 0.22 \text{ torr}, \quad P^0_{H_2O} = 23.8 \text{ torr}.$$

$$\Delta P_{soln} = P^0_{solv} X_{solute}$$

$$X_{solute} = \frac{\Delta P_{soln}}{P^0_{H_2O}} = \frac{0.22 \text{ torr}}{23.8 \text{ torr}} = 0.0092$$

But

$$X_{solute} = \frac{n_{solute}}{n_{solute} + n_{H_2O}}$$

$$n_{H_2O} = 25 \text{ g H}_2\text{O} \times \frac{1 \text{ mole H}_2\text{O}}{18.0 \text{ g H}_2\text{O}} = 1.389 \text{ moles}$$

$$0.0092 = \frac{n_{solute}}{n_{solute} + 1.389}$$

$$(0.0092)(n_{solute} + 1.389) = n_{solute}$$

$$(0.0092)(1.389) = n_{solute} - (0.0092) n_{solute}$$

$$0.9908 \, n_{solute} = 0.0128$$

$$n_{solute} = 0.0129 \text{ mole}$$

$$\text{Molecular weight of solute} = \frac{\text{Mass of solute in grams}}{\text{Number of moles of solute}} = \frac{2.25 \text{ g}}{0.0129 \text{ mole}} = 174.2 \text{ g/mole}$$

Thus, *we can determine the molecular weight of a substance without knowing anything about its composition*; this is one reason colligative properties are so useful.

15.2.3 Ideal solutions with two or more volatile components

A solution may sometimes contain two or more volatile components. If the solution behaves ideally, then we can still apply Raoult's law. For example, if an ideal solution contains two components, A and B, the partial pressures of A and B vapors above the solution are given by Raoult's law:

$$P_A = X_A P_A^0 \text{ and } P_B = X_B P_B^0$$

The total vapor pressure over the solution is obtained as the sum of the partial pressures of each volatile component. Thus:

$$P_{total} = P_A + P_B = X_A P_A^0 + X_B P_B^0$$

Example 15.3

Consider an ideal solution with a mole fraction of benzene, C_6H_6, equal to 0.30, and mole fraction of toluene, C_7H_8, equal to 0.70. If the vapor pressure of pure C_6H_6 is 68 torr and that of pure C_7H_8 is 21 torr at 22°C, what is the total vapor pressure at this temperature?

Solution

Calculate the partial pressure of each component:

$$P_{C_6H_6} = P_{C_6H_6}^0 X_{C_6H_6} = 68 \times 0.30 = 20.4 \text{ torr}$$

$$P_{C_7H_8} = P_{C_7H_8}^0 X_{C_7H_8} = 21 \times 0.70 = 14.7 \text{ torr}$$

And the total pressure:

$$P_{total} = P_{C_6H_6} + P_{C_7H_8} = 20.4 + 14.7 = 35.1 \text{ torr}$$

15.3 Elevation of Boiling Point

Besides lowering the vapor pressure, adding a nonvolatile solute to a solvent raises the boiling point of the resulting solution relative to that of the pure solvent. This increase, the *boiling point elevation*, is given by the equation:

$$\Delta T_b = i k_b m$$

where k_b is the boiling point constant, which depends on the solvent, and m is the molality of solute, and $\Delta T_b = T_{b(soln)} - T_{b(solvent)}$. The factor i, called the van't Hoff factor, is equal to the

Table 15.1   Freezing point depression and boiling point elevation constants

Solvent	Formula	Freezing point (°C)	k_f (°C Kg/mol)	Boiling point (°C)	k_b (°C Kg/mol)
Acetic acid	CH_3COOH	16.6	3.59	118.1	3.08
Benzene	C_6H_6	5.5	5.07	80.1	2.61
Camphor	$C_{10}H_{16}O$	179.8	40	208	5.95
Carbon disulfide	CS_2	−111.5	3.83	46.2	2.34
Carbon tetrachloride	CCl_4	−22.8	30	76.6	5.03
Chloroform	$CHCl_3$	−63.5	4.7	61.2	3.63
Cyclohexane	C_6H_{12}	6.6	20	80.7	2.79
Diethyl ether	$C_4H_{10}O$	−116.2	1.79	34.5	2.02
Ethyl alcohol	C_2H_5OH	−117.3	1.99	78.4	1.22
Naphthalene	$C_{10}H_8$	80.2	6.92	218	6.34
Water	H_2O	0	1.86	100	0.51

number of moles of solute ions per mole of the solute. Ionic compounds (or electrolytes) dissolve to give more than one particle per solute; in general they ionize completely, providing as many particles as there are ions in the compound. Thus, for NaCl, which ionizes to give Na^+ and Cl^-, i is 2; similarly, $Al(NO_3)_3$ ionizes in aqueous solution to give an i value of 4. For nonelectrolytes, that is, substances which do not ionize (such as glucose) i is 1. For these solutes, the boiling point elevation (ΔT_b) is simply

$$\Delta T_b = k_b m$$

For a solution containing a nonvolatile solute, ΔT_b is always positive since the boiling points of such solutions are always higher than those of pure solvents. Table 15.1 shows values of k_b for some common solvents.

Example 15.4

The boiling point of pure carbon tetrachloride, CCl_4 is 76.6°C. If the boiling point of a solution made by dissolving 2.036 g of naphthalene, $C_{10}H_8$, in 25 g of CCl_4 is 79.8°C, calculate the molar boiling point elevation constant. Assume that the solute is nonvolatile.

Solution

First, compute the molality:

$$\text{Moles of } C_{10}H_8 = \frac{\text{grams of } C_{10}H_8}{\text{Molar mass of } C_{10}H_8} = \frac{2.036 \text{ g}}{128 \text{ g/mol}} = 0.0159 \text{ mol}$$

$$\text{Molality}, m, = \frac{\text{moles of } C_{10}H_8}{\text{kg of } CCl_4} = \frac{0.0159 \text{ mol}}{0.025 \text{ kg}} = 0.636 \text{ m}$$

Now, calculate k_b from the expression:

$$\Delta T_b = k_b m$$

$$k_b = \frac{\Delta T_b}{m} = \frac{(79.8 - 76.6)\,^{\circ}\text{C}}{0.636\,\frac{\text{mol}}{\text{kg}}} = \frac{3.2\,^{\circ}\text{C}}{0.636\,\frac{\text{mol}}{\text{kg}}} = 5.03\,^{\circ}\text{C kg/mol}$$

Example 15.5

What is the boiling point of an aqueous solution containing 25 g of the nonelectrolyte urea, $CO(NH_2)_2$, per 500 g of water? k_b for water is 0.51°C kg/mol.

Solution

First, calculate the molality of the urea solution:

$$n_{urea} = \frac{\text{g of urea}}{\text{MW of urea}} = \frac{25\text{ g}}{60\text{ g/mol}} = 0.4167\text{ mol}$$

$$\text{Molality, } m, = \frac{n_{urea}}{\text{kg of urea}} = \frac{0.4167\text{ mol}}{0.5\text{ kg}} = 0.833\text{ m}$$

Next, calculate the boiling point of the solution:

$$\Delta T_b = k_b m$$

$$\Delta T_b = k_b m = \left(0.51\,^{\circ}\text{C}\frac{\text{kg}}{\text{mol}}\right)\left(0.833\frac{\text{mol}}{\text{kg}}\right) = 0.42\,^{\circ}\text{C}$$

But

$$\Delta T_b = T_{b(soln)} - T_{b(solvent)}$$

$$T_{b(soln)} = T_{b(solvent)} + \Delta T_b = 100\,^{\circ}\text{C} + 0.42\,^{\circ}\text{C} = 100.42\,^{\circ}\text{C}$$

Example 15.6

Calculate the boiling point of a solution prepared by dissolving 50 g of $Ca(NO_3)_2$ in 500 g H_2O.

Solution

In this case, the solute is an electrolyte and will ionize completely in water. So, the following expression will be used to calculate the boiling point elevation:

$$\Delta T_b = i k_b m$$

For $Ca(NO_3)_2$, $i = 3$ (1 Ca^{2+}, 2 NO_3^-).

First, calculate moles of $Ca(NO_3)_2$:

$$Mol\ Ca(NO_3)_2 = 50\ g \times \frac{1\ mol\ Ca(NO_3)_2}{102g} = 0.490\ mol\ Ca(NO_3)_2$$

Next, calculate molality of the solution:

$$Molality\ m = \frac{moles\ of\ solute}{kg\ of\ solvent}$$

$$= \frac{0.490\ mol}{0.500\ kg} = 0.980\ m$$

Now, calculate the ΔT_b and $T_{b(soln)}$

$$\Delta T_b = ik_b m = 3 \times \left(0.51\ ^\circ C\ \frac{kg}{mol}\right)\left(0.980\ \frac{mol}{kg}\right) = 1.5\ ^\circ C$$

$$T_{b(soln)} = T_{b(solvent)} + \Delta T_b = 100\ ^\circ C + 1.5\ ^\circ C = 101.5\ ^\circ C$$

15.4 Depression of Freezing Point

The addition of a nonvolatile solute to a solvent lowers the freezing point of the resulting solution below that of the pure solvent. The magnitude of the freezing-point depression (ΔT_f) is proportional to the molality of the solute (i.e. the number of moles of the solute per kg of solvent). In mathematical terms:

$$\Delta T_f = ik_f m$$

The constant k_f is the molar freezing point constant of the solvent. The factors i and m have the same meaning as in the previous section. When dealing with solutes that are nonelectrolytes, the expression simplifies to:

$$\Delta T_f = k_f m$$

Similarly, $\Delta T_f = T_{f(solvent)} - T_{f(soln)}$. For a solution containing a nonvolatile solute, ΔT_f is always positive since the freezing points of such solutions are always lower than those of pure solvents. Values of k_f for some common solvents are given in Table 15.1.

Example 15.7

Calculate the freezing point of a solution made by dissolving 50 g of glucose in 500 mL of water. Assume that the solute is nonvolatile. (The molecular weight of glucose is 180 g/mol.)

Solution

First, calculate the molality:

$$\text{Moles of solute } (C_6H_{12}O_6) = (50 \text{ g } C_6H_{12}O_6)\left(\frac{1 \text{ mol } C_6H_{12}O_6}{180\text{g } C_6H_{12}O_6}\right) = 0.2778 \text{ mol}$$

$$\text{Molality} = \frac{\text{moles of solute}}{\text{kg of solvent}} = \left(\frac{0.2778 \text{ mol } C_6H_{12}O_6}{0.500 \text{ kg } H_2O}\right) = 0.556 \text{ m}$$

Then, calculate the freezing point depression from the expression:

$$\Delta T = k_f m = \left(1.86\,^\circ C\frac{\text{kg}}{\text{mol}}\right)\left(0.56\,\frac{\text{mol}}{\text{kg}}\right) = 1.04\,^\circ C$$

But $\Delta T_f = T_{f(\text{solvent})} - T_{b(\text{soln})}$

i.e. $1.04\,^\circ C = 0.00\,^\circ C - T_{f(\text{soln})}$

$$T_{f(\text{soln})} = (0.00 - 1.04)\,^\circ C = -1.04\,^\circ C$$

Example 15.8

Ethylene glycol, $C_2H_6O_2$, is a nonvolatile nonelectrolyte commonly used in automobile antifreeze. How much ethylene glycol must be added to 1.00 L of water so that the resulting solution will not freeze at $-30\,^\circ C$?

Solution

First, calculate molality from the expression:

$$\Delta T_f = k_f m$$

$$\Delta T = 30\,^\circ C, k_f = 1.86\,^\circ C \frac{\text{kg}}{\text{mol}}$$

$$m = \frac{\Delta T}{k_f} = \frac{30\,^\circ C}{1.86\,^\circ C \frac{\text{kg}}{\text{mol}}} = 16.13 \frac{\text{mol}}{\text{kg}}$$

Since 1 L of water has a mass of 1 kg, the actual amount of solute is

$$1 \text{ kg H2O} \times \frac{16.13 \text{ mol ethylene glycol}}{\text{kg water}} = 16.13 \text{ mol}$$

Now, convert to mass in grams.

$$\text{MW of } C_2H_6O_2 = 62 \text{ g/mol}$$

$$\text{Mass of ethylene glycol used} = (16.13 \text{ mol})\left(\frac{62\text{g}}{\text{mol}}\right) = 1,000 \text{ g}$$

Example 15.9

A solution made by dissolving 1.75 g of an organic compound in 100 g of nitrobenzene freezes at 4.88°C. What is the molecular weight of the compound? The freezing point of pure nitrobenzene is 5.70°C and k_f is 7.00°C/m.

Solution

First, calculate ΔT and m:

$$\Delta T = T_{f(solv)} - T_{f(soln)} = 5.70 - 4.88 = 0.82\,^\circ C$$

$$m = \frac{\Delta T_f}{k_f} = \frac{0.82\,^\circ C}{7.00\,^\circ C/m} = 0.12\ m\ \text{or}\ 0.12\ \frac{\text{moles of solute}}{\text{kg of nitrobenzene}}$$

Now calculate the molar mass:

Moles of solute = molality $(m) \times$ kg of solvent

$$= \left(0.12\ \frac{\text{mol of solute}}{1\ \text{kg nitrobenzene}}\right)(0.10\ \text{kg nitrobenzene}) = 0.012\ \text{mol}$$

$$\text{Molar mass} = \frac{\text{mass in grams}}{\text{no. of moles}} = \frac{1.75\ \text{g}}{0.012\ \text{mol}} = 145.8\ \text{g/mol}$$

15.5 Osmosis and Osmotic Pressure

It is a common observation that when a solution and its pure solvent are separated by a semi-permeable membrane, the pure solvent will diffuse across the membrane to dilute the solution. This process is referred to as *osmosis*. Osmosis may be defined as the movement of solvent molecules from a less-concentrated to a more-concentrated solution across a semipermeable membrane so as to equalize concentrations. Like any reversible process, the diffusion of solvent molecules through the membrane occurs in both directions. But the rate of diffusion from the low-concentration region to the high-concentration region is higher than the reverse rate. The pressure that must be applied to equalize the flow of solvent molecules in both directions across the membrane is known as osmotic pressure, Π. For a dilute solution, the following relationship holds:

$$\Pi = \frac{nRT}{V}$$

where n is the number of moles of solute in V liters of the solution, R is the gas constant (0.0821 L.atm.mol^{-1}K^{-1}), and T is temperature (in K). When the concentration is expressed in terms of molarity, we have

$$\Pi = MRT$$

$$\text{since } M = \frac{n}{V}.$$

For dilute solution, the molarity (M) is approximately equal to molality (m). Thus,

$$\Pi = mRT$$

Also, for an electrolyte solution, the expression for osmotic pressure is modified to contain the van't Hoff factor, i:

$$\Pi = iMRT$$

Example 15.10

The osmotic pressure of a solution prepared by dissolving 38 g of insulin in 600 mL of water at 25°C is 0.62 atm. What is the molar mass of insulin?

Solution

First, calculate the molarity of the solution from the osmotic pressure:

$$\Pi = \left(\frac{n}{V}\right) RT = MRT$$

$$M = \frac{\Pi}{RT} = \frac{0.62 \text{ atm}}{\left(0.0821 \dfrac{\text{L atm}}{\text{mol.K}}\right)(298 \text{ K})} = 0.0253 \frac{\text{mol}}{\text{L}}$$

Now, calculate molar mass from the mass and molarity of insulin solution:

$$\text{Molar mass} = \left(\frac{38 \text{ g}}{600 \text{ mL}}\right)\left(\frac{1,000 \text{ mL}}{1 \text{ L}}\right)\left(\frac{1\text{L}}{0.0253 \text{ mol}}\right) = 2,500 \text{ g/mol}$$

Example 15.11

An intravenous (IV) solution prepared by adding 13.475 g of glucose per 250 mL of solution is isotonic with human blood at 37°C. Determine the osmotic pressure of a patient's blood after receiving this IV.

Solution

First, calculate the amount of glucose (moles) and molarity of the solution from the given information:

$$\text{MW of Glucose} = 180 \text{ g/mol}$$

$$V = 250 \text{ mL} = 0.250 \text{ L}$$

$$\text{Mass of glucose} = 13.475 \text{ g}$$

$$T = 310 \text{ K}$$

$$R = 0.0821 \text{ L.atm.mol}^{-1}\text{K}^{-1}$$

$$\text{mol Glucose } (C_6H_{12}O_6) = 13.475 \text{ g } C_6H_{12}O_6 \times \frac{1 \text{ mol } C_6H_{12}O_6}{180 \text{ g } C_6H_{12}O_6} = 0.075 \text{ mol } C_6H_{12}O_6$$

$$\text{Molarity (M)} = \frac{\text{mol Glucose } (C_6H_{12}O_6)}{\text{L solution}} = \frac{0.075 \text{ mol } C_6H_{12}O_6}{250 \text{ mL} \times \dfrac{1 \text{ L soln}}{1{,}000 \text{ mL}}} = 0.300 \text{ M } C_6H_{12}O_6$$

Next, use the molarity to calculate the osmotic pressure in the expression for Π:

$$\Pi = MRT$$
$$\Pi = (0.300 \text{ mol L}^{-1})\,(0.0821 \text{ L atm mol}^{-1}\text{ K}^{-1})\,(310 \text{ K}) = 7.64 \text{ atm.}$$

Example 15.12

What would be the osmotic pressure at 50°C of an aqueous solution containing 45 g of glucose ($C_6H_{12}O_6$) per 250 mL of solution?

Solution

$$\Pi = \left(\frac{n}{V}\right)RT = \left(\frac{m}{MW.V}\right)RT$$

MW of Glucose $= 180 \text{ g/mol}$

$V = 250 \text{ mL} = 0.250 \text{ L}$

Mass of glucose $= 45 \text{ g}$

$T = 323 \text{ K}$

$R = 0.0821 \text{ L atm.mol}^{-1}\text{K}^{-1}$

$$\Pi = \left(\frac{m}{MW.V}\right)RT$$

$$= \left(\frac{45 \text{ g}}{(180 \text{ g mol}^{-1})\,(0.25L)}\right)(0.0821 \text{ L atm mol}^{-1}\text{ K}^{-1})\,(323 \text{ K}) = 26.5 \text{ atm.}$$

Example 15.13

An aqueous solution of sucrose ($C_{12}O_{22}O_{11}$) has an osmotic pressure of 2.70 atm at 298 K.

(a) How many moles of sucrose were dissolved per liter of solution?
(b) How many grams of sucrose were dissolved per liter of solution?

Solution

$$\Pi = MRT$$

$$\text{Molarity } M = \frac{\Pi}{RT}$$

Molar mass of $C_{12}H_{22}O_{11} = 342$ g/mol

Osmotic pressure $\Pi = 2.70$ atm

$R = 0.0821$ L atm/mol K

$T = 298$ K

(a) The number of moles of sucrose dissolved in one liter of solution can be obtained from the molarity:

$$M = \frac{\Pi}{RT} = \frac{2.70 \text{ atm}}{(0.0821 \text{ L atm K}^{-1}\text{mol}^{-1})(298\text{K})} = 0.110 \text{ mol/liter}$$

(b) Mass in grams per liter = molarity × molar mass

Mass in grams per liter = molarity × molar mass

$$= \left(0.110 \frac{\text{mole}}{\text{liter}}\right)\left(342 \frac{\text{g}}{\text{mole}}\right) = 37.74 \text{ g/L}$$

15.6 Problems

1. Calculate the vapor pressure of water above a solution at 25°C prepared by dissolving 35 g of glucose ($C_6H_{12}O_6$) in 1,000 g of water. The vapor pressure of pure water is 23.8 mmHg.

2. What is the vapor pressure at 25°C of a solution prepared by dissolving 115 g of sucrose (molar mass = 342.3 g/mol) in 700 g of water? The vapor pressure of pure water is 23.8 mmHg and its density is 0.997 g/mL at 25°C.

3. The vapor pressure of water at 25°C is 23.76 mmHg. How many grams of sucrose must be added to 250 g of water to lower the vapor pressure by 1.25 mmHg at 25°C? The molar mass of sucrose is 342.3g/mol.

4. At 45°C the vapor pressure of pure water is 71.88 mmHg and that of a urea solution (NH_2CONH_2) solution is 71.20 mmHg. Calculate the molality of the solution.

5. What is the vapor pressure of a solution prepared by mixing 45.5 g of Na_3PO_4 with 255 g of water at 25°C? The vapor pressure of pure water at 25°C is 23.76 mmHg. (Hint: Note that Na_3PO_4 is a strong electrolyte. When 1 mole of Na_3PO_4 dissolves in water, it produces 3 moles of Na^+ ions and 1 mole of PO_4^{3-} ions. Therefore, the number of solute particles present in the solution will be four times the concentration of the original solute.)

6. A solution is prepared by dissolving 35 g of $NaNO_3$ in 500 g of water. What is the vapor pressure of this solution at 25°C? The vapor pressure of pure water at 25°C is 23.76 mmHg.

7. Consider an ideal solution with the mole fraction of methanol (CH_3OH) equal to 0.23, and mole fraction of ethanol (C_2H_5OH) equal to 0.77. If at 20°C the vapor pressure of pure CH_3OH is 94 mmHg and that of pure C_2H_5OH is 44 mmHg, what is the total vapor pressure at this temperature?

8. Assuming ideal behavior, calculate the mole fraction of toluene in the vapor phase that is in equilibrium with a solution having the following composition at 20°C:

$$X_{benzene}^l = 0.75 \text{ mmHg}, \quad X_{toluene}^l = 0.25 \text{ mmHg}, \quad P_{Total} = 8.20 \text{ mmHg},$$

$$P_{benzene}^o = 9.90 \text{ mmHg}, \quad \text{and} \quad P_{toluene}^o = 2.80 \text{ mmHg}$$

9. The freezing point depression constant, k_f, for camphor is 40°C.kg/mol. Calculate the freezing point of a solution made by dissolving 15.75 g of a nonvolatile solute (of a molecular weight 130) in 500 g of camphor. The normal freezing point of pure camphor is 178.4°C.

10. What is the freezing point of a solution made by dissolving 25 g of sucrose (MW = 342) in 800 g of water? ($k_f = 1.86$°C.kg/mol.)

11. The molecular weight of an unknown compound may be determined from freezing point depression data. A solution containing 30 g of a nonvolatile organic solute dissolved in 750 g of water freezes at −0.576°C. Calculate the molecular weight of the solute.

12. A solution made by dissolving 12.5 g of an unknown compound in 30 g of benzene has a freezing point of −3.375°C. Determine the molar mass of the compound.

13. The boiling point elevation constant, k_b, for nitrobenzene is 5.24°C kg/mol. The normal boiling point of nitrobenzene is 210.8°C. Calculate the boiling point (in °C) of a solution made by dissolving 5.15 g of naphthalene ($C_{10}H_8$, $M_w = 128.2$ g/mol) in 50 g of nitrobenzene.

14. An unknown organic compound extracted from a mango is found to contain 39.6% C, 7.75% H, and 52.7% O. A 1.005-g sample of this compound was dissolved in 100 g of water. The freezing point of this solution was −0.101°C.

    (a) What is the molecular mass of this compound?
    (b) What are its empirical and molecular formulas?

15. Estimate the freezing points of the following aqueous solutions:

    (a) 0.15 m NaCl
    (b) 0.25 m Na_2SO_4, and
    (c) 0.05 m $Al_2(SO_4)_3$.

16. A solution made by dissolving 1.500 g of an unknown compound Z in 5.00 g benzene had a freezing point of 2.07°C. Given that the freezing point of pure benzene is 5.46°C, determine the molecular weight of the unknown compound. The molal freezing point depression constant of benzene is 5.12°C/molal.

17. A 20% (w/w) aqueous solution of an unknown organic compound has a freezing point of −0.07°C. If the empirical formula of the compound is $C_6H_{10}O_5$, determine the molecular formula.

18. The average osmotic pressure of ten seawater samples taken from different ocean sites is approximately 30.15 atm at 25°C. What is the molar concentration of an aqueous solution of hemoglobin that is isotonic with seawater?

19. A solution is prepared by dissolving 15.25 g of a protein fraction in sufficient amount of water and making up to volume in a 1 L flask. What is the molar mass of the protein if the osmotic pressure of the solution measured at 25°C is 76 mmHg?

20. What is the molar concentration of solute particles in human blood if the osmotic pressure is 8.77 atm at 37°C?

21. An aqueous solution contains 4.5 g of bovine insulin per 500 mL of solution at 25°C. What is the osmotic pressure of the solution in mmHg? The molecular weight of the insulin is 5,700 g/mol.

22. What osmotic pressure (in atmospheres) would you expect for each of the following solutions?

(a) 10 g of LiCl in 250 mL of aqueous solution at 25°C
(b) 5.5 g of $CaCl_2$ in 500 mL at 10°C, and
(c) 0.125 M $Fe(NH_4)_2(SO_4)_2$ at 0°C.

16

Chemical Kinetics

. . . .

16.1 Rates of Reaction

Chemical kinetics is the aspect of chemistry that deals with the speed or rate of chemical reactions and the mechanisms by which they occur. The rate of a chemical reaction is a measure of how fast the reaction occurs, and it is defined as the change in the amount or concentration of a reactant or product per unit time. The mechanism of a reaction is the series of steps or processes through which it occurs.

16.2 Measurement of Reaction Rates

Most experimental techniques for measuring reaction rates involve the measurement of the rate of disappearance of a reactant, or the rate of appearance of a product. For a reaction in which the reactant Y is converted to some products,

$$\text{Rate} = \frac{\text{concentration of } Y \text{ at time } t_2 - \text{concentration of } Y \text{ at time } t_1}{t_2 - t_1}$$

$$\text{Rate} = \frac{\Delta [Y]}{\Delta t}$$

Where $[Y]$ indicates the molar concentration of the reactant of interest, and Δ refers to a change in the given amount. The rate for a reactant, by this definition, is a negative number because the amount decreases with time. For a product, it is positive.

Example 16.1

For the reaction

$$C_2H_6(g) \longrightarrow C_2H_4(g) + H_2(g)$$

the initial concentration of C_2H_6 gas is 2.500 mol/L and its concentration after forty-five seconds is 1.175 mol/L. What is the average rate of this reaction?

Chemistry in Quantitative Language: Fundamentals of General Chemistry Calculations. Second Edition.
Christopher O. Oriakhi, Oxford University Press. © Christopher O. Oriakhi 2021.
DOI: 10.1093/oso/9780198867784.003.0016

Solution

Assume that the reaction was initiated at time $t = 0$ and use the expression for average rate:

$$\text{Average rate} = -\frac{\Delta(\text{mol/L of } C_2H_6)}{\Delta t}$$

$$\text{Average rate} = -\frac{(\text{mol/L of } C_2H_6 \text{ at } t = 45 \text{ sec}) - (\text{mol of } C_2H_6 \text{ at } t = 0 \text{ sec})}{45 \text{ sec} - 0 \text{ sec}}$$

$$= -\frac{(1.175 - 2.500) \text{ mol/L}}{(45 - 0) \text{ sec}}$$

$$= 0.0294 \text{ mol/sec}$$

The rate of reaction always changes with time. For example, consider a reaction in which reactant A is converted to product B. The average rate may be derived as:

$$\text{Average rate} = \frac{\text{change in the concentration of B}}{\text{change in time}}$$

$$\text{Average rate} = \frac{\Delta(\text{moles of B})}{\Delta t}$$

Suppose 1.0 mol/L of A undergoes thermal decomposition to form B. No B is initially present. If after thirty minutes 0.65 mol/L of B is produced, the average rate of reaction will be:

$$\text{Average rate} = \frac{\Delta(\text{mol/L of B})}{\Delta t}$$

$$\text{Average rate} = \frac{(\text{mol/L of B at } t = 30 \text{ min}) - (\text{mol/L of B at } t = 0 \text{ min})}{30 \text{ min} - 0 \text{ min}}$$

$$= \frac{0.65 \text{ mol/L} - 0 \text{ mol/L}}{30 \text{ min} - 0 \text{ min}}$$

$$= 0.0217 \text{ mol/min}$$

We can also calculate the average rate of reaction with respect to A if we know how much of A is left after a given interval. The rate expression here will have a negative sign in front of it since the concentration of A will decrease with time. However, the value of the rate must be a positive quantity.

$$\text{Average rate} = -\frac{\Delta(\text{moles of A})}{\Delta t}$$

Example 16.2

The following data were obtained for the conversion (thermal electrocyclization) of 1,3,5-hexatriene to 1,3-cyclohexadiene (see Figure 16.1).

Using the data in the Table 16.1, determine the average rate for this reaction and tabulate your results.

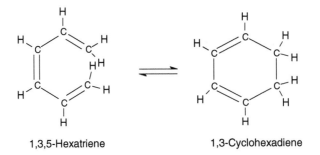

1,3,5-Hexatriene   1,3-Cyclohexadiene

Figure 16.1   The conversion of (thermal electrocyclization) of 1,3,5-hexatriene to 1,3-cyclohexadiene.

Table 16.1   Kinetic data for the conversion of
1,3,5-hexatriene to 1,3-cyclohexadiene

Time (h)	Concentration of 1,3,5-hexatriene (mol)
0	1.700
0.56	1.150
1.39	0.596
2.22	0.319
3.33	0.142
4.17	0.079

Solution

Calculate the average rate by taking two adjacent time intervals and their corresponding
concentrations. For example:

For the time interval 0 to 0.56 hr,

$$\text{Average rate} = -\frac{\Delta[1,3,5-\text{hexatriene}]}{\Delta t}$$

$$\text{Average rate} = -\frac{(1.150 - 1.700)\ \text{mol}}{(0.56 - 0)\ \text{hr}} = 0.982\ \text{mole/hr}$$

For the time interval 0.56 to 1.39 hr,

Table 16.2 Rate data for the conversion of 1,3,5-hexatriene to 1,3-cyclohexadiene

Time (h)	Concentration of 1,3,5-hexatriene (mol)	Average rate (mol/h)
0	1.700	–
0.56	1.150	0.982
1.39	0.596	0.668
2.22	0.319	0.334
3.33	0.142	0.160
4.17	0.079	0.075

$$\text{Average rate} = -\frac{(0.596 - 1.150) \text{ mol}}{(1.39 - 0.596) \text{ hr}} = 0.668 \text{ mole/hr}$$

For the time interval 1.39 to 2.22 hrs,

$$\text{Average rate} = -\frac{(0.319 - 0.596) \text{ mol}}{(2.22 - 1.39) \text{ hr}} = 0.334 \text{ mole/hr}$$

Complete the remaining calculations. The results are shown in Table 16.2.

16.2.1 Instantaneous rate

The value of the rate at a particular time is known as the instantaneous rate and it is different from the average rate. Its value can be obtained from the plot of concentration (mol/L) vs time (s) as the slope of a line tangent to the curve at the desired point.

Consider the following kinetic data for the decomposition of N_2O_5 to gaseous NO_2 and O_2 at 40°C (Table 16.3).

A plot of $[N_2O_5]$ vs time is shown in Figure 16.2. From this curve, the instantaneous rate of reaction at any time t can be obtained as the slope of the tangent to the curve. This corresponds to the value of $\Delta[N_2O_5]/\Delta t$ for the tangent at a given instant. The instantaneous rate at the beginning of the reaction ($t = 0$) is known as the *initial rate*. For this reaction, the initial rate is calculated as:

$$\text{Initial rate} = -\text{slope of tangent line}$$

$$= -\frac{\Delta[N_2O_5]}{\Delta t} = -\frac{(0 - 1.420) \text{ M}}{(200 - 0) \text{ min}} = 7.10 \times 10^{-3} \text{ M}$$

The instantaneous rate at 420 min is calculated in a similar way:

$$\text{Instantaneous rate} = -\text{slope of tangent line}$$

$$= -\frac{\Delta[N_2O_5]}{\Delta t} = -\frac{(0.100 - 0.260) \text{ M}}{(490 - 360) \text{ min}} = 1.23 \times 10^{-3} \text{ M}$$

Table 16.3   Kinetic data for the decomposition of N_2O_5 to gaseous NO_2 and O_2 at 40°C

Time (min)	Concentration of N_2O_5 (mol/L)
0	1.42
60	1.02
120	0.78
180	0.61
240	0.45
300	0.33
360	0.25
420	0.18
480	0.14
540	0.1
600	0.08
660	0.06
720	0.05

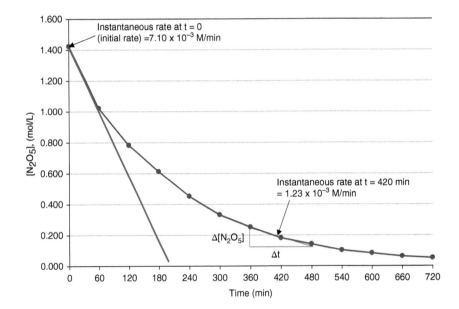

Figure 16.2   Concentration as a function of time for the decomposition of N_2O_5 at 40°C.

16.3 Reaction Rates and Stoichiometry

For the reaction

$$2 N_2O_5 (g) \longrightarrow 4 NO(g) + O_2 (g)$$

we can relate the rate of decomposition of N_2O_5 to the rates of formation of NO and O_2. The stoichiometry of the equation indicates that 2 moles of N_2O_5 produce 4 moles of NO and 1 mole of O_2. Therefore, the rate of decomposition of N_2O_5 is half the rate of formation of NO and twice the rate of formation of O_2.

$$\frac{\Delta [NO]}{\Delta t} = -2 \frac{\Delta [N_2O_5]}{\Delta t} \quad \text{and} \quad \frac{\Delta [O_2]}{\Delta t} = -\frac{1}{2} \frac{\Delta [N_2O_5]}{\Delta t}$$

Thus, if the rate of decomposition of N_2O_5 is known, the rates of formation NO and O_2 can be deduced. Consider the general reaction

$$aA + bB \longrightarrow cC + dD$$

Using the coefficients in the balanced equation, the rate is given by

$$\text{Rate} = -\frac{1}{a} \frac{\Delta [A]}{\Delta t} = -\frac{1}{b} \frac{\Delta [B]}{\Delta t} = \frac{1}{c} \frac{\Delta [C]}{\Delta t} = \frac{1}{d} \frac{\Delta [D]}{\Delta t}$$

and

$$\frac{\Delta [A]}{\Delta t} = \frac{a}{b} \frac{\Delta [B]}{\Delta t} = -\frac{a}{c} \frac{\Delta [C]}{\Delta t} = -\frac{a}{d} \frac{\Delta [D]}{\Delta t}$$

Example 16.3

Consider the reaction

$$2 NO(g) + O_2 (g) \longrightarrow NO_2 (g)$$

If the rate of formation of NO_2 within the reaction vessel is 2.5×10^{-5} M/s, what are the rates of disappearance of NO and O_2?

Solution

First, using the coefficients in the balanced equation, write the rate expression involving reactants and products.

$$\text{Rate} = -\frac{1}{2} \frac{\Delta [NO]}{\Delta t} = -\frac{\Delta [O_2]}{\Delta t} = \frac{\Delta [NO_2]}{\Delta t}$$

Then, use this expression to calculate the other rates.

$$-\frac{\Delta [NO]}{\Delta t} = 2 \frac{\Delta [NO_2]}{\Delta t} = 2(2.5 \times 10^{-5} \text{ M/s}) = 5.0 \times 10^{-5} \text{ M/s}$$

$$-\frac{\Delta [O_2]}{\Delta t} = \frac{\Delta [NO_2]}{\Delta t} = 2.5 \times 10^{-5} \text{ M/s}$$

16.4 Collision Theory of Reaction Rates

Collision theory stipulates the conditions necessary for a reaction to occur:

1. Reactant atoms, ions, or molecules must collide with each other for a reaction to occur.
2. For the collision to result in a reaction, the colliding particles must generate a minimum amount of total kinetic energy. The minimum energy necessary to cause colliding atoms, ions, or molecules of reactants to undergo specific reaction is known as the energy of activation, E_a.
3. The reactant particles must have an appropriate orientation upon collision for the reaction to occur.

Any conditions that increase the fraction of successful collisions will increase the rate of reaction.

16.4.1 Factors affecting reaction rates

1. *The nature of the reactants:* The type of reaction and the nature of bonds to be broken between reactants can affect the rate of reaction. Generally, ionic reactions are faster because they simply involve the formation of new chemical bonds by the union of oppositely charged ions. Reactions involving covalent bonds are slower since they involve the breaking of covalent bonds in reacting molecules. Also, gases react faster than liquids and liquids faster than solids. This follows the order of their molecular velocities.
2. *Concentration of the reactants:* An increase in the concentration of reactants will increase the rate of collision due to effective increase in the number of particles of reactants per unit volume. This results in an increase in the rate of reaction.
3. *Temperature:* An increase in temperature causes colliding molecules to have a higher velocity, and hence kinetic energy greater than or equal to the energy of activation. This increases the rate of reaction.
4. *Light:* The rates of many reactions are greatly increased by the presence of light of suitable wavelengths. Examples include photosynthesis in plants, certain halogenation reactions, and other organic photochemical reactions.
5. *Pressure:* A change in pressure is only applicable to reactions involving gases. It does not affect reactions involving only solids or liquids due to their low compressibility. An increase in total pressure of gaseous reactants will increase the rate of collision, and hence the rate of reaction.
6. *Surface Area:* The rate of a reaction is affected by the nature of the reactants. A solid will react faster when in powdered form than when in large pieces or lumps. Grinding the solid lumps into smaller particles will increase the surface area, thereby increasing the collision frequency and thus the rate of reaction.
7. *Catalysts:* A catalyst is a substance that is added to alter the rate of a reaction but remains chemically unchanged after the reaction is completed. Catalysts increase the rates of reactions by providing alternative reaction pathways that have a lower energy of activation relative to the uncatalyzed pathway (Figure 16.3).

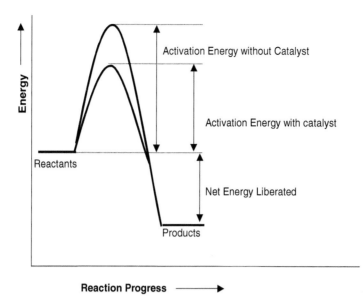

Reactants

Activation Energy without Catalyst

Activation Energy with catalyst

Net Energy Liberated

Products

Reaction Progress ⟶

Figure 16.3   Potential energy profile showing the activation energies for the catalyzed and uncatalyzed reactions.

16.5 Rate Laws and the Order of Reactions

The rates of many chemical reactions can be related to the concentrations of individual reactants. To express reaction rates in mathematical terms, the *Law of Mass Action* is used. According to this law, *the rate of a chemical reaction is directly proportional to the product of the concentration of the reactants*. For a one-step reaction of the general form $aA + bB \longrightarrow$ products, the law of mass action can be written as

$$\text{Rate} \propto [A]^x[B]^y$$

The rate equation may therefore be written as:

$$\text{Rate} = k[A]^x[B]^y$$

The proportionality constant k is called the rate constant (or specific rate constant) for the reaction; the [] represent the molar concentration of reactants A and B. The numerical values of the exponents x and y must be determined experimentally and are not necessarily related to the coefficients a, b.

The *order of the reaction* is governed by the exponents (x and y) to which the concentrations of the reactants are raised in the rate equation. Reaction order can be expressed either as order with respect to the individual reactant, or as the overall order of the reaction ($x + y$). If x is 1, the reaction is first-order with respect to A. If y is 1, the reaction is first-order with respect to B. If x is 1 and y is 2, the reaction is first order with respect to A, second order with respect to B, and third order overall.

16.6 Experimental Determination of Rate Law Using Initial Rates

The values of the exponents in a rate law can be determined by a series of experiments in which the initial rate of reaction is measured as a function of different initial concentrations. For a reaction that is zero order with respect to a particular reactant, a change in concentration will not affect the rate as long as the concentration of the reactant is not zero. This is because a zero-order reaction is concentration-independent. If the reaction is first order in a reactant, doubling the concentration of the reactant will double the reaction rate. Similarly, if a reaction is second order in a given reactant, doubling the concentration causes the rate to increase by a factor of $4\,(2^2)$. Let us illustrate this with an example.

Consider the reaction

$$2\,NO\,(g) + O_2\,(g) \longrightarrow NO_2\,(g)$$

The initial concentration and rate data are given in Table 16.4.

The general rate law for the reaction is

$$Rate = k[NO]^m[O_2]^n$$

Let us see how we can calculate the values of m and n. To determine the rate law, we need to evaluate the value of m from two different rates when $[O_2]$ is constant, and then to determine the value of n when $[NO]$ is constant. From the data in the table, we observe that as the concentration of NO is doubled from 0.03 M to 0.06 M, (comparing experiments 1 and 2 where $[O_2]$ is constant) the rate is increased four times (2^2) from 0.096 M/s to 0.384 M/s. This suggests the reaction rate is second order with respect to NO. On the other hand, we also observe that as the concentration of O_2 is doubled from 0.03 M to 0.06 M, (comparing experiments 1 and 3 where $[NO]$ is constant) the rate doubles, suggesting the reaction is first order with respect to O_2. The rate equation is therefore:

$$Rate = k[NO]^2[O_2]^1$$

An alternative method for finding m is to take the ratio of the initial rates for two different initial concentrations of NO that have the same initial concentrations of O_2.

$$\frac{Rate_2}{Rate_1} = \frac{k[NO]^m_{exp2}[O_2]^n_{exp2}}{k[NO]^m_{exp1}[O_2]^n_{exp1}}$$

$$\frac{0.384}{0.096} = \frac{k(0.06)^m(0.03)^n}{k(0.03)^m(0.03)^n}$$

$$4 = 2^m \implies 2^2 = 2^m \text{ and } m = 2$$

Table 16.4   Initial-rate data for the formation of NO_2

Experiment	Initial [NO], M	Initial [O₂], M	Initial rate of formation of NO₂, M/s
1	0.03	0.03	0.096
2	0.06	0.03	0.384
3	0.03	0.06	0.192
4	0.06	0.06	0.768

We can find the value of n in a similar way:

$$\frac{\text{Rate}_3}{\text{Rate}_1} = \frac{k[NO]_{exp3}^m [O_2]_{exp3}^n}{k[NO]_{exp1}^m [O_2]_{exp1}^n}$$

$$\frac{0.196}{0.098} = \frac{k(0.03)^m (0.06)^n}{k(0.03)^m (0.03)^n}$$

$$2 = 2^n \implies 2^1 = 2^n \text{ and } n = 1$$

Hence the rate for this reaction is proportional to the product of $[NO]^2$ and $[O_2]$ or

$$\text{Rate} = k[NO]^2[O_2]^1$$

Example 16.4

Under certain experimental conditions, the rate equation for the formation of nitrosyl bromide

$$2\,NO + Br_2 \longrightarrow 2\,NOBr$$

is

$$R = k[NO]^2[Br_2]$$

(a) What is the order of the reaction?
(b) Given that the rate of formation of NOBr is 4.24×10^{-5} M/s, calculate the rate constant for the reaction if $[NO] = 0.035$ M and $[Br_2] = 0.070$ M.

Solution

(a) Use the rate equation given. The overall order is the sum of the orders for each reaction. For this reaction $m + n = 3$. Therefore, the reaction is third order: second order in NO, and first order in Br_2.
(b) The value of k can be calculated by substituting the given data into the rate equation:

$$k = \frac{\text{rate}}{[NO]^2[Br_2]} = \frac{4.24 \times 10^{-5}\ \text{mol/L.sec}}{(0.035\ \text{mol/L})(0.070\ \text{mol/L})} = 0.0173\ \frac{L}{\text{mol.sec}}$$

Example 16.5

The rate law for the reduction of bromate ions (BrO_3^-) by bromide ions in acidic solution,

$$BrO_3^-\,(aq) + 5\,Br^-\,(aq) + 6\,H^+\,(aq) \longrightarrow 3\,Br_2\,(aq) + 3\,H_2O\,(l)$$

has the general form

$$\text{Rate} = k[BrO_3^-]^x[Br^-]^y[H^+]^z$$

Table 16.5 gives the results of experiments conducted at four different initial concentrations.
From these data, determine the

Table 16.5   Initial-rate data for the reduction of bromate ions ($BrO3^-$) by bromide ions in acidic solution

Experiment	Initial $[BrO_3^-]_0$, (M)	Initial $[Br^-]_0$, (M)	Initial $[H^+]_0$ (M)	Initial rate, mol (L − s)
1	0.1	0.1	0.1	0.00095
2	0.2	0.1	0.1	0.0019
3	0.2	0.2	0.1	0.0038
4	0.1	0.1	0.2	0.0038

(a) reaction order with respect to each reactant
(b) overall reaction order
(c) rate constant

Solution

The rate expression is

$$\text{Rate} = k\left[BrO_3^-\right]^x\left[Br^-\right]^y\left[H^+\right]^z$$

We can determine the values of x, y, and z by comparing the relevant experimental data. We can see from experiments 1 and 2 that the rate doubles as the concentration of BrO_3^- is increased from 0.10 M to 0.20 M while the other concentrations are held constant. Thus, the reaction is first order with respect to BrO_3^- ($x = 1$). To determine the order with respect to Br^-, consider experiments 2 and 3. When the concentration of Br^- is doubled from 0.10 M to 0.20 M, the reaction rate also doubles, indicating the reaction is again first order with respect to Br^- ($y = 1$). Finally, when the concentration of H^+ is increased from 0.10 M to 0.20 M (experiments 1 and 4), the reaction rate quadruples ($\times 2^2$). Therefore, the reaction is second order in H^+ ($z = 2$). The complete rate equation is

$$\text{Rate} = k\left[BrO_3^-\right]\left[Br^-\right]\left[H^+\right]^2$$

(a) The reaction is first order in BrO_3^-, first order in Br^-, and second order in H^+
(b) The reaction is fourth order overall, i.e. $x + y + z = 4$
(c) Having established the rate law, the value of k can be obtained from the results of any of the four experiments. Let us use the data for experiment 1.

$$\text{Rate} = k\left[BrO_3^-\right]\left[Br^-\right]\left[H^+\right]^2$$

$$k = \frac{\text{rate}_{\text{exp 1}}}{\left[BrO_3^-\right]_{\text{exp 1}}\left[Br^-\right]_{\text{exp 1}}\left[H^+\right]^2_{\text{exp 1}}}$$

$$= \frac{\left(9.5 \times 10^{-4}\ \text{mol}/(\text{L.sec})\right)}{(0.10\ \text{mol/L})(0.10\ \text{mol/L})(0.10\ \text{mol/L})^2} = 9.5\frac{\text{L}^3}{\text{mol}^3\ \text{sec}}$$

16.6.1 Alternate method

We can solve for the values of x, y, and z algebraically by comparing the rates from the four experiments. Substitute the appropriate initial concentrations and rates into the general rate equation.

16.6.2 Determining the value of x

Use the results from experiments 1 and 2, in which $\left[\text{Br}^-\right]$ and $\left[\text{H}^+\right]$ are constant and only $\left[\text{BrO}_3^-\right]$ changes. Substitute the values of the initial concentrations and rates into the general rate equation and compare the results.

$$\frac{\text{rate}_2}{\text{rate}_1} = \frac{k\left[\text{BrO}_3^-\right]_2^x\left[\text{Br}^-\right]_2^y\left[\text{H}^+\right]_2^z}{k\left[\text{BrO}_3^-\right]_1^x\left[\text{Br}^-\right]_1^y\left[\text{H}^+\right]_1^z}$$

$$\frac{1.9\times10^{-3}}{9.8\times10^{-4}} = \frac{k(0.20)^x\cancel{(0.10)^y}\cancel{(0.10)^z}}{k(0.10)^x\cancel{(0.10)^y}\cancel{(0.10)^z}}$$

$$2^1 = 2^x, \quad \Rightarrow x = 1$$

16.6.3 Determining the value of y

Use the results from experiments 2 and 3, in which $\left[\text{BrO}_3^-\right]$ and $\left[\text{H}^+\right]$ are constant and only $\left[\text{Br}^-\right]$ changes.

$$\frac{\text{rate}_3}{\text{rate}_2} = \frac{k\left[\text{BrO}_3^-\right]_3^x\left[\text{Br}^-\right]_3^y\left[\text{H}^+\right]_3^z}{k\left[\text{BrO}_3^-\right]_2^x\left[\text{Br}^-\right]_2^y\left[\text{H}^+\right]_2^z}$$

$$\frac{3.8\times10^{-3}}{1.9\times10^{-3}} = \frac{k\cancel{(0.20)^x}(0.20)^y\cancel{(0.10)^z}}{k\cancel{(0.20)^x}(0.10)^y\cancel{(0.10)^z}}$$

$$2^1 = 2^y, \quad \Rightarrow y = 1$$

16.6.4 Determining the value of z

Use the results from experiment 1 and 4 in which $\left[\text{BrO}_3^-\right]$ and $\left[\text{Br}^-\right]$ are constant and only $\left[\text{H}^+\right]$ changes.

$$\frac{\text{rate}_4}{\text{rate}_1} = \frac{k\left[\text{BrO}_3^-\right]_4^x\left[\text{Br}^-\right]_4^y\left[\text{H}^+\right]_4^z}{k\left[\text{BrO}_3^-\right]_1^x\left[\text{Br}^-\right]_1^y\left[\text{H}^+\right]_1^z}$$

$$\frac{3.8\times10^{-3}}{9.5\times10^{-4}} = \frac{k\cancel{(0.10)^x}\cancel{(0.10)^y}(0.20)^z}{k\cancel{(0.10)^x}\cancel{(0.10)^y}(0.10)^z}$$

$$4 = 2^2 = 2^z, \quad \Rightarrow z = 2$$

Thus, $x = 1$, $y = 1$, and $z = 2$.

(a) The reaction is first order in $[BrO_3^-]$ and $[Br^-]$ and second order in $[H^+]$.
(b) Overall, the reaction is fourth order $(x + y + z = 4)$.
(c) Here let us substitute the results of the second experiment into the rate law:

$$\text{rate} = k[BrO_3^-][Br^-][H^+]^2$$

$$k = \frac{\text{rate}_{\exp 2}}{[BrO_3^-]_{\exp 2}[Br^-]_{\exp 2}[H^+]^2_{\exp 2}}$$

$$= \frac{(1.9 \times 10^{-3} \text{ mol}/(\text{L.sec}))}{(0.20 \text{ mol/L})(0.10 \text{ mol/L})(0.10 \text{ mol/L})^2} = 9.5\frac{L^3}{\text{mol}^3 \text{ sec}}$$

16.7 The Integrated Rate Equation

16.7.1 First-order reactions

Consider a simple first-order reaction of the general type:

$$a \text{ A} \longrightarrow \text{Products}$$

The rate law is:

$$\text{Rate} = -\frac{\Delta[\text{A}]}{\Delta t} = k[\text{A}]$$

Using calculus, one can obtain the integrated rate law (equation):

$$\ln \frac{[\text{A}]_t}{[\text{A}]_0} = -kt \text{ or } \log \frac{[\text{A}]_t}{[\text{A}]_0} = \frac{-kt}{2.303}$$

Here, $[\text{A}]_t$ represents the concentration of the reactant A at time t, and $[\text{A}]_0$ is the initial concentration. The ratio $[\text{A}]_t/[\text{A}]_0$ is the fraction of reactant A that remains at time t.

Example 16.6

The thermal isomerization of cyclopropane to 1-propene follows first-order kinetics with a rate constant of 0.036 min^{-1} at 500°C. Calculate the concentration of cyclopropane after six hours if the initial concentration is 0.075 mol/L.

Solution

For a first-order reaction, the integrated law is

$$\log \frac{[A]_t}{[A]_0} = \frac{-kt}{2.303}$$

$$\log \frac{[A]_t}{[A]_0} = \frac{-(0.036 \text{ min}^{-1}) \left(6 \text{ h} \times 60 \frac{\text{min}}{\text{h}}\right)}{(2.303)} = -5.267$$

$$\frac{[A]_t}{[A]_0} = 10^{-5.267}$$

But $[A]_0 = 0.075 \text{ mol/L}$

$$[A]_{t=6h} = 10^{-5.267}[A]_0 = 10^{-5.267} \times 0.075 \text{ mol/L} = 4.0 \times 10^{-7} \text{ mol/L}$$

Example 16.7

The rate constant for the first-order decomposition of azomethane, $(CH_3)_2N_2$, is $1.5 \times 10^{-3} \text{s}^{-1}$. If the initial concentration is 0.25 M, calculate the time for it to decrease to 0.05 M.

Solution

$$\log \frac{[A]_t}{[A]_0} = \frac{-kt}{2.303}$$

$$[A]_0 = 0.25 \text{ M}$$

$$[A]_t = 0.05 \text{ M}$$

$$k = 1.5 \times 10^{-3} \text{s}^{-1}$$

$$\log \frac{0.05}{0.25} = \frac{-1.5 \times 10^{-3} \text{s}^{-1} t}{2.303}$$

$$-0.6990 = -0.000651 \text{ s}^{-1} \text{ t}$$

$$t = 1,073.7 \text{ s or } 17.9 \text{ min}.$$

16.7.2 Graphing first-order data

By using the property of the logarithms that $\log(A/B) = \log A - \log B$, we can rewrite the above integrated rate equation as

$$\log [A]_t = \left(\frac{-k}{2.303}\right) t + \log [A]_0$$

This conforms to the equation of a straight line, $y = mx + b$:

$$\log [A]_t = \left(\frac{-k}{2.303}\right) t + \log[A]_0$$

$$\uparrow \qquad \uparrow \quad \uparrow \qquad \uparrow$$
$$y \qquad m \quad x \qquad b$$

- Therefore, a plot of log $[A]_t$ on the vertical axis versus time on the horizontal axis should yield a straight line with slope $m = \frac{-k}{2.303}$ and intercept $b = \log [A]_0$. In terms of natural logarithms, the simplified expression would be:

$$\ln [A]_t = (-k)\, t + \ln [A]_0$$

$$\uparrow \qquad \uparrow \ \uparrow \qquad \uparrow$$
$$y \qquad m \ x \qquad b$$

The slope m is equal to $(-k)$, and the intercept b is equal to $\ln [A]_0$.

Example 16.8

Studies of the decomposition of N_2O_5 at 70°C yield the data in Table 16.6. From this information, show that the reaction is first order, and calculate the rate constant k.

Solution

The integrated rate equation for a first-order reaction is

$$\ln \frac{[A]_t}{[A]_0} = -kt$$

From the ideal gas equation, it can be shown that P is directly proportional to the number of moles of gas at constant temperature and volume:

$$PV = nRT$$

Since n/V equals the molar concentration of N_2O_5, we can substitute the partial pressure of N_2O_5 for molarity. Thus the rate equation becomes:

$$\ln \frac{P_t}{P_0} = -kt$$

Table 16.6   Rate data for the decomposition of N_2O_5 at 70°C

Time, min	0	1	2	3	4	5	6	7
P N_2O_5, torr	798	565	400	280	198	140	96	66

where P_0 is the partial pressure at time $t = 0$, and P_t is the partial pressure after time t. Rearranging this equation yields

$$\ln P_t = (-k)t + \ln P_0$$

Plot $\ln P_t$ against time t using the data in Table 16.6. If the graph yields a straight line, the reaction obeys first-order kinetics. Table 16.7 gives data for the decomposition of N_2O_5 at 70°C.

The plot of $\ln P$ vs t yields a straight line (Figure 16.4). Therefore, the decomposition of N_2O_5 is a first-order process with a slope of -0.3548 $(min)^{-1}$.

Table 16.7 Kinetic data for the decomposition of N_2O_5 at 70°C

Time, min	P (N_2O_5), torr	ln P
0	798	6.682
1	565	6.337
2	400	5.991
3	280	5.635
4	198	5.288
5	140	4.942
6	96	4.564
7	66	4.19

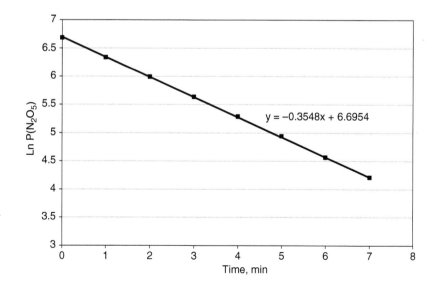

Figure 16.4   The plot of ln $[N_2O_5]$ as a function of time for the decomposition of N_2O_5 at 70°C.

The rate constant can be obtained from the slope as follows:

$$\text{Slope} = (-k)$$

$$-0.3548 \text{ min}^{-1} = -k$$

$$k = 0.3548 \text{ min}^{-1}$$

16.7.3 Second-order reactions

Consider a reaction of the general type:

$$a \text{ A} \longrightarrow \text{Products}$$

The second-order rate law is:

$$\text{Rate} = -\frac{\Delta[\text{A}]}{\Delta t} = k[\text{A}]^2$$

Using calculus, we can obtain the integrated rate law (equation):

$$\frac{1}{[\text{A}]_t} = kt + \frac{1}{[\text{A}]_0}$$

With this equation, we can calculate the concentration of A at any time t, if the initial concentration $[\text{A}]_0$ is known.

The integrated second-order rate equation is of the form $y = mx + b$. Therefore, a plot of $1/[\text{A}]_t$ against time should yield a straight line with slope $= k$, and intercept $= 1/[\text{A}]_0$.

Example 16.9

The gas-phase decomposition of ammonia into hydrogen and nitrogen in the presence of Pt catalyst is second order with a rate constant of $0.085 \text{ M}^{-1}\text{s}^{-1}$. If the initial concentration of ammonia was 0.15 M, what is the final concentration after thirty minutes?

Solution

$$\frac{1}{[\text{A}]_t} = kt + \frac{1}{[\text{A}]_0}$$

$$[\text{A}]_0 = 0.15 \text{ M}$$

$$k = 0.085 \text{ M}^{-1}\text{s}^{-1}$$

$$t = 30 \text{ min} \quad \text{or} \quad 180 \text{ s.}$$

$$[\text{A}]t = ?$$

$$\frac{1}{[\text{A}]_t} = 0.085 \text{ M}^{-1}\text{s}^{-1} \times 180 \text{ s} + \frac{1}{0.15 \text{ M}} = 21.9667 \text{ M}^{-1}$$

$$[\text{A}]_t = 0.046 \text{ M}$$

Example 16.10

The following data (Table 16.8) were obtained for the alkaline hydrolysis of ethyl ethanoate:

$$OH^- + CH_3COOC_2H_5 \longrightarrow CH_3COO^- + CH_3CH_2OH$$

Confirm that the reaction is second order and calculate the value of the rate constant k.

Solution

To confirm that the rate law is second order, construct a plot of $1/[CH_3COOC_2H_5]$ vs. time using the information in Table 16.9.

The plot of $1/[CH_3COOC_2H_5]$ versus time yields a straight line with a slope of 5.5477 M^{-1} min^{-1} (Figure 16.5). The reaction is thus second order.

Table 16.8 Rate data for the alkaline hydrolysis of ethyl ethanoate

Time (s)	Concentration of $CH_3COOC_2H_5$ (M)
0	0.02
5	0.013
10	0.0101
15	0.0077
20	0.0062
25	0.0055
35	0.0043
45	0.0033
55	0.00285
85	0.00195
100	0.00165
125	0.00135

Table 16.9 Kinetic data for the alkaline hydrolysis of ethyl ethanoate

Time (s)	Concentration of $CH_3COOC_2H_5$ (M)	$1/[CH_3 COOC_2H_5]$ $(M)^{-1}$
0	0.02	50
5	0.013	76.92
10	0.0101	99.01
15	0.0077	129.87
20	0.0062	161.29
25	0.0055	181.82
35	0.0043	232.56
45	0.0033	303.03
55	0.00285	350.88
85	0.00195	512.82
100	0.00165	606.06
125	0.00135	740.74

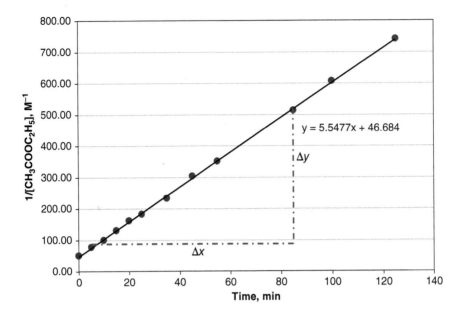

Figure 16.5   The plot of $1/[CH_3COOC_2H_5]$ versus time.

The rate constant k is equal to the slope of the straight line. To calculate the slope, use two widely separated coordinate points on the graph.

$$k = \text{slope} = \frac{\Delta y}{\Delta x} = \frac{(512.82 - 76.92)\ \text{M}^{-1}}{(85.0 - 5.0)\ \text{s}} = \frac{435.9}{80.0\ \text{s}} = 5.449\ \text{M}^{-1}\text{s}^{-1}$$

16.8 Half-Life of a Reaction

The half-life of a reaction, $t_{1/2}$, is the time taken for the reactant concentration to decrease to one-half of its original value.

Recall that the integrated rate equation for a first-order reaction is

$$\ln \frac{[A]_t}{[A]_0} = -kt$$

At time $t = t_{1/2}$, the fraction $\dfrac{[A]_t}{[A]_0}$ becomes ½. Substituting in the equation yields

$$\ln \frac{1}{2} = -kt_{1/2}\ \text{ or } t_{1/2} = \frac{-\ln 2}{k}$$

$$t_{1/2} = \frac{0.693}{k}$$

Thus the half-life of a first-order reaction is independent of the initial concentration of the reactant. This implies that the half-life is the same at any time during the reaction.

The half-life expression for a second-order reaction can be obtained in a similar manner. Substituting $t = t_{1/2}$, and $[A]_t = [A]_0/2$ in the integrated rate equation:

$$\frac{1}{[A]_t} = kt + \frac{1}{[A]_0}$$

yields

$$\frac{1}{[A_0]^2} = kt_{1/2} + \frac{1}{[A]_0}$$

$$\frac{2}{[A]_0} = kt_{1/2} + \frac{1}{[A]_0}$$

$$t_{1/2} = \frac{1}{k[A]_0}$$

Unlike the half-life of a first-order reaction, the half-life of a second-order reaction depends on the starting concentration of the reactant.

Example 16.11

For the first-order gas phase reaction

$$2NOCl(g) \longrightarrow 2NO(g) + Cl_2(g)$$

the rate constant is 9.5×10^{-6}/s at 77°C. What is the half-life of NOCl if 1.25 moles of NOCl are placed in a three-liter reactor?

Solution

The half-life of a first-order reaction is given by the equation

$$t_{1/2} = \frac{0.693}{k} = \frac{0.693}{9.5 \times 10^{-6}/s} = 7.30 \times 10^4 s$$

Note that $t_{1/2}$ does not depend on the initial concentration.

Example 16.12

The rate constant for the second-order hydrolysis of ethyl nitrobenzene at 25°C is $0.0795 \ M^{-1}min^{-1}$. Starting with an initial concentration of 0.05 M, calculate the half-life of ethyl nitrobenzene.

Solution

The half-life of a second-order process is given by the equation:

$$t_{1/2} = \frac{1}{k[A]_0}$$

where A represents ethyl nitrobenzene.

$$t_{1/2} = \frac{1}{k[A]_0} = \frac{1}{(0.0795 \ M^{-1}min^{-1})(0.05 \ M)} = 251.6 \ min$$

16.9 Reaction Rates and Temperature: The Arrhenius Equation

At higher temperatures, more molecules possess the necessary energy of activation and more collisions occur, so reaction rates increase significantly. The quantitative relationship between reaction rate, temperature, and energy of activation is given by the Arrhenius equation:

$$k = Ae^{-E_a/RT}$$

where k is the rate constant, A is a constant known as the pre-exponential factor, E_a is the energy of activation and R, the universal gas constant, is 8.314 J/mole-K. Taking the logarithm, the Arrhenius equation becomes

$$\ln k = \frac{-E_a}{RT} + \ln A \quad \text{or} \quad \log k = \log A - \frac{E_a}{2.303 \ RT}$$

The Arrhenius equation at two different temperatures can be written as

$$\log\frac{k_2}{k_1} = \frac{E_a}{2.303 \ R}\left(\frac{1}{T_1} - \frac{1}{T_2}\right) \quad \text{or} \quad \log\frac{k_2}{k_1} = \frac{E_a}{2.303 \ R}\left(\frac{T_2 - T_1}{T_1 T_2}\right)$$

Thus, measuring the rate constants at two different temperatures allows us to calculate E_a, which is one of the main uses of this equation.

Example 16.13

The kinetic data for temperature dependence of the rate constant for the reaction

$$CH_3COOC_2H_5\,(aq) + OH^-\,(aq) \longrightarrow CH_3COO^-\,(aq) + C_2H_5OH\,(aq)$$

are tabulated Table 16.10. Calculate the energy of activation, E_a, and the pre-exponential factor, A.

Solution

To determine E_a, plot $\ln k$ vs $1/T$ (Figure 16.6) using the data in Table 16.11. This should be a straight line with a slope $\frac{-E_a}{R}$.

$$\text{Slope} = -\frac{E_a}{R} = -5,345.3$$

$$E_a = -(-5,345.5) \times 8.314 \ J/mol = 4.44 \times 10^4 \ J/mol$$

Table 16.10 Kinetic data for temperature dependence of the rate constant for the hydrolysis of ethyl ethanoate

Temperature (°C)	k (M/s)
15	0.0511
25	0.104
35	0.187
45	0.332
55	0.533
65	0.783

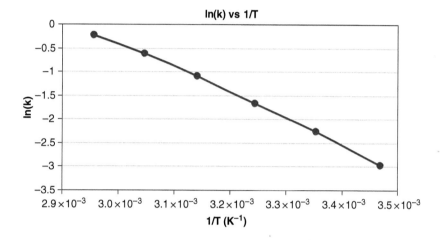

Figure 16.6   An Arrhenius plot for the alkaline hydrolysis of ethyl ethanoate.

Table 16.11   Determination of the energy of activation for the alkaline hydrolysis of ethyl ethanoate

T (K)	k (M/s)	ln k	1/T (K⁻¹)
288.2	0.0511	−2.974	0.00347
298.2	0.104	−2.264	0.00335
308.2	0.187	−1.677	0.00324
318.2	0.332	−1.103	0.00314
328.2	0.533	−0.629	0.00305
338.2	0.783	−0.245	0.00296

Example 16.14

At 600 K, the rate constant for the first-order decomposition of nitrogen pentoxide was found to be 3.0×10^{-3} s^{-1}. Determine the value of the rate constant at 750 K, given that $E_a = 1.5 \times 10^2$ kJ mol^{-1}, and $R = 8.314$ J mol^{-1}K^{-1}.

Solution

$$\ln \frac{k_2}{k_1} = \left(\frac{E_a}{R}\right)\left(\frac{1}{T_1} - \frac{1}{T_2}\right)$$

$$\ln \frac{k_2}{3.0 \times 10^{-3}} = \left(\frac{1.5 \times 10^2 \text{ kJ mol}^{-1} \times 1,000 \text{ J/kJ}}{8.314 \text{ J mol}^{-1}\text{K}^{-1}}\right)\left(\frac{1}{600\text{K}} - \frac{1}{750 \text{ K}}\right)$$

$$\ln \frac{k_2}{3.0 \times 10^{-3}} = (1,8041.85)(0.000333) = 6.0140$$

$$\frac{k_2}{3.0 \times 10^{-3}} = e^{6.0140} = 409.1$$

$$k_2 = 1.23 \text{ s}^{-1}$$

16.10 Problems

1. Write the rate expression in terms of the disappearance of the reactants and the appearance of the products for the following decomposition reactions:

   (a) $2 H_2O_2 (g) \longrightarrow 2 H_2O (l) + O_2 (g)$

   (b) $2 N_2O_5 (g) \longrightarrow 4 NO_2 (g) + O_2 (g)$

   (c) $SO_2Cl_2 (g) \longrightarrow SO_2 (g) + Cl_2 (g)$

   (d) $4 PH_3 (g) \longrightarrow P_4 (g) + 6 H_2 (g)$

2. For the following reactions, write the rate expression in terms of the disappearance of the reactants and the appearance of the products:

   (a) $2 NH_3 (g) + 5 O_2 (g) \longrightarrow 4 NO (g) + H_2O (g)$

   (b) $2 NO (g) + 2 H_2 (g) \longrightarrow N_2 (g) + H_2O (g)$

   (c) $CH_3CHO (g) \longrightarrow CH_4 (g) + CO (g)$

   (d) $2 SO_2 (g) + O_2 (g) \longrightarrow 2 SO_3 (g)$

3. The formation of nitrogen from the high-temperature reduction of nitric oxide in a hydrogen gas atmosphere is given by the equation:

   $$2 NO (g) + H_2 (g) \longrightarrow N_2 (g) + 2 H_2O (g)$$

   If NO had an initial concentration of 0.025 mol/L, what is the initial rate of production of N_2 gas?

4. The formation of hydrogen iodide from hydrogen and iodine gas is given by:

   $$H_2 (g) + I_2 (g) \longrightarrow 2 HI (g)$$

   If iodine gas is being consumed at a rate of 0.012 M/s, what is the rate of formation of hydrogen iodide?

5. Ozone decomposes according to:

$$2 \, O_3 \, (g) \longrightarrow 3 \, O_2 \, (g)$$

The initial concentration of oxygen is 0.050 M, and 0.130 M after two hours of reaction. Calculate the average rate of this reaction.

6. Initial rate data (Table 16.12) were collected for the reaction $X + Y \longrightarrow Z$.

   (a) What is the rate law for this reaction?
   (b) Determine the rate constant.

7. Initial rate data for the reaction $NH_4^+ \, (aq) + CNO^- \, (aq) \longrightarrow CO(NH_2)_2 \, (s)$ at 80°C are given in Table 16.13.

   (a) Determine the rate law.
   (b) Calculate the value of the rate constant.

8. The reaction

$$I^- \, (aq) + OCl^- \, (aq) \longrightarrow IO^- \, (aq) + Cl^- \, (aq)$$

was investigated in a basic solution. Table 16.14 shows the results. Determine:

   (a) the rate law, and
   (b) the order of the reaction.

9. Initial rate data for the reaction

$$2 \, S_2O_3^{2-} \, (aq) + I_2 \, (aq) \longrightarrow S_4O_6^{2-} \, (aq) + 2I^- \, (aq)$$

were collected at 25°C. The results are summarized in Table 16.15. Determine:

Table 16.12   Data for Problem 6

Experiment	[X] (M)	[Y] (M)	Initial Rate (M/min)
1	0.02	0.025	0.0036
2	0.01	0.025	0.0018
3	0.02	0.0125	0.0009

Table 16.13   Data for Problem 7

Experiment	$[NH_4^+]$ (M)	$[CNO^-]$ (M)	Initial Rate (M/s)
1	0.014	0.02	0.002
2	0.028	0.02	0.008
3	0.014	0.01	0.001

Table 16.14   Data for Problem 8

Experiment	$[I^-]$ (M)	$[OCl^-]$ (M)	$[OH^-]$ (M)	Initial Rate (M/min)
1	0.005	0.005	0.5	0.000275
2	0.0025	0.005	0.5	0.000138
3	0.0025	0.0025	0.5	0.000007
4	0.0025	0.0025	0.25	0.00014

Table 16.15   Data for Problem 9

Experiment	$[I_2]_0$ (M)	$[S_2O_3^-]_0$ (M)	Initial Rate (M/s)
1	0.01	0.01	0.0004
2	0.01	0.02	0.0004
3	0.02	0.01	0.0008

Table 16.16   Data for Problem 10

Experiment	$[H_2O_2]_0$ (M)	$[I^-]_0$ (M)	$[H^+]_0$ (M)	Initial Rate (M/s)
1	0.15	0.15	0.05	0.00012
2	0.15	0.3	0.05	0.00024
3	0.3	0.15	0.05	0.00024
4	0.15	0.15	0.1	0.00048

(a) the order with respect to each reactant
(b) the overall rate law, and
(c) the value of the rate constant.

10. The initial rates given in Table 16.16 were determined at 25°C for the reaction:

$$H_2O_2 \text{(aq)} + 3I^- \text{(aq)} + 2H^+ \text{(aq)} \longrightarrow I_3^- \text{(aq)} + 2H_2O \text{(l)}$$

(a) What is the order of the reaction with respect to each reactant?
(b) Determine the overall rate law.
(c) Calculate the value of the rate constant.
(d) What is the rate of reaction when all reactant concentrations are 0.025 M?

11. The decomposition of dinitrogen pentoxide according to the equation

$$2N_2O_5 \text{(g)} \xrightarrow{50°C} 4NO_2 \text{(g)} + O_2 \text{(g)}$$

follows first-order kinetics with a rate constant of $0.0065s^{-1}$. If the initial concentration of N_2O_5 is 0.275 M, determine:

(a) The concentration of N_2O_5 after three minutes.
(b) How long it will take for 98% of the N_2O_5 to decompose.

12. At 1200°C, CS_2 decomposes according to

$$CS_2 \text{(g)} \longrightarrow CS \text{(g)} + S \text{(g)}$$

The rate law is rate $= k[CS_2]$, where $k = 1.6 \times 10^{-6}s^{-1}$.

(a) What is the initial rate of decomposition if the concentration of CS_2 is 0.25 M?
(b) If the initial concentration of CS_2 is 0.25 M, calculate the residual concentration after a reaction time of five hours.

13. The first-order rate constant for the reaction $H_2O_2 \text{(aq)} \longrightarrow 2H_2O \text{(l)} + O_2 \text{(g)}$ is $2.6 \times 10^{-4}s^{-1}$ at 45°C.

(a) Calculate the half-life for the reaction.

(b) Determine the concentration of H_2O_2 after four half-lives, given that the initial concentration is 0.005000 M.

14. At 450°C, cyclopropane (C_3H_6) gas isomerizes to propene with first-order kinetics. The half-life for the reaction is thirteen hours.

(a) What is the rate constant for this reaction?

(b) Calculate the concentration of C_3H_6 remaining after one hour if the initial concentration is 0.015 M.

(c) What fraction of C_3H_6 remains after twenty-four hours?

(d) How many hours does it take for 80% of the C_3H_6 to react?

15. The thermal decomposition of N_2O_5, $2N_2O_5 (g) \longrightarrow 4NO_2 (g) + O_2 (g)$, is a first-order reaction. From the following kinetic data (Table 16.17), determine the value of the rate constant.

16. The following data were collected for the gas phase isomerization of methyl isonitrile (CH_3NC) to acetonitrile (CH_3CN) at 230°C (see Table 16.18).

$$CH_3NC \xrightarrow{230°C} CH_3CN$$

(a) Use the kinetic data above to determine if the isomerization reaction is first or second order.

(b) What is the value of the rate constant?

17. The thermal decomposition of NO_2 to yield NO and O_2 is a second-order reaction with a rate constant of 0.55 $M^{-1}s^{-1}$ at 300°C.

$$2NO_2 (g) \longrightarrow 2NO(g) + O_2 (g)$$

If the initial concentration of NO_2 gas is 0.0500 M:

(a) What is the concentration of NO_2 after five hours?

(b) How long does it take for the NO_2 concentration to decrease to 0.0050 M?

(c) What is the half-life of the reaction?

18. The kinetic data shown in Table 16.19 were obtained for the decomposition of HI at 400°C:

Table 16.17 Data for Problem 15

Time (s)	0	50	100	150	200	250	300	350	400
$[N_2O_5]$ (M)	0.075	0.072	0.07	0.067	0.065	0.064	0.062	0.061	0.059

Table 16.18 Data for Problem 16

Time (s)	0	300	600	900	1200	1500	1800
$[CH_3NC]$ (M)	0.085	0.067	0.053	0.041	0.031	0.023	0.018

Table 16.19 Data for Problem 18

Time (s)	0	20	40	60	80	100	120
$[HI]$ (M)	0.07	0.062	0.056	0.052	0.0475	0.0446	0.042

$$2HI(g) \xrightarrow{400°C} H_2(g) + I_2(g)$$

(a) Use these data to determine if the reaction is first order or second order.

(b) Calculate the value of the rate constant.

(c) What is the half-life of the reaction if the initial concentration of HI is 0.15 M?

19. The rate constant for the first-order decomposition of gaseous N_2O_5 at 55°C is $1.73 \times 10^{-4} s^{-1}$. If the activation energy is 121 kJ/mol, determine the rate constant at 150°C.

20. The rate constant for the second-order decomposition of gaseous NO_2 at 380°C is $10.50 \ M^{-1}s^{-1}$. If the activation energy is 233 kJ/mol, calculate the temperature at which the rate constant is $1.050 \times 10^3 \ M^{-1}s^{-1}$.

21. Rate constants for the decomposition of an organic peroxide, ROOR', are $2.5 \times 10^{-7}s^{-1}$ at 300 K and $1.25 \times 10^{-4}s^{-1}$ at 425 K, respectively. Calculate:

(a) The activation energy for the reaction in kJ/mol.

(b) The rate constant at 355 K.

22. Rate constants for the decomposition of azomethane, $(CH_3)_2N_2$, are $2.5 \times 10^{-3}s^{-1}$ at 300°C and $3.25 \times 10^{-2}s^{-1}$ at 333°C, respectively. Calculate:

(a) The activation energy for the reaction in kJ/mol.

(b) The rate constant at 25°C.

23. The hydrolysis of a certain sugar in 0.05 M HCl follows a first-order rate equation. Variation of the rate constant with temperature is given in Table 16.20. Determine graphically the activation energy and the pre-exponential factor, A, for the reaction.

24. The temperature dependence of the rate constant for the following reaction is given in Table 16.21:

$$CH_3Br(aq) + OH^-(aq) \longrightarrow CH_3OH(aq) + Br^-(aq)$$

From an appropriate graph of the data, calculate the activation energy, E_a, (in kJ/mol) and the pre-exponential or frequency factor, A.

Table 16.20   Data for Problem 23

Temperature (K)	273	298	308	318
$k \ (s^{-1})$	2.1×10^{-5}	6.38×10^{-4}	1.94×10^{-3}	5.84×10^{-3}

Table 16.21   Data for Problem 24

Temperature (°C)	15	25	35	45
$k \ (M^{-1}s^{-1})$	0.104	0.202	0.368	0.664

17

Chemical Equilibrium

. . . .

17.1 Reversible and Irreversible Reactions

Many chemical reactions go to completion, i.e. all the reactants are converted to products. A good example is the reaction of calcium with cold water.

$$Ca(s) + 2\,H_2O(l) \longrightarrow Ca(OH)_2(s) + H_2(g)$$

There is no evidence that the reverse reaction occurs. Such reactions are said to be *irreversible*. On the other hand, many reactions are *reversible*: the process can be made to go in the opposite direction. This means that both the reactants and products will be present at any given time. *A reversible reaction* is defined as *one in which the products formed react to give the original reactants*. A double arrow is used to indicate that the reaction is reversible, as illustrated by the general equation:

$$aA + bB \rightleftharpoons cC + dD$$

At the start of the reaction, the reactants convert more quickly to products than products turn back to reactants. The reaction is said to reach *equilibrium* when the net change in the products and reactants is zero, i.e. the rate of forward reaction equals the rate of reverse reaction. Chemical equilibria are *dynamic equilibria* because, although nothing appears to be happening, opposing reactions are occurring at the same rate.

   Figure 17.1 illustrates that for a reaction in chemical equilibrium, the rate of forward reaction equals the rate of reverse reaction.

17.2 The Equilibrium Constant

When a chemical reaction is at equilibrium, the concentrations of reactants and products are constant. The relationship between the concentrations of reactants and products is given by the equilibrium expression, also known as the law of mass action. For the general reaction

$$aA + bB \rightleftharpoons cC + dD$$

at a constant temperature, the equilibrium constant expression is written as

Chemistry in Quantitative Language: Fundamentals of General Chemistry Calculations. Second Edition.
Christopher O. Oriakhi, Oxford University Press. © Christopher O. Oriakhi 2021.
DOI: 10.1093/oso/9780198867784.003.0017

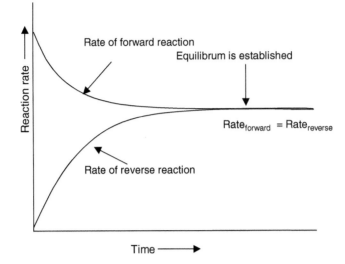

Figure 17.1   Rates of forward and reverse reactions for chemical reactions at equilibrium.

$$K_c = \frac{[C]^c[D]^d}{[A]^a[B]^b}$$

where [A], [B], [C], and [D] are the molar concentrations or partial pressures of A, B, C, and D at equilibrium. The exponents a, b, c, and d in the equilibrium expression are the coefficients in the balanced equation; K_c is the equilibrium constant and is not given units. The subscript c shows that K is in terms of concentration. The numerical value for K_c is usually determined experimentally. Note that the equilibrium constant expression depends only on the reaction stoichiometry, and not on the mechanism.

Definition: The equilibrium constant, K, is defined as the product of the molar concentrations (or partial pressure, if gaseous) of the products of a chemical reaction, each raised to the power of its coefficient in the balanced equation, divided by the product of the molar concentrations (or partial pressure, if gaseous) of the reactants, each raised to the power of its coefficient in the balanced equation.

Example 17.1

Write the equilibrium constant expressions for the following reactions:

(a) $N_2(g) + 3\,H_2(g) \rightleftharpoons 2\,NH_3(g)$
(b) $CO(g) + Cl_2(g) \rightleftharpoons COCl_2(g)$
(c) $H_2(g) + I_2(g) \rightleftharpoons 2\,HI(g)$

Solution

(a) $K_c = \dfrac{[NH_3]^2}{[N_2][H_2]^3}$

(b) $K_c = \dfrac{[COCl_2]}{[CO][Cl_2]}$

(c) $K_c = \dfrac{[HI]^2}{[H_2][I_2]}$

Example 17.2

Phosgene, $COCl_2$, is a very toxic gas that was used extensively during the First World War. It decomposes to carbon monoxide and chlorine on heating, according to

$$COCl_2(g) \rightleftharpoons CO(g) + Cl_2(g)$$

0.75 mole of $COCl_2$ was heated in a sealed 1-liter vessel at 373 K. The equilibrium concentration of Cl_2 was found to be 0.25 M. Calculate the equilibrium constant for the reaction. (Note: 0.75 mole in 1 L is 0.75 M.)

Solution

At the start of the reaction:

$$[COCl_2] = 0.75 \text{ M}$$
$$[CO] = 0$$
$$[Cl_2] = 0$$

At equilibrium:

$$[COCl_2] = 0.75 - 0.25 = 0.5 \text{ M}$$
$$[CO] = 0.25 \text{ M}$$
$$[Cl_2] = 0.25 \text{ M}$$
$$K_c = \dfrac{[Cl_2][CO]}{[COCl]}$$
$$K_c = \dfrac{[0.25][0.25]}{[0.5]} = 0.125$$

17.2.1 Equilibrium constant in terms of pressure

For reactions where the reactants and products are gases, the equilibrium constant expression can be written in terms of partial pressures. For the general equation

$$aA(g) + bB(g) \rightleftharpoons cC(g) + dD(g)$$

The equilibrium constant expression, K_p, can be written as:

$$K_p = \dfrac{(P_C)^c (P_D)^d}{(P_A)^a (P_B)^b}$$

where P_A, P_B, P_C, and P_D are the equilibrium partial pressures. The subscript p is used here because K is determined from partial pressures of products and reactants.

Example 17.3

Write the equilibrium constant expression K_p, in terms of partial pressure, for the decomposition of sulfur trioxide.

$$2\ SO_3(g) \rightleftharpoons 2\ SO_2(g) + O_2(g)$$

Solution

$$K_p = \frac{P_{SO_2}^2 P_{O_2}}{P_{SO_3}^2}$$

Example 17.4

Write balanced chemical equations for completely gaseous equilibrium reactions resulting in the following expressions for the equilibrium constant.

1. $\dfrac{[N_2][O_2]}{[NO]^2}$

2. $\dfrac{[NO]^2[Cl_2]}{[NOCl]^2}$

3. $\dfrac{[CH_4][H_2S]^2}{[CS_2][H_2]^4}$

4. $\dfrac{[N_2]^2[H_2O]^6}{[NH_3]^4[O_2]^3}$

Solution

1. $N_2(g) + O_2(g) \rightleftharpoons 2\ NO(g)$
2. $2\ NOCl(g) \rightleftharpoons 2\ NO(g) + Cl_2(g)$
3. $CS_2(g) + 4\ H_2(g) \rightleftharpoons CH_4(g) + 2\ H_2S(g)$
4. $4\ NH_3(g) + 3\ O_2(g) \rightleftharpoons 2\ N_2(g) + 6\ H_2O(g)$

17.2.2 Relationship between K_p and K_c

The relationship between K_p and K_c is given by the expression

$$K_p = K_c(RT)^{\Delta n_{gas}}$$

where Δn_{gas} is the sum of the coefficients of the gaseous products, minus the sum of the coefficients of the gaseous reactants, in the balanced equation.

Δn = (number of moles of gaseous products - number of moles of gaseous reactants).

Example 17.5

Calculate the value of K_c at 1000 K for the following reaction:

$$CH_4(g) + 2\,H_2S(g) \rightleftharpoons CS_2(g) + 4\,H_2(g); \quad K_p = 4.2 \times 10^{-3} \text{ at 1000 K}$$

Solution

1. Calculate the change in the number of moles of gases.

$$\Delta n = 5 \text{ moles of produce} - 3 \text{ moles of reactants} = 2$$

2. Substitute Δn, R, and T into the equation $K_p = K_c(RT)^{\Delta n_{gas}}$ and solve for K_p. Use the value of R consistent with the units given in the problem. Here, $R = 0.08206 \left(\dfrac{L.atm}{K.mol} \right)$

$$K_p = K_c(RT)^{\Delta n_{gas}}$$

$$4.2 \times 10^{-3} = K_c(0.08206 \times 1000)^2$$

$$K_c = \frac{4.2 \times 10^{-3}}{(0.08206 \times 1000)^2} = 6.55 \times 10^{-7}$$

17.3 The Reaction Quotient

For the general reaction

$$aA + bB \rightleftharpoons cC + dD$$

the mass action expression is given as

$$K = \frac{[C]^c[D]^d}{[A]^a[B]^b}$$

If the reaction is not at equilibrium, the above ratio is different from the equilibrium constant expression. In such case the ratio is called the reaction quotient Q.

$$Q = \frac{[C]^c[D]^d}{[A]^a[B]^b}$$

The concentrations [A], [B], [C], and [D] are not necessarily equilibrium concentrations; they are the actual concentrations at any point in time. When $Q = K$ the reaction is at equilibrium. Note that we use K here without the subscript c or p for simplicity.

17.4 Predicting the Direction of Reaction

The direction of a reaction can be predicted by comparing the magnitude of Q with the value of K_c. For the general reaction

$$aA + bB \rightleftharpoons cC + dD$$

1. If $Q < K_c$, the reaction will continue to move from left to right (forward) towards the product until equilibrium is achieved.
2. If $Q > K_c$, the reaction moves from right to left (reverse) towards the reactants until equilibrium is achieved.
3. If $Q = K_c$, the reaction is at equilibrium.

17.5 Position of Equilibrium

The position of equilibrium, that is, the magnitude of K, indicates in qualitative terms the extent to which a reaction at equilibrium has moved towards completion. The relationship between K, relative concentration of products and reactants, and equilibrium position is shown in Table 17.1.

Example 17.6

For the following reactions, give a qualitative description of the position of equilibrium. Also, predict the relative concentrations of the products and reactants at 298 K.

1. $N_2(g) + O_2(g) \rightleftharpoons 2\, NO(g)$ \qquad $K_{eq} = 2.45 \times 10^{45}$
2. $2\, NOCl(g) \rightleftharpoons 2\, NO(g) + Cl_2(g)$ \qquad $K_{eq} = 1.125 \times 10^{-33}$
3. $CS_2(g) + 4\, H_2(g) \rightleftharpoons CH_4(g) + 2\, H_2S(g)$ \qquad $K_{eq} = 9.95 \times 10^{-1}$
4. $4\, NH_3(g) + 3\, O_2(g) \rightleftharpoons 2\, N_2(g) + 6\, H_2O(g)$ \qquad $K_{eq} = 12$

Solution

1. Equilibrium position is far to the right. The reaction has gone to completion. Mainly NO exists.
2. Equilibrium position is far to the left. The reaction will not proceed to any significant extent. Very little of NO and Cl_2 exist in the gas phase.
3. The equilibrium is neither to the left nor to the right. Significant concentrations of reactants (CS_2 and H_2) and products (CH_4 and H_2S) exist.

Table 17.1 Relationship between the magnitude of K, equilibrium position, and the concentrations of reactants and products

Value of K_{eq}	Range of K_{eq}	Equilibrium position	Relative concentration of products and reactants
Very large	$10^{20} - 10^{30}$	Far to the right. Reaction proceeds almost 100% to completion	Mainly all products
Large	$> 10^3$	To the right	Products > reactants
Unity	$10^{-3} - 10^3$	Neither to the left nor to right	Almost the same
Small	$< 10^{-3}$	To the left	Reactants > products
Very small	$10^{-20} - 10^{-30}$	Far to the left	Mainly all reactants

4. The equilibrium is neither to the left nor to the right. Significant concentrations of reactants (NH_3 and O_2) and products (N_2 and H_2O) exist.

17.6 Homogeneous vs Heterogeneous Equilibria

Homogeneous equilibria involve reactions in which all the products and reactants are in the same phase. For example, in the Haber process for ammonia synthesis,

$$N_2(g) + 3\,H_2(g) \rightleftharpoons 2\,NH_3(g)$$

the reactants and product are all gaseous.

Heterogeneous equilibra, on the other hand, involve reactions in which the substances in equilibrium are in different phases. Consider the decomposition of solid phosphorus pentachloride to liquid phosphorus trichloride and chlorine gas:

$$PCl_5(s) \rightleftharpoons PCl_3(l) + Cl_2(g)$$

The system involves solid PCl_5 in equilibrium with liquid PCl_3 and Cl_2 gas. The equilibrium constant expression is:

$$K_c = \frac{[PCl_3][Cl_2]}{[PCl_5]}$$

The concentrations (or activities) of pure solids and liquids do not change much with temperature or pressure, so they are taken to be 1 in these calculations. Consequently, for heterogeneous equilibria, terms for liquids and solids are omitted in the K expression. Thus, the K_c expression for the decomposition of PCl_5 becomes simply

$$K_{eq} = [Cl_2] \text{ or } P_{Cl_2}$$

Example 17.7

Write the equilibrium constant expressions for the following reversible reactions:

1. $NH_4Cl(s) \rightleftharpoons NH_3(g) + HCl(g)$
2. $CO_2(g) + NaOH(s) \rightleftharpoons NaHCO_3(s)$
3. $NH_3(g) + H_2SO_4(l) \rightleftharpoons (NH_4)_2SO_4(s)$
4. $2\,Pb(NO_3)_2(s) \rightleftharpoons 2\,PbO(s) + 4\,NO_2(g) + O_2(g)$

Solution

Note that the concentrations of pure liquids and solids do not appear in the K expression.

1. $K_c = [NH_3][HCl]$
2. $K_c = \dfrac{1}{[CO_2]}$
3. $K_c = \dfrac{1}{[NH_3]}$
4. $K_c = [NO_2]^4[O_2]$

17.7 Calculating Equilibrium Constants

The equilibrium constant K_c can be calculated from a set of equilibrium concentrations. Once K_c is known, its value can be used to calculate equilibrium concentrations, since the initial concentrations are always known.

In general we need to know two basic things in order to calculate the value of the equilibrium constant:

1. A balanced equation for the reaction system. Be sure to include the physical states of each species so that the appropriate equilibrium expression for calculating K_c or K_p can be derived.
2. The equilibrium concentrations or pressures of each reactant or product included in the equilibrium expression. Substituting these values into the equilibrium expression allows the value of the equilibrium constant to be calculated.

17.7.1 Calculating K_c or K_p from known equilibrium amounts

1. Write a balanced chemical equation for the reaction at equilibrium, if not already given.
2. Write the equilibrium expression for the reaction.
3. Determine the molar concentrations or partial pressures of each species involved.
4. Plug the value of the molar concentrations or partial pressures of each species into the equilibrium expression and solve for K.

Example 17.8

The equilibrium concentrations of H_2, I_2, and HI in the following reaction at 700 K are 0.0058 M, 0.105 M, and 0.190 M, respectively. Calculate the value of the equilibrium constant, K_c:

$$H_2(g) + I_2(g) \rightleftharpoons 2\,HI(g)$$

Solution

1. We already have a balanced equation.
2. Write the equilibrium expression for the reaction.

$$K_c = \frac{[HI]^2}{[H_2][I_2]}$$

3. Skip this step since all concentrations are known and expressed in moles per liter (M).
4. Plug the value of the molar concentration or partial pressure of each species into the equilibrium expression and solve for K_c:

$$K_c = \frac{[HI]^2}{[H_2][I_2]} = \frac{(0.190)^2}{(0.0058)(0.105)} = 59.3$$

17.7.2 Calculating K from initial concentration and one equilibrium concentration

1. Write a balanced chemical equation for the reaction at equilibrium, if not already given.
2. Write the equilibrium constant expression for the reaction.
3. Determine the initial molar concentrations or partial pressures of each species involved.
4. Determine changes from initial concentrations or partial pressures, based on stoichiometric ratios. If unknown, assign x to it.
5. Use (possibly unknown) changes to calculate final concentrations.
6. Plug the values of the molar concentration or partial pressure of each species into the equilibrium expression and solve for K.

Example 17.9

Consider the following:

$$CH_4(g) + H_2O(g) \rightleftharpoons CO(g) + 3\,H_2(g)$$

The reaction starts with a mixture of 0.15 M CH_4, 0.30 M H_2O, 0.40 M CO, and 0.60M H_2, and is allowed to reach equilibrium. At equilibrium, the concentration of CH_4 is found to be 0.05 M. Calculate K_c for the reaction.

Solution

1. Write the equilibrium constant expression for the reaction.

$$K_c = \frac{[CO][H_2]^3}{[CH_4][H_2O]}$$

2. Determine the initial molar concentrations of each species involved.

$$[CH_4] = 0.15\ M;\ [H_2O] = 0.30\ M;\ [CO] = 0.40\ M;\ [H_2] = 0.60\ M.$$

3. Determine changes from initial concentrations based on stoichiometric ratios, assigning x to unknown changes. Use Table 17.2a to simplify your calculations. We know the equilibrium concentration of CH_4 to be 0.05 M. Thus, we can determine what the change is for the other species.
4. Substitute the value of the molar concentration or partial pressure of each species into the equilibrium expression and solve for K:

$$K_c = \frac{[CO][H_2]^3}{[CH_4][H_2O]} = \frac{(0.50)(0.90)^3}{(0.05)(0.20)} = 36.45$$

17.8 Calculating Equilibrium Concentrations from K

If we know the equilibrium constant for a given reaction, we can calculate the concentrations in the mixture provided the initial concentrations are known.

1. Write down a balanced chemical equation for the reaction at equilibrium, if not already given.
2. Write the equilibrium constant expression for the reaction.
3. Determine the initial molar concentrations or partial pressures of each species involved.
4. Determine changes from initial concentrations or partial pressures, based on stoichiometric ratios. Assign x to an unknown concentration.
5. Substitute concentrations in the K expression and solve for the unknown.
6. Use the value of the unknown to determine the desired concentrations.
7. Check values as needed.

The following examples illustrate the steps involved in the calculation of equilibrium concentrations of reaction mixtures.

Example 17.10

One mole of PCl_5 was decomposed to gaseous PCl_3 and Cl_2 in a 1-liter sealed vessel. If K_c at the decomposition temperature is 1×10^{-3}, what are the equilibrium concentrations of the reactant and products?

Solution

Step 1:

$$PCl_5(g) \rightleftharpoons PCl_3(g) + Cl_2(g)$$

Step 2:

$$K_c = \frac{[PCl_3][Cl_2]}{[PCl_5]} = 1.0 \times 10^{-3}$$

Step 3:

$$[PCl_5]_0 = \frac{1.0 \text{ mole}}{1.0 \text{ L}} = 1.0 \text{ M}$$

$$[PCl_3]_0 = 0$$

$$[Cl_2]_0 = 0$$

Table 17.2   Data for Example 17.9

(a)	$[CH_4]$	$[H_2O]$	$[CO]$	$[H_2]$
Initial concentration (M)	0.15 M	0.30 M	0.40 M	0.60 M
Change (M)	$(-)x$	$(-)x$	$(+)x$	$(+)x$
Equilibrium concentration (M)	$(0.15 - x) = 0.05$	$(0.30 - x)$	$(x + 0.4)$	$(x + 0.6)$

(b)	$[CH_4]$	$[H_2O]$	$[CO]$	$[H_2]$
Initial concentration (M)	0.15 M	0.30 M	0.40 M	0.60 M
Change (M)	$(-)0.10$	$(-0.10$	0.10	0.30
Equilibrium concentration (M)	0.05	0.20	0.50	0.90

Table 17.3   Data for Example 17.10

	[PCl₅]	[PCl₃]	[Cl₂]
Initial concentration (M)	1	0	0
Change (M)	$(-)x$	$(+)x$	$(+)x$
Equilibrium concentration (M)	$1-x$	x	x

Step 4: Let x equal the amount of PCl₅ that reacted to reach equilibrium. Table 17.3 shows the equilibrium concentrations of all the species.

Steps 5 & 6:

$$K_c = \frac{[PCl_3][Cl_2]}{[PCl_5]} = 1.0 \times 10^{-3}$$

$$1.0 \times 10^{-3} = \frac{(x)(x)}{(1-x)} = \frac{x^2}{1-x}$$

$$x^2 + 1.0 \times 10^{-3}x + 1.0 \times 10^{-3} = 0$$

This yields a quadratic equation of the general form $ax^2 + bx + c = 0$. The roots can be obtained by using 'the (almighty) quadratic formula':

$$x = \frac{-b \pm \sqrt{b^2 - 4ac}}{2a}$$

Comparing our equation with the general form, $a = 1.0$, $b = 1.0 \times 10^{-3}$, and $c = 1.0 \times 10^{-3}$. The value for x is obtained by substituting these values into the quadratic formula. This gives $x = 0.031$ M or $x = -0.032$ M. The acceptable value of x is taken to be 0.031 M since concentration cannot be negative.
Therefore,

$$[PCl_3]_{eq} = 0.031 \text{ M}$$

$$[Cl_2]_{eq} = 0.031 \text{ M}$$

$$[PCl_5]_{eq} = 1.0 - 0.031 = 0.968 \text{ M}$$

Step 7:

$$K_c = \frac{[PCl_3][Cl_2]}{[PCl_5]} = \frac{(0.031)(0.031)}{0.968} = 1.0 \times 10^{-3}$$

This compares favorably with the given K_c of 1.0×10^{-3}.

17.8.1 A faster method for solving equilibrium problems

Note that we could have simplified the calculation from step 7 by assuming x to be very small compared to the initial concentration of PCl₅.

In other words, x is very small compared to 1, or $1 - x$ is approximately equal to 1, so that the equation:

$$1.0 \times 10^{-3} = \frac{x^2}{1-x}$$

becomes:

$$x^2 = 1.0 \times 10^{-3}.$$

$$x = \pm\sqrt{1.0 \times 10^{-3}},$$

and

$$x = +0.0315 \text{ or } -0.0315$$

17.8.2 When to use the approximation method

If the answer from the approximation method is less than 10% of the original concentration, then it is reasonably valid. However, if the approximation method gives an answer outside this range, then you will have to use the quadratic equation. For example, x above was found to be 0.0315 M, compared to the initial concentration of 1.0 M; that is, x is 3.2 percent of 1.0 M, so this answer is good.

Example 17.11

For the equilibrium

$$H_2(g) + I_2(g) \rightleftharpoons 2\,HI(g)$$

at 763 K, K_{eq} is 46.5. Calculate the composition of the equilibrium mixture obtained if 1.0 mole each of H_2 and I_2 were equilibrated in a closed 1-liter vessel at 763 K.

Solution

1. $H_2(g) + I_2(g) \rightleftharpoons 2\,HI(g)$

2. $K_c = \dfrac{[HI]^2}{[H_2][I_2]}$

3. $[H_2]_0 = 1.0\,M$
   $[I_2]_0 = 1.0\,M$
   $[HI]_0 = 0\,M$

4. Assuming x equals the amount of H_2 and I_2 that reacts, then the equilibrium concentrations are as shown in Table 17.4.
   Note that the change in [HI] is $2x$, which reflects the reaction stoichiometry.

Table 17.4   Data for Example 17.11

	$[H_2]$	$[I_2]$	$[HI]$
Initial concentration (M)	1	1	0
Change (M)	$(-)x$	$(-)x$	$(+)2x$
Equilibrium concentration (M)	$1-x$	$1-x$	$2x$

5.
$$K_c = \frac{[HI]^2}{[H_2][I_2]} = 46.5$$

$$46.5 = \frac{(2x)^2}{(1-x)(1-x)}$$

The right side of this equation is a perfect square, so we can simply take the square root of both sides. (We could also use the quadratic equation, but that would be more work.) Taking the square root of both sides we have:

$$6.82 = \frac{(2x)}{(1-x)}$$

$$2x + 6.82x = 6.82$$

$$8.82x = 6.82 \text{ or } x = 0.773 \ M$$

Thus,

$$[HI]_{eq} = 1.55 \text{ M}; \ [H_2]_{eq} = 0.227 \text{ M}; \ [I_2]_{eq} = 0.227 \text{ M}$$

6. Check for K_c.

$$K_c = \frac{[HI]^2}{[H_2][I_2]} = \frac{(1.55)^2}{(0.227)(0.227)} = 46.62$$

Example 7.12

The equilibrium constant K_p for the reaction

$$NH_4CO_2NH_2 (g) \rightleftharpoons 2 \ NH_3 (g) + CO_2 (g)$$

is 7.1×10^{-3} at 40°C. Calculate the partial pressures of all species present after 5.0 moles of ammonium carbamate, $NH_4CO_2NH_2$, is decomposed in an evacuated 2.5-liter vessel and equilibrium is reached at 40°C.

Solution

1. The equation given is balanced
2. $K_p = \dfrac{P_{NH_3} \times P_{CO_2}}{P_{NH_4CO_2NH_2}}$
3. Calculate the initial partial pressure of $NH_4CO_2NH_2$, using the ideal gas equation.

$$P_{NH_4CO_2NH_2} = \frac{nRT}{V} = \frac{(5 \text{ mol})\left(0.082\dfrac{\text{L.atm}}{\text{mol.K}}\right)(313 \text{ K})}{2.5 \text{ L}} = 51.33 \text{ atm.}$$

Table 17.5   Data for Example 17.12

(a)	[$NH_4CO_2NH_2$]	[NH_3]	[CO_2]
Initial concentration (M)	51.33	0	0
Change (M)	$(-)x$	$(+)2x$	$(+)x$
Equilibrium concentration (M)	$(51.33 - x)$	$2x$	x

(b)	[$NH_4CO_2NH_2$]	[NH_3]	[CO_2]
Initial concentration (M)	51.33	0	0
Change (M)	$(-)\,0.714$	$(+)\,1.428$	$(+)\,0.714$
Equilibrium concentration (M)	50.62	1.428	0.714

4. Determine the equilibrium partial pressure for each species present (Table 17.5a).
5. Calculate the partial pressures of all species present at equilibrium:

$$K_p = \frac{P_{NH_3} \times P_{CO_2}}{P_{NH_4CO_2NH_2}} = \frac{(2x)^2 \times x}{(51.33 - x)} = \frac{4x^3}{(51.33 - x)}$$

$$7.1 \times 10^{-3} = \frac{4x^3}{(51.33 - x)}$$

6. This equation would be difficult to solve, so we try the approximation method first:

$$7.1 \times 10^{-3} = \frac{4x^3}{51.33}$$

$$4x^3 = (7.1 \times 10^{-3})(51.33) = 0.36444$$

$$x = \sqrt[3]{0.36444} = 0.714$$

The equilibrium concentrations are summarized in Table 17.5b.

$$P_{NH_4CO_2NH_2} = 50.73 \text{ atm}; P_{NH_3} = 1.428 \text{ atm}; P_{CO_2} = 0.714 \text{ atm}.$$

7. Our assumption is valid since 0.714 is much less than 51.55.

17.9 Qualitative Treatment of Equilibrium: Le Chatelier's Principle

In 1888, Le Chatelier, a French industrial chemist, proposed a qualitative principle for predicting the behavior of a chemical system in equilibrium:

When a chemical system in equilibrium is disturbed by imposing a stress (e.g. a change in temperature, pressure, or concentration of reactants or products) on it, the position of the equilibrium will shift so as to counteract the effect of the disturbance.

17.9.1 Factors affecting a chemical reaction at equilibrium

There are three ways we can alter a chemical system in equilibrium. We can change:

1. the concentration of one of the reactants or products.
2. the pressure or volume.
3. the reaction temperature.

When any of the above factors in an equilibrium system is altered, Le Chatelier's principle can be used to predict how the system can regain equilibrium.

17.9.1.1 Changes in concentration

If a system is in equilibrium and we increase the concentration of one of the reactants or products, the reaction will shift in the forward or reverse direction to consume added reactant or product and restore equilibrium. Consider the Haber process for ammonia synthesis:

$$N_2(g) + 3 H_2(g) \rightleftharpoons 2 NH_3(g)$$

An increase in the concentration of nitrogen or hydrogen will cause the formation of more ammonia. This will undo the added stress of increased hydrogen or nitrogen. Therefore, the equilibrium will shift forward, towards the products. Similarly, a decrease in the concentration of ammonia (i.e., removing ammonia as soon as it is formed), will shift the reaction to the right so that more ammonia will be produced. Again the equilibrium will favor more product formation. On the other hand, if we remove some of the hydrogen gas at equilibrium, the reverse reaction, which favors the decomposition of ammonia to increase the concentration of nitrogen and hydrogen, will be favored.

The reaction quotient Q is very helpful when analyzing the effect of an imposed stress on a system in equilibrium. When a reaction is at equilibrium, $Q = K_c$. For the Haber process,

$$K_c = \frac{[NH_3]^2}{[N_2][H_2]^3} = Q$$

If we disturb the equilibrium by adding more nitrogen gas to the system, the denominator will increase relative to the numerator. Once this happens, Q becomes less than K_c. To restore the system to equilibrium (make $Q = K_c$), the concentration of ammonia in the numerator will have to increase. For this to happen, the overall reaction must proceed from left to right to form more ammonia. On the other hand, if we disturb the equilibrium by adding more ammonia to the system, Q will become larger than K_{eq} because the concentration of ammonia in the numerator has become larger. To restore equilibrium, the concentration of hydrogen, nitrogen or both will have to increase. This means the reverse reaction, favoring the decomposition of ammonia, will be favored.

Thus, without any memorization, we can use Q to predict how a reaction in equilibrium will respond to added stress and arrive at the same conclusion that Le Chatelier's principle will give us.

17.9.1.2 Changes in pressure

Changes in pressure or volume are only important for equilibrium reactions in which one or more reactants or products are gaseous. There are three ways to change the pressure of such a system. These include:

1. addition or removal of a gaseous reactant or product.
2. changing the volume of the container.
3. addition of an inert gas to the system.

An increase in pressure, like concentration, does not affect the value of K_{eq}, but the value of Q may be altered. The same principles outlined earlier for change in concentration can be used to predict how the system will deal with an added pressure stress. For equilibrium reactions involving gases, we can also use Le Chatelier's principles in combination with Avogadro's and Gay-Lussac's laws to predict how a system in equilibrium will handle a stress.

You may recall that at constant temperature and pressure, the volume of a given mass of gas is directly proportional to the number of moles of gas present. Generally, an increase in pressure (or a decrease in volume) will cause the equilibrium condition to shift in the direction where there are fewer numbers of moles of gas. A decrease in pressure (or an increase in volume) of an equilibrium mixture of gases will shift the equilibrium in the direction where there are more moles of gas.

Consider the following system at equilibrium at constant temperature.

$$2\,SO_3\,(g) \quad \rightleftharpoons 2\,SO_2\,(g) \quad + O_2\,(g)$$

| 2 molecules | 2 molecules | 1 molecule |
| 2 volumes | 2 volumes | 1 volume | (by Avogadro's law) |

The total moles of reactants and products are 2 and 3, respectively. Therefore, an increase in pressure will favor the reverse reaction since the equilibrium will shift in the direction where there are fewer moles of gas: 3 moles products (2 mol $SO_2 + 1$ mol O_2) $\longrightarrow$ 2 moles reactants (SO_3).

In cases where the total number of gas molecules remains constant, pressure change will have no effect on the equilibrium. Consider the reaction in which one mole of hydrogen reacts with one mole of iodine to form two moles of hydrogen iodide.

$$H_2\,(g) \;+ I_2\,(g) \quad \rightleftharpoons 2\,HI(g)$$

| 1 mole | 1 mole | 2 moles |
| 1 vol. | 1 vol. | 2 vol. | (by Avogadro's law) |

The number of moles on both sides of the equation is the same. A change in P does not favor either side, so will have no effect on the position of equilibrium.

The addition of an inert gas that is not involved in the equilibrium reaction has no effect on the equilibrium position. The total pressure is increased when an inert gas is added, but the concentrations or partial pressures of the reactants and products remain unchanged. Hence the original equilibrium position is maintained.

17.9.1.3 Changes in temperature

If the temperature remains constant, imposing a stress (change in concentration, pressure, or volume) on a reaction at equilibrium affects only the composition of the equilibrium mixture and hence the reaction quotient Q. The value of the equilibrium constant K_{eq} is not affected.

On the other hand, a change in temperature will alter the value of the equilibrium constant itself. There is a general relationship between temperature and K_{eq}. For an exothermic reaction (negative $\Delta H°$) K_c decreases as the temperature rises, but for an endothermic reaction (positive

$\Delta H°$), K_c increases as temperature rises. To use Le Chatelier's principle to describe the effect of temperature changes on a system at equilibrium, treat the heat of reaction as a product of an exothermic reaction, or as a reactant in an endothermic reaction.

The Haber synthesis of ammonia is an exothermic reaction, liberating about 92.2 kJ of heat energy:

$$N_2(g) + 3\,H_2(g) \rightleftharpoons 2\,NH_3(g) + 92.2\,kJ \qquad \Delta H° = -92.2\,kJ$$

If the temperature is increased, the equilibrium will shift to the left in favor of reactants to absorb the heat and thus annul the stress. If the temperature is decreased, the equilibrium will be displaced to the right, favoring the formation of more ammonia.

Let us consider another example. The transformation of graphite to diamond is an endothermic process represented as:

$$C(graphite) + 2\,kJ \rightleftharpoons C(diamond) \qquad \Delta H = +2\,kJ$$

If the temperature is increased, the equilibrium shifts in the direction that absorbs heat. This is an endothermic reaction; raising the temperature will shift the equilibrium to the right in favor of the formation of more diamond.

17.9.2 Addition of a catalyst

The condition for a chemical system to be at equilibrium is that the rates of the forward and reverse reactions be equal. The addition of a catalyst affects the rates of both the forward and reverses reactions equally. A catalyst increases the rate at which equilibrium is reached. It neither affects the equilibrium concentrations of reactants and products nor the equilibrium constant for the reaction.

Example 17.13

The formation of sulfur trioxide is a key step in the "contact process" for manufacturing sulfuric acid. The equilibrium reaction is described by the equation below:

$$2\,SO_2(g) + O_2(g) \rightleftharpoons 2\,SO_3(g) \quad \Delta H = -196\,kJ/mol$$

What will be the effect on the equilibrium if

(a) The temperature is increased?
(b) The pressure is increased?
(c) The pressure is decreased?
(d) SO_3 is removed as soon as it is formed?
(e) O_2 is removed from the reaction mixture?

Solution

(a) Since the reaction is exothermic, an increase in temperature favors the formation of reactants. This will cause the decomposition of SO_3 to yield more SO_2 and O_2, thus shifting the equilibrium towards the reactants.
(b) An increase in pressure will force the equilibrium to shift in the direction where there are fewer moles of gas. There are 3 moles of reactants and 2 moles of the products.

Therefore, the equilibrium will shift from left to right, in the direction of SO_3, as the pressure is increased.

(c) A decrease in pressure will force the equilibrium to shift in the direction where there are more gaseous molecules. There are 3 moles of reactant molecules and 2 moles of the products. Therefore, the equilibrium will shift from right (product) to left, in the direction of reactants, as the pressure is decreased.

(d) The removal of products as soon as they are formed results in the formation of more products. Therefore, the removal of SO_3 causes more SO_2 and O_2 to react to form additional SO_3, thus shifting the equilibrium from left to right.

(e) The removal of one or more reactants causes one or more products to decompose to form additional reactants. Thus, removal of O_2 will shift the equilibrium from right (products) to left (reactants).

Example 17.14

Nitric oxide, a notable air pollutant, is produced in automobile engines at elevated temperatures according to the reaction:

$$N_2(g) + O_2(g) \rightleftharpoons 2\,NO(g) \quad \Delta H^\circ = +180.5 \text{ kJ}$$

What is the effect on equilibrium of increasing the pressure, assuming no change in temperature?

Solution

An increase in pressure will shift the equilibrium in the direction where there are fewer gaseous molecules. Since the reaction contains an equal number of moles of gas on both sides of the equation, an increase or decrease in pressure will have no effect on the position of equilibrium.

Example 17.15

Consider the decomposition of limestone, $CaCO_3$, represented by

$$CaCO_3(s) \rightleftharpoons CaO(s) + CO_2(g) \quad \Delta H = +178.5 \text{ kJ}$$

What is the effect on the equilibrium position of

(a) an increase in pressure?
(b) adding a catalyst?

Solution

(a) Since one of the products is a gas, an increase in pressure is equivalent to an increase in the concentration of the gas involved. Increasing the pressure will shift the position of equilibrium to the left, in favor of the reactants.

(b) A catalyst does not alter the equilibrium position of a reaction. Therefore the addition of a catalyst to this reaction will have no effect.

17.10 Problems

1. Write the equilibrium constant expression (K_c) for the following reactions:

   (a) $2\,SO_2\,(g) + O_2\,(g) \rightleftharpoons 2\,SO_3\,(g)$

   (b) $CO\,(g) + H_2O\,(g) \rightleftharpoons CO_2\,(g) + H_2\,(g)$

   (c) $CH_4\,(g) + 2H_2S\,(g) \rightleftharpoons CS_2\,(g) + 4\,H_2\,(g)$

   (d) $N_2O_4\,(g) \rightleftharpoons 2\,NO_2\,(g)$

2. Write the equilibrium expression for K_p for the reactions in problem 1.

3. Write the equilibrium constant expression (K_c) for the following reactions:

   (a) $CO\,(g) + 2H_2\,(g) \overset{\text{Catalyst}}{\rightleftharpoons} CH_3OH\,(g)$

   (b) $CH_3Cl\,(aq) + OH^-\,(aq) \rightleftharpoons CH_3OH\,(aq) + Cl^-\,(aq)$

   (c) $Ag^+\,(aq) + 2NH_3\,(aq) \rightleftharpoons Ag(NH_3)_2^+\,(aq)$

   (d) $CH_3COOH\,(aq) + C_2H_5OH\,(aq) \rightleftharpoons CH_3COOC_2H_5\,(aq) + H_2O(l)$

4. Write the equilibrium constant (K_p) expression for the following reactions:

   (a) $PCl_3\,(g) + Cl_2\,(g) \rightleftharpoons PCl_5\,(g)$

   (b) $2\,PCl_3\,(g) + O_2\,(g) \rightleftharpoons 2\,POCl_3\,(g)$

   (c) $N_2H_4\,(g) + 6\,H_2O_2\,(g) \rightleftharpoons 2\,NO_2\,(g) + 8\,H_2O\,(g)$

   (d) $NO_2\,(g) + SO_2\,(g) \rightleftharpoons NO\,(g) + SO_3\,(g)$

5. The following are the equilibrium constant expressions (K_c) for totally gaseous equilibrium reactions. Write a balanced chemical equation for each of them.

   (a) $\dfrac{[H_2][F_2]}{[HF]^2}$ (b) $\dfrac{[CO_2]^4[H_2O]^2}{[C_2H_2]^2[O_2]^5}$

   (c) $\dfrac{[CO_2][H_2]}{[CO][H_2O]}$ (d) $\dfrac{[HCl]^4[O_2]}{[Cl_2]^2[H_2O]^2}$

6. The following are the equilibrium constant expression (K_c) for totally gaseous equilibrium reactions. Write a balanced chemical equation for each of them.

   (a) $\dfrac{[N_2][H_2]^3}{[NH_3]^2}$ (b) $\dfrac{[CS_2][H_2]^4}{[CH_4][H_2S]^2}$

   (c) $\dfrac{[H_2][CO]}{[CH_2O]}$ (d) $\dfrac{[NO]^2}{[N_2][O_2]}$

7. Rewrite the following K_c expressions as K_p.

   (a) $\dfrac{[N_2][H_2]^3}{[NH_3]^2}$ (b) $\dfrac{[CS_2][H_2]^4}{[CH_4][H_2S]^2}$

   (c) $\dfrac{[H_2][CO]}{[CH_2O]}$ (d) $\dfrac{[NO]^2}{[N_2][O_2]}$

8. For the following reactions, give a qualitative description of the position of equilibrium. Also, predict the relative concentration of the products and reactants at 298 K.

   (a) $2Cl_2(g) + 2H_2O(g) \rightleftharpoons 4HCl(g) + O_2(g)$ $K_p = 3.2 \times 10^{-14}$
   (b) $H_2(g) + I_2(g) \rightleftharpoons 2HI(g)$ $K_{eq} = 51$

9. For the reaction $2\,SO_2(g) + O_2(g) \rightleftharpoons 2\,SO_3(g)$, the equilibrium constant, K_c, is 3.5×10^6 at 400°C.

   (a) At this temperature, does the equilibrium favor the product, SO_3, or the reactants, SO_2 and O_2?
   (b) Calculate the value of K_p for this reaction.
   (c) Calculate K_c for the reaction $2SO_3(g) \rightleftharpoons 2SO_2(g) + O_2(g)$.

10. A 1-liter reaction vessel containing SO_2 gas and O_2 gas was equilibrated at 300°C. The equilibrium concentrations of SO_2, O_2, and SO_3 in the gaseous mixture were 0.229 M, 0.115 M, and 0.0373 M, respectively. Determine K_c for the reaction $2\,SO_2(g) + O_2(g) \rightleftharpoons 2\,SO_3(g)$.

11. The following reaction was carried out in a 7.14-L vessel at 690°C.

$$2\,H_2(g) + S_2(g) \rightleftharpoons 2\,H_2S(g)$$

Upon analysis the equilibrium mixture consisted of the following: $[H_2] = 1.5$ mol; $[S_2] = 7.8 \times 10^{-6}$ mol; and $[H_2S] = 5.35$ mol. Calculate the equilibrium constant K_c.

12. For the reaction

$$CH_3COOH(aq) + C_2H_5OH(aq) \rightleftharpoons CH_3COOC_2H_5(aq) + H_2O(g) \quad K_c = 4$$

calculate the equilibrium composition of reactants and products obtained by reacting one mole of ethanoic acid (CH_3COOH) with one mole of ethanol (C_2H_5OH) in a 1.0-L flask.

13. The equilibrium constant, K_c, for the reaction

$$H_2(g) + I_2(g) \rightleftharpoons 2HI(g)$$

is 60 at 425°C. What is the composition of the equilibrium mixture if 8 g of hydrogen react with 508 g of iodine vapor in a sealed vessel?

14. A sample of phosgene, $COCl_2$, was allowed to decompose by heating it in a sealed 1 dm^3 reactor. The reaction is shown below:

$$COCl_2(g) \rightleftharpoons CO(g) + Cl_2(g) \quad K_c = 2.2 \times 10^{-10}$$

Determine the equilibrium concentrations of the products and residual reactant assuming an initial concentration of $COCl_2$ of 0.10 M.

15. Write the K_p expressions for the following equilibrium reactions:

   (a) $SrC_2O_4 \cdot 2H_2O(s) \rightleftharpoons SrC_2O_4(s) + 2\,H_2O(g)$
   (b) $CaCl_2 \cdot 6H_2O(s) \rightleftharpoons CaCl_2(s) + 6\,H_2O(g)$
   (c) $Hg_2O \cdot H_2O(s) \rightleftharpoons Hg_2O(s) + H_2O(g)$
   (d) $Na_2SO_4 \cdot 10H_2O(s) \rightleftharpoons Na_2SO_4(s) + 10H_2O(g)$

16. At 25°C, the values of K_p for problem 15(b) and (d) are 5.1×10^{-44} and 1.0×10^{-16}, respectively. Calculate the water vapor pressure above each of these hydrated salts.

17. 0.1125 mole of Br_2 gas is heated in a 1.50-L flask at 1800 K. At equilibrium, the bromine is 5.0% dissociated. Determine the equilibrium constant, K_c.

18. NH_4HS decomposes into ammonia and hydrogen sulfide at low temperatures:

$$NH_4HS(g) \rightleftharpoons NH_3(g) + H_2S(g)$$

(a) In a 1.0-L vessel, a 0.05-mole sample of NH_4HS was heated at 80°C. Calculate K_c for the reaction if the vessel contained 0.03 moles of H_2S at equilibrium.
(b) Using the ideal gas equation, calculate the pressure in the vessel at equilibrium.
(c) What is the partial pressure of each gas in the reaction vessel?
(d) Calculate K_p for the reaction.

19. With reference to the following reaction:

$$2\ POCl_3(g) \rightleftharpoons 2\ PCl_3(g) + O_2(g) \quad \Delta H = +572\ kJ$$

(a) What is the effect of increasing the temperature?Assume no change in pressure.
(b) What is the effect of increasing the pressure?Assume no change in temperature.
(c) Describe the effect of removing PCl_3.
(d) How does the equilibrium amount of O_2 vary if a platinum catalyst is added?

20. The initial step of the Oswald process for the industrial production of nitric acid is the oxidation of ammonia to nitric oxide

$$4\ NH_3(g) + 5\ O_2(g) \rightleftharpoons 4\ NO(g) + 6\ H_2O(g) \quad \Delta H^o = -905.6\ kJ$$

What happens to the amount of NO when

(a) the temperature is increased?
(b) a platinum catalyst is added?
(c) the pressure is increased by adding helium gas?
(d) some ammonia gas is pumped out of the reactor?

18

Ionic Equilibria and pH

. . . .

18.1 The Ionization of Water

Water is a weak acid. At 25°C, pure water ionizes to form a hydrogen ion and a hydroxide ion:

$$H_2O \rightleftharpoons H^+ + OH^-$$

Hydration of the proton (hydrogen ion) to form hydroxonium ion is ignored here for simplicity. The equilibrium lies mainly to the left; that is, the ionization only happens to a slight extent. We know that 1 L of pure water contains 55.6 moles. Of this, only 10^{-7} mole actually ionizes into equal amounts of $[H^+]$ and $[OH^-]$, i.e.

$$\left[H^+\right] = \left[OH^-\right] = 10^{-7} \, M$$

This is why pure water is neutral. It contains equal amounts of hydrogen ions and hydroxide ions.
A solution is acidic if it contains more hydrogen ions than hydroxide ions. Similarly, a solution is basic if it contains more hydroxide ions than hydrogen ions.

18.2 Definition of Acidity and Basicity

Acidity is defined as the concentration of hydrated protons (hydrogen ions); basicity is the concentration of hydroxide ions.

18.2.1 Ionic product of water

Pure water ionizes at 25°C to produce 10^{-7} M of $[H^+]$ and 10^{-7} M of $[OH^-]$. The product

$$\left[H^+\right] \times \left[OH^-\right] = 10^{-7} \, M \times 10^{-7} M = 10^{-14} \, M$$

is known as the ionic product of water. Note that this is simply the equilibrium expression for the dissociation of water. This equation holds for any dilute aqueous solution of acid, base, and salt.

Chemistry in Quantitative Language: Fundamentals of General Chemistry Calculations. Second Edition.
Christopher O. Oriakhi, Oxford University Press. © Christopher O. Oriakhi 2021.
DOI: 10.1093/oso/9780198867784.003.0018

Example 18.1

The hydrogen ion concentration of a solution is 10^{-5} M. What is the hydroxide ion concentration? Is the solution acidic or basic?

Solution

$$[H^+] = 10^{-5}M$$
$$[OH^-] = ?$$
$$[H^+][OH^-] = 10^{-14}$$
$$[OH^-] = \frac{10^{-14}}{[H^+]} = \frac{10^{-14}}{10^{-5}} = 10^{-9}M$$

The solution is acidic, because $[H^+] = 10^{-5}M > [OH^-] = 10^{-9}M$

18.3 The pH of a Solution

The pH of a solution is defined as *the negative logarithm of the molar concentration of hydrogen ions*. The lower the pH, the greater the acidity of the solution. Mathematically

$$pH = -\log_{10}[H^+] \text{ or } -\log_{10}[H_3O^+]$$

This can also be written as:

$$pH = \log_{10}\frac{1}{[H^+]} \text{ or } \log_{10}\frac{1}{[H_3O^+]}$$

Taking the antilogarithm of both sides and rearranging gives

$$[H_3O^+] = 10^{-pH}$$

This equation can be used to calculate the hydrogen ion concentration when the pH of the solution is known.

Example 18.2

Calculate the pH of a 0.0010 M solution of nitric acid.

Solution

$$[H_3O^+] = 10^{-3} M$$
$$pH = -\log_{10}[H_3O^+]$$
$$= -\log_{10}\left[10^{-3}\right]$$
$$= 3$$

Example 18.3

Calculate the $[H_3O^+]$ of a solution having a pH of 5.

Solution

$$[H_3O^+] = 10^{-pH}$$
$$= 10^{-5} \, M$$

Example 8.4

What is the $[H_3O^+]$ in a cup of coffee with a pH of 5.05?

Solution

$$[H_3O^+] = 10^{-pH}$$
$$= 10^{-5.05} = 8.9 \times 10^{-6} \, M$$

18.4 The pOH of a Solution

The pOH of a solution is defined as the negative logarithm of the molar concentration of hydroxide ions. The lower the pOH, the greater the basicity of the solution. Mathematically

$$pOH = -\log_{10}[OH^-]$$

The relationship between pH and pOH is

$$pH + pOH = 14$$

If the pH of a solution is known, the pOH can be calculated using the above equation.

Example 18.5

Tomato juice has a pH of 4.2. What is the pOH of the solution?

Solution

$$pH = 4.2$$
$$pOH = ?$$
$$pH + pOH = 14$$
$$4.2 + pOH = 14$$
$$pOH = 9.8$$

Example 18.6

Lemon juice has a pOH of 11.7. Calculate the pH of the lemon juice.

Solution

$pOH = 11.7$

$pH = ?$

$pH + pOH = 14$

$pH + 11.7 = 14$

$pOH = 14 - 11.7 = 2.3$

Example 18.7

Calculate the pH of each of the following solutions:

(a) A 0.25 M solution of KOH
(b) A 0.0025 M solution of $Mg(OH)_2$

Solution

(a) KOH is a strong base and so it is 100% ionized in aqueous solution. Therefore, a 0.25 M KOH solution will provide one OH^- per molecule. $[OH^-] = 0.25$ M.

$$[H^+][OH^-] = 1.0 \times 10^{-14}$$

$$[H^+] = \frac{1.0 \times 10^{-14}}{[OH^-]} = \frac{1.0 \times 10^{-14}}{0.25 \text{ M}} = 4 \times 10^{-14} \text{M}$$

$$pH = -\log_{10}[H^+] = -\log (4.0 \times 10^{-14}) = 13.4$$

(b) $Mg(OH)_2$ is a strong base and so it is 100% ionized in aqueous solution. Therefore a 0.0025 M $Mg(OH)_2$ solution will provide two OH^- per molecule. $[OH^-] = 2 \times 0.0025$ M $= 0.005$ M.

$$[H^+][OH^-] = 1.0 \times 10^{-14}$$

$$[H^+] = \frac{1.0 \times 10^{-14}}{[OH^-]} = \frac{1.0 \times 10^{-14}}{0.005 \text{ M}} = 2 \times 10^{-12} \text{ M}$$

$$pH = -\log_{10}[H^+] = -\log (2.0 \times 10^{-12}) = 11.7$$

18.5 The Acid Ionization Constant, K_a

Protonic acids undergo ionization in aqueous solution. The ionization process may be represented by:

$$HA\,(aq) + H_2O\,(l) \rightleftharpoons H_3O^+ + A^-$$

or simply

$$HA\,(aq) \rightleftharpoons H^+ + A^-$$

The equilibrium expression is then

$$K_a = \frac{[H^+][A^-]}{[HA]}$$

K_a is called the ionization constant and is a measure of the strength of an acid. The larger the K_a, the stronger the acid. Strong acids undergo complete or 100% ionization and have K_a values greater than 1. Weak acids, on the other hand, are only partially ionized, with K_a values less than 1. For example, HCl is a strong acid, which ionizes in water according to

$$HCl\,(aq) \longrightarrow H^+\,(aq) + Cl^-\,(aq)$$

The equilibrium expression for the reaction is

$$K_a = \frac{[H^+][Cl^-]}{[HCl]} = 2 \times 10^6$$

Since HCl dissociates fully, [HCl] will be nearly zero, making K_a very large. Acetic acid (CH_3COOH), on the other hand, is a weak acid and dissociates according to

$$CH_3COOH\,(aq) \rightleftharpoons H^+\,(aq) + CH_3COO^-\,(aq)$$

A significant amount of CH_3COOH is still present, so the equilibrium constant measured at 25°C is 1.8×10^{-5}.

18.5.1 Definition of pK_a

Like pH, pK_a is defined as the negative logarithm of the acid ionization constant, K_a. That is

$$pK_a = -\log K_a$$

A large K_a value will have a corresponding small pK_a, and vice versa.

18.6 Calculating pH and Equilibrium Concentrations in Solutions of Weak Acids

Acid–base equilibrium problems are solved using the same methods adopted for any other equilibrium problems.

Table 18.1   Data for Example 18.8

	[HCN]	[H⁺]	[CN⁻]
Initial concentration (M)	0.1	0	0
Change (M)	(−) 0.0003	(+) 0.0003	(+) 0.0003
Equilibrium concentration (M)	0.0997	0.0003	0.0003

Example 18.8

The pH of 0.10 M HCN is 3.5. Calculate the value of K_a for hydrocyanic acid (Table 18.1).

Solution

1. First, write the balanced equation for the dissociation reaction:

$$HCN\,(aq) \rightleftharpoons H^+\,(aq) + CN^-\,(aq)$$

2. Then, write the expression for K_a:

$$K_a = \frac{[H^+][CN^-]}{[HCN]}$$

3. Next, calculate the hydrogen ion concentration $[H^+]$ from the given pH.

$$pH = -\log[H^+]$$
$$[H^+] = 10^{-pH} = 10^{-3.5} = 3.0 \times 10^{-4}M$$
$$\text{But } [CN^-] = [H^+] = 3.0 \times 10^{-4}M$$

4. Then, calculate the concentrations in the equilibrium mixture, i.e., [HCN], [H⁺], and [CN⁻].
5. Finally, calculate K_a by substituting the various equilibrium concentrations in the expression for K_a.

$$K_a = \frac{[H^+][CN^-]}{[HCN]} = \frac{(0.0003)(0.0003)}{(0.0997)} = 9.03 \times 10^{-7}$$

Example 18.9

Calculate the equilibrium concentrations of the various species present in a 0.1 M solution of HF ($K_a = 6.5 \times 10^{-4}$). What is the pH of the solution (Table 18.2)?

Solution

1. Write the balanced equation for the equilibrium dissociation reaction:

$$HF\,(aq) \rightleftharpoons H^+\,(aq) + F^-\,(aq)$$

Table 18.2   Data for Example 18.9

	[HF]	[H⁺]	[F⁻]
Initial concentration (M)	0.1	0	0
Change (M)	$(-)x$	$(+)x$	$(+)x$
Equilibrium concentration (M)	$0.1 - x$	x	x

2. Write an expression for K_a:

$$K_a = \frac{[\text{H}^+][\text{F}^-]}{[\text{HF}]}$$

3. Set up a table listing the initial and equilibrium concentrations of the species present.
4. Substitute the equilibrium concentrations into the K_a expression and then solve for the unknown, x.

$$K_a = \frac{(x) \times (x)}{0.1 - x} = \frac{x^2}{0.1 - x} = 6.5 \times 10^{-4}$$

$$x^2 + 6.5 \times 10^{-4}x - 6.5 \times 10^{-5} = 0$$

This is a quadratic equation. Solve for x using the quadratic formula

$$x = \frac{-b \pm \sqrt{b^2 - 4ac}}{2a}, \text{ where } a = 1, \ b = 6.5 \times 10^{-4}, \text{ and } c = -6.5 \times 10^{-5}$$

$$x = \frac{-\left(6.5 \times 10^{-4}\right) \pm \sqrt{\left(6.5 \times 10^{-4}\right)^2 - 4(1)\left(-6.5 \times 10^{-5}\right)}}{2(1)}$$

$$x = \frac{-\left(6.5 \times 10^{-4}\right) \pm 0.0173}{2}$$

$$x = 0.0083 \text{ or } -0.0090$$

Since x is the concentration of H^+, it must have a positive value. Thus $[\text{H}^+] = 0.008$ M. We can also solve this problem by using the approximation method.
Here we assume that $x \ll 0.1$

$$0.1 - x \approx 0.1 \text{ and hence } \frac{x^2}{0.1 - x} = \frac{x^2}{0.1} = 6.5 \times 10^{-4}$$

On solving, $x^2 = \left(6.5 \times 10^{-4}\right)(0.1)$ or $x = 0.008$ (see below for a discussion of how to decide if this answer is valid).

Therefore equilibrium $[\text{H}^+] = [\text{F}^-] = 0.008$ M and $[\text{HF}] = 0.092$ M

5. Now calculate the pH of the solution.

$$pH = -\log[\text{H}^+]$$
$$= -\log(0.008)$$
$$= 2.1$$

18.6.1 When to use the approximation method

The approximation method works best when x is less than 10% of the original concentration. In the preceding example, x was found to be equal to 0.008 compared to the initial concentration of 0.10 M: That is, x is 8% of 0.10. If this criterion is not met, use the quadratic equation to solve for x.

18.7 Percent Dissociation of Weak Acids

Since weak acids are only partially dissociated in aqueous solution, the percent dissociation or ionization can be used in conjunction with K_a to measure the strength of an acid. It is calculated from the initial concentration of the acid and the concentration of hydrogen ion at equilibrium. The percent dissociation is defined as follows:

$$\text{Percent dissociation} = \frac{\text{Amount of acid dissociated (mol/L)}}{\text{Initial concentration (mol/L)}}$$

For a monoprotic acid HA,

$$\%\text{Dissociation} = \frac{[H^+]_{eq}}{[HA]_{initial}} \times 100\%$$

Example 18.10

An etching solution is prepared by dissolving 1.0 mole of HF in 1 liter of water. If this solution contains 0.025 M of hydrogen ion concentration, calculate the percent dissociation of HF.

Solution

$$\%\text{Dissociation} = \frac{[H^+]_{eq}}{[HF]_{initial}} \times 100\%$$

$$= \frac{0.025\ M}{1.0\ M} \times 100 = 2.5\%$$

Example 18.11

Calculate the pH and the degree of dissociation of a 0.10 M benzoic acid ($HC_7H_5O_2$) solution. K_a for benzoic acid is 6.6×10^{-5} (Table 18.3).

Solution

1. First, write equation for the reaction.

$$HC_7H_5O_2\,(aq) \rightleftharpoons H^+\,(aq) + C_7H_5O_2^-\,(aq)$$

Table 18.3   Data for Example 18.11

	$[HC_7H_5O_2]$	$[H^+]$	$[C_7H_5O_2^-]$
Initial concentration (M)	0.1	0	0
Change (M)	$(-)x$	$(+)x$	$(+)x$
Equilibrium concentration (M)	$0.10 - x$	x	x

2. Then, write an expression for the dissociation constant.

$$K_a = \frac{[H^+][C_7H_5O_2^-]}{[HC_7H_5O_2]}$$

3. Next, calculate the equilibrium concentrations of all species using the table.
4. Then, substitute in the equation for K_a. Determine x and hence $[H^+]$.

$$K_a = \frac{(x)(x)}{0.10 - x} = \frac{x^2}{0.10 - x} = 6.6 \times 10^{-5}$$

Assume $0.10 << x$

$$6.6 \times 10^{-5} = \frac{x^2}{0.10}$$

$$x = 2.6 \times 10^{-3} = [H^+] \text{ (Assumption valid)}$$

$$pH = -\log(2.6 \times 10^{-3}) = 2.59$$

5. Finally, calculate the percent dissociation.

$$\%\text{Dissociation} = \frac{[H^+]}{[HA]} \times 100 = \frac{2.6 \times 10^{-3}}{0.10} \times 100 = 2.6\%$$

Example 18.12

The accumulation of lactic acid ($HC_3H_5O_3$) in muscle tissue during exertion or "workout" often results in pain and tiredness. Calculate pK_a for lactic acid, if a 0.0500 M solution is 4% ionized.

Solution

1. First, write the equation for the reaction.

$$HC_3H_5O_3 \text{ (aq)} \rightleftharpoons H^+ \text{ (aq)} + C_3H_5O_3^- \text{ (aq)}$$

2. Next, write an expression for the dissociation constant.

$$K_a = \frac{[H^+][C_3H_5O_3^-]}{[HC_3H_5O_3]}$$

3. Then, from the percent dissociation given, calculate the equilibrium concentration of hydrogen ions.

$$\%\text{Dissociation} = \frac{[H^+]_{eq}}{[HC_3H_5O_3]_0} \times 100 = 4$$

$$[H^+]_{eq} = \frac{4 \times [HC_3H_5O_3]_0}{100}$$

But $[HC_3H_5O_3]_0 = 0.0500$ M

Hence $[H^+]_{eq} = 0.0020$ M

4. Next, solve for K_a by substituting the equilibrium concentrations of the species in solution.

   At equilibrium,

$$[H^+]_{eq} = [C_3H_5O_3^-]_{eq} = 0.0020 \text{ M}$$

$$[HC_3H_5O_3]_{eq} = [HC_3H_5O_3]_0 - [H^+]_{eq} = 0.050 - 0.0020 = 0.0480 \text{ M}$$

$$K_a = \frac{[H^+][C_3H_5O_3^-]}{[HC_3H_5O_3]} = \frac{(0.0020)(0.0020)}{0.0480} = 1.0 \times 10^{-4}$$

5. Finally, solve for pK_a.

$$pK_a = -\log K_a$$

$$= -\log(0.0001) = 4$$

18.8 The Base Dissociation Constant, K_b

Generally, weak bases (B) accept a proton from water to form the conjugate acid of the base (BH^+) and OH^- ions.

$$B\,(aq) + H_2O\,(l) \rightleftharpoons BH^+\,(aq) + OH^-\,(aq)$$

The base dissociation constant for this equilibrium reaction is

$$K_b = \frac{[BH^+][OH^-]}{[B]}$$

The weak base B may be a neutral molecule such as ammonia or an organic amine, or an anion such as fluoride (F^-) or cyanide (CN^-).

Problems involving equilibria in solutions of weak bases are solved by the same procedure used for weak acid problems.

Example 18.13

Aniline $(C_6H_5NH_2)$ is an organic base used in the synthesis of the electronically conducting poly(aniline). Calculate the pH and the concentrations of all species present in a 0.25 M solution of aniline $(K_b = 4.3 \times 10^{-10})$ (Table 18.4).

Table 18.4   Data for Example 18.13

	$[C_6H_5NH_2]$	$[C_6H_5NH_3{}^+]$	$[OH^-]$
Initial concentration (M)	0.25	0	0
Change (M)	$(-)x$	$(+)x$	$(+)x$
Equilibrium concentration (M)	$0.25 - x$	x	x

Solution

1. Write an equation for the reaction and an expression for K_b.

$$C_6H_5NH_2\,(aq) + H_2O\,(l) \rightleftharpoons C_6H_5NH_3^+\,(aq) + OH^-\,(aq)$$

$$K_b = \frac{\left[C_6H_5NH_3^+\right]\left[OH^-\right]}{[C_6H_5NH_2]}$$

2. Set up a table showing the species and their concentrations at equilibrium.
3. Solve for the value of x from the equilibrium equation to find the concentrations of all species at equilibrium.

$$K_b = 4.3 \times 10^{-10} = \frac{\left[C_6H_5NH_3{}^+\right]\left[OH^-\right]}{[C_6H_5NH_2]} = \frac{(x)(x)}{(0.25 - x)}$$

If we assume that $x << 0.25$,

$$x^2 = (0.25)(4.3 \times 10^{-10}) = 1.075 \times 10^{-10}$$

$$x = 1.037 \times 10^{-5}$$

So our assumption was justified, Thus, $\left[C_6H_5NH_3{}^+\right] = \left[OH^-\right] = 1.037 \times 10^{-5}$ M and

$$[C_6H_5NH_2]_{eq} = 0.25 - 0.00001037 \text{ or } 0.24999 \text{ M} \approx 0.25 \text{ M}$$

4. Now calculate the hydrogen ion concentration and the pH. Note that the small $[H^+]$ concentration is produced from the subsidiary equilibrium dissociation of water:

$$\left[H^+\right]\left[OH^-\right] = 10^{-14}$$

$$\left[H^+\right] = \frac{10^{-14}}{\left[OH^-\right]} = \frac{10^{-14}}{1.0 \times 10^{-5}} = 1.0 \times 10^{-9} \text{ M}$$

$$pH = -\log\left[H^+\right] = -\log(1.0 \times 10^{-9}) = 9$$

So the solution is mildly basic as expected.

18.9 Relationship Between K_a and K_b

The relationship between K_a for a weak acid and K_b for its conjugate base can be deduced by considering the equilibrium conditions of the acidic ionization of an acid HA and the basic ionization of its conjugate base A^- in aqueous solution. The derivation is shown in the following steps:

1. $HA \rightleftharpoons H^+ + A^-$

   $$K_a = \frac{[H^+][A^-]}{[HA]}$$

2. $A^- + H_2O \rightleftharpoons HA + OH^-$

   $$K_b = \frac{[HA][OH^-]}{[A^-]}$$

3. Multiply the expressions for K_a and K_b and eliminate common terms:

   $$K_a \times K_b = \frac{[H^+][\cancel{A^-}]}{\cancel{[HA]}} \times \frac{\cancel{[HA]}[OH^-]}{\cancel{[A^-]}} = [H^+][OH^-]$$

4. Recall that $[H^-][OH^-] = K_w$

   $$K_a K_b = K_w$$

Of course this also holds between K_b for a weak base and K_a for its conjugate acid.

Example 18.14

Calculate the pH of a 0.10 M KF solution. The K_a for HF is 6.4×10^{-4}. ($K_w = 1.0 \times 10^{-14}$.)

Solution

1. Write an equation for the reaction of KF in water (specifically, for the F^- ion).

   $$F^-(aq) + H_2O(l) \rightleftharpoons HF(aq) + OH^-(aq)$$

2. Write an expression for K_b for the above reaction.

   $$K_b = \frac{K_w}{K_a} = \frac{1 \times 10^{-14}}{6.4 \times 10^{-4}} = 1.56 \times 10^{-11} = \frac{[HF][OH^-]}{[F^-]}$$

3. Solve for $[OH^-]$, and hence for $[H^+]$.

   $$K_b = 1.56 \times 10^{-11} = \frac{[HF][OH^-]}{[F^-]}$$

   Let $[HF] = [OH^-] = x$ at equilibrium.
   Initial $[F^-] = 0.10$ M. Equilibrium value $[F^-]_{eq} = 0.10 - x$

Substitute these in the K_b expression:

$$1.56 \times 10^{-11} = \frac{(x)(x)}{0.10 - x} = \frac{x^2}{0.10 - x}$$

Assume that $x << 0.10$ and solve the resulting equation:

$$x^2 = 1.56 \times 10^{-12} \text{ or } x = 1.25 \times 10^{-6}$$

$$[OH^-] = 1.25 \times 10^{-6} M$$

But

$$[H^+] = \frac{K_w}{[OH^-]} = \frac{1.0 \times 10^{-14}}{1.25 \times 10^{-6}} = 8.0 \times 10^{-9} \text{ M}$$

4. Now we can find the pH of the solution.

$$pH = -\log[H^+] = -\log(8.0 \times 10^{-9}) = 8.10$$

Example 18.15

The dissociation constant of benzoic acid, C_6H_5COOH, is 6.46×10^{-5}. Calculate K_b and pK_b for the benzoate ion, $C_6H_5COO^-$.

Solution

$$K_a K_b = K_w$$

$$K_b = \frac{K_w}{K_a} = \frac{1.0 \times 10^{-14}}{6.46 \times 10^{-5}} = 1.55 \times 10^{-10}$$

$$pK_b = -\log K_b = -\log(1.55 \times 10^{-10}) = 9.8$$

18.10 Salt Hydrolysis: Acid–Basis Properties of Salts

In a neutralization reaction, an acid reacts with a base to form water and a salt, a substance composed of an anion and a cation. Characteristically, salts are strong electrolytes that completely dissociate into anions and cations in water. Salt hydrolysis refers to the reaction of the anionic or cationic component of a salt (or both) with water. It is essentially a proton transfer reaction between salt and water and is the same process as ionization of weak acid or base. Therefore the mathematical treatment of the equilibrium process is the same.

There are three different kinds of salts:

1. Acidic salts which ionize in water to produce acidic solutions. These are typically salts of a weak base and a strong acid. Examples include NH_4Cl, $AlCl_3$, and $Fe(NO_3)_3$.
2. Basic salts which ionize in water to produce basic solutions. These are typically salts of a strong base and a weak acid. Examples include KNO_2, CH_3COONa, and Na_2S.

3. Neutral salts that ionize in water to produce a neutral (or nearly neutral) solution. These are typically salts of a strong acid and a strong base. Examples include $NaNO_3$, KI, and $BaCl_2$.

18.10.1 The hydrolysis constant, K_h

Consider the hydrolysis reaction of ammonium chloride.

$$NH_4Cl(s) \longrightarrow NH_4^+(aq) + Cl^-(aq) \tag{18.1}$$

$$NH_4^+(aq) + H_2O(l) \rightleftharpoons NH_3(aq) + H_3O^+(aq) \tag{18.2}$$

We can write the equilibrium expression for the hydrolysis reaction

$$K_h = \frac{[NH_3][H_3O^+]}{[NH_4^+]} \tag{18.3}$$

The equilibrium constant K_h is known as the hydrolysis constant.

18.10.2 Relationship between K_h and K_w

For the weak acid HA

$$HA + H_2O \rightleftharpoons H_3O^+ + A^- \text{ and } K_a = \frac{[H_3O^+][A^-]}{[HA]}$$

For the hydrolysis of the conjugate base A^-

$$A^- + H_2O \rightleftharpoons [HA] + OH^- \text{ and } K_h = \frac{[HA][OH^-]}{[A^-]}$$

Now multiply the expressions for K_a and K_h and eliminate common terms.

$$K_a K_h = \frac{[H_3O^+][\cancel{A^-}]}{\cancel{[HA]}} \times \frac{\cancel{[HA]}[OH^-]}{\cancel{[A^-]}} = [H_3O^+][OH^-] \tag{18.4}$$

We know that

$$K_w = [H_3O^+][OH^-]$$

Therefore,

$$K_a K_h = K_w \text{ or } K_h = \frac{K_w}{K_a}$$

Example 8.16

Calculate the hydroxide ion concentration, the degree of hydrolysis, and the pH of a 0.05 M solution of sodium benzoate. Assume that sodium benzoate salt ionizes completely in water. $K_h\ (C_6H_5CO_2^-) = 1.5 \times 10^{-10}$ at $25°C$ (Table 18.5).

Table 18.5   Data for Example 18.16

	$[C_6H_5CO_2^-]$	$[C_6H_4CO_2H]$	$[OH^-]$
Initial concentration (M)	0.05	0	0
Change (M)	$(-)x$	$(+)x$	$(+)x$
Equilibrium concentration (M)	$0.05 - x$	x	x

Solution

1. First, write an equation for the reaction and an expression for K_h.

$$C_6H_5O_2^- (aq) + H_2O(l) \rightleftharpoons HC_6H_5O_2(aq) + OH^-(aq)$$

$$K_h = \frac{[HC_6H_5O_2][OH^-]}{[C_6H_5O_2^-]} = 1.5 \times 10^{-10}$$

2. Determine the equilibrium concentration of all species in solution, including $[H^+]$. Substituting in the equilibrium expression yields:

$$1.5 \times 10^{-10} = \frac{(x)(x)}{(0.050 - x)} = \frac{x^2}{(0.050 - x)}$$

Since K $<<$ 0.050 (the initial concentration), we can assume x will be very small, hence we can ignore it relative to 0.05, and solve without using the quadratic equation:

$$1.5 \times 10^{-10} = \frac{x^2}{(0.050)}$$

$$x = \sqrt{(1.5 \times 10^{-10})(0.05)} = 2.7 \times 10^{-6} \text{ M} = [OH^-] = [C_6H_5CO_2H]$$

$$[C_6H_5CO_2^-] = 0.05 - 0.0000027 = 0.049997 \text{ M}$$

1. Calculate the degree of hydrolysis:

$$\text{Degree of hydrolysis} = \frac{[C_6H_5CO_2H]}{[C_6H_5CO_2^-]} \times 100 = \frac{2.7 \times 10^{-6}}{0.05} \times 100 = 5.4 \times 10^{-3}\%$$

2. Calculate the pH from $[OH^-]$.

$$pOH = -\log[OH^-] = -\log(2.7 \times 10^{-6}) = 5.57$$

$$pH = 14.00 - 5.57 = 8.43$$

Example 18.17

How many grams of ammonium bromide are needed per liter of solution to give a pH of 5.0? $K_a = 5.7 \times 10^{-10}$.

Solution

1. First, write the equation for the reaction and an expression for K_a.

$$NH_4^+ \, (aq) + H_2O \, (l) \rightleftharpoons NH_3 \, (aq) + H_3O^+ \, (aq)$$

$$K_a = \frac{[NH_3][H_3O^+]}{[NH_4^+]} = 5.7 \times 10^{-10}$$

2. Next, determine the equilibrium concentrations of ammonia and $[H^+]$ in solution from the given pH.

$$pH = 5.5 = -\log[H_3O^+]$$

$$[H_3O^+] = 10^{-5.5} = 3.1 \times 10^{-6} \, M$$

$$[H_3O^+] = [NH_3] = 3.1 \times 10^{-6} \, M$$

3. Then, calculate the concentration of ammonium ion.

$$K_a = \frac{[NH_3][H_3O^+]}{[NH_4^+]} = 5.7 \times 10^{-10}$$

$$\frac{(3.1 \times 10^{-6})(3.1 \times 10^{-6})}{[NH_4^+]} = 5.7 \times 10^{-10}$$

$$[NH_4^+] = 0.0169 \, M$$

$$\text{Mass of } NH_4Br \text{ needed per liter} = 0.0169 \, \frac{mol}{L} \times 98.0 \, \frac{g}{mol} = 1.65 \, g.$$

Example 18.18

The pH of 0.02 M $Al(NO_3)_3$ is 3.3. Calculate the acidic ionization constant for the metal ion hydrolysis equilibrium reaction:

$$Al(H_2O)_6^{3+}(aq) \rightleftharpoons Al(OH)(H_2O)_5^{2+}(aq) + H^+(aq)$$

Solution

1. First, write an equation for the reaction and an expression for K_a.

$$Al(H_2O)_6^{3+}(aq) \rightleftharpoons Al(OH)(H_2O)_5^{2+}(aq) + H^+(aq)$$

$$K_a = \frac{[Al(OH)(H_2O)_5^{2+}][H^+]}{[Al(H_2O)_6^{3+}]}$$

2. Next, determine the equilibrium concentrations of $Al(H_2O)_6^{3+}$, $Al(OH)(H_2O)_5^{2+}$ and $[H^+]$ in solution from the given pH.

$$pH = 3.3 = -\log[H_3O^+]$$

$$[H_3O^+] = 10^{-3.3} = 5.01 \times 10^{-4} \text{ M}$$

$$[H_3O^+] = [Al(OH)(H_2O)_5^{2+}] = 5.01 \times 10^{-4} \text{ M}$$

$$[Al(H_2O)_6^{3+}] = 0.02 \text{ M} - 0.000501 \text{ M} = 0.0195 \text{ M}$$

3. Then, calculate the equilibrium constant, K_a.

$$K_a = \frac{[Al(OH)(H_2O)_5^{2+}][H^+]}{[Al(H_2O)_6^{3+}]} = \frac{(5.01 \times 10^{-4})(5.01 \times 10^{-4})}{(0.0195)} = 1.3 \times 10^{-5}$$

18.11 The Common Ion Effect

The *common ion effect* is an application of Le Chatelier's principle. The dissociation of a weak acid or base is decreased when a solution of strong electrolyte that has an ion in common with the weak acid or base is added. For example, the addition of the strong electrolyte CH_3COONa to the weak acid CH_3COOH has the effect of decreasing the dissociation of the acid in solution. CH_3COOH is a weak acid and undergoes partial dissociation as follows:

$$CH_3COOH(aq) \rightleftharpoons CH_3COO^-(aq) + H^+(aq)$$

Sodium acetate, on the other hand, is a strong electrolyte and dissociates completely according to:

$$CH_3COONa(aq) \longrightarrow CH_3COO^-(aq) + Na^+(aq)$$

When sodium acetate is added to a solution of acetic acid, the acetate ion from the sodium acetate will increase the concentration of acetate ion. This will drive the equilibrium

$$CH_3COOH(aq) \rightleftharpoons CH_3COO^-(aq) + H^+(aq)$$

to the left in accordance with Le Chatelier's principle. Therefore the degree of dissociation of CH_3COOH will be reduced. Such a shift in equilibrium due to the addition of an ion already involved in the reaction is referred to as the common ion effect.

The pH of a solution containing a weak acid and an added common ion can be calculated in a manner similar to solutions containing only weak acids as previously described.

Example 18.19

Find the acetate ion $(C_2H_3O_2^-)$ concentration and the pH of a solution containing 0.10 mol of HCl and 0.20 mol acetic acid $(HC_2H_3O_2)$ in 1 liter of solution (Table 18.6).

Table 18.6   Data for Example 18.19

	$[HC_2H_3O_2]$	$[H_3O^+]$	$[C_2H_3O_2]$
Initial concentration (M)	0.20 M	0.10 M	0
Change (M)	$(-)x$	$(+)x$	$(+)x$
Equilibrium concentration (M)	$(0.20-x)$	$(0.1+x)$	x

Solution

1. Write an equation for the reaction and an expression for K_a.

$$HC_2H_3O_2\,(aq) + H_2O(l) \rightleftharpoons C_2H_3O_2^-\,(aq) + H_3O^+\,(aq)$$

$$K_a = \frac{\left[C_2H_3O_2^-\right]\left[H_3O^+\right]}{[HC_2H_3O_2]}$$

2. Determine the equilibrium concentrations of all species in solution including $[H^+]$. Substituting in the equilibrium expression yields:

$$1.8 \times 10^{-5} = \frac{(0.10+x)(x)}{(0.20-x)}$$

Since HCl is much stronger than $HC_2H_3O_2$, the H^+ in solution will mainly be from HCl. Therefore $[H^+]$ will be approximately equal to 0.10 M, and x will be very small. Therefore,

$$1.8 \times 10^{-5} = \frac{(0.10)(x)}{(0.20)} \text{ or } x = 3.6 \times 10^{-5}\,M = \left[C_2H_3O_2^-\right]$$

$$\left[H_3O^+\right] = (0.10+x)\,M \simeq 0.10\,M$$

3. Calculate the pH from $[H^+]$.

$$pH = -\log\left[H_3O^+\right] = -\log(0.10) = 1.0$$

18.12 Buffers and pH of Buffer Solutions

A buffer is a solution that resists changes in pH upon the addition of small amounts of acid or base. That is, a buffer solution keeps its pH almost constant. Buffers play a significant role in helping to maintain a constant pH in many biological and chemical processes. Human blood has a pH of about 7.4. If this value drops below 7.0, fatality can result by a process known as acidosis. On the other hand, if the value rises above 7.7, death can also result by a process known as alkalosis. The good news is that our blood has a buffer system that maintains the pH at the proper level.

A buffer can be prepared by mixing either a weak acid with a salt of the weak acid or by mixing a weak base with a salt of the weak base. Some examples of buffers are:

- acetic acid and sodium acetate (CH_3COOH/CH_3COONa)
- carbonic acid and sodium carbonate (H_2CO_3/Na_2CO_3)
- ammonium hydroxide and ammonium chloride (NH_4OH/NH_4Cl).

The reason buffers resist changes in pH is that they contain both an acidic species to neutralize OH^- ions and a basic species to neutralize H^+ ions.

18.12.1 How does a buffer work?

To understand the mechanism of buffering, consider the following system of a weak acid HA and its salt, MA. The dissociation equilibrium is:

$$HA\,(aq) \rightleftharpoons H^+\,(aq) + A^-\,(aq)$$

For this reaction, K_a is:

$$K_a = \frac{[H^+][A^-]}{[HA]} \quad \text{and} \quad [H^+] = K_a\frac{[HA]}{[A^-]}$$

This expression indicates that the $[H^+]$ and hence the pH depends on K_a and the ratio $[HA]/[A^-]$.

If a small amount of a base (OH^-) is added to the buffer, the OH^- ions will react with the acid, HA. As some of the HA is consumed, more of the conjugate base $[A^-]$ will be produced.

$$OH^-\,(aq) + HA\,(aq) \rightleftharpoons H_2O\,(l) + A^-\,(aq)$$

The hydrogen ion concentration $[H^+]$, and the pH will not change as long as the value of the ratio $[HA]/[A^-]$ does not deviate greatly from the original value.

Similarly, if a small amount of acid (H^+) is added to the buffer, the H^+ will react with the conjugate base, A^-, and increase the concentration of the weak acid.

$$H^+\,(aq) + A^-\,(aq) \rightleftharpoons HA\,(aq)$$

Again, the $[H^+]$ and hence the pH will not change as long as the value of the ratio $[HA]/[A^-]$ stays close to the original value.

18.12.2 Buffer capacity and pH

No buffer has unlimited capacity to continuously take abuse before it is rendered useless. At some point a buffer loses its ability to function. The buffer capacity of a buffer is the amount of acid or base that the solution can absorb before a significant change in pH will occur. The capacity of a buffer is directly related to the number of moles of weak acid and weak base in the system. The buffering capacity is destroyed once these species (weak acid and weak base) are consumed.

In mathematical terms, we can define buffer capacity (β) as

$$\text{Buffer Capacity } (\beta) = \frac{\Delta C_b}{\Delta pH} = -\frac{\Delta C_a}{\Delta pH}$$

The minus sign is used because the addition of an acid causes a decrease in pH. A quick and handy formula for calculating buffer capacity is

$$\beta = \frac{2.303\,C_{HA}C_{A^-}}{(C_{HA} + C_{A^-})}$$

The derivation of this formula is beyond the scope of this book.

Example 18.20

What is the buffer capacity of a solution containing 0.20 M benzoic acid and 0.20 M sodium benzoate? If KOH is added such that the solution becomes 0.01 M in KOH, what will be the change in pH?

Solution

For the first part of the question, substitute directly into the equation

$$\beta = \frac{2.303 \times C_{HA} \times C_{A^-}}{(C_{HA} + C_{A^-})} = \frac{2.303(0.2)(0.2)}{(0.2 + 0.2)} = 0.23$$

For the second part of the question, calculate the change in pH using the equation

$$\text{Buffer Capacity } (\beta) = \frac{\Delta C_b}{\Delta pH}$$

$$0.23 = \frac{\Delta C_b}{\Delta pH} = \frac{0.01}{\Delta pH}$$

$$\Delta pH = \frac{0.010}{0.230} = 0.043$$

Buffer capacity depends on the number of moles of the weak acid and conjugate base present in the solution.

18.12.3 The Henderson–Hasselbalch equation

The pH of a buffer solution depends on the K_a for the acid as well as the relative concentrations of the weak acid and its conjugate base in the solution. We can derive an equation that shows the relationship between conjugate acid–base concentrations, pH, and K_a.

Consider the buffer consisting of a weak acid HA and its salt, MA. The dissociation equilibrium is

$$HA\,(aq) \rightleftharpoons H^+\,(aq) + A^-\,(aq)$$

For this reaction, K_a is

$$K_a = \frac{[H^+][A^-]}{[HA]} \quad \text{and} \quad [H^+] = K_a \frac{[HA]}{[A^-]}$$

In general form,

$$[H^+] = K_a \frac{[Acid]}{[Conjugate\ base]}$$

Taking the negative log of both sides gives the so-called Henderson–Hasselbalch equation

$$pH = pK_a + \log \frac{[Base, A^-]}{[Acid, HA]}$$

or

$$pH = pK_a - \log \frac{[Acid, HA]}{[Base, A^-]}$$

Table 18.7   Data for Example 18.21

	$[H_3BO_3]$	$[H_3O^+]$	$[H_2BO_3^-]$
Initial concentration (M)	0.1	0	0.2
Change (M)	$(-)x$	$(+)x$	$(+)x$
Equilibrium concentration (M)	$(0.10 - x)$	x	$(0.2 + x)$

The Henderson–Hasselbalch equation can also be written in terms of pOH and pK_b as

$$pOH = pK_b + \log \frac{\left[\text{Conjugate cation}\right]}{[\text{Base}]}$$

or

$$pOH = pK_b - \log \frac{[\text{Base}]}{\left[\text{Conjugate cation}\right]}$$

Example 18.21

What is the pH of a solution containing 0.10 M H_3BO_3 and 0.20 M NaH_2BO_3? For boric acid, $K_a = 7.3 \times 10^{-10}$ (Table 18.7).

Solution

There are two ways of solving this problem. We can use the Henderson–Hasselbalch equation or solve for $[H^+]$ and then the pH.

Method 1

We will first determine the pK_a from the given K_a.

$$pK_a = -\log K_a = -\log(7.3 \times 10^{-10}) = 9.14$$

Now calculate the pH by substituting the pK_a value and known concentrations into the Henderson–Hasselbalch equation.

$$pH = pK_a + \log \frac{\left[\text{Base}, H_2BO_3^-\right]}{[\text{Acid}, H_3BO_3]}$$

$$pK_a = 9.14, \ \left[H_2BO_3^-\right] = 0.20 \text{ M}, [H_3BO_3] = 0.10 \text{ M}$$

$$pH = 9.14 + \log \frac{(0.20)}{(0.10)} = 9.14 + 0.30 = 9.44$$

Method 2

1. First, we the write equation for the reaction and an expression for K_a.

$$H_3BO_3 \, (aq) + H_2O \, (l) \rightleftharpoons H_2BO_3^- \, (aq) + H_3O^+ \, (aq)$$

$$K_a = \frac{\left[H_2BO_3^-\right]\left[H_3O^+\right]}{[H_3BO_3]}$$

2. Next, determine the equilibrium concentrations of all species in solution including $[H^+]$.

$$K_a = \frac{(x)(x+0.2)}{(0.10-x)} = 7.3 \times 10^{-10}$$

Since we have a common ion in the solution, x will be very small compared to 0.10 or 0.20 M. Therefore we can use the approximation method to solve for x.

$$\frac{(x)(0.2)}{(0.10)} = 7.3 \times 10^{-10} \text{ or } x = 3.6 \times 10^{-10}$$

(So the assumption is justified.)

$$[H_3O^+] = 3.6 \times 10^{-10} M$$

3. Then, calculate the pH from $[H^+]$.

$$pH = -\log[H_3O^+] = -\log(3.6 \times 10^{-10}) = 9.4$$

Example 18.22

The pH of a solution containing an HCN and NaCN buffer system is 9.50. What is the ratio of $[CN^-]/[HCN]$ in the buffer? K_a for HCN is 6.2×10^{-10}.

Solution

First, determine the pK_a from the given K_a.

$$pK_a = -\log K_a = -\log(6.2 \times 10^{-10}) = 9.2$$

Now, calculate the [base]/[acid] ratio by substituting the pH and pK_a values into the Henderson–Hasselbalch equation.

$$pH = pK_a + \log\frac{[\text{Base, CN}^-]}{[\text{Acid, HCN}]}$$

$$pH = 9.5, pK_a = 9.2$$

$$9.5 = 9.2 + \log\frac{[CN^-]}{[HCN]}$$

$$\log\frac{[CN^-]}{[HCN]} = 0.30$$

$$\frac{[CN^-]}{[HCN]} = 10^{0.30} = 2.0$$

18.13 Polyprotic Acids and Bases

A polyprotic acid such as phosphoric acid or tartaric acid has more than one ionizable hydrogen atom per molecule. In solution, polyprotic acids undergo successive dissociations, or react with a base stepwise for each proton. For example, phosphoric acid contains three protons (hence is a triprotic acid), and it dissociates as follows:

$$H_3PO_4\,(aq) \rightleftharpoons H^+\,(aq) + H_2PO_4^-\,(aq) \quad K_{a_1} = \frac{[H^+][H_2PO_4^-]}{[H_3PO_4]} = 7.5 \times 10^{-3}$$

$$H_2PO_4^-\,(aq) \rightleftharpoons H^+\,(aq) + HPO_4^-\,(aq) \quad K_{a_2} = \frac{[H^+][HPO_4^-]}{[H_2PO_4^-]} = 6.2 \times 10^{-8}$$

$$HPO_4^-\,(aq) \rightleftharpoons H^+\,(aq) + PO_4^-\,(aq) \quad K_{a_3} = \frac{[H^+][PO_4^-]}{[HPO_4]} = 4.8 \times 10^{-13}$$

Here, K_{a_1}, K_{a_2}, and K_{a_3} represent the first, second, and third ionization constants. For any polyprotic acid, the ionization constants always decrease in the order $K_{a_1} > K_{a_2} > K_{a_3}$.
The overall K_a value for a polyprotic acid can be found by multiplying the K_a values for the individual ionization steps. For phosphoric acid

$$K_a = K_{a_2} K_{a_2} K_{a_3} = \frac{[H^+][H_2PO_4^-]}{[H_3PO_4]} \times \frac{[H^+][HPO_4^-]}{[H_2PO_4^-]} \times \frac{[H^+][PO_4^-]}{[HPO_4^-]}$$

$$K_a = \frac{[H^+]^3 [PO_4^{3-}]}{[H_3PO_4]} = 2.23 \times 10^{-22}$$

This represents the overall reaction

$$H_3PO_4\,(aq) + H_2O(l) \rightleftharpoons 3\,H_3O^+\,(aq) + PO_43^-\,(aq)$$

Similarly, a diprotic base such as Na_2CO_3 (or CO_3^{2-}) will have two ionization constants:

$$HCO_3^-\,(aq) + H_2O\,(l) \rightleftharpoons H_2CO_3\,(aq) + OH^-\,(aq) \quad K_{b_1} = \frac{[H_2CO_3][OH^-]}{[HCO_3^-]} = 2.33 \times 10^{-8}$$

$$CO_3^{2-}\,(aq) + H_2O\,(l) \rightleftharpoons HCO_3^-\,(aq) + OH^-\,(aq) \quad K_{b_2} = \frac{[HCO_3^-][OH^-]}{[CO_3^{2-}]} = 1.78 \times 10^{-4}$$

Example 18.23

Calculate the concentrations of H^+ and S^{2-} in a solution of 0.100 M of H_2S. K_a^1 and K_a^2 for H_2S are 1.0×10^{-7} and 1.2×10^{-13}, respectively.

Solution

1. Write equations for the dissociation reactions and expressions for the equilibrium dissociation constants.

$$H_2S \rightleftharpoons H^+ + HS^- \qquad K_{a_1} = \frac{[H^+][HS^-]}{[H_2S]} = 1.0 \times 10^{-7}$$

$$HS^- \rightleftharpoons H^+ + S^{2-} \qquad K_{a_2} = \frac{[H^+][S^{2-}]}{[HS^-]} = 3.0 \times 10^{-13}$$

2. To calculate the hydrogen ion concentration, we assume that all the hydrogen ions come from the first or primary reaction, since K_{a_1} is much larger than K_{a_2}.
   At equilibrium, let $[H^+] = [HS^-] = x$. Then $[H_2S] = 0.10 - x$.
   Using the approximation method, $[H_2S] \simeq 0.10$ M since K_{a_1} is very small.
3. Substitute these values in the K_{a_1} expression and solve for x.

$$K_{a_1} = \frac{[H^+][HS^-]}{[H_2S]} = \frac{x^2}{(0.10)} = 1.0 \times 10^{-7}$$

$$x = \sqrt{(1.0 \times 10^{-7})} = 0.00010 \text{ or } 1.0 \times 10^{-4}\text{M}$$

Thus the concentration of H^+ in the solution is 1.0×10^{-4}M.
To calculate the concentration of S^{2-} in the solution, use the second equilibrium equation above.

$$K_{a_2} = \frac{[H^+][S^{2-}]}{[HS^-]} = 3.0 \times 10^{-13}$$

From part 1 above, $[H^+] = [HS^-] = 1.0 \times 10^{-4}$M

$$3.0 \times 10^{-13} = \frac{1.0 \times 10^{-4}[S^{2-}]}{1.0 \times 10^{-4}}$$

$$[S^{2-}] = 3.0 \times 10^{-13}\text{M}$$

Thus, the concentration of S^{2-} in the solution is 3.0×10^{-13} M.

Example 18.24

Calculate the concentrations of all species present at equilibrium in 0.10 M H_3PO_4. K_{a_1}, K_{a_2}, and K_{a_3} for H_3PO_4 are 7.5×10^{-3}, 6.2×10^{-8}, and 4.8×10^{-13}, respectively.

Solution

1. Write chemical equations for all the dissociation reactions:

$$H_3PO_4\,(aq) \rightleftharpoons H^+\,(aq) + H_2PO_4^-\,(aq) \quad K_{a_1} = \frac{[H^+][H_2PO_4^-]}{[H_3PO_4]} = 7.5 \times 10^{-3}$$

$$H_2PO_4^-\,(aq) \rightleftharpoons H^+\,(aq) + HPO_4^-\,(aq) \quad K_{a_2} = \frac{[H^+][HPO_4^-]}{[H_2PO_4^-]} = 6.2 \times 10^{-8}$$

$$HPO_4^-\,(aq) \rightleftharpoons H^+\,(aq) + PO_4^-\,(aq) \quad K_{a_3} = \frac{[H^+][PO_4^-]}{[HPO_4]} = 4.8 \times 10^{-13}$$

2. Assume that H^+ comes mainly from the initial stage of the dissociation. Also assume that the concentration of any anion formed at one step of dissociation is not decreased by subsequent dissociation steps. Therefore, we can say from the first equation

$$H_3PO_4\,(aq) \rightleftharpoons H^+\,(aq) + H_2PO_4^-\,(aq) \quad K_{a_1} = \frac{[H^+][H_2PO_4^-]}{[H_3PO_4]} = 7.5 \times 10^{-3}$$

$$[H^+] = [H_2PO_4^-] = x$$

Substitute in the K_{a_1} expression:

$$7.5 \times 10^{-3} = \frac{x^2}{(0.10 - x)} \simeq \frac{x^2}{(0.10)} \quad \text{or } x = 0.0274 \text{ M}$$

If you solve for x in the equation $x^2 + 7.5 \times 10^{-4}x - 7.5 \times 10^{-4} = 0$ using the quadratic formula

$$x = \frac{-b \pm \sqrt{b^2 - 4ac}}{2a}$$

with $a = 1, b = 7.5 \times 10^{-4}$, and $c = -7.5 \times 10^{-4}$ you will get the same answer, i.e. $x = 0.027$ M. Thus

$$[H^+] = [H_2PO_4^-] = 0.027 \text{ M}$$

3. Use the values for $[H^+]$ and $[H_2PO_4^-]$ above to solve for $\left[HPO_4^{2-}\right]$:

$$H_2PO_4^-\,(aq) \rightleftharpoons H^+\,(aq) + HPO_4^{2-}\,(aq) \quad K_{a_2} = \frac{[H^+][HPO_4^{2-}]}{[H_2PO_4^-]} = 6.2 \times 10^{-8}$$

$$\left[HPO_4^{2-}\right] = K_{a_2}\frac{[H_2PO_4^-]}{[H^+]} = \frac{(6.2 \times 10^{-8})(0.027)}{(0.027)} = 6.2 \times 10^{-8}M$$

4. Finally, use the values for $[H^+]$ and $\left[HPO_4^{2-}\right]$ above to solve for $\left[PO_4^{3-}\right]$.

$$HPO_4^{2-}\,(aq) \rightleftharpoons H^+\,(aq) + PO_4^{3-}\,(aq) \quad K_{a_3} = \frac{[H^+][PO_4^{3-}]}{[HPO_4^{2-}]} = 4.8 \times 10^{-13}$$

$$\left[PO_4^{3-}\right] = K_{a_3}\frac{[HPO_4^{2-}]}{[H^+]} = \frac{(4.8 \times 10^{-13})(6.2 \times 10^{-8})}{(0.027)} = 1.10 \times 10^{-18}M$$

Table 18.8   Selected acid/base indicators

Indicator	Acid color	Base color	pH Range	pK_a
Methyl orange	Red	Yellow	3.2–4.4	3.4
Bromophenol blue	Yellow	Blue	3.0–4.6	3.9
Methyl red	Red	Yellow	4.8–6.0	5
Bromothymol blue	Yellow	Blue	6.0–7.6	7.1
Litmus	Red	Blue	5.0–8.0	6.5
Thymol blue	Yellow	Blue	8.0–9.6	8.9
Phenolphthalein	Colorless	Pink	8.2–10.0	9.4
Alizarin yellow	Yellow	Red	10.1–12.0	11.2

18.14 More Acid–Base Titration

The progressive addition of a measured volume of an acid of known concentration (standard acid) to a solution of a base of unknown concentration (or vice versa) is called a *titration* (see Chapter 14). When a solution of an acid such as HCl is titrated with a base such NaOH, the reaction that occurs is

$$HCl\,(aq) + NaOH\,(aq) \longrightarrow NaCl\,(aq) + H_2O\,(l)$$

or simply,

$$H^+\,(aq) + OH^-\,(aq) \longrightarrow H_2O\,(l)$$

18.14.1 Acid–base indicators

Indicators are organic compounds used to detect equivalence points, i.e. the point at which the reaction is complete in an acid–base titration. Indicators change colors over a characteristic short range of pH. Common acid–base indicators include methyl orange, methyl red, bromothymol blue, litmus, and phenolphthalein. Table 18.8 lists some common acid–base indicators, their colors in acidic and basic media, the pH ranges, and pK_a values.

Table 18.8 is a guide to selecting a suitable indicator for an acid–base titration experiment. First, determine or estimate the pH of the solution (titration mixture) at the equivalence point. Then, select an indicator that changes color at or close to this pH.

Acid–base indicators are weak organic acids, HIn, or, to a lesser degree, weak organic bases, InOH. The "In" in the formula represents the complex organic group. For a weak indicator, the following equilibrium is established with its conjugate base at equilibrium:

$$HIn\,(aq) + H_2O\,(l) \rightleftharpoons H_3O^+\,(aq) + In^-\,(aq)$$

(color A)   (color B)

The acid (HIn) and its conjugate base (In$^-$) have different colors. At low pH values the concentration of H_3O^+ is high. Consequently the equilibrium position lies on the left side of the equation and the equilibrium solution exhibits color A. On the other hand, at high pH values, the concentration of H_3O^+ is low and as a result the equilibrium position lies to the right, with the equilibrium solution exhibiting color B.

The equilibrium constant for the above acid–base indicator equilibrium is

$$K_{In} = \frac{[H_3O^+][In^-]}{[HIn]}$$

K_{In} is known as the *indicator dissociation constant*. Let us assume that the change between the two colored forms HIn and In$^-$ occurs when we have equal amounts of each. The transition or turning point is when $[HIn] = [In^-]$. At this point, substituting in the K_{In} expression yields

$$K_{In} = \frac{[H_3O^+][\cancel{In^-}]}{\cancel{[HIn]}} = [H_3O^+]$$

$$pK_{In} = -\log K_{In} = -\log[H_3O^+] = pH$$

The pH of the solution at its turning point is known as the pK_{In}. At this pH, half of the indicator is in its acid form and the other half in the form of its conjugate base.
The equilibrium constant expression

$$K_{In} = \frac{[H_3O^+][In^-]}{[HIn]}$$

can be rearranged into a ratio form:

$$\frac{[In^-]}{[HIn]} = \frac{K_{In}}{[H_3O^+]}$$

Thus the ratio $[In^-]/[HIn]$ is inversely proportional to $[H_3O^+]$ and is responsible for the color exhibited by the indicator.
The detailed behavior of acid–base indicators depends on several considerations besides the pH of the solution, including:

- the nature of the solvent
- ionic strength
- temperature
- colloidal particulates, which can interfere through surface adsorption of the indicator.

18.14.2 Perception of color change of indicators

The human eye typically responds to only dramatic color changes. Color changes that are less than 10% are hardly seen. Hence, for the color change of an indicator to be seen clearly, the molar concentrations of the indicator species must be 90% or more of the indicator. The pK_a of an indicator plays a significant role in color perception. At the pK_a, acid–base indicators are 50% ionized. If the pH of the solution is one unit above the pK_a, 90% of the ionizable indicator will be in its basic form. On the other hand, if the pH of the solution is one unit below the pK_a, 90% of the ionizable indicator will be in its acidic form. Hence, indicators show a color transition 1 pH unit above or below their pK_a. This is why indicators are generally selected based on the proximity of their pK_a to pH at the titration endpoint. To see the HIn color

$$\frac{[In^-]}{[HIn]} \leq 0.10$$

and to see In^-

$$\frac{[In^-]}{[HIn]} \geq 0.10$$

Example 18.25

A certain indicator "HIn" has $K_{In} = 5.25 \times 10^{-10}$. Calculate the pH of the solution when the ratio $[In^-]/[HIn]$ is ½.

Solution

1. Write an equation for the ionization of the indicator and an expression for the hydrogen ion concentration in terms of K_{In}.

$$HIn + H_2O(l) \rightleftharpoons H_3O^+ + In^-$$

$$K_{In} = \frac{[H_3O^+][In^-]}{[HIn]} \quad \text{or} \quad [H_3O^+] = K_{In}\frac{[HIn]}{[In^-]}$$

2. Solve for the hydrogen ion concentration by making appropriate substitutions:

$$[H_3O^+] = K_{In}\frac{[HIn]}{[In^-]} = (5.25 \times 10^{-10})\left(\frac{2}{1}\right) = 1.05 \times 10^{-9}M$$

3. Calculate the pH:

$$pH = -\log[H_3O^+] = -\log(1.05 \times 10^{-9}) = 9.0$$

18.15 pH Titration Curves

An acid–base titration curve is a plot of the pH of a solution of acid (or base) versus the volume of base (or acid) added. Depending on the strength and concentrations of the acid and base involved, pH curves can have different characteristic shapes. For example, the titration curve for the titration of a strong acid by a strong base is different from the curve for the titration of a weak acid by a strong base. Titration curves can be used in selecting a good indicator for a given titration process. While a pH meter can determine changes in pH, it is also possible to calculate the pH of each at point on the titration curve.

18.15.1 Titration of strong acid against a strong base

The net reaction between a strong acid and a strong base is the formation of water:

$$H^+(aq) + OH^-(aq) \rightarrow H_2O(l)$$

The hydrogen ion concentration at any given point in the titration can be calculated from the amount of hydrogen ion remaining at that point divided by the total volume of the solution.

- The pH of the solution before the addition of the strong base is determined by the initial concentration of the strong acid.
- As the strong base is added to the strong acid, the pH values between the initial pH and the equivalence point are determined by the concentration of excess acid (unneutralized acid).
- The pH at the equivalence point is easy to determine. No calculation is required because at the equivalence point, equal amounts or numbers of moles of the acid and the base have reacted, resulting in the formation of a neutral salt. The pH is 7.0.
- After the equivalence point has been reached, the pH of the solution is determined by the concentration of excess OH^- in the solution.

The example below illustrates the steps involved in the calculation.

Example 18.26

A 50-mL sample of 0.10 M HCl is titrated with 0.10 M NaOH. Calculate the pH

(a) at the start of the titration
(b) after 10 mL of NaOH has been added
(c) after 25 mL of NaOH has been added
(d) after 50 mL of NaOH has been added
(e) after 70 mL of NaOH has been added.

Solution

(a) At the start of the titration, before NaOH is added, we have:

$$HCl + H_2O \xrightarrow{100\%} H_3O^+ + Cl^-$$

$$[H_3O^+] = 0.10 \text{ M}, pH = 1.0$$

(b) After 10 mL of NaOH has been added:

$$HCl + NaOH \rightarrow NaCl + H_2O$$

$$\text{Mol } OH^- \text{ added } = 0.010 \text{ L} \times 0.10\frac{\text{mol}}{\text{L}} = 0.0010 \text{ mol}$$

$$\text{Original mol of } H^+ \text{ added} = 0.050 \text{ L} \times 0.10\frac{\text{mol}}{\text{L}} = 0.0050 \text{ mol}$$

All the OH^- is completely reacted leaving excess H^+.
Note that the volume of solution has increased by 10 mL from the original 50 mL.

$$\text{Excess } H^+ = (0.005 - 0.001) = 0.004 \text{ mol}$$

$$[H^+] = \frac{0.004 \text{ mol}}{60 \text{ mL}} \times \frac{1000 \text{ mL}}{1 \text{ L}} = 0.0667 \text{ M}$$

$$pH = -\log(0.0667) = 1.18$$

(c) Calculate pH after 25 mL NaOH has been added. HCl and NaOH react in 1:1 ratio. Repeat calculation as above remembering the total volume.

$$HCl + NaOH \rightarrow NaCl + H_2O$$

$$Mol\ OH^- \text{ added} = 0.025\ L \times 0.10\frac{mol}{L} = 0.0025\ mol$$

$$\text{Original mol of } H^+ = 0.050\ L \times 0.10\frac{mol}{L} = 0.0050\ mol$$

All the OH^- is completely reacted leaving excess H^+.
Excess $H^+ = (0.005 - 0.0025) = 0.0025\ mol$

$$[H^+] = \frac{0.0025\ mol}{75\ mL} \times \frac{1000\ mL}{1\ L} = 0.033\ M$$

$$pH = -\log(0.033) = 1.48$$

(d) After 50 mL of NaOH has been added,

$$HCl + NaOH \rightarrow NaCl + H_2O$$

$$Mol\ OH^- \text{ added} = 0.050\ L \times 0.10\frac{mol}{L} = 0.005\ mol$$

$$\text{Original mol of } H^+ = 0.050\ L \times 0.10\frac{mol}{L} = 0.0050\ mol$$

At this point both OH^- and H^+ are completely consumed.

$$\text{Excess } H^+ = (0.005 - 0.005) = 0$$

The equivalence point is reached when 50 mL of NaOH has been added. NaCl will be formed and $pH = pOH = 7$.

(e) After 70 mL of NaOH has been added, the excess NaOH will be calculated as follows:

$$\text{Excess } OH^- = 20\ mL \times \frac{0.10\ mol}{1L} \times \frac{1L}{1000\ mL} = 0.0020\ mol$$

This amount is present in 120 mL of solution.

$$[OH^-] = \frac{0.0020\ mol}{0.120\ L} = 0.0167\ M$$

$$pOH = -\log[OH^-] = -\log(0.0167) = 1.78$$

$$pH = 12.22$$

18.15.2 Titration of weak acid against a strong base

- At the start of the titration of a weak acid HA versus a strong base such as NaOH, the initial pH is the pH of the weak acid solution. As the strong base is gradually added to the weak acid, partial neutralization of the acid occurs, resulting in a mixture of the acid HA and its conjugate base A^-.

- To calculate the pH prior to the equivalence point, we first determine the concentrations of HA and A^-. From the values of K_a, [HA], and $[A^-]$, we can determine $[H^+]$, and hence the pH. This can be readily done with the Henderson–Hasselbalch equation.
- The equivalence point is reached after equal number of moles of NaOH and HA have reacted, resulting in the formation of sodium salt of the acid. Since A^- is a weak base, the pH at the equivalence point is normally above 7.
- After the equivalence point, the $[OH^-]$ derived from the hydrolysis of A^- is negligible compared to the $[OH^-]$ from excess NaOH. Hence, the pH is determined by the concentration of OH^- from the excess NaOH.

The following example will illustrate the steps involved in the calculation.

Example 18.27

50 mL of 0.100 M acetic acid, $HC_2H_3O_2$, is titrated with 0.100 M NaOH. Calculate the pH of the solution when

(a) 40 mL of 0.10 M NaOH has been added
(b) 50 mL of 0.10 M NaOH has been added
(c) 70 mL of 0.10 M NaOH has been added

(K_a for acetic acid $= 1.8 \times 10^{-5}$.)

Solution

(a) Acetic acid is a weak acid. Use the following procedure to calculate the pH:

1. Write the equation for the reaction.

$$HC_2H_3O_2 + OH^- \rightleftharpoons C_2H_3O_2^- + H_2O$$

2. Calculate the moles of reactants before reaction.

$$\text{mol NaOH} = (40 \text{ mL}) \left(\frac{0.100\text{mol}}{1\text{L}} \right) \left(\frac{1\text{L}}{1000\text{mL}} \right) = 4.0 \times 10^{-3}\text{mol}$$

$$\text{mol } HC_2H_3O_2 = (50 \text{ mL}) \left(\frac{0.100\text{mol}}{1\text{L}} \right) \left(\frac{1\text{L}}{1000\text{mL}} \right) = 5.0 \times 10^{-3}\text{mol}$$

Once the NaOH solution is added, 0.004 mol of acetic acid is consumed by 0.004 mol of NaOH, and 0.004 mol of acetate ion is produced. The excess acid is calculated as follows:

$$\text{Excess acid} = (0.005 - 0.004) = 0.001 \text{ mol}$$

$$\text{Total vol} = 40 \text{ mL} + 50 \text{ mL} = 90 \text{ mL} = 0.090 \text{ L}$$

$$[HC_2H_3O_2] = \left(\frac{0.001 \text{ mol}}{0.090 \text{ L}} \right) = 0.011 \text{ M}$$

$$[C_2H_3O_2^-] = \left(\frac{0.004 \text{ mol}}{0.090 \text{ L}} \right) = 0.044 \text{ M}$$

3. Using equilibrium conditions, calculate hydrogen ion concentration and then pH.

$$K_a = 1.8 \times 10^{-5} = \frac{[H^+][C_2H_3O_2^-]}{[HC_2H_3O_2]} \quad \text{or}$$

$$[H^+] = (1.8 \times 10^{-5}) \frac{[HC_2H_3O_2]}{[C_2H_3O_2^-]}$$

$$[H^+] = (1.8 \times 10^{-5}) \left(\frac{0.011}{0.044}\right) = 4.5 \times 10^{-6} \text{ M}$$

$$pH = -\log\left(4.5 \times 10^{-6}\right) = 5.35$$

(b) Calculate the pH after 50 mL NaOH has been added.

At the equivalence point, moles of acetic acid = moles of NaOH, and the solution will contain essentially 5.0×10^{-3} mol sodium acetate. The following hydrolysis reaction will occur:

$$C_2H_3O_2^- + H_2O \rightleftharpoons HC_2H_3O_2 + OH^- \quad \text{and } K_b = \frac{K_w}{K_a} = \frac{[OH^-][HC_2H_3O_2]}{[C_2H_3O_2^-]}$$

$$K_b = \frac{[OH^-][HC_2H_3O_2]}{[C_2H_3O_2^-]} = \frac{1 \times 10^{-14}}{1.8 \times 10^{-5}} = 5.56 \times 10^{-10}$$

$$[HC_2H_3O_2] = [OH^-] = x$$

$$[C_2H_3O_2^-] = \frac{0.005 \text{ mol}}{0.100 \text{ L}} = 0.050 \text{ M}$$

$$5.56 \times 10^{-10} = \frac{x^2}{0.050} \text{ or } x = 5.27 \times 10^{-6} \text{M}$$

Thus $[OH^-] = 5.27 \times 10^{-6}$M and pOH $= 5.28$

$$pH = 8.72$$

(c) Calculate the pH after 70 mL NaOH has been added.

After the addition of 70 mL of 0.100 M NaOH:

$$\text{mol of NaOH added} = \left(\frac{70 \text{ mL} \times 0.100 \text{ mol}}{1 \text{ L}}\right)\left(\frac{1 \text{ L}}{1000 \text{ mL}}\right) = 0.0070 \text{ mol.}$$

This OH⁻ concentration is more than enough to neutralize the 0.0050 mol of acid initially present. The total volume of the solution is 50 mL + 70 mL = 120 mL = 0.120 L.

$$[CH_3COO^-] = \frac{0.005 \text{ mol}}{0.12 \text{ L}} = 0.042 \text{ M}$$

$$[OH^-] = \frac{(0.007 \text{ mol} - 0.005 \text{ mol})}{0.12 \text{ L}} = 0.067 \text{ M}$$

At this stage of the titration $[OH^-]$ from the hydrolysis of CH_3COO^- is negligible compared with $[OH^-]$ from the excess NaOH. Therefore, the $[H_3O^+]$, and hence the pH, of the solution after the equivalence point can be calculated from the $[OH^-]$ derived from excess NaOH.

$$[H_3O^+][OH^-] = K_w$$

$$[H_3O^+] = \frac{K_w}{[OH^-]} = \frac{1.0 \times 10^{-14}}{0.0167} = 6.0 \times 10^{-13}M$$

$$pH = -\log[H_3O^+] = 12.2$$

18.16 Problems

18.16.1 pH, pOH, and percent ionization

1. Calculate the pH of each of the following solutions:

   (a) $[H^+] = 1.0 \times 10^{-5}M$
   (b) $[H^+] = 5.5 \times 10^{-2}M$
   (c) $[H^+] = 2.1 \times 10^{-11}M$
   (d) $[H^+] = 9.8 \times 10^{-7}M$

2. Calculate the pH of each of the following solutions:

   (a) $[OH^-] = 1.0 \times 10^{-5}M$
   (b) $[OH^-] = 5.5 \times 10^{-2}M$
   (c) $[OH^-] = 2.1 \times 10^{-11}M$
   (d) $[OH^-] = 9.8 \times 10^{-7}M$

3. Calculate the hydrogen ion concentration in solutions with the following pH values:
   (a) 11.8, (b) 2, (c) 7.9, (d) 6.5, (e) 1

4. Calculate the hydroxide ion concentration in solutions with the following pH values:
   (a) 11.8, (b) 2, (c) 7.9, (d) 6.5, (e) 1

5. What is the pH of the following?

   (a) 6.3 g of HNO_3 in 1 liter of solution
   (b) 0.56 g of KOH in 500 mL of solution
   (c) 4.9 g of H_2SO_4 in 250 mL of solution
   (d) 3.65 g of HCl in 2 liters of solution

6. Calculate the range of hydrogen ion concentration in the following substances from their typical pH ranges:

   (a) Gastric juice (1.6–1.8)
   (b) Human urine (4.8–8.8)
   (c) Cow's milk (6.3–6.6)
   (d) Lime juice (1.8–2.1)
   (e) Human blood (7.35–7.45)
   (f) Milk of magnesia (9.9–10.1)
   (g) Tomatoes (4.0–4.4)

7. The hydrogen ion concentration in a sample of black coffee is 1.2×10^{-5} M. What is the hydroxide ion concentration in this coffee?

8. Commercial concentrated sulfuric acid labeled 98 percent (w/w) H_2SO_4 has a density of 1.84.g/mL.

   (a) Calculate the molarity of this (stock) solution.
   (b) Calculate the pH of a solution prepared by diluting 1.50 mL of the stock solution to 2.00 L, assuming 100% ionization.

9. A chemistry student from Ambrose Alli University in a final year project decided to investigate the temperature dependence of pH in water. He collected water samples from Ikpoba River for his experiment. The hydrogen ion concentration in the water samples was determined at 25 and 100°C, respectively. To his surprise, at 100°C the hydrogen ion concentration in the water sample was found to be 1.05×10^{-6} M. This is about ten times higher than the result he obtained at 25°C. What can he conclude from these data?

10. Determine the pH and pOH of water at 60°C given that at 60°C K_w has a value of 9.62×10^{-14}.

11. Calculate the pH of the following solutions:

    (a) 5% (w/w) of HNO_3
    (b) 0.03% (w/w) of HBr
    (c) 1.0% (w/w) of NaOH
    (d) 0.035% (w/w) of RbOH

12. What are the equilibrium concentrations of benzoic acid $(C_6H_5CO_2H)$, the benzoate ion, and H_3O^+ for a 0.020 M benzoic acid solution $(K_a = 6.3 \times 10^{-5})$? What is the pH of the solution?

13. The pH of 0.0009 M formic acid (HCO_2H) is 3.5. Determine the K_a and the equilibrium concentration of HCO_2H.

14. Calculate the pH of 0.25 M aqueous acetic acid $(CH_3CO_2H, K_a = 1.8 \times 10^{-5})$.

15. Calculate the pH and pOH of a 0.10 M nitrous acid (HNO_2) solution $(K_a = 4.6 \times 10^{-4})$.

16. Calculate the pH of 0.20M CH_3NH_2 $(K_b = 4.2 \times 10^{-4})$.

17. A solution is prepared by dissolving 0.34 g of ammonia gas in 1 L of water. Calculate the pH and pOH, given that K_b for ammonia is 1.8×10^{-5}.

18. Calculate the concentration of OH^- in a 0.01 M solution of pyridine $(K_b = 1.7 \times 10^{-9})$.

19. What is the percent ionization of a 0.0010 M solution of phenol $(C_6H_5OH, K_a = 1.3 \times 10^{-10})$?

20. Calculate the molar concentration of a benzoic acid $(C_6H_5CO_2H)$ solution if it is 5% ionized $(K_a = 6.3 \times 10^{-5})$.

21. In a 9.0×10^{-4}M HCO_2H solution, the concentration of HCO_2H is 3.16×10^{-4}M. Determine the degree of ionization.

22. At 25°C, a 0.20 M solution of vitamin C (ascorbic acid) is 2.0% dissociated. Calculate the equilibrium constant, K_a, for the reaction:

$$C_6H_8O_6 + H_2O \rightleftharpoons C_6H_7O_6^- + H_3O^+$$

18.16.2 Salt hydrolysis

23. Calculate the pH of a 0.10 M solution of sodium acetate. $K_a = 1.8 \times 10^{-5}$. (Note: you will need to determine the hydrolysis constant K_h from $K_h = \frac{K_w}{K_a}$.)

24. Calculate the pH and the degree of hydrolysis of a 0.20 M solution of sodium acetate $(CH_3CO_2Na, K_b = 5.6 \times 10^{-10})$.

25. Calculate the pH of a 0.0350 M solution of $Al(NO_3)_3$ given that K_a for $Al(H_2O)_6^{3+}$ is 1.35×10^{-5}. Assume that the only reactions of importance are:

    (1) $Al(NO_3)_3 (s) \rightarrow Al^{3+} (aq) + 3\ NO_3^- (aq)$

    (2) $Al(H_2O)_6^{3+} (aq) + H_2O\,(l) \rightleftharpoons Al(OH)(H_2O)_5^{2+} (aq) + H_3O^+ (aq)$

26. The pH of a 0.30 M $Co(NO_3)_2$ solution is 3.0. Assuming the only significant hydrolysis is

$$Co(H_2O)_6^{2+} (aq) + H_2O\,(l) \rightleftharpoons Co(OH)(H_2O)_5^+ (aq) + H_3O^+ (aq),$$

    what is the hydrolysis constant?

18.16.3 Common ion effect

27. What is the hydrogen ion concentration in a solution made by adding 6.66 g of HCO_2Na to 1.00 L of 0.100 M formic acid $(HCO_2H, K_a = 1.8 \times 10^{-4})$?

28. Calculate the pH of a solution containing 0.50 M acetic acid (CH_3CO_2H) and 0.50 M sodium acetate (CH_3CO_2Na). For acetic acid, CH_3CO_2H, $K_a = 1.8 \times 10^{-5}$.

29. Calculate the pH and the percent ionization of ammonia in a solution that is 0.10 M in NH_3 and 0.30 M NH_4Cl. $K_{b,NH_3} = 1.8 \times 10^{-5}$.

30. Calculate the concentration of all the species and the pH of a solution made by adding 0.065 mol $CH_3NH_3^+Cl^-$ to 1.0 L of 0.055 M CH_3NH_2. Note: $pK_b\,(CH_3NH_2) = 3.44$.

18.16.4 Buffers

31. What is the pH of the buffer prepared by making up a solution that is 0.25 M CH_3CO_2H and 0.30 M CH_3CO_2Na? $K_a\,(CH_3CO_2H) = 1.8 \times 10^{-5}$.

32. What $[CH_3CO_2H]/[CH_3CO_2^-]$ ratio is required for the buffer solution in problem 18.3 to maintain a pH of 7?

33. Provide a recipe for preparing a phosphate buffer with a pH of approximately 7.50. Which of the following equations will offer the most effective buffer system and why?

$$H_3PO_4 (aq) \rightleftharpoons H^+ (aq) + H_2PO_4^- (aq), pK_{a1} = 2.12$$

$$H_2PO_4^- (aq) \rightleftharpoons H^+ (aq) + HPO_4^{2-} (aq), pK_{a2} = 7.21$$

$$HPO_4^{2-} (aq) \rightleftharpoons H^+ (aq) + PO_4^{3-} (aq), pK_{a3} = 12.32$$

34. A buffer solution was prepared by mixing equal volumes of a 0.2 M acid and its 0.75 M salt. The pH of the resulting buffer was measured to be 6.1.

   (a) Calculate K_a for the acid.
   (b) Is this a good buffer system for this pH?

35. The pK_a of benzoic acid is 4.18. How many moles of the salt sodium benzoate must be added to 1.0 L of 0.05 M benzoic acid to bring the pH of the solution to 3.8? (Assume no volume change by the addition of the salt.)

36. Calculate the pH of a solution prepared by adding 7.5 g of NaOH to 1.0 L of 0.50 M phenylboric acid ($C_6H_5H_2BO_3$). The pK_a of phenylboric acid is 8.86.

37. Determine the pH, $[HCO_3^-]$, and $[CO_3^{2-}]$ of a 0.05M aqueous CO_2 solution at 25°C. The equilibrium reactions are:

$$CO_2\,(aq) + 2\,H_2O\,(l) \rightleftharpoons H_3O^+\,(aq) + HCO_3^-\,(aq), K_{a1} = 2.0 \times 10^{-4}$$

$$HCO_3^-\,(aq) + H_2O\,(l) \rightleftharpoons H_3O^+\,(aq) + CO_3^{2-}\,(aq), K_{a2} = 5.6 \times 10^{-11}$$

38. The ionization of hydrogen sulfide, H_2S, is shown below.

$$H_2S \rightleftharpoons H^+ + HS^-, \quad K_{a1} = 1.0 \times 10^{-7}$$

$$HS^- \rightleftharpoons H^+ + S^{2-}, \quad K_{a2} = 1.0 \times 10^{-19}$$

   What are the pH, $[HS^-]$, and $[S^{2-}]$ of 0.20 M aqueous H_2S at 25°C?

39. A solution saturated with tellurous acid undergoes the following equilibrium reactions:

$$H_2TeO_3 \rightleftharpoons H^+ + HTeO_3^-, \quad K_{a1} = 2.0 \times 10^{-3}$$

$$HTeO_3^- \rightleftharpoons H^+ + TeO_3^{2-}, \quad K_{a2} = 1.0 \times 10^{-8}$$

   (a) What are the approximate values of $[H^+]$ and $[HTeO_3^-]$ in a 0.200 M solution of H_2TeO_3?
   (b) Calculate the concentration of TeO_3^{2-} present at equilibrium.

18.16.5 Titration

40. Using Table 18.1 as a guide, recommend indicators for the following titrations:

   (a) 0.10 M NaOH titrated with 0.10 M HCl.
   (b) 0.10 M NaOH titrated with formic acid 0.10 M (HCO_2H).
   (c) 0.10 M NH_3 titrated with 0.10 M HCl.

41. At the equivalence point in the titration of 50.00 mL of 0.100 M (CH_3CO_2H) with 0.100 M NaOH, 0.0050 mole of CH_3CO_2Na ($K_b = 5.6 \times 10^{-10}$) have been formed in 100.00 mL of solution.

   (a) Calculate the pH for this point on the titration curve.
   (b) Recommend a suitable indicator for the titration.

42. For Alizarin Yellow (HAY), $K_a = 6.3 \times 10^{-12}$. When the ratio $\frac{[AY^-]}{[HAY]} \leq \frac{1}{10}$, the human eye can just detect the red color. Calculate the highest pH at which the solution of the acid form (HAY) appears red.

43. In a titration experiment, 25.00 mL of 0.10 M HCl is titrated with 0.10 M NaOH. Calculate the pH of the solution formed when

   (a) 0.00 mL of NaOH is added
   (b) 5.00 mL of NaOH is added
   (c) 15.00 mL of NaOH is added
   (d) 25.00 mL of NaOH is added
   (e) 35.00 mL of NaOH is added

44. In the titration of 25.00 mL of 0.10 M HCl by a 0.10 M KOH solution, calculate the pH of the titration mixture

   (a) before the addition of any acid, and
   (b) after the addition of 20 mL HCl.

45. In the titration of 25 mL of 0.10 M acetic acid with 0.10 M NaOH, what is the pH of the titration mixture after

   (a) 0.00 mL of 0.10 M NaOH is added
   (b) 10.00 mL of 0.10 M NaOH is added
   (c) 25.00 mL of 0.10 M NaOH is added
   (d) 35.00 mL of 0.10 M NaOH is added

46. In the titration of 25 mL of 0.10 M NH_3 with 0.10 M HCl, what is the pH of the titration mixture after

   (a) 0.00 mL of 0.10 M HCl is added
   (b) 10.00 mL of 0.10 M HCl is added
   (c) 20.00 mL of 0.10 M HCl is added
   (d) 35.00 mL of 0.10 M HCl is added

19

Solubility and Complex-Ion Equilibria

. . . .

19.1 Solubility Equilibria

Solubility equilibria are heterogeneous equilibria that describe the dissolution and precipitation of slightly soluble ionic compounds. Examples are commonly encountered in many chemical and biological processes. This chapter deals with the application of chemical equilibrium to heterogeneous systems involving saturated solutions of slightly soluble ionic compounds.

19.2 The Solubility Product Principle

The solubility product constant is equal to the product of the concentration of the constituent ions involved in the equilibrium, each raised to the power corresponding to its coefficient in the balanced equilibrium equation.

When a slightly soluble salt is dissolved in water, it eventually reaches a point of saturation where equilibrium is established between the undissolved salt and its solution.

For example, a saturated solution of $BaSO_4$ involves the equilibrium

$$BaSO_4\,(s) \rightleftharpoons Ba^{2+}\,(aq) + SO_4^{2-}\,(aq)$$

We can apply the law of chemical equilibrium to this system and obtain the expression

$$K_{eq} = \frac{\left[Ba^{2+}\right]\left[SO_4^{2-}\right]}{[BaSO_4]}$$

Since the concentration of solid $BaSO_4$ remains essentially constant, the product $K_{eq} \times [BaSO_4]$ will also be constant. Therefore, we can write

$$K_{eq}\,[BaSO_4] = \left[Ba^{2+}\right]\left[SO_4^{2-}\right] = K_{sp}$$

The equilibrium constant K_{sp} is called the *solubility product constant*, or simply solubility product.

For the general solubility equilibrium

$$A_xB_y\,(s) \rightleftharpoons xA^{n+}(aq) + yB^{m-}\,(aq)$$

Chemistry in Quantitative Language: Fundamentals of General Chemistry Calculations. Second Edition.
Christopher O. Oriakhi, Oxford University Press. © Christopher O. Oriakhi 2021.
DOI: 10.1093/oso/9780198867784.003.0019

The equilibrium constant expression is

$$K_{sp} = \left[A^{n+}(aq)\right]^x \left[B^{m-}(aq)\right]^y$$

Example 19.1

Write expressions for the solubility product constant, K_{sp}, for the following salts:

(a) $CaF_2(s)$
(b) $Ag_2CrO_4(s)$
(c) $Bi_2S_3(s)$

Solution

(a) $CaF_2(s) \rightleftharpoons Ca^{2+}(aq) + 2F^-(aq)$ $K_{sp} = \left[Ca^{2+}\right]\left[F^-\right]^2$

(b) $Ag_2CrO_4(s) \rightleftharpoons 2\,Ag^+(aq) + CrO_4^{2-}(aq)$ $K_{sp} = \left[Ag^+\right]^2 \left[CrO_4^{2-}\right]$

(c) $Bi_2S_3(s) \rightleftharpoons 2\,Bi^{3+}(aq) + 3S^{2-}(aq)$ $K_{sp} = \left[Bi^{3+}\right]^2\left[S^{2-}\right]^3$

19.3 Determining K_{sp} from Molar Solubility

The value of K_{sp} for a salt whose ions do not react appreciably with water can be determined if the solubility of the salt is known. Note that sometimes the concentration of the ions in solution or solubility of a salt is expressed in mass per unit volume of water, but all mass concentrations must be converted to moles per liter before substituting in the K_{sp} expression.

Example 19.2

The solubility of magnesium carbonate at room temperature is 0.2666 g/L. Calculate the solubility product for this salt.

Solution

1. Write the chemical equation and K_{sp} expression for the dissolution of $MgCO_3$.

$$MgCO_3(s) \rightleftharpoons Mg^{2+}(aq) + CO_3^{2-}(aq)$$

$$K_{sp} = \left[Mg^{2+}\right]\left[CO_3^{2-}\right]$$

2. A saturated solution of $MgCO_3$ contains equal amounts of Mg^{2+} and CO_3^{2-}. This implies that $\left[Mg^{2+}\right] = \left[CO_3^{2-}\right] =$ molar solubility, x.

3. Calculate the solubility of $MgCO_3$ in moles per liter.

$$\text{Solubility } (x) \text{ in moles per liter} = \left(\frac{0.2666 \text{ g } MgCO_3}{1 \text{ L}}\right)\left(\frac{1 \text{ mol } MgCO_3}{84.3 \text{ g } MgCO_3}\right)$$

$$= 3.2 \times 10^{-3}\text{M}$$

Hence, $\left[Mg^{2+}\right] = \left[CO_3^{2-}\right] = 3.2 \times 10^{-3}\text{M}$

4. Now substitute for x in the K_{sp} expression

$$K_{sp} = \left[Mg^{2+}\right]\left[CO_3^{2-}\right] = (x)(x) = \left(x^2\right)$$

$$= \left(3.2 \times 10^{-3}\right)^2 = 1.0 \times 10^{-5}$$

Example 19.3

A saturated solution of BaF_2 prepared by dissolving solid BaF_2 in water has $\left[F^-\right] = 1.0 \times 10^{-3}$ M. What is the value of K_{sp} for BaF_2?

Solution

1. Write the chemical equation and K_{sp} expression for the dissolution of BaF_2:

$$BaF_2 \text{ (s)} \rightleftharpoons Ba^{2+} \text{ (aq)} + 2 \, F^- \text{ (aq)}$$

$$K_{sp} = \left[Ba^{2+}\right]\left[F^-\right]^2$$

2. Two moles of fluoride ions are formed for each mole of BaF_2 that dissolves. Thus

$$\left[F^-\right] = 2\left[Ba^{2+}\right]$$

$$\text{Since } \left[F^-\right] = 1.0 \times 10^{-3}\text{M}, \left[Ba^{2+}\right] = 5.0 \times 10^{-4} \text{ M}$$

3. Substitute these values in the K_{sp} expression.

$$K_{sp} = \left[Ba^{2+}\right]\left[F^-\right]^2 = \left(5.0 \times 10^{-4}\right)\left(1.0 \times 10^{-3}\right)^2$$

$$= 5.0 \times 10^{-10}$$

Example 19.4

The solubility of silver sulfate is 0.020 M at 25°C. Calculate the solubility product constant.

Solution

1. Write the chemical equation and K_{sp} expression for the dissolution of Ag_2SO_4:

$$Ag_2SO_4\,(s) \rightleftharpoons 2Ag^+\,(aq) + SO_4^{2-}\,(aq)$$

$$K_{sp} = [Ag^+]^2 [SO_4^{2-}]$$

2. Two moles of silver ions are formed for each mole of Ag_2SO_4 that dissolves. Thus

$$[Ag^+] = 2[Ag_2SO_4] = 2[SO_4^{2-}]$$

$$\text{Since } [SO_4^{2-}] = 2.0 \times 10^{-2}M, [Ag^+] = 4.0 \times 10^{-2}\ M$$

3. Substitute these values in the K_{sp} expression.

$$K_{sp} = [Ag^+]^2 [SO_4^{2-}] = \left(4.0 \times 10^{-2}\right)^2 \left(2.0 \times 10^{-2}\right)$$

$$= 3.2 \times 10^{-5}$$

19.4 Calculating Molar Solubility from K_{sp}

The solubility of a slightly soluble salt in moles per liter can be readily calculated if the solubility product constant is known.

Example 19.5

The solubility product constant of lead iodide in water is 7.1×10^{-9}. Calculate the solubility in

(a) moles per liter, and
(b) grams per liter.

Solution

1. Write the chemical equation and K_{sp} expression for the dissolution of PbI_2

$$PbI_2\,(s) \rightleftharpoons Pb^{2+}\,(aq) + 2\,I^-\,(aq)$$

$$K_{sp} = [Pb^{2+}][I^-]^2 = 7.1 \times 10^{-9}$$

2. Let x be the solubility of PbI_2 in moles per liter. Two moles of iodide ions and one mole of lead ions are formed for each mole of the salt that dissolves. Therefore,

$\left[Pb^{2+} \right] = x$, and $\left[I^- \right] = 2x$

$$PbI_2 \, (s) \rightleftharpoons \underset{x}{Pb^{2+}} \, (aq) + \underset{2x}{2I^-} \, (aq)$$

$$K_{sp} = \left[Pb^{2+} \right]\left[I^- \right]^2 = (x)(2x)^2 = (4x)^3 = 7.1 \times 10^{-9}$$

$$x = 1.2 \times 10^{-3} M$$

(a) The solubility of PbI_2 is 1.2×10^{-3} moles per liter.

(b) Solubility in g/liter $= \left(\dfrac{1.2 \times 10^{-3} \text{ mol } PbI_2}{1 \text{ Liter}} \right) \left(\dfrac{334.2 \text{ g } PbI_2}{1 \text{ mol } PbI_2} \right) = 0.40$ g/liter

Example 19.6

Magnesium hydroxide is used as an antacid and laxative.

(a) How many moles of $Mg(OH)_2$ would dissolve in 1.0 liter of water?
(b) If $Mg(OH)_2$ is the only active ingredient in the antacid, what is the pH of the resulting medicinal emulsion? The K_{sp} for $Mg(OH)_2$ is 8.8×10^{-12}.

Solution

(a) Write the equation for the dissolution reaction and an expression for K_{sp}.

$$Mg(OH)_2 \, (s) \rightleftharpoons Mg^{2+} \, (aq) + 2 \, OH^- \, (aq)$$

$$K_{sp} = \left[Mg^{2+} \right]\left[OH^- \right]^2 = 8.8 \times 10^{-12}$$

Let x = moles of $Mg(OH)_2$ which dissolves.

$\left[Mg^{2+} \right] = x$, and $\left[OH^- \right] = 2x$

$$K_{sp} = (x)(2x)^2 = 4x^3 = 8.8 \times 10^{-12} \text{ or } x^3 = 2.2 \times 10^{-12}$$

$$x = \sqrt[3]{(2.2 \times 10^{-12})} = 1.3 \times 10^{-4} \text{ M}.$$

Therefore the solubility of $Mg(OH)_2$ in moles per liter is $1.3 \times 10^{-4} M$ Also, $\left[OH^- \right] = 2.6 \times 10^{-4}$ M.

(b) Calculate the hydrogen ion concentration using the known value of hydroxide ion concentration.

$$\left[H^+ \right]\left[OH^- \right] = 10^{-14}$$

$$\left[H^+ \right] = \frac{10^{-14}}{\left[OH^- \right]} = \frac{10^{-14}}{2.6 \times 10^{-4}} = 3.8 \times 10^{-11} \text{ M}$$

$$pH = - \log\left[H^+ \right] = - \log\left(3.8 \times 10^{-11} \right) = 10.4$$

19.5 K_{sp} and Precipitation

Generally, a precipitate will form whenever the solubility limit of a substance is exceeded. To predict whether a precipitate will form when solutions are mixed, we calculate the quantity known as ion product Q, and compare its value with the K_{sp} of a given solid. The ion product Q is defined just like K_{sp}, except that the ion concentrations used for Q are not necessarily equilibrium ion concentrations. For example, the Q expression for Ag_2CrO_4 is

$$Q = [Ag^+]_0^2 [CrO_4^{2-}]_0$$

If $Q > K_{sp}$, the solution is supersaturated, and precipitation will occur until $Q = K_{sp}$.
If $Q = K_{sp}$, the solution is saturated and is already in equilibrium.
If $Q < K_{sp}$, the solution is unsaturated, so more solid will dissolve until $Q = K_{sp}$.

Example 19.7

A solution is prepared by mixing 100 mL of 0.010 M $CaCl_2$ solution and 50 mL of 0.005 M NaF solution at 25°C. Is it possible for CaF_2 to precipitate? K_{sp} for CaF_2 is 1.5×10^{-10}.

Solution

1. First calculate the $[Ca^{2+}]$ and $[F^-]$ in the mixture, assuming no precipitation.

The ion product $Q = [Ca^{2+}][F^-]^2$

Total after mixing $= 150$ mL or 0.150 L

$$[Ca^{2+}] = \frac{(0.100 \text{ L} \times 0.010 \text{ M})}{0.150 \text{ L}} = 0.0067 \text{ M}$$

$$[F^-] = \frac{(0.050 \text{ L} \times 0.005 \text{ M})}{0.150 \text{ L}} = 0.0017 \text{ M}$$

2. Now calculate Q after mixing:

$$Q = (0.0067 \text{ M})(0.0017 \text{ M})^2$$

$$= 1.94 \times 10^{-8}$$

3. Compare Q to K_{sp} and decide if precipitation will occur.
   Since K_{sp} is 1.5×10^{-10}, Q is greater than K_{sp}, so CaF_2 will precipitate.

Example 19.8

A student had a solution of lead dichromate in water; she wanted to bring the Pb^{2+} concentration to 0.20 M exactly. What concentration of CrO_4^{2-} (added as chromic acid) would accomplish this? $\left(K_{sp} \text{ for } PbCrO_4 = 2.0 \times 10^{-16}\right)$

Solution

1. Write an equation for the dissolution reaction.

$$PbCrO_4 \rightleftharpoons Pb^{2+} + CrO_4^{2-}$$

2. When the solution is saturated with $PbCrO_4$, the expression

$$\left[Pb^{2+}\right]\left[CrO_4^{2-}\right] = K_{sp} = 2.0 \times 10^{-16} \text{ holds.}$$

3. Calculate the chromate ion concentration from the expression for K_{sp}.

$$\left[Pb^{2+}\right]\left[CrO_4^{2-}\right] = K_{sp} = 2.0 \times 10^{-16}$$

$$\left[CrO_4^{2-}\right] = \frac{2.0 \times 10^{-16}}{\left[Pb^{2+}\right]} = \frac{2.0 \times 10^{-16}}{0.20} = 1.0 \times 10^{-15} \text{ M}$$

To bring the lead ion concentration to 0.20 M will require a chromate concentration of 1.0×10^{-15} M.

Example 19.9

A solution contains 0.010 M silver ion (Ag^+) and 0.0010 M lead ion (Pb^{2+}). If an iodide ion solution is slowly added, which solid will precipitate first, AgI or PbI_2? K_{sp} for $AgI = 1.5 \times 10^{-16}$, and for $PbI_2 = 7.1 \times 10^{-8}$.

Solution

1. Write equations for both reactions:

$$AgI \rightleftharpoons Ag^+ + I^-, K_{sp} = \left[Ag^+\right]\left[I^-\right] = 1.5 \times 10^{-16}$$

$$PbI_2 \rightleftharpoons Pb^{2+} + 2I^-, K_{sp} = \left[Pb^{2+}\right]\left[I^-\right]^2 = 7.1 \times 10^{-8}$$

2. PbI_2 will precipitate when $\left[Pb^{2+}\right]\left[I^-\right]^2 = K_{sp} = 7.1 \times 10^{-8}$. From this expression, calculate the iodide ion concentration required for precipitation to occur.

$$\left[Pb^{2+}\right]\left[I^-\right]^2 = K_{sp} = 7.1 \times 10^{-8}$$

$$\left[Pb^{2+}\right] = 0.0010 \text{ M}$$

$$\left[I^-\right] = \sqrt{\frac{7.1 \times 10^{-8}}{\left[Pb^{2+}\right]}} = \sqrt{\frac{7.1 \times 10^{-8}}{0.0010}} = 0.0084 \text{ M}$$

Thus PbI_2 will precipitate when $[I^-] = 0.0084$ M.

4. Similarly, AgI will precipitate when $\left[Ag^+\right]\left[I^-\right] = K_{sp} = 1.50 \times 10^{-16}$. From this expression, calculate the iodide ion concentration required for precipitation to occur.

$$\left[Ag^+\right]\left[I^-\right] = 1.50 \times 10^{-16}$$

$$\left[Ag^+\right] = 0.010 \text{ M}$$

$$\left[I^-\right] = \frac{1.50 \times 10^{-16}}{\left[Ag^+\right]} = \frac{1.50 \times 10^{-16}}{0.010} = 1.50 \times 10^{-14} \text{ M}$$

So AgI will precipitate when the iodide ion concentration equals 1.50×10^{-14} M.

4. The solid that requires the lowest $[I^-]$ will be the first to precipitate. Therefore AgI will precipitate first since 1.50×10^{-14} M is reached before 3.70×10^{-3} M.

19.6 Complex-Ion Equilibria

19.6.1 Formation of complex ions

A complex ion is one which contains a positive metal ion (usually a transition metal) bonded to one or more uncharged small molecules (e.g. H_2O or NH_3) or negative ions (e.g. OH^-, CN^-, or Cl^-) called *ligands*. The metal ion is an electron-acceptor (or Lewis acid), while the ligands are electron-donating groups (or Lewis bases). All the ligands surrounding the central ion need not be the same, and solvent molecules can occupy some positions. Ligands are linked to the central metal ion by coordinate (dative covalent) bonds. The number of coordinate bonds formed by the central metal ion (or the number of ligands bonded to the central metal ion) is known as the *coordination number*.

The cations in many slightly soluble electrolytes can be dissolved through complex ion formation. For example, consider the dissolution of AgCl precipitate by the addition of aqueous ammonia. This results in the formation of the complex $\left[Ag(NH_3)_2\right]^+$. The net reaction is shown by equation (19.3) below.

$$AgCl(s) \rightleftharpoons Ag^+(aq) + Cl^-(aq) \tag{19.1}$$

$$Ag^+ + 2NH_3(aq) \rightleftharpoons Ag(NH_3)_2^+(aq) \tag{19.2}$$

$$AgCl(s) + 2NH_3(aq) \rightleftharpoons Ag(NH_3)_2^+ + Cl^-(aq) \tag{19.3}$$

According to Le Chatelier's principle, the added ammonia removes Ag^+ ions from solution and shifts the position of equilibrium. More AgCl(s) will dissolve to furnish new Ag^+ ions in solution to restore the equilibrium. These are further converted to $[Ag(NH_3)_2]^+$ by the NH_3 present in the solution. The entire precipitate can be dissolved if excess ammonia is present in the solution.

An equilibrium constant expression can be written for the formation of the complex ion (equation 19.2) as:

$$K = \frac{[Ag(NH_3)^+_2]}{[Ag^+][NH_3]^2} = K_f$$

This equilibrium constant is known as the *formation or stability constant*, K_f. K_f measures how tightly or loosely the ligands hold the metal ions.

The dissociation of the complex ion $[Ag(NH_3)_2]^+$ may be represented as

$$Ag(NH_3)_2^+ \rightleftharpoons Ag^+ + 2NH_3$$

The equilibrium constant for this reaction is known as the *dissociation constant* K_d, and is given by the expression:

$$K_d = \frac{[Ag^+][NH_3]^2}{[Ag(NH_3)_2^+]}$$

Like polyprotic acids, complex ions usually exhibit stepwise dissociation equilibria. Equilibrium constants can be written for each reaction step. The net dissociation constant K_d will be the product of the K values for the successive steps.

K_d is related to K_f by

$$K_d = \frac{1}{K_f}$$

These constants can be used to calculate the solubility and concentration of an aqueous metal ion in equilibrium with a complex ion.

Example 19.10

A solution was prepared by adding 0.10 mole of $AgNO_3$ to 1.0 L of 2.5 M NH_3. Calculate the concentrations of Ag^+ and $Ag(NH_3)_2^+$ ions in the solution. The K_f value for $Ag(NH_3)_2^+$ is 1.70×10^7.

Solution

1. First write a balanced equation for the complex ion formation.

$$Ag^+ (aq) + 2NH_3 (aq) \rightleftharpoons Ag(NH_3)_2^+ (aq)$$

2. Write the equilibrium or K_f expression for the reaction.

$$K_f = \frac{[Ag(NH_3)_2^+]}{[Ag^+][NH_3]^2}$$

3. Assuming complete conversion of Ag^+ to $Ag(NH_3)_2^+$ (i.e. K_f is large), make a table showing initial concentration and equilibrium concentrations of all species (Table 19.1).

Table 19.1   Data for Example 19.10

	$[Ag^+]$	$[NH_3]$	$[Ag(NH_3)_2^+]$
Initial concentration (M)	0.10 M	2.5	0
After complete reaction of Ag^+ (M)	0	2.3	0.1
Equilibrium concentration (M)	x	$2.3 + 2x$	$0.1 - x$

4. Substitute the equilibrium concentrations into the K_f expression and solve for x:

$$K_f = 1.70 \times 10^7 = \frac{\left[Ag(NH_3)_2^+\right]}{\left[Ag^+\right][NH_3]^2} = \frac{(0.10 - x)}{(x)(2.3 + 2x)}$$

The expression can be simplified by making the approximation that x is small compared to 0.10 and 2.30.

$$1.70 \times 10^7 = \frac{(0.10)}{(x)(2.3)} \text{ or } x = 2.56 \times 10^{-9}$$

Hence, $\left[Ag^+\right] = 2.56 \times 10^{-9} M$, and

$$\left[Ag(NH_3)_2^+\right] = (0.10 - x) = \left(0.10 - 2.56 \times 10^{-9}\right) = 0.10 \text{ M}$$

5. The concentrations of Ag^+ and $Ag(NH_3)_2^+$ ions are 2.56×10^{-9} M and 0.10 M, respectively.

Example 19.11

The dissociation constant of $Ag(NH_3)_2^+$ is 6.2×10^{-8}. For a 0.15 M solution prepared by dissolving solid $[Ag(NH_3)_2]NO_3$ in water, calculate the equilibrium concentration of Ag^+ and NH_3.

Solution

1. First write a balanced equation for the dissociation of the complex ion.

$$Ag(NH_3)_2^+ (aq) \rightleftharpoons Ag^+ (aq) + 2NH_3 (aq)$$

2. Write the equilibrium or K_d expression for the reaction:

$$K_d = \frac{\left[Ag^+\right][NH_3]^2}{\left[Ag(NH_3)_2^+\right]}$$

3. For the dissociation of $Ag(NH_3)_2^+$, make a table showing initial and equilibrium concentrations of all species (Table 19.2).

Table 19.2   Data for Example 19.11

	$[Ag(NH_3)_2^+]$	$[Ag^+]$	$[NH_3]$
Initial concentration (M)	0.15 M	0	0
Change (M)	$(-)x$	$(+)x$	$(+)2x$
Equilibrium concentration (M)	$(0.15 - x)$	x	$2x$

4. Substitute the equilibrium concentrations into the K_f expression and solve for x:

$$K_d = 6.20 \times 10^{-8} = \frac{[Ag^+][NH_3]^2}{[Ag(NH_3)_2^+]} = \frac{x(2x)^2}{(0.15-x)} = \frac{4x^3}{(0.15-x)}$$

The expression can be simplified by making the approximation that x is small compared to 0.15.

$$6.20 \times 10^{-8} \simeq \frac{4x^3}{(0.15)} \quad \text{or } x = 1.30 \times 10^{-3}$$

Hence, $[Ag^+] = 1.3 \times 10^{-3}M$, $[NH_3] = 2.6 \times 10^{-3}M$ and

$$[Ag(NH_3)_2^+] = (0.15 - x) = \left(0.15 - 1.30 \times 10^{-3}\right) \simeq 0.15 \text{ M}$$

5. The concentrations of Ag^+ ions and NH_3 at equilibrium are 1.30×10^{-3} M and 2.60×10^{-3} M, respectively.

19.7 Problems

1. Write the expression for the solubility product, K_{sp}, of the following:
   (a) Ag_2CrO_4 (b) CaF_2 (c) $CaCO_3$ (d) $Ca_3(PO_4)_2$ (e) $Pb_3(AsO_4)_2$

2. Write the solubility product expression for the following salts:
   (a) AgI (b) CaC_2O_4 (c) Hg_2Cl_2 (d) PbI_2 (e) $Fe(OH)_3$

3. At 25°C, a 1.0 L solution saturated with magnesium oxalate, MgC_2O_4, is evaporated to dryness, resulting in 1.0345 g of MgC_2O_4 residue. Calculate the K_{sp} at this temperature.

4. A saturated solution prepared by dissolving excess MgF_2 in water has $[Mg^{2+}] = 2.55 \times 10^{-4}M$ at 25°C. Calculate the K_{sp}.

5. Calculate the molar solubility of aluminum hydroxide, $Al(OH)_3$, in water at 25°C. $K_{sp} = 1.9 \times 10^{-33}$.

6. The solubility product constant, K_{sp}, of calcium phosphate, $Ca_3(PO_4)_2$, is 1.0×10^{-26} at 25°C. Calculate the solubility of $Ca_3(PO_4)_2$ (a) in mol/L and (b) in g/L.

7. The solubility product constant, K_{sp}, of CaF_2 is 3.9×10^{-11}. What is the solubility of CaF_2 in a 0.030M LiF solution?

8. A student mixed 50 mL of 0.200 M $AgNO_3$ with 70 mL of 0.10 M K_2CrO_4. Calculate the concentration of each ion $(Ag^+, NO_3^-, K^+, \text{and } CrO_4^{2-})$ in solution at equilibrium if K_{sp} of $Ag_2CrO_4 = 1.2 \times 10^{-13}$.

9. The concentration of Ba^{2+} ions in a given aqueous solution is $5 \times 10^{-4}M$. What is the SO_4^{2-} ion concentration that must be exceeded before $BaSO_4$ can precipitate? The K_{sp} of $BaSO_4$ at 25°C is 1.5×10^{-9}.

10. A 100 mL solution containing 0.0025 mol Ag^+ ions is mixed with 100 mL of 0.250 M HBr to precipitate $AgBr$. Calculate the concentration of Ag^+ ions remaining in solution. The K_{sp} of $AgBr$ at 25°C is 5×10^{-13}.

11. A solution containing 0.01 M mercuric (I) ion is mixed with another solution containing 0.001 M lead (II) ions. A dilute NaI solution is slowly added. Which solid will precipitate first: Hg_2I_2 ($K_{sp} = 4.5 \times 10^{-29}$) or PbI_2 ($K_{sp} = 1.4 \times 10^{-8}$)?

12. A solution of $AgNO_3$ is gradually added to one containing 0.025 mol of NaBr and 0.0025 mol of NaI. Determine which salt, AgBr ($K_{sp} = 5.0 \times 10^{-13}$) or AgI ($K_{sp} = 1.5 \times 10^{-16}$), will precipitate first.

13. Determine the maximum concentration of hydrogen ion at which FeS ($K_{sp} = 3.7 \times 10^{-19}$) will not precipitate from a solution which is 0.01 M in $FeCl_2$ and saturated with H_2S ($K_{sp} = 1.3 \times 10^{-21}$).

14. A solution is prepared by adding 450 mL of 0.05 M $Pb(NO_3)_2$ to 250 mL of 0.125 M Na_2CO_3. Would you expect $PbCO_3$ ($K_{sp} = 1.5 \times 10^{-15}$) to precipitate from this solution?

15. A solution contains 0.0010 M Ba^{2+} and 0.100 M Ca^{2+}. A dilute NaF solution is slowly added. Will BaF_2 ($K_{sp} = 2.4 \times 10^{-5}$) or CaF_2 ($K_{sp} = 4.0 \times 10^{-11}$) precipitate first? Indicate the concentration of fluoride ion necessary to begin precipitation of each salt.

16. Calculate the number of moles of ammonia that must be added to 1.0 liter of 0.50 M $AgNO_3$ so as to reduce the silver ion concentration $[Ag^+]$ to 2.50×10^{-6} M. The formation constant K_f for the amine complex ion at 25°C is 1.7×10^7.

17. Determine the number of moles of ammonia that must be added to 1.0 L of 0.025 M $Cu(NO_3)_2$ so as to reduce the copper ion concentration $[Cu^{2+}]$ to 2.50×10^{-12} M. The dissociation constant K_d for $Cu(NH_3)_4^{2+}$ at 25°C is 4.35×10^{-13}. (Remember $K_f = 1/K_d$.)

18. What is the molar solubility of AgCl at 25°C in 1.0 M NH_3?

19. The solubility product constant, K_{sp}, of $Zn(OH)_2$ is 1.8×10^{-14}. Calculate the concentration of aqueous ammonia required to initiate the precipitation of $Zn(OH)_2$ from a 0.050 M solution of $Zn(NO_3)_2$.

20. Nickel forms a complex with ammonia according to the following equation:

$$Ni(NH_3)_4 \rightleftharpoons Ni^{2+} + 4\,NH_3 \quad K_d = 1.0 \times 10^{-8}$$

If excess powdered NiS is added to 1.0 M NH_3 solution, only a small amount will dissolve. Calculate the exact amount of NiS that will dissolve in 1.0 M NH_3 solution. The K_{sp} of NiS is 1×10^{-22}.

20

Thermochemistry

. . . .

20.1 Introduction

All chemical reactions involve energy changes. Some reactions liberate heat to the surroundings, while others absorb it. The breaking of chemical bonds in reactants and the formation of new ones in the products is the source these of energy changes.

20.2 Calorimetry and Heat Capacity

Calorimetry is the experimental determination of the amount of heat transferred during a chemical reaction. This measurement is carried out in a device called a *calorimeter*, which allows all the heat entering or leaving the reaction to be accounted for. This is done by observing the temperature change within the calorimeter as the reaction takes place; if we know how much energy is needed to change the calorimeter's temperature by a given amount, we can calculate the amount of energy involved in the reaction.

The relation between energy and temperature change for the calorimeter or for any other physical object is known as its *heat capacity* (C), which is the amount of heat energy required to raise the temperature of that object by 1°C (or 1 K). This can be expressed in mathematical terms as:

$$C = \frac{q}{\Delta T}$$

where q is the quantity of heat transferred and ΔT is the change in temperature, calculated as $\Delta T = T_f - T_i$. The larger the heat capacity of a body, the larger the amount of heat required to produce a given rise in temperature. The heat capacity of 1 mol of a substance is known as the *molar heat capacity*. Also, the heat capacity of 1 g of a substance is known as the *specific heat capacity*. To determine the specific heat of a substance, we measure the temperature change, ΔT, that a known mass, m, of a substance undergoes as it gains or loses a known quantity of heat, q. That is:

Chemistry in Quantitative Language: Fundamentals of General Chemistry Calculations. Second Edition.
Christopher O. Oriakhi, Oxford University Press. © Christopher O. Oriakhi 2021.
DOI: 10.1093/oso/9780198867784.003.0020

$$\text{Specific heat } (c) = \frac{\text{Quantity of heat gained or lost}}{\text{Mass of substance (in grams)} \times \text{Temperature change } (\Delta T)} \quad \text{or}$$

$$c = \frac{q}{m \times \Delta T}$$

The unit of specific heat is J/g-K.

Example 20.1

A 20-g sample of water absorbs 250.5 J of heat in a calorimeter. If the temperature rises from 25°C to 28°C, what is the specific heat capacity of water?

Solution

Specific heat, $c = \dfrac{q}{m \times \Delta T}$

$m = 20 \text{ g}; q = 250.5 \text{ J}; \Delta T = 28°C - 25°C = 3°C \times \dfrac{1K}{1°C} = 3K$

$c = \dfrac{q}{m \times \Delta T} = \dfrac{250.5 \text{ J}}{20\text{g} \times 3\text{K}} = 4.18 \text{ Jg}^{-1}\text{K}^{-1}$

Example 20.2

The specific heat capacity of NaCl is 0.864 J/g-K. Calculate the quantity of heat necessary to raise the temperature of 100 g of NaCl by 7.5 K.

Solution

Substitute known values into the equation and solve for q directly.

$$\text{Specific heat, } c = \frac{q}{m \times \Delta T}$$

or

$$q = mc\Delta T$$

$$m = 100 \text{ g}; c = 0.864 \text{ J/g-K}; \Delta T = 70.5 \text{ K}$$

$$q = mc\Delta T = 100 \text{ g} \times 0.864 \ \frac{\text{J}}{\text{g K}} \times 70.5 \text{ K} = 6,090 \text{ J}$$

Example 20.3

The specific heat of gold is 0.129 J/g-K. What will be the final temperature of 1500 g of gold at 60°C after 4.35 kJ of heat is transferred away from it?

Solution

Calculate ΔT and add the value to the initial temperature. This will give the final temperature.

$$\Delta T = \frac{q}{m \times c}$$

$m = 1500$ g; $q = 4.35$ kJ or $4,350$ J; $c = 0.129$ J/g-K

$$\Delta T = \frac{4,350 \text{J}}{1500 \text{ g} \times 0.129 \frac{\text{J}}{\text{g.K}}} = 22.5 \text{ K or } 22.5°\text{C} \left(\text{i.e. } 22.5 \text{ K} \times \frac{1°\text{C}}{1\text{K}} \right)$$

$$T_f = T_i + \Delta T = 60°\text{C} + 22.5°\text{C} = 82.5°\text{C}$$

20.3 Enthalpy

Enthalpy (H) is the total heat bound up in a substance; in practice, what is always measured in a reaction is the change in the heat energy, so we use the symbol ΔH. Note that we cannot measure the absolute amount of heat contained in either the products or the reactants; we can only measure the change that occurs during a reaction. The enthalpy change is also known as the heat of reaction, and has units of kJ/mol.

The change in the enthalpy or heat content in a chemical reaction is given by the equation:

$$\Delta H = H_2 - H_1$$

H_1 is the heat content before the reaction, and H_2 is the heat content after the reaction.

An *exothermic reaction* is one in which *energy is given off by the reacting substances*. This generally appears as an increase in temperature as internal energy is converted to heat (though it can take other forms). In these reactions, ΔH is negative because the products have less potential energy than the reactants. Consequently, there is an increase in the temperature of the reaction mixture. An exothermic reaction will continue to sustain itself, once it is initiated, until the reactants are completely consumed. Figure 20.1 illustrates the energy changes in an exothermic reaction.

An *endothermic reaction* is one in which *heat energy is absorbed from the surroundings*. Here, ΔH is positive because the potential energy of the products is higher than that of the reactants, as illustrated by Figure 20.2. Consequently, there is a decrease in the temperature of the reaction mixture as heat is converted into chemical bonds. Endothermic reactions are not self-sustaining. Once the source of energy is removed, the reaction will terminate.

The heat of reaction is the heat change when the number of moles of the reactants as represented by the chemical equation has reacted completely. The symbol for the heat of reaction is ΔH and it has the unit of kJ mol^{-1}.

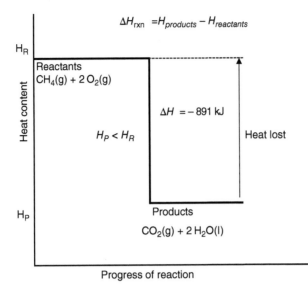

$$\Delta H_{rxn} = H_{products} - H_{reactants}$$

H_R

Reactants
$CH_4(g) + 2\,O_2(g)$

Heat content

$\Delta H = -891\ kJ$

$H_P < H_R$ Heat lost

H_P Products
                      $CO_2(g) + 2\,H_2O(l)$

Progress of reaction

Figure 20.1   Change in heat content during the combustion of methane gas.

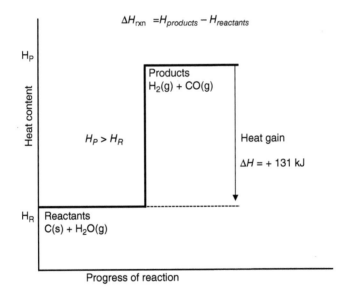

$$\Delta H_{rxn} = H_{products} - H_{reactants}$$

H_P

                   Products
                   $H_2(g) + CO(g)$

Heat content

$H_P > H_R$ Heat gain

                                           $\Delta H = +131\ kJ$

H_R Reactants
        $C(s) + H_2O(g)$

Progress of reaction

Figure 20.2   Change in heat content during the water gas reaction (formation of carbon monoxide).

20.3.1 Calculating ΔH of reaction

Consider the following general reaction:

$$a\text{A} + b\text{B} \longrightarrow c\text{C} + d\text{D}$$

reactants   products

ΔH of reaction = Total heat of formation of products − Total heat

of formation of reactants

$$\Delta H \text{ of reaction} = \Sigma \Delta H_{products} - \Sigma \Delta H_{reactants}$$

$$\Delta H \text{ of reaction} = (c \times \Delta H_C + d \times \Delta H_D) - (a \times \Delta H_A + b \times \Delta H_B)$$

The *heat of formation* of a compound is defined as the amount of heat liberated or absorbed when 1 mole of it is formed from its constituent elements. For example, the heat of formation of ammonia gas is

$$N_2\,(g) + 3H_2\,(g) \longrightarrow 2NH_3\,(g) \quad \Delta H = -100 \text{ kJ}$$

The *standard molar heat of formation* is the amount of heat absorbed or liberated when one mole of a substance in a specified state is formed from its elements in their standard states. It is given the symbol ΔH_f^0. The superscript zero in the ΔH_f^0 indicates standard temperature of 298 K and pressure of one atm. Under international convention, it is generally accepted that the standard heat of formation of all elements in their most stable form is zero. Table 20.1 provides some standard heats of formation.

Example 20.4

Syngas (from *synthesis gas*) consists primarily of carbon monoxide and hydrogen and has less than half the energy density of natural gas. Methanol can be produced by the direct conversion of Syngas according to the following equation:

$$CO\,(g) + 2H_2\,(g) \longrightarrow CH_3OH\,(l)$$

Calculate the heat of reaction, given the heats of formation of reactants and product as follows:

$$\Delta H_{CO(g)} = -110.5 \text{ kJmol}^{-1}$$

$$\Delta H_{H_2(g)} = 0$$

$$\Delta H_{CH_3OH(l)} = -239 \text{ kJmol}^{-1}$$

Solution

$$\Delta H \text{ of reaction} = \Sigma \Delta H_{products} - \Sigma \Delta H_{reactants}$$

$$\Delta H \text{ of reaction} = (-239 \text{ kJ mol}^{-1}) - (-110.5 \text{ kJ mol}^{-1} + 0)$$

$$\Delta H \text{ of reaction} = -128.5 \text{ kJ mol}^{-1}$$

Table 20.1   Standard Heat of Formation of Selected Compounds

Substance and State	ΔH_f^0 (kJ/mole)	Substance and State	ΔH_f^0 (kJ/mole)	Substance and State	ΔH_f^0 (kJ/mole)
$Al(s)$	0	$Cl_2(g)$	0	$HNO_3(aq)$	−207
$Al_2O_3(s)$	−1676	$Cl_2(aq)$	−23	$HNO_3(l)$	−174
$Al(OH)_3(s)$	−1277	$Cl^-(aq)$	−167	$O_2(g)$	0
$Ba(s)$	0	$HCl(g)$	−92	$O(g)$	249
$Ba(OH)_2(s)$	−946	$F_2(g)$	0	$O_3(g)$	143
$BaSO_4(s)$	−1465	$F^-(aq)$	−333	$H_3PO_4(s)$	−1279
$Br_2(l)$	0	$HF(g)$	−271	$H_3PO_4(l)$	−1267
$Br_2(g)$	31	$H_2(g)$	0	$H_3PO_4(aq)$	−1288
$Br_2(aq)$	−3	$H(g)$	217	$K(s)$	0
$Br^-(aq)$	−121	$H^+(aq)$	0	$KCl(s)$	−436
$HBr(aq)$	−36	$OH^-(aq)$	230	$KOH(s)$	−425
$Ca(s)$	0	$H_2O(l)$	−286	$KOH(aq)$	−481
$CaCO_3(s)$	−63	$H_2O(g)$	−242	$Na(s)$	0
$CaO(s)$	−635	$I_2(s)$	0	$Na^+(aq)$	−240
$Ca(OH)_2(s)$	−987	$I_2(g)$	62	$NaCl(s)$	−411
$C(s)$ (graphite)	0	$I_2(aq)$	23	$NaOH(s)$	−427
$C(s)$ (diamond)	2	$I^-(aq)$	−55	$NaOH(aq)$	−470
$CO(g)$	−110.5	$N_2(g)$	0	$S(s)$ (rhombic)	0
$CO_2(g)$	−393.5	$NH_3(g)$	−46	$S(s)$ (monoclinic)	0.3
$CH_4(s)$	−75	$NH_3(aq)$	−80	$S_8(g)$	102
$CH_3OH(g)$	−201	$NH_4^+(aq)$	−132	$SF_6(g)$	−1209
$CH_3OH(l)$	−239	$NO(g)$	90	$H_2S(g)$	−21
$C_2H_2(g)$	227	$NO_2(g)$	94	$SO_2(g)$	−297
$C_2H_4(g)$	52	$N_2O(g)$	82	$SO_3(g)$	−396
$C_2H_5OH(l)$	−278	$N_2O_4(g)$	10	$SO_4^{2-}(aq)$	−909
$C_2H_6(g)$	−84.7	$N_2O_4(l)$	−20	$H_2SO_4(l)$	−814
$C_6H_{12}O_6(s)$	−1275	$N_2O_5(s)$	−42	$H_2SO_4(aq)$	−909

Heat of combustion is defined as the heat evolved when one mole of a substance is burned completely in oxygen at constant temperature. For example, the heat of combustion of methane is:

$$CH_4(g) + 2\ O_2(g) \longrightarrow CO_2(g) + 2\ H_2O(l) \quad \Delta H = -891 \text{ kJ}$$

Combustion is always an exothermic process.

The *heat of neutralization* is defined as the amount of heat liberated when one mole of hydrogen ion, H^+, from an acid reacts with one mole of hydroxide ion, OH^-, from a base to form one mole of water.

$$H+(aq) + OH^-(aq) \longrightarrow H_2O(l) \quad \Delta H = -57.4 \text{ kJ}$$

The heat of neutralization is always negative.

The *heat of solution* of a substance is the quantity of heat liberated or absorbed when one mole of it is dissolved in a large quantity of solvent. For example, the heat of solution of sodium chloride is:

$$NaCl(s) + H_2O(l) \longrightarrow NaCl(aq) \quad \Delta H + 4 \text{ kJ}$$

20.4 Hess's Law of Heat Summation

Hess's law states that if a reaction is carried out in a series of steps, the overall enthalpy change is equal to the sum of the enthalpy changes for the individual steps in the reaction. This implies that the overall enthalpy change for the process is independent of the number or type of steps through which the reaction is carried out.

When solving problems using Hess's law, it is important to know that reactants and products in the individual steps can be added or subtracted just like algebraic terms to obtain the overall equation. In general, the law may be represented as:

$$\Delta H^0_{rxn} = \Delta H^0_1 + \Delta H^0_2 + \Delta H^0_3 + ...$$

Where 1, 2, 3, represent the equations that can be added to give the target equation for the reaction.

20.4.1 Hints for using Hess's law

1. The enthalpy change for any reaction is equal in magnitude but opposite in sign to that for the backward reaction. When an equation is reversed, the associated ΔH must be multiplied by -1.
2. An equation can be multiplied by any necessary coefficient; when this is done, the ΔH value must be multiplied by the same coefficient.

The following worked examples will illustrate Hess's law.

Example 20.5

A mixture of CO and H_2, known as 'Syngas', is prepared by passing steam over charcoal at $1000°C$ according to the following equation:

$$C(s) + H_2O(g) \longrightarrow CO(s) + H_2(g) \quad \Delta H_{rxn} = ?$$

Use the following data to calculate ΔH^0 in kJ for the reaction.

$$C(s) + O_2(g) \longrightarrow CO_2(g) \qquad \Delta H_1 = -393.5 \text{ kJ} \qquad (20.1)$$
$$2 H_2(g) + O_2(g) \longrightarrow 2 H_2O(g) \qquad \Delta H_2 = -483.6 \text{ kJ} \qquad (20.2)$$
$$2 CO(g) + O_2(g) \longrightarrow 2 CO_2(g) \qquad \Delta H_3 = -566.0 \text{ kJ} \qquad (20.3)$$

Solution

The target equation is

$$C(s) + H_2O(g) \longrightarrow CO(s) + H_2(g) \qquad \Delta H_{rxn} = ?$$

To produce the overall equation, we need to multiply equation 1 (including the associated enthalpy change) by 2, which becomes equation (20.4). Equations (20.2) and (20.3) must be written backward from the given direction, resulting in equations (20.5) and (20.6). Also the signs of ΔH_2^0 and ΔH_3^0 must be changed. The target equation and the enthalpy change for the reaction can be obtained by adding the three equations, and then dividing the final result by 2.

$$2C(s) + 2O_2(g) \longrightarrow 2CO_2(g) \qquad 2 \times \Delta H_1 = 2 \times -393.5 \text{ kJ} = -787 \text{ kJ} \qquad (20.4)$$

$$2H_2O(g) \longrightarrow 2H_2(g) + O_2(g) \qquad \Delta H_2 = -1 \times -483.6 \text{ kJ} = 483.6 \text{ kJ} \qquad (20.5)$$

$$2CO_2(g) \longrightarrow 2CO(g) + O_2(g) \qquad \Delta H_3 = -1 \times -566.0 \text{ kJ} = 566 \text{ kJ} \qquad (20.6)$$

$$2 C(s) + 2 H_2O(g) \longrightarrow 2 CO(g) + 2 H_2(g) \quad \Delta H_{rxn} = 2\, \Delta H_2 + (-\Delta H_2) + (-\Delta H_3)$$

$$= 262.6 \text{ kJ}$$

$$C(s) + H_2O(g) \longrightarrow CO(g) + H_2(g) \quad \Delta H_{rxn} = \frac{1}{2}[2\, \Delta H_2 + (-\Delta H_2) + (-\Delta H_3)]$$

$$= 131.3 \text{ kJ}$$

Thus the heat of reaction for water gas synthesis is $+131.3$ kJ.

Example 20.6

Ethanol can be produced by reacting ethylene with water according to the following equation.

$$C_2H_4(g) + H_2O(l) \longrightarrow C_2H_5OH(l)$$

From the following data, calculate the heat of reaction at 298 K.

$$C_2H_5OH(l) + 3O_2(g) \longrightarrow 2 CO_2(g) + 3 H_2O(l) \qquad \Delta H = -1367 \text{ kJ} \qquad (20.7)$$

$$C_2H_4(g) + 3 O_2(g) \longrightarrow 2 CO_2(g) + 2 H_2O(l) \qquad \Delta H = -1411 \text{ kJ} \qquad (20.8)$$

Solution

To solve this problem, reverse equation (20.7) and multiply its ΔH by -1. Add the result to equation (20.8):

$$2 CO_2(g) + 3 H_2O(l) \longrightarrow C_2H_5OH(l) + 3O_2(g) \qquad \Delta H = 1367 \text{ kJ}$$

$$C_2H_4(g) + 3O_2(g) \longrightarrow 2 CO_2(g) + 2 H_2O(l) \qquad \Delta H = -1411 \text{ kJ}$$

$$C_2H_4(g) + H_2O(l) \longrightarrow C_2H_5OH(l) \qquad \Delta H_{rxn}^0 = -44 \text{ kJ}$$

Thus ΔH_{rxn}^0 per mole of $C_2H_5OH(l)$ formed is -44 kJ.

20.5 Lattice Energy and the Born–Haber Cycle

The concept of lattice energy was introduced in Chapter 9 to explain the stability of ionic solids. Because it is experimentally difficult to isolate gaseous ions, the lattice energy of an ionic solid cannot be measured directly. Born and Haber first introduced a method for determining lattice energy using what is now known as the Born–Haber Thermochemical Cycle, a calculation that allows one to move from any starting point, go through a series of changes, and then return to the starting point. Essentially, the Born–Haber Cycle is the application of Hess's Law of heat summation to indirectly estimate the lattice energy. Therefore, to use the cycle to determine the lattice energy several important concepts including ionization energy, electron affinity, dissociation energy, sublimation energy, heat of formation, and Hess's Law must be understood.

Let us go through a Born–Haber cycle for the formation of an ionic compound MX from the reaction of an alkali metal M (M = Li, Na, K, Rb, Cs) with a gaseous halogen X_2 (X = F, Cl).

- Step 1 is the formation of the gaseous metal from the solid metal. The enthalpy change is the sublimation energy of the metal.

$$M(s) \longrightarrow M(g) \quad \Delta H^0_{step1} \text{(Sublimation energy of M)}$$

- Step 2 is the formation of the halogen atoms from the molecule, where the enthalpy change is the bond dissociation energy of the molecule.

$$1/2 \ X_2(g) \longrightarrow X(g) \quad \Delta H^0_{step2} (1/2 \times \text{Bond energy of } X_2)$$

- Step 3 represents the ionization of the gaseous metal, and the enthalpy change is the ionization energy of the metal.

$$M(g) \longrightarrow M^+(g) + e^- \quad \Delta H^0_{step3} \text{(Ionization energy of M)}$$

- Step 4 represents the addition of an electron to gaseous halogen. The associated energy change is the electron affinity of the halogen.

$$X(g) + e^- \longrightarrow X^-(g) \quad \Delta H^0_{step4} \text{(Electron affinity of X)}$$

- Step 5 is the formation of the ionic solid from gaseous ions. The enthalpy change for this step is the unknown and by definition the lattice energy.

$$M^+(g) + X^-(g) \longrightarrow MX(s) \quad \Delta H^0_{step\ 5} = -\Delta H^0_{lattice} \text{ of MX}$$

The Born–Haber cycle for the formation of MS from M(s) and $X_2(g)$ is shown below:

$$
\begin{array}{ccccc}
M(s) & + & 1/2X_2(g) & \xrightarrow{\ \Delta H_f\ } & MX(s) \\
\end{array}
$$

M(s)	+	1/2X₂(g)	ΔH_f → MX(s)
ΔH^0_{step1} Sublimation		ΔH^0_{step2} Dissociation	
M(g)		X(g)	$\Delta H^0_{step5} = \Delta H^0_{lattice}$
ΔH^0_{step3} Ionization energy(IE)		ΔH^0_{step4} Electron affinity (EA)	Formation of crystalline solid MX(s) from gaseous ions
$M^+(g)$	+	$X^-(g)$	

$$\Delta H^0_f = \Delta H^0_{step1} + \Delta H^0_{step2} + \Delta H^0_{step3} + \Delta H^0_{step4} + \Delta H^0_{step5} \text{ (or Lattice energy)}$$

Example 20.7

Draw the Born–Haber cycle for the data given in the table below and calculate the lattice energy of LiF(s).

		ΔH^0
Li (s) $+ \frac{1}{2} F_2$ (g) $\rightarrow$ LiF (s)	Standard enthalpy of formation of LiF(s)	−617 kJ
Li (s) $\rightarrow$ Li (g)	Formation of Li atom	161 kJ
Li (g) $\rightarrow$ Li$^+$ (g) $+ e^-$	First ionization energy of Li	520 kJ
$\frac{1}{2} F_2$ (g) $\rightarrow$ F (g)	Dissociation energy of F_2	159 kJ
F (g) $+ e^- \rightarrow$ F$^-$ (g)	Electron affinity for fluorine	−328 kJ
Li+ (g) $+ F^-$ (g) $\rightarrow$ LiF (s)	Lattice energy for LiF	??

Solution

First draw the Born–Haber cycle diagram following the example given for MX(s), substituting the values given for the various enthalpies. Remember to divide the dissociation enthalpy for F_2 by 2. Next solve for the lattice energy in the equation:

$$\Delta H_f^0 = \Delta H_{step1}^0 + \Delta H_{step2}^0 + \Delta H_{step3}^0 + \Delta H_{step4}^0 + \Delta H_{step5}^0 \text{ (or Lattice energy)}$$

The Born–Haber cycle for the formation of LiF(s) is

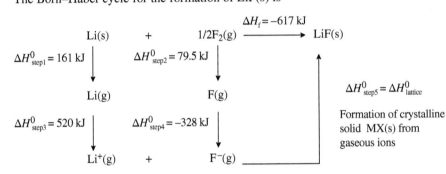

The lattice energy is calculated as follows:

$$\Delta H_f^0 = \Delta H_{step1}^0 + \Delta H_{step2}^0 + \Delta H_{step3}^0 + \Delta H_{step4}^0 + \Delta H_{step5}^0 \text{ (or lattice energy)}$$

$$- 617 \text{ KJ} = 161 \text{ KJ} + 79.5 \text{ KJ} + 520 \text{ KJ} + (-328 \text{ KJ}) + \text{LE}$$

Lattice energy for LiF $= -617 \text{ KJ} - 432.5 \text{ KJ} = 1049.5 \text{ KJ}$

Example 20.8

For calcium, the enthalpy of sublimation is +178 kJ and the first and second ionization energies are +590 kJ and +1150 kJ, respectively. The dissociation energy of fluorine is +154 kJ, and its electron affinity is −328 kJ. If the lattice energy of CaF_2 is 2590 kJ/mole, calculate the net enthalpy change for the reaction

$$Ca(s) + F_2(g) \longrightarrow CaF_2(s) \quad \Delta H_f^0 = ?$$

Solution

The information needed to solve this problem is provided. Outline the individual reactions/steps of the Born–Haber cycle. The summation of the ΔH values for each of these steps gives the enthalpy change for the formation of $CaF_2(s)$.

		ΔH^0/mol
1. Formation of Ca atoms:	$Ca(s) \longrightarrow Ca(g)$	+178 kJ
2. 1st ionization energy of Ca	$Ca(g) \longrightarrow Ca^+(g) + e^-$	+590 kJ
3. 2nd ionization energy of Ca	$Ca^+(g) \longrightarrow Ca^{2+}(g) + e^-$	+1150 kJ
4. Dissociation energy of $F_2(g)$	$F_2(g) \longrightarrow 2F(g)$	+154 kJ
5. Electron affinity for F	$2F(g) + 2e^-(g) \longrightarrow 2F^-(g)$	$2 \times (-328$ kJ)
6. Formation of crystal lattice	$Ca^{2+}(g) + 2F^-(g) \longrightarrow CaF_2(s)$	-2590 kJ

Net Reaction: $Ca(s) + F_2(g) \longrightarrow CaF_2(s)$

$\Delta H_f^0 = 178$ kJ $+ 590$ kJ $+ 1150$ kJ $+ 154$ kJ $+ 2(-328$ kJ$) + (-2590$ kJ$) = -1{,}174$ kJ

$Ca(s) + F_2(g) \longrightarrow CaF_2(s)$ $\Delta H_f^0 = -1{,}174$ kJ

20.6 Bond Energies and Enthalpy

Chemical reactions involve the breaking and formation of chemical bonds. To break a bond, energy must be added to the system; hence this is an endothermic process. The formation of a bond liberates energy, that is, it is exothermic.

The average amount of energy necessary to dissociate one mole of bonds in a covalent substance in the gaseous state into atoms in the gaseous state is known as the bond energy (BE).

Bond energy (BE) always has a positive sign. The larger the bond energy, the larger is the energy liberated when the bond is formed, and the larger the energy needed to break the bond. Thus bond-energy data can serve as a measure of bond stability. For example, the H—F bond, with energy of $+569$ kJ/mol, has higher bond stability than C—C with a bond energy of $+339$ kJ/mol.

The enthalpy change for a reaction can be written as

$$\Delta H_{rxn}^0 = \sum BE \text{ (bonds broken)} - \sum BE \text{ (bonds formed)}$$

$$\text{(Energy needed)} \qquad\qquad \text{(Energy liberated)}$$

where $\sum$ represents the sum of terms while BE represents the bond energy per mole of bonds. Since bonds are always broken in reactants and formed in products, the above expression can also be written as:

$$\Delta H_{rxn}^0 = \sum BE \text{ (reactants)} - \sum BE \text{ (products)}$$

Consider the reaction between hydrogen and iodine to form hydrogen iodide:

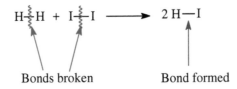

Bonds broken   Bond formed

One H—H bond and one I—I bond are broken. On the other hand, two H—I bonds are formed. The approximate overall heat of reaction can be calculated as follows:

$$\Delta H^0_{rxn} = \sum BE \ (bonds \ broken) - \sum BE \ (bonds \ formed)$$

$$= (BE_{H_2} + BE_{I_2}) - (2 \times BE_{HI})$$

Using BE data from Table 20.2, we have

$$= \left(1 \ mol \times 432 \frac{kJ}{mol}\right) + \left(1 \ mol \times 149 \frac{kJ}{mol}\right) - \left(2 \ mol \times 295 \frac{kJ}{mol}\right)$$

$$= 581 \ kJ - 590 \ kJ$$

$$= -9 \ kJ$$

Thus the reaction is slightly exothermic.

Example 20.9

Using the bond energies listed in Table 20.2, estimate ΔH^0 for the following reaction:

$$H_2 \,(g) + Br_2 \,(g) \longrightarrow 2 \, HBr \,(g)$$

Table 20.2   Average bond energies (kJ/mol) of some common bonds

Bond	Bond energy (kJ/mol)	Bond	Bond energy (kJ/mol)	Bond	Bond energy (kJ/mol)	Bond	Bond energy (kJ/mol)
H—H	432	F—F	154	S—H	347	Si—H	323
H—F	565	Cl—Cl	239	S—F	327	Si—C	301
H—Cl	427	Cl—Br	218	S—Cl	253	Si—O	368
H—Br	365	Br—Br	193	S—Br	218	Si—Si	226
H—I	295	I—I	149	S—S	266		
C—H	413	N—H	391	O—H	467		
C—C	347	N—N	160	O—O	146		
C=C	614	N=N	418	O=O	495		
C≡C	839	N≡N	941	O—F	190		
C—N	305	N—F	272	O—Cl	203		
C≡N	891	N—Cl	200	O—I	234		
C—O	358	N—Br	243	O—P	351		
C=O	799	N—P	209	O—N	201		
C≡O	1072						
C—Cl	339						
C—Br	276						

Is the reaction endothermic or exothermic?

Solution

To solve this problem, we can use the following expression:

$$\Delta H_{rxn}^0 = \sum BE \text{ (reactants)} - \sum BE \text{ (products)}$$

1. Obtain the bond dissociation values for all the bonds involved:

$$D(H-H) = 432 \text{ kJ/mol}; D(Br-Br) = 193 \text{ kJ/mol}; D(H-Br) = 365 \text{ kJ/mol}$$

2. Write the expression for the heat of reaction and substitute the bond energy values. Remember to multiply each BE by the corresponding number of moles.

$$\Delta H_{rxn} = [D(H-H) + D(Br-Br)] - [2 \times D(H-Br)]$$
$$= [(1 \text{ mol})(432 \text{ kJ/mol}) + (1 \text{ mol})(193 \text{ kJ/mol})] - [(2 \text{ mol})(365 \text{ kJ/mol})]$$
$$= -105 \text{ kJ}$$

The reaction is exothermic since ΔH_{rxn} is negative. This suggests that the bonds formed are more stable than those that are broken.

Example 20.10

Calculate the approximate heat of reaction for the hydrogenation of ethylene using the bond energies listed in Table 20.2.

$$H_2C = CH_2 \text{ (g)} + H_2 \text{ (g)} \longrightarrow CH_3CH_3 \text{ (g)}$$

Is the reaction endothermic or exothermic?

Solution

To solve this problem, we can use the following expression:

$$\Delta H_{rxn}^0 = \sum BE \text{ (reactants)} - \sum BE \text{ (products)}$$

1. Obtain the bond dissociation values for all the bonds involved:

Bonds broken

$1 D(H-H) = 432 \text{ kJ/mol}$; and $1 D(C=C) = 614 \text{ kJ/mol}$

Bonds formed

$1 D(C-C) = 347 \text{ kJ/mol}$, and $2 D(C-H) = 413 \text{ kJ/mol}$

2. Write the expression for the heat of reaction and substitute the bond energy values. Remember to multiply each BE by the corresponding number of moles.

$$\Delta H_{rxn} = [D(H-H) + D(C=C)] - [2 \times D(C-H) + D(C-C)]$$
$$= \{(1\ mol)(432\ kJ/mol) + (1\ mol)(614\ kJ/mol)\}$$
$$- \{(2\ mol)(413\ kJ/mol) + (1\ mol)(347\ kJ/mol)\}$$
$$= -127\ kJ$$

The reaction is exothermic since ΔH_{rxn} is negative.

20.7 Problems

1. Determine the amount of heat absorbed when 500 g of ethyl alcohol is heated from 30°C to 65°C. The specific heat capacity for ethyl alcohol is 2.2202 J/g.⁻°C.

2. An industrial chemistry student was asked to measure the heat of combustion of benzene, C_6H_6, a commercially important hydrocarbon for enhancing the octane rating of gasoline. To determine the heat of combustion, she ignited a 1.2575-g sample of benzene in a bomb calorimeter in the presence of excess oxygen. Calculate the heat of combustion per mole for benzene if the temperature increase of the calorimeter is 17°C. The heat capacity of the bomb calorimeter is 8.23 kJ/°C.

3. Determine the heat capacity of a 25-g sample that has absorbed 4200 J of heat over a temperature change of 50°C.

4. A 75-g sample of an alloy was heated to 100°C and dropped into a beaker containing 150 g of water at 25°C. Calculate the final temperature of the water if the specific heat capacity of the metal is 0.55 J/°C. (Specific heat capacity for water is 4.2 J/g.°C.)

5. Find the amount of heat required to raise the temperature of 550g of nickel (Ni) from 15°C to 110°C. The specific heat capacity of Ni is 0.4452 J/g.⁻°C.

6. The thermochemical equation for the hydrogenation of ethene is

$$C_2H_4(g) + H_2(g) \longrightarrow C_2H_6(g) \quad \Delta H_{rxn}^0 = -138\ kJ$$

   (a) Is the reaction endothermic or exothermic?
   (b) Calculate the quantity of heat liberated when 5.0 moles of ethene react with excess hydrogen.
   (c) Estimate the mass of ethene required to produce 420 kJ of heat.

7. The thermochemical equation for the combustion of ethane is

$$C_2H_6(g) + \frac{7}{2}O_2(g) \longrightarrow 2\ CO_2(g) + 2\ H_2O(g) \quad \Delta H_{rxn}^0 = -1560\ kJ$$

   Find the amount of heat in kJ given off by burning 0.50 kg of ethane.

8. Consider the reaction

$$2\ CH_4(g) + 2\ NH_3(g) + 3\ O_2(g) \longrightarrow 2\ HCN(g) + 6\ H_2O(g) \quad \Delta H_{rxn}^0 = -930\ kJ$$

(a) Is the reaction endothermic or exothermic?

(b) Rewrite the above reaction showing the heat energy appropriately as a reactant or a product.

(c) What is the heat released per mole of CH_4?

(d) Find the amount of heat liberated when 48.0 g of CH_4 are burned in excess oxygen.

9. Classify each of the following reactions as endothermic or exothermic.

(a) $2 NH_3 (g) + CO_2 (g) \longrightarrow (H_2N)_2CO(g) + H_2O(g) + 133.6 kJ$

(b) $C_{(graphite)} + 1.895 kJ \longrightarrow C_{(Diamond)}$

(c) $Sn(s) + 2 Cl_2 (g) \longrightarrow SnCl_4 (l) + 545.2 kJ$

(d) $HCl(aq) + NaOH(aq) \longrightarrow NaCl(aq) + H_2O(l) + 57.3 kJ$

(e) $Ca^{2+} (aq) + CO_3^{2-} (aq) + 26.6 kJ \longrightarrow CaCO_3 (s)$

10. The formation of limestone, $CaCO_3$, is represented by the reaction

$$CaO(s) + CO_2 (g) \longrightarrow CaCO_3 (s)$$

(a) Determine the heat of reaction, given that

$$\Delta H_{f(CaO)} = -635 \text{ kJ/mol}$$

$$\Delta H_{f(CO_2)} = -393.5 \text{ kJ/mol}$$

$$\Delta H_{f(CaCO_3)} = -1205.9 \text{ kJ/mol}$$

(b) Is the reaction endothermic or exothermic?

(c) What is the effect of adding heat to the reaction vessel?

11. Dissolving sulfur trioxide gas in water can produce sulfuric acid.

$$H_2O(l) + SO_3 (g) \longrightarrow H_2SO_4 (aq)$$

(a) Determine the heat of reaction from the following data:

$$\Delta H_{f(H_2O)} = -286 \text{ kJ/mol}$$

$$\Delta H_{f(SO_3)} = -396 \text{ kJ/mol}$$

$$\Delta H_{f(H_2SO_4)} = -909 \text{ kJ/mol}$$

(b) Is the reaction endothermic or exothermic?

(c) What is the effect of adding heat to the reaction vessel?

12. The enthalpy change at 25°C for the reaction

$$2 N_2O(g) + 3 O_2 (g) \longrightarrow 2 N_2O_4 (g)$$

$\Delta H_f^0 (kJ/mol):$ 81.6   0   ?

is −143.9 kJ. Determine the standard heat of formation of $N_2O_4 (g)$.

13. The combustion of glucose, $C_6H_{12}O_6$, is represented by the equation:

$$C_6H_{12}O_6\,(s) + 6\,O_2\,(g) \longrightarrow 6\,CO_2\,(g) + 6H_2O\,(g)$$

Calculate ΔH^0_{rxn} at 25°C for the combustion of 1 mole of glucose from the following standard heat of formation data.

$$\Delta H_{f(H_2O)} = -241.6 \text{ kJ/mol}$$
$$\Delta H_{f(CO_2)} = -393.5 \text{ kJ/mol}$$
$$\Delta H_{f(O_2)} = 0$$
$$\Delta H_{f(C_6H_{12}O_6)} = -1275.1 \text{ kJ/mol}$$

14. The enthalpy change for the reaction

$$2\,KClO_3\,(s) \longrightarrow 2\,KCl\,(s) + 3\,O_2\,(g)$$

is equal to –89.3 kJ. What is the enthalpy of formation for $KClO_3$?

$$\Delta H^0_{f\,(KClO_3)} = ?$$
$$\Delta H^0_{f\,(KCl)} = -435.5 \text{ kJ/mol}$$
$$\Delta H^0_{f\,(O_2)} = 0$$

15. The first step in the manufacture of nitric acid, HNO_3, is

$$4\,NH_3\,(g) + 7\,O_2\,(g) \longrightarrow 4\,NO_3\,(g) + 6\,H_2O\,(l)$$

The ΔH^0_{rxn} is – 1396 kJ. What is the enthalpy of formation of NO_2?

$$\Delta H_{f(H_2O(l))} = -286 \text{ kJ/mol}$$
$$\Delta H_{f(NO_2)} = ? \text{ kJ/mol}$$
$$\Delta H_{f(O_2)} = 0$$
$$\Delta H_{f(NH_3)} = -46 \text{ kJ/mol}$$

16. Calculate the heat of reaction for

$$2\,CO\,(g) + O_2\,(g) \longrightarrow 2\,CO_2\,(g)$$

from the enthalpy changes for the following reactions:

$$C\,(s) + O_2\,(g) \longrightarrow CO_2\,(g) \qquad\qquad \Delta H = -393.3 \text{ kJ}$$
$$CO\,(g) \longrightarrow C\,(s) + \frac{1}{2}O_2(g) \qquad \Delta H = +110.5 \text{ kJ}$$

17. Calculate the enthalpy change for the formation of solid PCl_5 from the enthalpy changes for the following reactions:

$$P_4(s) + 6\,Cl_2(g) \longrightarrow 4\,PCl_3(l) \quad \Delta H = -1270.7\ kJ$$

$$PCl_3(l) + Cl_2(g) \longrightarrow PCl_5(s) \quad \Delta H = -137.1\ kJ$$

18. The formation of diborane from its constituent elements is represented by the chemical equation:

$$2\,B(s) + 3\,H_2(g) \longrightarrow B_2H_6(g)$$

Use the following thermochemical data to calculate the enthalpy change in kJ for this reaction.

$$B_2H_6(g) + 3\,O_2(g) \longrightarrow B_2O_3(g) + 3\,H_2O(g) \quad \Delta H = -1941\ kJ$$

$$2\,B(s) + \frac{3}{2}O_2(g) \longrightarrow B_2O_3(g) \quad \Delta H = -2368\ kJ$$

$$H_2(g) + \frac{1}{2}O_2(g) \longrightarrow H_2O(g) \quad \Delta H = -242\ kJ$$

19. Methylene chloride is an important industrial solvent, commonly prepared by the chlorination of methane gas.

$$CH_4(g) + 2\,Cl_2(g) \longrightarrow CH_2Cl_2(g) + 2\,HCl(g)$$

From the following thermochemical data calculate the enthalpy change for this chlorination reaction:

$$CH_4(g) + Cl_2(g) \longrightarrow CH_3Cl(g) + HCl(g) \quad \Delta H = -98.3\ kJ$$

$$CH_3Cl(g) + Cl_2(g) \longrightarrow CH_2Cl_2(g) + HCl(g) \quad \Delta H = -104\ kJ$$

20. In the laboratory, acetylene gas, C_2H_2, is prepared by the action of cold water on calcium carbide:

$$CaC_2(s) + 2H_2O(l) \longrightarrow Ca(OH)_2(aq) + C_2H_2(g)$$

(a) Calculate the enthalpy change for the reaction using the following thermochemical data:

$$Ca(s) + 2\,C_{(graphite)} \longrightarrow CaC_2(s) \quad \Delta H = -62.8\ kJ$$

$$Ca(s) + \frac{1}{2}O_2(g) \longrightarrow CaO(s) \quad \Delta H = -635.5\ kJ$$

$$CaO(s) + H_2O(l) \longrightarrow Ca(OH)_2(aq) \quad \Delta H = -653.1\ kJ$$

$$C_2H_2(g) + \frac{5}{2}O_2(g) \longrightarrow 2\,CO_2(g) + H_2O(l) \quad \Delta H = -1300.1\ kJ$$

$$C_{(graphite)} + O_2(g) \longrightarrow CO_2(g) \quad \Delta H = -393.5\ kJ$$

(b) Is the reaction endothermic or exothermic?

21. The enthalpy of sublimation of magnesium is +147.7 kJ, and its first and second ionization energies are +736 kJ and +1450 kJ, respectively. The dissociation energy of bromine is +198 kJ, and its electron affinity is −325 kJ. If the standard enthalpy of formation of MgBr$_2$ is −524.3 kJ/mole, calculate the lattice enthalpy for MgBr$_2$.

22. With the aid of a Born–Haber cycle, calculate the lattice energy for MgO(s) from the following data:

Sublimation enthalpy for Mg(s) = +147.7 kJ/mol
First ionization energy for Mg(g) = +736 kJ/mol
Second ionization energy for Mg$^+$(g) = +1450 kJ/mol
Bond dissociation energy of O$_2$(g) = +498 kJ/mol
Electron affinity for O(g) + 1e$^-$ ⟶ O$^-$(g) = −141 kJ/mol
Electron affinity for O$^-$(g) + 1e$^-$ ⟶ O^{2-}(g) = +844 kJ/mol
Lattice energy for Mg^{2+}(g) + O^{2-}(g) ⟶ Mg^{2+}O^{2-}(s) = ΔH_{LE} = ?
Standard enthalpy of formation of MgO(s): Mg(s) + 1/2O$_2$(g) ⟶ MgO(s) = −602 kJ/mol

23. For the reaction

$$H_2(g) + F_2(g) \longrightarrow 2\, HF(g)$$

(a) Estimate ΔH_{rxn}^0 using the bond energy values given below:

H—H	436.4 kJ/mol
F—F	156.9 kJ/mol
H—F	568.2 kJ/mol

(b) Calculate ΔH_{rxn}^0 from the standard enthalpy of formation data.

$$\Delta H_{f(H_2)} = 0$$
$$\Delta H_{f(F_2)} = 0$$
$$\Delta H_{f(HF)} = -268.6 \text{ kJ/mol}$$

24. Using the bond dissociation energy data below, calculate the ΔH^0 for the following reactions:

(a) $C_2H_6(g) \longrightarrow C_2H_4(g) + H_2(g)$

(b) $C_4H_{10}(g) + \dfrac{13}{2} O_2(g) \longrightarrow 4\, CO_2(g) + 5\, H_2O(g)$

(c) $2\, H_2(g) + O_2(g) \longrightarrow 2\, H_2O(g)$

Bond	kJ/mol
C—C	331
C=C	590
C—H	414
H—H	436.4
O=O	498
C=O	803
O—H	464

25. Esters have pleasant aromas and are commonly used in perfumes, and also as flavoring agents in the food industry. They are typically formed by the reaction of a carboxylic acid with an alcohol:

$$
\begin{array}{ccc}
\underset{\substack{| \\ H}}{\overset{\substack{H \quad O \\ | \quad \|}}{H-C-C}}-OH & + \quad HO-\underset{\substack{| \\ H \ \ H}}{\overset{\substack{H \ \ H \\ | \ \ |}}{C-C}}-H & \longrightarrow \quad \underset{\substack{| \\ H}}{\overset{\substack{H \quad O \\ | \quad \|}}{H-C-C}}-O-\underset{\substack{| \\ H}}{\overset{\substack{H \\ |}}{C}}-H + H_2O
\end{array}
$$

Using bond dissociation energy data, calculate the standard enthalpy change for the reaction.

Bond	kJ/mol
C—C	331
C—H	414
C—O	351
C=O	803
O—H	464

26. The hydrogenation of ethylene is represented by the reaction:

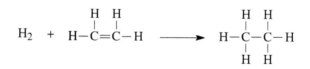

ΔH^0 for the reaction is −184 kJ. Calculate the bond energy of the carbon-to-carbon double bond in ethylene, given that the C—C = 347 kJ/mol, C—H = 413 kJ/mol and H—H = 432 kJ/mol.

21

Chemical Thermodynamics

. . . .

21.1 Definition of Terms

Chemical thermodynamics is the study of the energy changes and transfers associated with chemical and physical transformations.

Energy is the ability to do work or to transfer heat.

A *spontaneous process* is one that occurs on its own without any external influence. A spontaneous process always moves a system in the direction of equilibrium.

When a process or reaction cannot occur under the prescribed conditions, it is *nonspontaneous*. The reverse of a spontaneous process or reaction is always nonspontaneous.

Heat (q) is the energy transferred between a system and its surroundings due to a temperature difference.

Work (W) is the energy change when a force (F) moves an object through a distance (d). Thus $W = F \times d$.

A *system* is a specified part of the universe, e.g. a sample or a reaction mixture we are studying. Everything outside a system is referred to as the *surroundings*.

The *universe* is a system plus its surroundings.

A *state function* is a thermodynamic quantity that defines the present state or condition of a system. Changes in state function quantities are independent of the path used to arrive at the final state from the initial state. Examples of state functions include enthalpy change, ΔH, entropy change, ΔS, and free energy change, ΔG.

The *internal energy* of a system is the sum of the kinetic and potential energies of the particles making up the system. While it is not possible to determine the absolute internal energy of a system, we can easily measure changes in internal energy (which correspond to energy given off or absorbed by the system). The change in internal energy, ΔE, is:

$$\Delta E = E_{final} - E_{initial}$$

21.2 The First Law of Thermodynamics

The first law of thermodynamics is also called the *law of conservation of energy*, and states that *the total amount of energy in the universe is constant*. Energy can neither be created nor

Chemistry in Quantitative Language: Fundamentals of General Chemistry Calculations. Second Edition.
Christopher O. Oriakhi, Oxford University Press. © Christopher O. Oriakhi 2021.
DOI: 10.1093/oso/9780198867784.003.0021

destroyed. It can only be converted from one form into another. In mathematical terms, the law states that the change in internal energy of a system, ΔE, equals $q + w$. That is,

$$\Delta E = q + w$$

In other words, the change in E is equal to the heat absorbed (or emitted) by the system, plus work done on (or by) the system.

Example 21.1

What is the change in internal energy in a process in which the system absorbs 550 J of heat energy, while at the same time 300 J of work are done on the system?

Solution

We are told that 300 J of work are done on the system and 550 J of heat absorbed. Therefore, w has a positive value ($w = +300$) and $q = +550$ J. We can calculate the change in internal energy as follows:

$$\Delta E = q + w = 550 + 300 = 850 \text{ J}$$

Example 21.2

In a certain chemical process, the system lost 1300 J of heat while doing work by expanding against the surrounding atmosphere. What is the work done if the change in internal energy is -2500 J?

Solution

Here heat is transferred from the system to the surroundings, and work is done by the system. Therefore q is negative, and the value of w should also be negative (the system is giving up energy to the surroundings in both cases).

$$\Delta E = q + w$$
$$w = \Delta E - q = (-2500 \text{ J}) - (-1300 \text{ J}) = -1200 \text{ J}$$

21.3 Expansion Work

When a force causes a mass to move, or when energy is transferred from one point to another, work is done. Consider a cylinder with a movable piston and filled with a gas. Work is done whenever the gas is compressed or expands due to the influence of external pressure. The work done by a constant pressure during expansion or contraction is given by

$$\text{Work } (w) = \text{Pressure } (P) \times \text{change in volume } (\Delta V)$$

or

$$w = P\Delta V = P(V_2 - V_1)$$

If the piston pushes back the surrounding atmosphere, the system (the gas within the piston) is giving up energy to do so; that is, if V increases ($\Delta V > 0$) energy is lost by the system, so that $w < 0$. On the other hand, if V decreases ($\Delta V < 0$) positive work is done on the system. Hence, we define:

$$w_{expansion} = -P\Delta V$$

Now, if we substitute $-P\Delta V$ for w in the expression $\Delta E = q + w$, we obtain

$$\Delta E = q - P\Delta V$$

In a constant-volume process, $\Delta V = 0$, and no $P\Delta V$ work is done. Hence,

$$\Delta E = q_v$$

Thus, in a constant-volume reaction, no work is done, and the heat evolved is equal to the change in internal energy. Note, q_v refers to the heat generated at constant volume.

When a gas is compressed, ΔV is a negative quantity due to the decrease in volume. This makes w a positive quantity.

$$w_{compression} = P\Delta V$$

Liquids and solids are relatively incompressible, and so for these components of a system, $P\Delta V$ is usually zero, i.e. $\Delta E = q$.

For reactions involving gases, the work done can be calculated from the expression

$$w = P\Delta V = (\Delta n)R\,T$$

Here, Δn equals the total moles of gaseous products minus the total moles of gaseous reactants; R is the ideal gas constant, and T is the absolute temperature.

Example 21.3

During a chemical decomposition reaction the volume of a reaction vessel expanded from 20.0 L to 35.0 L against an external pressure of 3.5 atm. What is the work done in kJ? ($1\,L - atm = 101\,J$.)

Solution

1. Work done in expansion or contraction is $w = -P\Delta V$. Here, $P = 3.5$ atm, and
   $\Delta V = V_f - V_i = (35.0 - 20.0) = 15.0\,L$.

   $$w = -P\Delta V = -(3.5\ \text{atm})(15.0\ \text{L}) = -52.5\ \text{L-atm}$$

2. Convert L-atm to kJ

   $$(-52.5\ \text{L-atm})\left(101\frac{J}{\text{L-atm}}\right)\left(\frac{1\ \text{kJ}}{1000\ \text{J}}\right) = -5.3\ \text{kJ}$$

Example 21.4

The reaction of nitric oxide with oxygen to form nitrogen dioxide is an important step in the industrial synthesis of nitric acid via the Ostwald process.

$$2\,NO(g) + O_2(g) \rightleftharpoons 2\,NO_2(g)$$

Calculate the work done if the volume contracts from 25.5 L to 18.0 L at a fixed pressure of 20 atm.

Solution

1. Since the volume is changing at constant pressure, use the formula $w = -P\Delta V$ to calculate work done. $P = 20.0$ atm and $\Delta V = V_f - V_i = (18.0 - 25.5) = -7.5$ L.

$$w = -P\Delta V = -(20.0\text{ atm})(-7.5\text{ L}) = 150.0\text{ L-atm}$$

2. Convert L-atm to kJ

$$(150.0\text{ L.atm})\left(101\frac{\text{J}}{\text{L-atm}}\right)\left(\frac{1\text{ kJ}}{1000\text{ J}}\right) = 15.15\text{ kJ}$$

The sign is positive because work is done on the system, that is, energy flows into the system.

Example 21.5

The reaction between gaseous carbon monoxide and oxygen to form carbon dioxide generated 566.0 kJ of heat.

$$2\,CO(g) + O_2(g) \longrightarrow 2\,CO_2(g) \quad \Delta H = -566.0\text{ kJ}$$

How much P–V work is done if the reaction is carried out at a constant pressure of 30 atm and the volume change is -2.5 L? Calculate the change in internal energy, ΔE, of the system.

Solution

1. First calculate the work done in kJ:

$$w = -P\Delta V = -(30.0\text{ atm})(-2.5\text{ L}) = 75.0\text{ L.atm}$$

$$w = (75.0\text{ L-atm})\left(101\frac{\text{J}}{\text{L-atm}}\right)\left(\frac{1\text{ kJ}}{1000\text{ J}}\right) = 7.58\text{ kJ}$$

2. Then calculate the value of ΔE:

$$\Delta E = q + w \text{ or } \Delta E = q - P\Delta V$$

From the question, $q = -566.0$ kJ, and $w = 7.58$ kJ

$$\Delta E = q + w = -566.0 + 7.58 = -558.4\text{ kJ}$$

Thus we can see that the system has given off energy.

21.4 Entropy

Entropy is a measure of the degree of disorder or randomness of a system and is denoted by the symbol S. Entropy can also be regarded as a measure of the statistical probability of a system. The entropy of the universe always increases in a spontaneous event. The larger the value of the entropy, the greater is the disorder or randomness of the atoms, ions, or molecules of the system. For example, the particles in a solid are more closely packed has than those in a liquid or gas; hence, they are less disordered. In other words, they have lower entropy. As a system changes phase from solid to liquid to gas, the entropy of the system increases.

A change in entropy is represented by ΔS and can be evaluated from the expression:

$$\Delta S = S_{final} - S_{intial}$$

If $S_{final} > S_{initial}$, ΔS is positive, indicating the system has become more disordered.
If $S_{final} < S_{initial}$, ΔS is negative, indicating the system has become less disordered.
A process is likely to be favored, that is, spontaneous, if S for the system increases; it is probably nonspontaneous if S decreases. The unit of entropy is joules per Kelvin (J/K).

21.5 The Second Law of Thermodynamics

The second law states that *in any spontaneous process the universe tends toward a state of higher entropy*. This implies that every spontaneous chemical or physical change will increase the entropy of the universe. In any process, spontaneous or not, the following condition holds:

$$\Delta S_{universe} = \Delta S_{system} + \Delta S_{surrounding}$$

If $\Delta S_{univ.} > 0$, the process is spontaneous. If $\Delta S_{univ.} = 0$, the process can go equally well in either direction (it is reversible). If $\Delta S_{univ.} < 0$, the process is nonspontaneous; it will not proceed as written, but it will be spontaneous in the reverse direction.

21.6 Calculation of Entropy Changes in Chemical Reactions

The standard state entropy change, ΔS^0, for a chemical reaction can be calculated from the absolute standard state entropies of the reactants and products. Values of S^0 at 298 K are usually tabulated in units of J/K instead of kJ/mol as used for enthalpy changes. Table 21.1 gives the standard molar enthalpies, entropies, and free energies of formation of some common substances. A comprehensive list can be found in most general chemistry textbooks, or a chemistry handbook.

The entropy change is given by the sum of the entropies of the products minus the sum of the entropies of the reactants. This calculation works just like the one for ΔH^0 of reaction. That is:

$$\Delta S^0 = \sum m S^0 \text{ (produdts)} - \sum n S^0 \text{ (reactants)}$$

The coefficients m and n represent the stoichiometric coefficients in the balanced chemical equation.

Table 21.1 Standard state free energies, enthalpies, and entropies of formation for selected substances at 25°C(298 K)

Substance	Formula	ΔH_f^0 (kJ/mol)	ΔG_f^0 (kJ/mol)	S_f^0 (J/mol–K)
Aluminum	Al(s)	0	0	28.32
	Al_2O_3(s)	−1668.8	−1576.5	51
Bromine	Br_2(l)	0	0	152.3
	Br_2(g)	30.71	3.14	245.3
	Br^-(aq)	−121	−102.8	80.71
	HBr(aq)	−36.23	−53.2	198.5
Carbon	C(s) (graphite)	0	0	5.69
	C(s) (diamond)	2	2.84	2.43
	CO(g)	−110.5	−137.2	198
	CO_2(g)	−393.5	−394.4	213.6
	CH_4(s)	−75	−50.8	186.3
	CH_3OH(g)	−201	−162	237.6
	C_2H_4(g)	52.3	68.1	219.4
	C_2H_6(g)	−84.7	−32.9	229.5
Chlorine	Cl_2(g)	0	0	223
	Cl^-(aq)	−167	−131.2	56.5
	HCl(g)	−92.3	−95.3	186.7
Fluorine	F_2(g)	0	0	202.7
	HF(g)	−271.1	−273.2	173.7
Hydrogen	H_2(g)	0	0	130.6
	H^+(aq)	0	0	0
Iron	Fe(s)	0	0	27.15
	$FeCl_2$(S)	−342	−302.2	118
	$FeCl_3$(S)	−400	−334	142.3
	Fe_2O_3(s)	−822.2	−741	90
	Fe_3O_4(s)	−1117.1	−1014.2	146
Iodine	I_2(s)	0	0	116.73
	I_2(g)	62.25	19.37	260.6
	I^-(aq)	−55.2	−51.6	111.3
	HI(g)	25.94	1.3	206.3
Nitrogen	N_2(g)	0	0	191.5
	NH_3(g)	−46	−16.7	192.5
	N_2H_4(g)	95.4	159.4	238.5
	NO(g)	90.37	86.7	210.6
	NO_2(g)	33.8	51.84	240.45
	N_2O_4(g)	10	98.28	304.3
Oxygen	O_2(g)	0	0	205
	O_3(g)	143	163.4	237.6
	OH^-(aq)	−230	−157.3	−10.7
	H_2O(l)	−286	−237.2	69.9
	H_2O(g)	−242	−228.6	188.8
Sulfur	S(s, rhombic)	0	0	31.8
	H_2S(g)	−21	−33.6	205.7
	SO_2(g)	−297	−300.4	248.5
	SO_3(g)	−396	−370.4	256.2

Unlike the enthalpies of formation, the standard molar entropies of formation of the elements are not zero and must be included in all calculations.

Example 21.6

Considering the reaction

$$C(s) + O_2(g) \longrightarrow CO_2(g)$$

calculate the standard state entropy of formation of gaseous carbon dioxide at 25°C, given the following standard state entropies at 25°C:

$$S_C^0 = 158 \text{ Jmol}^{-1} \text{ K}^{-1}; \quad S_{O_2}^0 = 205 \text{ Jmol}^{-1} \text{ K}^{-1}; \quad S_{CO_2}^0 = 214 \text{ Jmol}^{-1} \text{ K}^{-1}$$

Solution

For the formation of 1 mol of CO_2 from 1 mol of C and 1 mol of O_2,

$$\Delta S^0 = \sum n\Delta S_{product}^0 - \sum n\Delta S_{reactant}^0$$

$$= (1 \text{ mol})\left(214 \text{ Jmol}^{-1}\text{K}^{-1} - \left\{(1 \text{ mol})\left(158 \text{ Jmol}^{-1}\text{K}^{-1}\right)\right.\right.$$

$$\left.+ (1 \text{ mol})\left(205 \text{ Jmol}^{-1}\text{K}^{-1}\right)\right\}$$

$$= -149 \text{ JK}^{-1}$$

This reaction is spontaneous if considered from the standpoint of entropy alone.

Example 21.7

Using data from the standard entropy table, calculate ΔS^0 for the following reactions:

1. $N_2(g) + H_2(g) \rightleftharpoons 2NH_3(g)$
2. $C_2H_6(g) \longrightarrow C_2H_4(g) + H_2(g)$
3. $H_2(g) + F_2(g) \longrightarrow 2 HF(g)$
4. $2H_2(g) + 2C(s) + O_2(g) \longrightarrow CH_3COOH(1)$

In each case, decide if the reaction is spontaneous on the basis of entropy alone.

Solution

For each problem estimate ΔS^0 by using the expression:

$$\Delta S^0 = \sum nS^0 \text{ (product)} - \sum nS^0 \text{ (reactants)}$$

Then substitute the appropriate ΔS^0 values from the Table 21.1 to obtain ΔS^0.

1. $N_2(g) + H_2(g) \rightleftharpoons 2\,NH_3(g)$

$$\Delta S^0 = 2S^0_{NH_3} - \left[S^0_{N_2} + 3S^0_{H_2}\right]$$

$$\Delta S^0 = (2\text{ mol})\left(192.5\text{ Jmol}^{-1}\text{ K}^{-1}\right) - \left\{(1\text{ mol})\left(191.5\text{ Jmol}^{-1}\text{ K}^{-1}\right)\right.$$

$$\left. + (3\text{ mol})\left(130.6\text{ Jmol}^{-1}\text{ K}^{-1}\right)\right\}$$

$$= -198.3\text{ JK}^{-1}$$

The reaction is nonspontaneous from entropy considerations alone.

2. $C_2H_6(g) \longrightarrow C_2H_4(g) + H_2(g)$

$$S^0 = \left[S^0_{C_2H_4} + S^0_{H}\right] - \left[S^0_{C_2H_6}\right]$$

$$= \left\{(1\text{ mol})\left(219.4\text{ Jmol}^{-1}\text{ K}^{-1}\right) + (1\text{ mol})\left(130.6\text{ Jmol}^{-1}\text{ K}^{-1}\right)\right.$$

$$\left. - (1\text{ mol})\left(229.5\text{ Jmol}^{-1}\text{ K}^{-1}\right)\right\}$$

$$= +120.5\text{ JK}^{-1}$$

This reaction is spontaneous from entropy considerations alone.

3. $H_2(g) + F_2(g) \longrightarrow 2HF(g)$

$$\Delta S^0 = 2S^0_{HF} - \left[S^0_{H_2} + 3S^0_{F_2}\right]$$

$$\Delta S^0 = (2\text{ mol})\left(173.51\text{ Jmol}^{-1}\text{ K}^{-1}\right) - \left\{(1\text{ mol})\left(130.58\text{ Jmol}^{-1}\text{ K}^{-1}\right)\right.$$

$$\left. + (1\text{ mol})\left(202.7\text{ Jmol}^{-1}\text{ K}^{-1}\right)\right\}$$

$$= +13.74\text{ JK}^{-1}$$

The reaction is spontaneous from entropy considerations alone.

4. $2H_2(g) + 2C(s) + O_2(g) \longrightarrow CH_3COOH(1)$

$$\Delta S^0 = S^0_{CH_3COOH} - \left[2S^0_{H_2} + 2S^0_{C} + S^0_{O_2}\right]$$

$$\Delta S^0 = (1\text{ mol})\left(159.8\text{ Jmol}^{-1}\text{ K}^{-1}\right) - \left\{(2\text{ mol})\left(130.58\text{ Jmol}^{-1}\text{ K}^{-1}\right)\right.$$

$$\left. + (2\text{ mol})\left(5.69\text{ Jmol}^{-1}\text{ K}^{-1}\right) + (1\text{ mol})\left(205\text{ Jmol}^{-1}\text{ K}^{-1}\right)\right\}$$

$$= -317.74\text{ JK}^{-1}$$

The reaction is nonspontaneous considering only entropy changes.

21.7 Free Energy

The Gibbs free energy (G) of a system is a thermodynamic measure that incorporates enthalpy and entropy changes. Neither of those alone is enough to determine absolutely whether a process is spontaneous; G combines them in a way that provides a single simple answer to the question "Is this process spontaneous?" In mathematical terms, G is defined by

$$G = H - TS$$

Like enthalpy and entropy, the Gibbs free energy of a system is a state function. For a given reaction at constant temperature, changes in H and S result in a change in G, which can be expressed by the equation:

$$\Delta G = \Delta H - T\Delta S$$

The sign of ΔG is a general criterion for the spontaneity of a chemical reaction. If

$$\Delta G < 0 \,(\text{negative})\text{, the reaction is spontaneous}$$

$$\Delta G = 0\text{, the reaction is at equilibrium} - \text{no change}$$

$$\Delta G > 0 \,(\text{positive})\text{, the reaction is not spontaneous}$$

If a reaction is not spontaneous as written, the reverse reaction is always spontaneous. A larger magnitude for ΔG implies that the process is more strongly spontaneous (if negative) or nonspontaneous (if positive).

21.8 The Standard Free Energy Change

The standard free energy change, ΔG^0, for a given reaction is the free energy change that occurs as reactants in their standard states are converted to products in their standard states. The standard conditions include 1 atm pressure, 1 M concentration for solution, and 25°C (298 K) temperature.

21.8.1 Calculating the standard free energy change

There are several ways of calculating ΔG^0 for a given chemical reaction. Three of the common methods are described here.

First, we can calculate it by direct substitution into the equation $\Delta G^0 = \Delta H^0 = T\Delta S^0$. If we know the values of T, ΔH^0 and ΔS^0, then we can calculate ΔG^0 directly.

Example 21.8

Calculate ΔG^0 in kJ per mole of SO_3 formed for the following reaction at 298 K, given that the values of ΔH^0 and ΔS^0 are -198 kJ and -200 J/K.

$$2\,SO_2\,(g) + O_2\,(g) \longrightarrow 2\,SO_3\,(g)$$

Is the reaction spontaneous under these conditions?

Solution

Substitute the given values directly into the expression for ΔG^0 (watch the units: ΔH^0 is usually in kJ but ΔS^0 is usually in J).

$$\Delta G^0 = \Delta H^0 - T\Delta S^0$$

$$= (-198,000 \text{ J}) - (298 \text{ K})\left(-200\frac{\text{J}}{\text{K}}\right) = -138,400 \text{ J}$$

$$= 138.4 \text{ kJ}$$

$$\Delta G^0 \text{ in kJ/mol SO}_3 = \frac{-138.4 \text{ kJ}}{2 \text{ mol SO}_3} = -69.2 \text{ kJ/mol}$$

The reaction is spontaneous since ΔG^0 is negative.

Second, we can use Hess's Law: ΔG^0 is a state function, therefore, if we know the values of ΔG^0 for the series of reactions given, we can use Hess's law to combine the ΔG^0 values and the equations to get the net ΔG^0 for the reaction.

Example 21.9

Using the following equations at 298 K

$$2 \text{ FeO}(s) + \frac{1}{2} \text{O}_2(g) \longrightarrow \text{Fe}_2\text{O}_3(s) \quad \Delta G^0 = -252.5 \text{ kJ} \tag{21.1}$$

$$3 \text{ FeO}(s) + \frac{1}{2} \text{O}_2(g) \longrightarrow \text{Fe}_3\text{O}_4(s) \quad \Delta G^0 = -281.2 \text{ kJ} \tag{21.2}$$

calculate ΔG^0 for the reaction

$$3 \text{ Fe}_2\text{O}_3(s) \longrightarrow 2 \text{ Fe}_3\text{O}_4(s) + \frac{1}{2} \text{O}_2(g)$$

Solution

To solve this problem, we multiply equation (21.1) by 3 and equation (21.2) by 2. Then we reverse equation (21.1) to make iron (III) oxide a reactant and then add both equations. (Remember that the ΔG^0 values for both equations must be multiplied by the coefficients as well; and when an equation is reversed, the sign for its ΔG^0 must be changed.)

$$3 \times \left[2\text{FeO}(s) + \frac{1}{2}\text{O}_2(g) \longrightarrow \text{Fe}_2\text{O}_3(s) \quad \Delta G^0 = -252.5 \text{ kJ} \right] \tag{21.1}$$

$$2 \times \left[3\text{FeO}(s) + \frac{1}{2}\text{O}_2(g) \longrightarrow \text{Fe}_3\text{O}_4(s) \quad \Delta G^0 = -281.2 \text{ kJ} \right] \tag{21.2}$$

Reversed eqn 21.1

$$3\ \text{Fe}_2\text{O}_3(\text{s}) \quad\longrightarrow\ \cancel{6\ \text{FeO(s)}} + \cancel{\tfrac{3}{2}\tfrac{1}{2}}\text{O}_2(\text{g}) \qquad \Delta G^0 = +757.5\ \text{kJ} \qquad (21.3)$$

$$\cancel{6\ \text{FeO(s)}} + \cancel{\text{O}_2(\text{g})} \longrightarrow 2\ \text{Fe}_3\text{O}_4(\text{s}) \qquad\qquad \Delta G^0 = -562.4\ \text{kJ} \qquad (21.4)$$

$$3\ \text{Fe}_2\text{O}_3(\text{s}) \quad\longrightarrow 2\ \text{Fe}_3\text{O}_4(\text{s}) + \tfrac{1}{2}\ \text{O}_2(\text{g}) \qquad \Delta G^0 = +195.1\ \text{kJ} \qquad (21.5)$$

Since ΔG^0 is positive, the transformation from hematite (Fe_2O_3) to magnetite (Fe_3O_4) is not spontaneous at 298 K.

Lastly, we can do this by calculation from tabulated values, using the following expression:

$$\Delta G^0 = \sum \Delta G_f^0\ (\text{products}) - \sum \Delta G_f^0\ (\text{reactants})$$

If we know the standard free energies of formation of reactants and products, then we can calculate ΔG^0.

Example 21.10

Is it feasible to synthesize KClO_3 from solid KCl and gaseous O_2 under standard conditions?

$$2\ \text{KCl(s)} + 3\ \text{O}_2\,(\text{g}) \longrightarrow 2\ \text{KClO}_3\,(\text{s})$$

The values of ΔG_f^0 (in kJ/mol) are:

$$\text{KCl(s)} = -408.8,\, \text{O}_2\,(\text{g}) = 0,\, \text{and KClO}_3\,(\text{s}) = -289.9.$$

Solution

To determine whether the given reaction is feasible (or spontaneous) under standard conditions, we will need to determine ΔG^0 for the reaction. Then use the ΔG^0 criteria for spontaneity of a chemical reaction. Calculate ΔG^0 by substituting the values of the ΔG_f^0 for the reactants and product into the expression:

$$\Delta G^0 = \sum \Delta G_f^0\ (\text{products}) - \sum \Delta G_f^0\ (\text{reactants})$$

$$= 2\Delta G_f^0\ (\text{KClO}_3\,(\text{s})) - \left[2\Delta G_f^0\ (\text{KCl(s)}) + 3\Delta G_f^0(\text{O}_2(\text{g})) \right]$$

$$= 2\ \text{mol} \times \left(-289.9\frac{\text{kJ}}{\text{mol}} \right) - \left[2\ \text{mol} \times -408.8\frac{\text{kJ}}{\text{mol}} + 3\ \text{mol} \times 0\frac{\text{kJ}}{\text{mol}} \right]$$

$$= -579.8\ \text{kJ} + 817.6\ \text{kJ}$$

$$= +237.8\ \text{kJ}$$

The reaction is not feasible as written since ΔG^0 is positive and fairly large. However, the reverse reaction will be spontaneous.

21.9 Enthalpy and Entropy Changes During a Phase Change

A *phase change* is a process in which a substance changes physical form but not chemical composition or identity. Common phase changes include:

- *melting* (fusion)—change from a solid to a liquid
- *sublimation*—change from a solid to a gas
- *freezing*—change from a liquid to a solid
- *vaporization*—change from a liquid to a gas
- *condensation*—change from a gas to a liquid
- *deposition*—change from a gas to a solid.

All phase changes are accompanied by changes in enthalpy and entropy, as well as a change in the free energy of the system. If we know the values of ΔH and ΔS for a phase transition, we can calculate the temperature at which a given phase change will occur. Since the two phases are at equilibrium, set $\Delta G = 0$, and solve for T. That is:

$$\Delta G = \Delta H - T\Delta S$$

$$0 = \Delta H - T\Delta S$$

$$\Delta H = T\Delta S$$

$$T = \frac{\Delta H}{\Delta S}$$

Hence the molar entropy change for any phase transition is given by

$$\Delta S = \frac{\Delta H}{T}$$

In this equation, ΔH represents the enthalpy change for the phase change and T is the absolute temperature of the change of state.

Example 21.11

The heat of vaporization of benzene is $\Delta H_{vap} = 30.8$ kJ/mol and the entropy change for the vaporization is $\Delta S_{vap} = 87.2$ J/K. What is the boiling point of benzene?

Solution

To determine the boiling point, use the formula:

$$T_b = \frac{\Delta H_{vap}}{\Delta S_{vap}} = \frac{30,800 \text{ J/mol}}{87.2 \text{ J/K·mol}} = 353.2 \text{ K or } 80.2°C$$

The trick here is to remember to convert your units. You cannot operate on J and kJ at the same time. You must express both units in J or kJ before carrying out any mathematical operation involving them.

Example 21.12

The molar heat of fusion of ice is 333.15 J/g. What is the entropy change for the melting of 1 mole of ice to liquid water at 0°C?

Solution

1. First, convert molar heat of fusion from J/g to J/mol.

$$\Delta H_{fus} = \left(333.15\frac{J}{g}\right)\left(18.0\frac{g}{mol}\right) = 5,996.6 \text{ Jmol}^{-1}$$

2. Calculate ΔS_{fus} using the expression: $\Delta S = \frac{\Delta H}{T}$

$$\Delta S = \frac{\Delta H}{T} = \frac{5,996.6 \text{ Jmol}^{-1}}{273 \text{ K}} = 21.97 \text{ Jmol}^{-1} \text{ K}^{-1}$$

Example 21.13

At 2.6°C, gallium undergoes a solid-solid phase transition. If ΔS^0 for this change is +7.62 J/mol-K, what is ΔH^0 for the transition at this temperature?

Solution

The entropy change for this phase transition is

$$\Delta S^0_{275.6 \text{ K}} = \frac{\Delta H_{s-s \text{ phase change}}}{T}$$

$$\Delta H_{s-s \text{ phase change}} = T\Delta S^0_{275.6 \text{ K}} = (275.6 \text{ K})\left(7.62 \text{ Jmol}^{-1} \text{ K}^{-1}\right)$$

$$= 2100 \text{ J mol}^{-1}$$

21.10 Free Energy and the Equilibrium Constant

For a reaction under standard conditions, the standard-state free energy ΔG^0 and the equilibrium constant K are related by

$$\Delta G^0 = -RT \ln K \text{ or } \Delta G^0 = -2.303 \ RT \log K$$

R is the gas constant (8.314 J/K mol) and T is the thermodynamic or absolute temperature in Kelvin.

In actual reactions where the reactants and products are present at nonstandard pressures and concentrations, the free energy change ΔG is related to the standard state free energy ΔG^0 and the composition of the reaction mixture as follows:

$$\Delta G = \Delta G^0 + RT \ln Q \text{ or } \Delta G = \Delta G^0 + 2.303 \ RT \log Q$$

Table 21.2   Relationship between standard free energy change and equilibrium constant

ΔG^0	$\ln K$	K	Comments on Products and Reactants
$\Delta G^0 < 0$	$\ln K > 0$	$K > 1$	Products are favored over reactants at equilibrium
$\Delta G^0 = 0$	$\ln K = 0$	$K = 1$	Amount of products is approximately equal to the amount of reactants at equilibrium
$\Delta G^0 > 0$	$\ln K < 0$	$K < 1$	Reactants are favored over products at equilibrium

where ΔG is the free energy change under nonstandard conditions, and Q is the reaction quotient. For more details, see Chapter 17, which deals with chemical equilibrium. Q can be expressed in terms of partial pressure for reactions involving gases, or in terms of molar concentration for reaction involving solutes in solution. Table 21.2 describes the relationship between ΔG^0 and K.

Example 21.14

Calculate the value of the equilibrium constant at 298 K for the reaction:

$$N_2\,(g) + 3H_2\,(g) \longrightarrow 2NH_3\,(g) \quad \Delta G^0 = -33.1 \text{ kJ}$$

Solution

We can calculate K from the expression $\Delta G^0 = -2.303 \, RT \log K$. Rearrange this equation and express it in terms of $\log K$.

$$\log K_p = \frac{-\Delta G^0}{2.303RT}$$

Now solve for K by substituting the values of R, T, and ΔG^0 into the equation.

$$\log K_p = \frac{-\Delta G^0}{2.303RT} = \frac{-(-33.1 \times 10^3 \text{ J})}{(2.303) \times \left(8.314\frac{J}{K}\right) \times (298 \text{ K})} = 5.8$$

$$K = \text{antilog}(5.8) = 10^{5.8} = 6.32 \times 10^5$$

Example 21.15

At 25°C, the solubility product constant K_{sp} for $CuCO_3$ is 2.5×10^{-10}. Calculate ΔG^0 at 25°C for the reaction

$$CuCO_3\,(s) \rightleftharpoons Cu^{2+}\,(aq) + CO_3^2\,(aq)$$

Solution

Calculate ΔG^0 from the expression $\Delta G^0 = -2.303RT \log K_p$. Note that $K_{sp} = K$.

$$\Delta G^0 = -2.303RT \log K = (-2.303)(8.314 \text{ J/K})(298 \text{ K})(\log 2.5 \times 10^{-10})$$

$$= 54{,}788 \text{ J} = 54.8 \text{ kJ}$$

Example 21.16

For the gaseous reaction

$$4\,NH_3\,(g) + 5\,O_2\,(g) \rightleftharpoons 4\,NO\,(g) + 6\,H_2O\,(g) \quad \Delta G^0 = -957.4\;kJ$$

the equilibrium partial pressures are

$$P_{NH_3} = 0.2\;atm, P_{O_2} = 15\;atm, P_{NO} = 2.5, and\; P_{H_2O} = 1.5\;atm.$$

Calculate ΔG at 298 K for this reaction.

Solution

To calculate ΔG for this reaction we use the expression

$$\Delta G = \Delta G^0 + 2.303\,RT\,\log Q$$

First, we need to calculate Q. From the given reaction,

$$Q = \frac{(P_{NO})^4\,(P_{H_2O})^6}{(P_{NH_3})^4\,(P_{O_2})^5} = \frac{(2.5)^4(1.5)^6}{(0.2)^4(15)^5} = 0.3662$$

Now, solve for ΔG:

$$\Delta G = \Delta G^0 + 2.303\,RT \log Q$$
$$= (-957,400\;J) + (2.303)(8.314\;J/K)(298\;K)(\log 0.3662)$$
$$= -9.60 \times 10^5\;J = -960\;kJ$$

This reaction is spontaneous since ΔG is negative.

21.11 Variation of ΔG^0 and Equilibrium Constant with Temperature

The dependence of ΔG^0 on the reaction temperature can be seen from the quantitative relationship between ΔG^0, ΔH^0, and ΔS^0, as well as between ΔG^0 and the equilibrium constant K. If we know ΔG^0 at one temperature, we can calculate ΔG^0 at another temperature if we assume that ΔH^0 and ΔS^0 are constant. Let us derive the relationship:

$$\Delta G_1^0 = \Delta H^0 - T_1 \Delta S \text{ and } \Delta G_2^0 = \Delta H^0 - T_2 \Delta S$$

Divide through the expressions for ΔG^0 by the temperature term. This gives:

$$\frac{\Delta G_1^0}{T_1} = \frac{\Delta H^0}{T_1} - \Delta S \text{ and } \frac{\Delta G_2^0}{T_2} = \frac{\Delta H^0}{T_2} - \Delta S$$

Combining these two equations yields:

$$\frac{\Delta H^0}{T_1} - \frac{\Delta G^0}{T_1} = \frac{\Delta H^0}{T_2} - \frac{\Delta G^0}{T_2}$$

$$\frac{\Delta G_2^0}{T_2} - \frac{\Delta G_1^0}{T_1} = \Delta H \left(\frac{1}{T_2} - \frac{1}{T_1} \right)$$

or

$$\frac{\Delta G_2^0}{T_2} = \frac{\Delta G_1^0}{T_1} + \Delta H \left(\frac{1}{T_2} - \frac{1}{T_1} \right)$$

Example 21.17

The standard free energy change at 25°C is −57.7 kJ for the following reaction:

$$FeCl_2 (aq) + 1/2\, Cl_2 (g) \longrightarrow FeCl_3 (aq)$$

Given that ΔH^0 is −127.2 kJ and ΔS^0 is −233 J/K, calculate ΔG^0 at 100°C. Comment on your answer.

Solution

Note that ΔS^0 is given. However, it is not relevant in solving this problem. To calculate ΔG^0 at 100°C or 373 K, use the equation:

$$\frac{\Delta G_2^0}{T_2} = \frac{\Delta G_1^0}{T_1} + \Delta H \left(\frac{1}{T_2} - \frac{1}{T_1} \right)$$

$$\frac{\Delta G_2^0}{373\ K} = \frac{-57.7\ kJ}{298\ K} + (-127.2\ kJ) \left(\frac{1}{373\ K} - \frac{1}{298\ K} \right)$$

$$= \left(-0.1936 \frac{kJ}{K} \right) + \left(0.0858 \frac{kJ}{K} \right)$$

$$\frac{\Delta G_2^0}{373\ K} = -0.1078 \frac{kJ}{K}$$

$$\Delta G_2^0 = \left(-0.1078 \frac{kJ}{K} \right) (373\ K) = -40.2\ kJ$$

The value of ΔG^0 at 25°C is more negative than that at 100°C. Therefore, the reaction is less favorable at elevated temperature.

21.11.1 Relationship between ΔG^0 and K at different temperatures

The expressions for ΔG^0 we have so far encountered are:

$$\Delta G^0 = \Delta H^0 - T\Delta S^0$$

$$\Delta G^0 = -2.303\, RT \log K$$

If we combine these two equations (i.e. $\Delta H^0 - T\Delta S^0 = -2.303RT\log K$) and solve for $\log K$, we will have:

$$\log K = -\frac{\Delta H^0}{2.303\,RT} + \frac{T\Delta S^0}{2.303\,RT} \quad \text{or} \quad \log K = -\frac{\Delta H^0}{2.303\,RT} + \frac{\Delta S^0}{2.303\,R}$$

Now assuming that a reaction is carried out at two temperatures, T_1, and T_2, and that ΔH^0 and ΔS^0 are constant at these temperatures, the above equation will give:

$$\log K_1 = -\frac{\Delta H^0}{2.303RT_1} + \frac{\Delta S^0}{2.303R}$$

and

$$\log K_2 = -\frac{\Delta H^0}{2.303RT_2} + \frac{\Delta S^0}{2.303R}$$

Subtracting the first equation from the second and rearranging gives:

$$\log\left(\frac{K_2}{K_1}\right) = \frac{-\Delta H^0}{2.303R}\left(\frac{1}{T_2} - \frac{1}{T_1}\right)$$

Example 21.18

The following reaction plays a significant role in the refining of copper and is usually carried out at high temperatures.

$$CuS\,(s) + H_2\,(g) \longrightarrow Cu\,(s) + H_2\,S\,(g)$$

$K = 3.14 \times 10^{-4}$ at 25°C. What is K at 727°C, given that $\Delta H^0 = 33$ kJ? Assume that ΔH^0 is constant over this temperature range.

Solution

To calculate K at 727°C, use the equation:

$$\log\left(\frac{K_2}{K_1}\right) = \frac{-\Delta H^0}{2.303R}\left(\frac{1}{T_2} - \frac{1}{T_1}\right)$$

$$\log\left(\frac{K_2}{3.14 \times 10^{-4}}\right) = \frac{(-33,000\ \text{J})}{(2.303)(8.314\ \text{J/K})}\left(\frac{1}{1000} - \frac{1}{298}\right)$$

$$\log\left(\frac{K_2}{3.14 \times 10^{-4}}\right) = 4.060$$

$$K_2 = \left(3.14 \times 10^{-4}\right)(\text{antilog } 4.060) = 3.6$$

The equilibrium constant at 727°C is 3.6. Thus the process is more favorable at higher temperature.

21.12 Problems

1. 150 J of heat was added to a biological system. Determine the change in internal energy for the system if it does 250 J of work on its surroundings.

2. The reaction of nitrogen with hydrogen to make ammonia yields −92.2 kJ.

$$N_2(g) + 3H_2(g) \longrightarrow 2NH_3(g)$$

If the change in internal energy is 80.5 kJ, calculate the maximum amount of work that can be done by this chemical process.

3. The combustion of methane gas in a cylindrical reactor causes a volume expansion of 7.5 L against an external pressure of 10 atm. Calculate the work done in kJ, during the reaction. (Note: 1 L-atm = 101 J.)

4. The highly exothermic fluorination of methane is represented by

$$CH_4(g) + 4\,F_2(g) \longrightarrow CF_4(g) + 4HF(g) \quad \Delta H = -1935 \text{ kJ}$$

If the reaction is carried out at a constant pressure of 25 atm, what is the change in internal energy, ΔE, if the reaction resulted in a volume change of −0.95 L?

5. The industrial production of ammonia is represented by

$$N_2(g) + 3H_2(g) \longrightarrow 2NH_3(g) \quad \Delta H = -92.2 \text{ kJ}$$

Calculate the change in internal energy, ΔE, for the reaction at 298K. ($R = 8.314$ J/K-mol.)

6. Oxygen is often prepared in the laboratory by the thermal decomposition of potassium chlorate in the presence of manganese dioxide,

$$2KClO_3(s) \xrightarrow{MnO_2} 2KCl(s) + 3O_2(g)$$

Calculate the work done when the above reaction is conducted at STP.

7. Using the appropriate S^0 values in Table 21.1, calculate the standard entropy change for the following reaction:

$$N_2O_4(g) \longrightarrow 2NO_2(g)$$

8. Using the appropriate S^0 values in Table 21.1, calculate the entropy change that accompanies the formation of 2 moles of nitrogen dioxide from nitrogen monoxide and oxygen.

$$2NO(g) + O_2(g) \longrightarrow 2NO_2(g)$$

9. Determine whether the following processes are accompanied by an increase or a decrease in the entropy of the system. No calculations are needed.

(a) $I_2(s) \longrightarrow I_2(g)$

(b) $CaCO_3(s) \longrightarrow CaO(s) + CO_2(g)$

(c) $H_2(g) + I_2(g) \longrightarrow 2HI(g)$

(d) $2SO_2(g) + O_2(g) \longrightarrow 2SO_3(g)$

(e) $H_2O(s) \longrightarrow H_2O(l)$

(f) $2H_2(g) + CO(g) \longrightarrow CH_3OH(l)$

10. Without doing any calculations, predict whether the entropy change in each of the following systems is positive or negative.

(a) $3O_2(g) \longrightarrow 2O_3(g)$

(b) $NH_4Cl(s) \longrightarrow NH_3(g) + HCl(g)$

(c) $H_2(g) \longrightarrow 2H(g)$

(d) $2HgO(s) \longrightarrow 2Hg(l) + O_2(g)$

(e) $N_2O_4(g) \longrightarrow 2NO_2(g)$

(f) $H_2O(l, 90^0C) \longrightarrow H_2O(s, 0^0C)$

11. Calculate the standard entropy changes for transformations of 1 mole of the following at 25°C:

(a) S_8 (rhombic) $\longrightarrow$ S_8 (monoclinic)

    $(S_r^0 = 254.4 \text{ J/Kmol})$ $(S_m^0 = 260.8 \text{ J/Kmol})$

(b) SiO_2 (quartz) $\longrightarrow$ SiO_2 (tridymite)

    $(S_q^0 = 10 \text{ J/Kmol})$ $(S_t^0 = 11.2 \text{ J/Kmol})$

12. Calculate the standard entropy changes for each of the following reactions:

(a) $2H_2 S(g) + SO_2(g) \longrightarrow 3 S(s) + 2H_2O(g)$

(b) $SO_3(g) + H_2O(l) \longrightarrow H_2SO_4(aq)$

(c) $2H_2O_2(l) \longrightarrow 2H_2O(l) + O_2(g)$

    S^0 in J/K.mol : $H_2O_2(l) = 110$, $SO_2(g) = 248.1$, S (rhombic) $= 31.8$

    $SO_3(g) = 256.6$, $H_2O(l) = 69.9$, $H_2O(g) = 188.7$, $H_2SO_4(aq) = 20$, $O_2(g) = 250$,

    $H_2 S(g) = 205.7$

13. The heat of vaporization of ethanol is 39.3 kJ/mol and its normal boiling point is 78°C. Calculate the entropy change associated with the vaporization of ethanol.

14. The molar heat of sublimation of solid carbon dioxide (dry ice) is 25.2 kJ/mol, and it sublimes at 194.8 K. What is the entropy change for the sublimation of dry ice?

15. The molar heat of fusion and vaporization of NaCl are 28.5 and 170.7 kJ/mol, respectively. Determine the entropy changes for the solid-to-liquid and liquid-to-gas transformations. The melting and boiling point are 1081 K and 1738 K at 1 atm pressure.

16. A certain polymerization process has $\Delta H = 55.8$ kJ and $\Delta S = 335$ J/K at 298 K.

   (a) Calculate ΔG.
   (b) Is the reaction spontaneous under these conditions?
   (c) Is the reaction spontaneous at 200°C?

17. Calculate ΔG^0 in kJ per mole of H_2 formed for the following reaction at 298 K using the data below:

$$H_2(g) + Cl_2(g) \longrightarrow 2HCl(g)$$

Is the reaction spontaneous under these conditions?

	ΔH_f^0/kJmol^{-1}	S^0/JK^{-1} mol^{-1}
$H_2(g)$	0.0	130.6
$Cl_2(g)$	0.0	130.6
$HCl(g)$	−92.3	186.8

18. An engineer without adequate chemistry background boasted he could synthesize ammonia by dissolving nitric oxide in water at 80°C according to the following equation:

$$4NO(g) + 6H_2O(g) \longrightarrow 4NH_3(g) + 5O_2(g)$$

Using the following thermodynamic data predict if the engineer will have any luck synthesizing ammonia under these conditions.

	ΔH_f^0/kJmol^{-1}	S^0/JK^{-1} mol^{-1}
$NO(g)$	90.2	210.7
$H_2O(g)$	241.8	188.7
$O_2(g)$	0	205.0
$NH_3(g)$	−46.1	192.3

19. Using the given enthalpy and entropy changes, estimate the temperature range for which each of the following standard reactions is spontaneous.

   (a) $NH_4Cl(s) \longrightarrow NH_3(g) + HCl(g)$
   $\left(\Delta H_{rxn}^0 = +176 \text{ kJ; } \Delta S_{rxn}^0 = +285 \text{ J/K}\right)$
   (b) $N_2H_4(g) + 2H_2O_2(1) \longrightarrow N_2(g) + 4H_2O(g)$
   $\left(\Delta H_{rxn}^0 = -642.2 \text{ kJ; } \Delta S_{rxn}^0 = +606 \text{ J/K}\right)$

20. For the reaction

$$Fe_2O_3(s) + 3CO(g) \longrightarrow 2Fe(s) + 3CO_2(g)$$

ΔG^0 and ΔH^0 were experimentally determined to be −29.5 kJ and −24.9 kJ, respectively.

(a) Calculate $\Delta G^0_{CO_2}$ and $\Delta H^0_{CO_2}$ from the thermodynamic data given below.

(b) Determine the change in entropy for the reaction at 25°C.

	$\Delta G^0_f/kJmol^{-1}$	$\Delta H^0_f/kJmol^{-1}$
CO(g)	−137.7	−110.5
CO$_2$(g)	?	?
Fe$_2$O$_3$(s)	−742.2	−824.2
Fe(s)	0	0
NH$_3$(g)	−46.1	192.3

21. Hydrogen peroxide decomposes according to the equation

$$H_2O_2(1) \longrightarrow H_2O(g) + \tfrac{1}{2}O_2(g)$$
$$\Delta G^0_f (kJ/mol) \quad -187.9 \qquad\qquad -241.8 \qquad 0$$

Calculate ΔG^0 for the reaction and determine if the reaction will occur as it is written (forward direction).

22. Calculate the equilibrium constant (K_p) for the reaction given in problem 21 at 25°C. ($R = 8.31$ J/K.)

23. Using the standard free energies of formation given below, calculate the equilibrium constant (K_p) at 25°C for the process

$$2NH_3(g) \rightleftharpoons N_2H_4(g) + H_2(g)$$
$$(\Delta G^0_f \text{ in kJ/mol} = -16.5 \qquad 159.3 \qquad 0$$

24. Calculate the free energy change for the oxidation of ammonia at 25°C from the following equilibrium partial pressures: 2.0 atm NH$_3$, 1.0 atm O$_2$, 0.5 atm NO, and 0.75 atm H$_2$O.

$$4NH_3(g) + 5O_2(g) \rightleftharpoons 4NO(g) + 6H_2O(g) \, \Delta G^0_{rxn} = -958 \text{ kJ/mol}$$

25. The standard free energies of formation for liquid and gaseous ethanol are −174.9 and −168.6 kJ/mol, respectively. Calculate the vapor pressure of ethanol (C$_2$H$_5$OH) at 25°C.

26. The vapor pressure of mercury at 25°C is 2.6×10^{-6} atm. Calculate the free energy change for the vaporization process.

27. The equilibrium constant for the formation of hydrogen iodide gas is 10.2 at 25°C, and 78 at 425°C. Calculate the enthalpy change for the reaction.

28. At 25°C, the equilibrium constant, K_c, for the process

$$N_2O_4(g) \rightleftharpoons 2NO_2(g) \quad \Delta H^0_{nxn} = 58 \text{ kJ}$$

Is 4.6×10^{-3}. What is the equilibrium constant at 100°C?

22

Oxidation and Reduction Reactions

. . . .

22.1 Introduction

Oxidation-reduction, or *redox*, reactions occur in many chemical and biochemical systems. The process may involve the complete or partial transfer of electrons from one atom to another. Oxidation and reduction processes are complementary. Every oxidation always has a corresponding reduction process. This is because, for a substance to gain electrons in a chemical reaction, another substance has to lose those electrons.

22.2 Oxidation and Reduction in Terms of Electron Transfer

Oxidation is defined as *a process by which an atom or ion loses electrons*. This can occur in the following ways:

- Addition of oxygen or other electronegative elements to a substance:

$$2 \, Mg \, (s) + O_2 \, (g) \longrightarrow 2 \, MgO \, (s)$$
$$Mg \, (s) + Cl_2 \, (g) \longrightarrow MgCl_2 \, (s)$$

- Removal of hydrogen or other electropositive elements from a substance:

$$H_2S \, (g) + Cl_2 \, (g) \longrightarrow 2 \, HCl \, (g) + S \, (s)$$

- The removal of electrons from a substance (loss of electrons):

$$2 \, FeCl_2 \, (s) + Cl_2 \, (g) \longrightarrow 2 \, FeCl_3 \, (s)$$
$$Fe^{2+} \longrightarrow Fe^{3+} + e^-$$

Chemistry in Quantitative Language: Fundamentals of General Chemistry Calculations. Second Edition. Christopher O. Oriakhi, Oxford University Press. © Christopher O. Oriakhi 2021. DOI: 10.1093/oso/9780198867784.003.0022

Reduction is defined as a process by which an atom or ion gains electrons. This can occur in the following ways:

- Removal of oxygen or other electronegative elements from a substance:

$$MgO(s) + H_2(g) \longrightarrow Mg(s) + H_2O(g)$$

- Addition of hydrogen or other electropositive elements to a substance:

$$H_2(g) + Br_2(g) \longrightarrow 2HBr(g)$$

$$2Na(s) + Cl_2(g) \longrightarrow 2NaCl(s)$$

- The addition of electrons to a substance (gain of electrons):

$$Fe^{3+} + e^- \longrightarrow Fe^{2+}$$

22.3 Oxidation Numbers (ON)

Oxidation number or *oxidation state* is a number assigned to atoms in a substance to describe their relative state of oxidation or reduction. The numbers are used to keep track of electron transfer in chemical reactions. Some general rules are used to determine the ON of an atom in free or combined state.

22.3.1 Rules for assigning oxidation numbers

1. Any atom in an uncombined (or free) element (e.g. N_2, Cl_2, S_8, O_2, O_3, and P_4) has an ON of zero.
2. Hydrogen has an ON of $+1$ except in metal hydrides (e.g. NaH, MgH_2) where it is -1.
3. Oxygen has an ON of -2 in all compounds except in peroxides (e.g. H_2O_2, Na_2O_2) where it is -1.
4. In simple monoatomic ions such as Na^+, Zn^{2+}, Al^{3+}, Cl^-, and C^{4-}, the ON is equal to the charge on the ion.
5. The algebraic sum of the ONs in a neutral molecule (e.g. $KMnO_4$, $NaClO$, H_2SO_4) is zero.
6. The algebraic sum of the ONs in a polyatomic ion (e.g. $SO4^{2-}$, $Cr_2O_7^{2-}$) is equal to the charge on the ion.
7. Generally, in any compound or ion, the more electronegative atom is assigned the negative ON while the less electronegative number is assigned the positive ON.

22.3.2 Oxidation numbers in formulas

The ON of an atom in a compound or ion can be determined using the applicable rules. For example, the ON of K in $K_2Cr_2O_7$ can be obtained as follows:

$2 \times ON[K] + 2 \times ON[Cr] + 7 \times ON[O] = 0$

ON of K $= +1$

ON of Cr $= ?$

ON of O $= -2$

$2 \times 1 + 2Cr - 7 \times 2 = 0$

$2 + 2Cr - 14 = 0$

$2Cr - 12 = 0$

$Cr = 6$

Hence, the ON of Cr in $K_2Cr_2O_7 = +6$. In a similar manner, the ON of Mn in MnO_4^- can be calculated as follows:

$1 \times ON[Mn] + 4 \times ON[O] = -1$

ON of Mn $= ?$

ON of O $= -2$

$Mn + 4 \times -2 = -1$

$Mn - 8 = -1$

$Mn = +7$

Therefore, the ON of Mn in MnO_4^- is $+7$.

Example 22.1

Calculate the ON for the element indicated in each of the following compounds or ions:

(a) As in Na_3AsO_4
(b) S in S_2F_{10}
(c) Br in $HBrO_3$
(d) P in $P_2O_7^{4-}$
(e) Pt in $[PtCl_6]^{2-}$
(f) Fe in $\left[Fe(CN)_6\right]^{3-}$

Solution

(a) As in Na_3AsO_4

The sum of the oxidation numbers should be zero:
$Na_3\,As\,O_4$
$3(+1) + As + 4(-2) = 0$
$3 + As - 8 = 0$
$As = +5$

(b) S in S_2F_{10}

The sum of the oxidation numbers should be zero:

$S_2\ F_{10}$

$2(S) + 10(-1) = 0$

$2S - 10 = 0$

$S = +5$

(c) Br in $HBrO_3$

The sum of the oxidation numbers should be zero:

$H\ Br\ O_3$

$1(+1) + Br + 3(-2) = 0$

$1 + Br - 6 = 0$

$Br = +5$

(d) P in $P_2O_7^{4-}$

The sum of the oxidation numbers should be -4:

$P_2\ O_7^{4-}$

$2(P) + 7(-2) = -4$

$2P - 14 = -4$

$P = +5$

(e) Pt in $[PtCl_6]^{2-}$

The sum of the oxidation numbers should be -2:

$[Pt\ Cl_6]^{2-}$

$(Pt) + 6(-1) = -2$

$Pt - 6 = -2$

$Pt = +4$

(f) Fe in $\left[Fe(CN)_6\right]^{3-}$

The sum of the oxidation numbers should be -3:

$\left[Fe\ (CN)_6\right]^{3-}$

$1(Fe) + 6(-1) = -3$

$Fe - 6 = -3$

$Fe = +3$

22.3.3 Oxidation number and nomenclature

The concept of ON is very important in naming complex molecular or ionic compounds. Binary chemical compounds such as CO, S_2Cl_2, NO_2, and CCl_4, are commonly named by indicating numbers of constituent atoms. The ON is often not stated. Thus, CO is named carbon *mono*xide, S_2Cl_2 is *di*sulfur *di*chloride; NO_2 is nitrogen *di*oxide and CCl_4 is carbon *tetra*chloride.

When naming complex compounds, or ions under the IUPAC convention, the ON of the component element that exhibits variable ON is indicated in Roman numerals. Table 22.1 shows some examples.

22.4 Oxidation and Reduction in Terms of Oxidation Number

In oxidation—reduction reactions, the reacting substances undergo changes in ON. Oxidation results in an increase in ON due to a loss of electrons. Reduction results in a decrease in ON

Table 22.1 Oxidation number and IUPAC nomenclature of some common compounds

Formula	IUPAC Name	Common Name
HNO_2	Nitric(III) acid	Nitrous acid
HNO_3	Nitric(V) acid	Nitric acid
H_2SO_3	Sulfuric(IV) acid	Sulfurous acid
H_2SO_4	Sulfuric(VI) acid	Sulfuric acid
$FeCl_2$	Iron(II) Chloride	Ferrous Chloride
$FeCl_3$	Iron(III) Chloride	Ferric Chloride

due to a gain of electrons. In the following reaction, the ON of Mg has increased from 0 to $+2$ and is therefore said to be oxidized. Hydrogen, on the other hand, is reduced because its ON has decreased from $+1$ to 0.

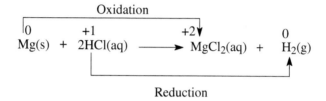

22.5 Disproportionation Reactions

Disproportionation is a special type of redox reaction in which a single reactant is simultaneously oxidized and reduced. A good example is the preparation of nitric acid from the action of nitrogen dioxide on water.

$$\overset{+4}{3NO_2(g)} + H_2O(l) \longrightarrow 2H\overset{+5}{N}O_3(aq) + \overset{+2}{N}O(g)$$

The nitrogen atom in nitrogen dioxide disproportionates from $+4$ oxidation state to both $+5$ and $+2$. Similarly, copper (I) ions disproportionate in solution according to:

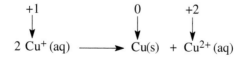

22.6 Oxidizing and Reducing Agents

Substances that bring about oxidation are called *oxidizing agents*. In a redox reaction, the oxidizing agent is the:

1. Substance that is reduced.
2. Substance that gains electrons.
3. Substance in which the ON has decreased (reduced).

Substances that bring about reduction are called *reducing agents*. In a redox reaction, the reducing agent is the:

1. Substance that is oxidized.
2. Substance that loses electrons.
3. Substance in which the ON has increased (oxidized).

Consider the following reaction:

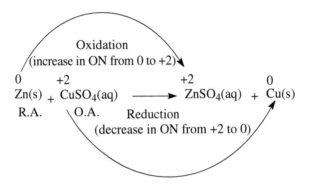

Zinc donates 2 electrons and goes into solution as zinc (Zn^{2+}) ions. Copper ions (Cu^{2+}) gain these 2 electrons and are deposited as elemental copper. Zinc is oxidized, and hence it is the reducing agent (R.A.). On the other hand, copper, which is reduced, is the oxidizing agent (O.A.).

In the industry, reducing agents are used in the extraction of metals from their ores. For example, carbon in the form of coke can be used to reduce copper (II) oxide and iron (III) oxide to the metals—copper and iron.

22.6.1 Identifying oxidizing and reducing agents

In redox reactions, oxidizing and reducing agents can be identified by calculating the ON for every atom in reactants and products, and then comparing changes in ON. A reactant molecule with a net increase in ON (oxidation) is the reducing agent. Similarly, a reactant molecule with a net decrease in ON (reduction) is the oxidizing agent.

Example 22.2

Assign ONs to all atoms in the following equation. Identify which substance is oxidized, and which is reduced. Identify the oxidizing and reducing agents.

$$2Fe_2O_3(s) + 3C(s) \longrightarrow 4Fe(s) + 3CO_2(g)$$

Solution

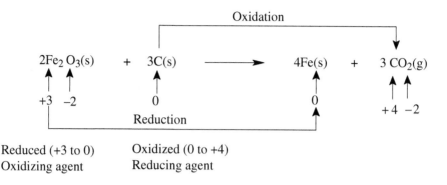

The uncombined C and Fe have ONs of zero; Fe in Fe_2O_3 is +3. O in Fe_2O_3 and CO_2 is –2. C is oxidized (from 0 to +4) while Fe_2O_3 is reduced (from +3 to 0). Fe_2O_3 is the oxidizing agent, while C is the reducing agent.

22.7 Half-Cell Reactions

The oxidation and reduction steps of a redox reaction can be described by two separate equations called *half-equations* or *half-cell reactions*. A half-cell reaction is largely a convenient calculating aid; one half-reaction never takes place in isolation. Each such equation represents half of a redox reaction, either the oxidation or the reduction.

The overall reaction is always the sum of two half-cell reactions—one for oxidation, and the other for reduction. For example, consider the redox reaction:

$$Zn(s) + Cu^{2+}(aq) \longrightarrow Zn^{2+}(aq) + Cu(s)$$

This can be separated into half-reactions for each metal:

Oxidation half-reaction $Zn(s)$ $\longrightarrow Zn^{2+}(aq) + 2e^-$

Reduction half-reaction $Cu^{2+}(aq) + 2e^-$ $\longrightarrow Cu(s)$

Net Reaction $Zn(s) + Cu^{2+}(aq) \longrightarrow Zn^{2+}(aq) + Cu(s)$

Example 22.3

Iron metal reacts with aqueous copper (II) ion to give iron (II) ion and copper metal according to the following equation:

$$Fe(s) + Cu^{2+}(aq) \longrightarrow Fe^{2+}(aq) + Cu(s)$$

Write the two half-cell reactions, labeling the oxidation and reduction reactions.

Solution

The two half-cell reactions are given below. When added, we obtain the original net reaction.

$$Fe\,(s) \longrightarrow Fe^{2+}\,(aq) + 2e^- \quad \text{Oxidation}$$
$$Cu^{2+}\,(aq) + 2e^- \longrightarrow Cu\,(s) \qquad \text{Reduction}$$

22.8 Balancing Redox Equations

Generally, many chemical reactions are represented by simple equations that can be balanced by mere inspection. But redox reactions are often very complex, and it can be quite difficult and time consuming to use inspection or trial and error to balance their equations. In general, redox equations are balanced by following the principle of conservation of charge, i.e. the total number of electrons lost by the reducing agent must be equal to the number of electrons gained by the oxidizing agent in a chemical reaction. Two common methods for balancing redox equations are discussed here:

1. The ON method
2. The half-reaction method.

22.8.1 The oxidation-number method

1. Write the unbalanced equation.
2. Assign ONs to all the atoms in the equation. Identify the atoms that are oxidized and reduced.
3. Write two new equations, one for the oxidation step and the other for the reduction step, using only species that change in ON. Add electrons to reflect the change in ON.
4. Multiply both equations by the smallest integers such that the total net increase in ON for the oxidized atoms equals the total net decrease in ON for the reduced atoms.
5. Assign these integers as coefficients of the oxidized and reduced species on both sides of the unbalanced equation.
6. Add H_2O to the side with less O until O is balanced. Then, if H is not balanced, add H^+ to the side with less H. If the reaction takes place in acidic solution, you are done, but if the reaction takes place in basic solution, there is one more step: add as many OH^- to each side as you added free H^+, remembering that $H^+ + OH^-$ can be written as water, H_2O.
7. Balance the remaining species that are not oxidized or reduced.
8. Check for atom and charge balance of the final equation.

Example 22.4

Use the ON method to balance the equation for the reaction of nitric acid with hydrogen sulfide.

Solution

Step 1: Write the unbalanced equation.

$$HNO_3\,(aq) + H_2S\,(g) \longrightarrow S\,(s) + NO\,(g) + H_2O\,(l)$$

Step 2: Assign ONs to all the atoms in the equation. Identify the atoms that are oxidized and reduced.

$$\begin{array}{ccccc} HNO_3\,(aq) & + H_2S\,(g) \longrightarrow & S\,(s) & + NO\,(g) & + H_2O\,(l) \\ +1+5-2 & +1-2 & 0 & +2-2 & +1-2 \end{array}$$

Nitric acid is reduced (gain of $3e^-$), sulfur is oxidized (loss of $2e^-$).

Step 3: Write two new equations, one for the oxidation step and the other for the reduction step, using only species that change in ON. Add electrons to reflect the change in ON.

$$\begin{array}{lccc} \text{Oxidation} & S^{2-} & \longrightarrow S^0 & + \quad 2e^- \\ \text{Reduction} & N^{5+} + 3e^- & \longrightarrow & N^{2+} \end{array}$$

Step 4: Multiply both equations by the smallest integers such that the total net increase in ON for the oxidized atoms equals the total net decrease in ON for the reduced atoms.

$$\begin{array}{lcc} \text{Oxidation} & 3S^{2-} & \longrightarrow 3S^0 + 6e^- \\ \text{Reduction} & 2N^{5+} + 6e^- & \longrightarrow \quad 2N^{2+} \end{array}$$

Step 5: Assign these integers as coefficients of the oxidized and reduced species on both side of the unbalanced equation.

$$2HNO_3\,(aq) + 3H_2S\,(g) \longrightarrow 3S\,(s) + 2NO\,(g) + H_2O\,(l) \text{ (unbalanced)}$$

Step 6: This is not necessary for the present problem.
Step 7: Balance the remaining species that are not oxidized or reduced.

$$2HNO_3\,(aq) + 3H_2S\,(g) \longrightarrow 3S\,(s) + 2NO\,(g) + 4H_2O\,(l) \text{ (balanced)}$$

Step 8: The final balanced equation contains equal number of atoms on both sides of the equation.

Example 22.5

Balance the equation below by the ON method, assuming it takes place in a basic solution.

$$MnO_4^-\,(aq) + SO_3^{2-}\,(aq) \longrightarrow MnO_4^{2-}\,(aq) + SO_4^{2-}\,(aq)$$

Solution

Step 1: Write the unbalanced equation.

$$MnO_4^-\,(aq) + SO_3^{2-}\,(aq) \longrightarrow MnO_4^{2-}\,(aq) + SO_4^{2-}\,(aq)$$

Step 2: Assign ONs to all the atoms in the equation. Identify the atoms that are oxidized and reduced.

$$MnO_4^- \text{ (aq)} + SO_3^{2-}\text{(aq)} \longrightarrow MnO_4^{2-} \text{ (aq)} + SO_4^{2-} \text{ (aq)}$$
$$+7-2 \qquad\quad +4-2 \qquad\qquad +6-2 \qquad\quad +6-2$$

MnO_4^- (aq) is reduced $(Mn^{7+} \longrightarrow Mn^{6+}, \text{gain of } 1e^-)$

SO_3^{2-} (aq) is oxidized $(S^{4+} \longrightarrow S^{6+}, \text{loss of } 2e^-)$

Step 3: Write two new equations, one for the oxidation step and the other for the reduction step, using only species that change in ON. Add electrons to reflect the change in ON.

$$\text{Oxidation} \quad S^{4+} \longrightarrow S^{6+} + 2e^-$$

$$\text{Reduction} \quad Mn^{7+} + 1e^- \longrightarrow Mn^{6+}$$

Step 4: Multiply both equations by the smallest integers such that the total net increase in ON for the oxidized atoms equals the total net decrease in ON for the reduced atoms.

$$\text{Oxidation} \quad S^{4+} \longrightarrow S^{6+} + 2e^-$$

$$\text{Reduction} \quad 2\,Mn^{7+} + 2e^- \longrightarrow 2\,Mn^{6+}$$

Step 5: Assign these integers as coefficients of the oxidized and reduced species on both side of the skeletal equation.

$$2MnO_4^- \text{ (aq)} + SO_3^{2-} \text{ (aq)} \longrightarrow 2MnO_4^{2-} \text{ (aq)} + SO_4^{2-} \text{ (aq)}$$

Step 6: Balance the equation for oxygen and hydrogen as follows: add H_2O to the side with less O and add H^+ to the side with less H.

$$2MnO_4^- \text{ (aq)} + SO_3^{2-} \text{ (aq)} + H_2O\,(l) \longrightarrow 2MnO_4^{2-} \text{ (aq)} + SO_4^{2-} \text{ (aq)} + 2H^+ \text{ (aq)}$$

Since the reaction occurs in a basic solution, we now add $2OH^-$ to both sides of the above equation (and cancel water if it appears on both sides) to give

$$2MnO_4^- \text{ (aq)} + SO_3^{2-} \text{ (aq)} + 2OH^- \text{ (aq)} \longrightarrow 2MnO_4^{2-} \text{ (aq)} + SO_4^{2-} \text{ (aq)} + H_2O\,(l)$$

Step 7: This is not required since all the atoms are already balanced.

Step 8: The final balanced equation contains equal numbers of atoms and equal charges on both sides of the equation.

22.8.2 The half-reaction method

1. Write the unbalanced ionic equation and decide which atoms are oxidized and which atoms are reduced.
2. Write unbalanced equations for the oxidation and reduction half-reactions.
3. Balance all the elements other than hydrogen and oxygen in each half-reaction using the inspection method.

4. Balance oxygen and hydrogen. For reactions in acidic solutions, add H^+ to balance hydrogen atoms and H_2O to balance oxygen. For reactions occurring in a basic solution, first treat as if the reaction is occurring in an acidic solution; then add as many OH^- ions to the reactants and products side of the reaction as there are H^+ ions in the equation. Combine the H^+ and OH^- ions to form H_2O. Rewrite the equation and eliminate equal numbers of water molecules appearing on opposite side of the equation.
5. Balance the charge in each half-reaction by adding electrons as reactants or products.
6. Multiply each half-reaction by an appropriate integer so that the number of electrons gained in the reduction process equals the number of electrons lost in the oxidation process.
7. Add the two half-reactions and eliminate any species that appears on both sides of the equation.

Example 22.6

Balance the following half-reactions in acidic solution:

$$Cr_2O_7^{2-}(aq) \longrightarrow Cr^{3+}(aq)$$
$$Ti^{3+}(aq) \longrightarrow TiO_2(s)$$

Solution

(a) Balance Cr
$$Cr_2O_7^{2-}(aq) \longrightarrow 2Cr^{3+}(aq)$$
Balance O by adding H_2O.
$$Cr_2O_7^{2-}(aq) + 2Cr^{3+}(aq) + 7H_2O(l)$$
Balance H by adding H^+.
$$Cr_2O_7^{2-}(aq) + 14H^+(aq) \longrightarrow 2Cr^{3+}(aq) + 7H_2O(l)$$
Balance charge by adding electrons. This gives the balanced equation.
$$Cr_2O_7^{2-}(aq) + 14H^+(aq) + 6e^- \longrightarrow 2Cr^{3+}(aq) + 7H_2O(l)$$

(b) $Ti^{3+}(aq) \longrightarrow TiO_2(s)$
Balance O with H_2O.
$$Ti^{3+}(aq) + 2H_2O(l) \longrightarrow TiO_2(s)$$
Balance H using H^+.
$$Ti^{3+}(aq) + 2H_2O(l) \longrightarrow TiO_2(s) + 4H^+(aq)$$
Balance charge by adding electrons. This gives the balanced equation.
$$Ti^{3+}(aq) + 2H_2O(l) \longrightarrow TiO_2(s) + 4H^+(aq) + e^-$$

Example 22.7

Balance the following ionic equation in acidic solution (add H^+ or H_2O as necessary).

$$Fe^{2+}(aq) + MnO_4^-(aq) \longrightarrow Fe^{3+}(aq) + Mn^{2+}(aq)$$

Solution

Step 1: Write the unbalanced ionic equation and decide which atoms are oxidized and which atoms are reduced.

$$Fe^{2+}(aq) + MnO_4^-(aq) \longrightarrow Fe^{3+}(aq) + Mn^{2+}(aq)$$

Oxidized   Reduced

Step 2: Write unbalanced equations for the oxidation and reduction half-reactions.

Oxidation half-reaction $Fe^{2+} \longrightarrow Fe^{3+}$

Reduction half-reaction $MnO_4^- \longrightarrow Mn^{2+}$

Step 3: Balance all the elements other than hydrogen and oxygen in each half-reaction using the inspection method. This is balanced.

Step 4: Balance oxygen and hydrogen. For reactions in acidic solutions, add H^+ to balance hydrogen and H_2O to balance oxygen. Balance oxygen in the second equation by adding 4 H_2O to the right side of the equation. Then balance hydrogen by adding 8 H^+ to the left side of the equation.

$$MnO_4^-(aq) + 8H^+(aq) \longrightarrow Mn^{2+}(aq) + 4H_2O(l)$$

Step 5: Balance the charge in each half-reaction by adding electrons as reactants or products. In the reduction, the total charge is +7 on the left side and +2 on the right side. Therefore, add electrons to the left side. For the oxidation, the total charge is +2 on the left side, and +3 on the right side. Therefore, add one electron to the right side.

$$MnO_4^-(aq) + 8H^+(aq) + 5e^- \longrightarrow Mn^{2+}(aq) + 4H_2O(l)$$
$$Fe^{2+}(aq) \longrightarrow Fe^{3+}(aq) + e^-$$

Step 6: Multiply each half-reaction by an appropriate integer so that the number of electrons gained in the reduction process equals the number of electrons lost in the oxidation process. The reduction reaction requires the transfer of 5 electrons while the oxidation reaction requires the transfer of only one. Therefore, multiply the oxidation half-equation by 5 to equalize the electron transfer.

$$MnO_4^-(aq) + 8H^+(aq) + 5e^- \longrightarrow Mn^{2+}(aq) + 4H_2O(l)$$
$$5Fe^{2+}(aq) \longrightarrow 5Fe^{3+}(aq) + 5e^-$$

Step 7: Add the two half-reactions and eliminate any species that appear on both sides of the equation. This yields the balanced equation.

$$MnO_4^-(aq) + 8H^+(aq) + 5e^- \longrightarrow Mn^{2+}(aq) + 4H_2O(l)$$
$$5Fe^{2+}(aq) \longrightarrow 5Fe^{3+}(aq) + 5e^-$$

$$MnO_4^-(aq) + 8H^+(aq) + 5Fe^{2+}(aq) \longrightarrow Mn^{2+}(aq) + 4H_2O(l) + 5Fe^{3+}(aq)$$

Example 22.8

Balance the following net ionic reaction in acidic solution:

$$Zn(s) + NO_3^- (aq) \longrightarrow NH_4^+ (aq) + Zn^{2+} (aq)$$

Solution

The unbalanced equation is $Zn(s) + NO_3^- (aq) \longrightarrow NH_4^+ (aq) + Zn^{2+} (aq)$.

Step 1: Write and balance the reduction half-equation.

$NO_3^- (aq) \longrightarrow NH_4^+ (aq)$

Balance O with H_2O and H with H^+.

$NO_3^- (aq) + 10 H^+ (aq) \longrightarrow NH_4^+ (aq) + 3 H_2O(l)$

Balance charge with electrons.

$NO_3^- (aq) + 10 H^+ (aq) + 8 e^- \longrightarrow NH_4^+ (aq) + 3 H_2O(l)$

Step 2: Write and balance the oxidation half-equation.

$$Zn(s) \longrightarrow Zn^{2+} (aq) + 2 e^-$$

Step 3: Multiply the redox half-equations by appropriate factors such that the number of electrons gained is equal to the number of electrons lost.

$$4 \times \left[Zn(s) \longrightarrow Zn^{2+} (aq) + 2 e^- \right] = 4 Zn(s) \longrightarrow 4 Zn^{2+} (aq) + 8 e^-$$

$$1 \times \left[NO_3^- (aq) + 10 H^+ (aq) + 8 e^- \longrightarrow NH_4^+ (aq) + 3 H_2O(l) \right]$$

$$NO_3^- (aq) + 10 H^+ (aq) + 8 e^- \longrightarrow NH_4^+ (aq) + 3 H_2O(1)$$

Step 4: Add the two half-equations to get the overall balanced equation. Be sure to cancel the electrons and other species that appear on both sides of the equation, and that the coefficients are reduced to the simplest whole numbers.

$$4 Zn(s) \longrightarrow 4 Zn^{2+}(aq) + 8 e^-$$

$$NO_3^- (aq) + 10 H^+(aq) + 8 e^- \longrightarrow NH_4^+(aq) + 3 H_2O(l)$$

$$\overline{}$$

$$4 Zn(s) + NO_3^- (aq) + 10 H^+(aq) \longrightarrow 4 Zn^{2+}(aq) + NH_4^+(aq) + 3 H_2O(l)$$

Example 22.9

Balance the following redox reaction in acidic solution using the half-reaction method.

$$I^- (aq) + Cr_2O_7^{2-} (aq) \longrightarrow Cr^{3+} (aq) + IO_3^- (aq)$$

Solution

The unbalanced equation is $I^- (aq) + Cr_2O_7^{2-} (aq) \longrightarrow Cr^{3+} (aq) + IO_3^- (aq)$.

Step 1: Write and balance the reduction half-equation.

$$Cr_2O_7^{2-} (aq) \longrightarrow 2Cr^{3+} (aq)$$

Balance O with H_2O and H with H^+.

$$Cr_2O_7^{2-} (aq) + 14H^+ (aq) \longrightarrow 2Cr^{3+} (aq) + 7H_2O(l)$$

Balance charge with electrons.

$$Cr_2O_7^{2-} (aq) + 14H^+ (aq) + 6e^- \longrightarrow 2Cr^{3+} (aq) + 7H_2O(l)$$

Step 2: Write and balance the oxidation half-equation.

$$I^- (aq) \longrightarrow IO_3^- (aq)$$

Balance O with H_2O and H with H^+.

$$I^- (aq) + 3H_2O(l) \longrightarrow IO_3^- (aq) + 6H^+ (aq)$$

Balance charge with electrons.

$$I^- (aq) + 3H_2O(l) \longrightarrow IO_3^- (aq) + 6H^+ (aq) + 6e^-$$

Step 3: Multiply the redox half-equations by appropriate factors such that the number of electrons gained is equal to the number of electrons lost. In this case, both half-equations contain equal numbers of electrons.

Step 4: Add the two half-equations to get the overall balanced equation. Be sure to cancel the electrons and other species that appear on both sides of the equation, and that the coefficients are reduced to the simplest whole numbers.

$$I^- (aq) + 3 H_2O(l) \longrightarrow IO_3^- (aq) + 6 H^+ (aq) + 6e^-$$
$$Cr_2O_7^{2-} (aq) + 14 H^+ (aq) + 6 e^- \longrightarrow 2 Cr^{3+} (aq) + 7 H_2O(l)$$
$$\overline{}$$
$$I^- (aq) + Cr_2O_7^{2-} (aq) + 8 H^+ (aq) \longrightarrow IO_3^- (aq) + 2 Cr^{3+} (aq) + 4 H_2O(l)$$

Example 22.10

Balance the following half-reactions in basic solution (add H^+, OH^-, or H_2O as necessary).

1. $Bi^{3+} (aq) \longrightarrow BiO_3^- (aq)$
2. $Br_2(aq) \longrightarrow BrO_3^- (aq)$
3. $CrO_4^{2-} (aq) \longrightarrow Cr(OH)_4^- (aq)$

Solution

First, balance the equation $Bi^{3+} (aq) \longrightarrow BiO_3^- (aq)$.

1. Balance the equation as if H^+ were present:

   $$Bi^{3+} (aq) \longrightarrow BiO_3^- (aq)$$

   Balance O with H_2O.

   $$3H_2O(l) + Bi^{3+} (aq) \longrightarrow BiO_3^- (aq)$$

   Balance H using H^+, and charge with electrons.

   $$3H_2O(l) + Bi^{3+} (aq) \longrightarrow BiO_3^- (aq) + 6H^+ (aq) + 2e^-$$

2. Since we know that the reaction is taking place in a basic solution, we must add 6 OH^- to both sides of the equation. This will neutralize the 6 H^+ on the right side.

$3 H_2O(l) + Bi^{3+}(aq) + 6 OH^-(aq) \longrightarrow BiO_3^-(aq) + 6 H^+(aq) + 6 OH^-(aq) + 2 e^-$

Combine $H^+ + OH^-$ to form H_2O on the right side of the equation.

$3 H_2O(l) + Bi^{3+}(aq) + 6 OH^-(aq) \longrightarrow BiO_3^-(aq) + 6 H_2O(aq) + 2 e^-$

Eliminate $3H_2O$ from each side of the equation to get the final balanced equation.

$Bi^{3+}(aq) + 6 OH^-(aq) \longrightarrow BiO_3^-(aq) + 3 H_2O(aq) + 2 e^-$

Then, balance the equation $Br_2(aq) \longrightarrow BrO_3^-(aq)$.

1. Balance the equation as if H^+ were present:

   Balance Br.

   $Br_2(aq) \longrightarrow 2 BrO_3^-(aq)$

   Balance O with H_2O.

   $6 H_2O(l) + Br_2(aq) \longrightarrow 2 BrO_3^-(aq)$

   Balance H using H^+, and charge with electrons.

   $6 H_2O(l) + Br_2(aq) \longrightarrow 2 BrO_3^-(aq) + 12 H^+(aq) + 10 e^-$

2. Since we know that the reaction takes place in a basic solution, we must add $12 OH^-$ to both sides of the equation. This will neutralize the $12 H^+$ on the right side.

   $6 H_2O(l) + Br_2(aq) + 12 OH^-(aq) \longrightarrow 2 BrO_3^-(aq) + 12 H^+(aq) + 12 OH^-(aq) + 10 e^-$

   Combine $H^+ + OH^-$ to form H_2O on the right side of the equation.

   $6 H_2O(l) + Br_2(aq) + 12 OH^-(aq) \longrightarrow 2 BrO_3^-(aq) + 12 H_2O(aq) + 10 e^-$

   Eliminate $3 H_2O$ from each side of the equation to get the final balanced equation.

   $Br_2(aq) + 12 OH^-(aq) \longrightarrow 2 BrO_3^-(aq) + 6 H_2O(aq) + 10 e^-$

Then, balance the equation $CrO_4^{2-}(aq) \longrightarrow Cr(OH)_4^-(aq)$.

1. Balance the equation as if H^+ were present:

   Cr and O are balanced.

   $CrO_4^{2-}(aq) \longrightarrow Cr(OH)_4^-(aq)$

   Balance H with H^+. Also, balance the charge.

   $CrO_4^{2-}(aq) + 4 H^+(aq) + 3 e^- \longrightarrow Cr(OH)_4^-(aq)$

2. We know that the reaction takes place in a basic solution. Therefore we must add $4 OH^-$ to both sides of the equation. This will neutralize the $12 H^+$ on the left side.

   $CrO_4^{2-}(aq) + 4 H^+(aq) + 4 OH^-(aq) + 3 e^- \longrightarrow Cr(OH)_4^-(aq) + 4 OH^-(aq)$

   Combine H^+ and OH^- to form H_2O. This yields the balanced equation.

   $CrO_4^{2-}(aq) + 4 H_2O(aq) + 3 e^- \longrightarrow Cr(OH)_4^-(aq) + 4 OH^-(aq)$

Example 22.11

Balance the following net ionic reaction in basic solution using the half-reaction method.

$$MnO_2(s) + Zn(s) \longrightarrow MnO(OH)(s) + Zn(OH)_4^{2-}(aq)$$

Solution

Step 1: Balance the reduction half-reaction as if it occurs in acidic solution:

$$MnO_2\,(s) \longrightarrow MnO\,(OH)\,(s)$$

Mn and O are both balanced.

Balance H with H^+.

$$MnO_2\,(s) + H^+\,(aq) \longrightarrow MnO\,(OH)\,(s)$$

Balance charge with electrons.

$$MnO_2\,(s) + H^+\,(aq) + e^- \longrightarrow MnO\,(OH)\,(s)$$

Step 2: Balance the oxidation half-reaction as if it occurs in acidic solution:

$$Zn\,(s) \longrightarrow Zn(OH)_4^{2-}\,(aq)$$

Zn is balanced.

Balance O with H_2O and H with H^+.

$$Zn\,(s) + 4\,H_2O\,(l) \longrightarrow Zn(OH)_4^{2-}\,(aq) + 4\,H^+\,(aq)$$

Balance charge with electrons.

$$Zn\,(s) + 4\,H_2O\,(l) \longrightarrow Zn(OH)_4^{2-}\,(aq) + 4\,H^+\,(aq) + 2\,e^-$$

Step 3: Multiply the balanced reduction half-equation by 2 to make the number of electrons transferred equal.

$$2\left[MnO_2\,(s) + H^+\,(aq) + e^- \longrightarrow MnO\,(OH)\,(s)\right]$$

$$2\,MnO_2\,(s) + 2\,H^+\,(aq) + 2\,e^- \longrightarrow 2\,MnO\,(OH)\,(s)$$

Step 4: Add the two half-reactions and eliminate any species that appears on both sides of the equation. This yields the balanced equation as if it were in acidic solution.

$$2\,MnO_2(s) + 2\,H^+(aq) + 2\,e^- \longrightarrow 2\,MnO(OH)(s)$$

$$Zn(s) + 4\,H_2O(l) \longrightarrow Zn(OH)_4^{2-}\,(aq) + 4\,H^+(aq) + 2\,e^-$$

$$Zn(s) + 4\,H_2O(l) + 2\,MnO_2(s) \longrightarrow Zn(OH)_4^{2-}\,(aq) + 2\,H^+(aq) + 2\,MnO(OH)(s)$$

Step 5: Since the reaction occurs in alkaline solution, we must add $2\,OH^-$ to both sides of the equation to neutralize the $2\,H^+$.

$$Zn(s) + 4\,H_2O(l) + 2\,MnO_2(s) + 2\,OH^-\,(aq) \longrightarrow$$
$$Zn(OH)_4^{2-}\,(aq) + 2\,H^+(aq) + 2\,OH^-\,(aq) + 2\,MnO(OH)(s)$$

Combine $2\,H^+$ and $2\,OH^-$ to form water

$$Zn(s) + 4\,H_2O(l) + 2\,MnO_2(s) + 2\,OH^-\,(aq) \longrightarrow$$
$$Zn(OH)_4^{2-}\,(aq) + 2\,H_2O(aq) + 2\,MnO(OH)(s)$$

Eliminate $2\,H_2O$ from both sides. This gives the balanced equation.

$$Zn(s) + 2\,H_2O(l) + 2\,MnO_2(s) + 2\,OH^-\,(aq) \longrightarrow$$
$$Zn(OH)_4^{2-}\,(aq) + 2\,MnO(OH)(s)$$

22.9 Oxidation-Reduction Titration

A redox titration is a volumetric analysis method just like an acid–base titration, except that it is based on a redox reaction in which electrons are transferred to an oxidizing agent from a reducing agent. The substance being titrated needs to be 100% in a single oxidation state. Where this is not the case, the substance has to be oxidized or reduced by adding excess reducing or oxidizing agent before initiating the titration. (In this case it is important to remove or destroy the excess agent before the titration to avoid interference.)

As in acid–base titration, the equivalence point of a redox titration can be determined visually, by the color change of an indicator, or graphically (from titration curves), by plotting the electrode potential measured by an electrode versus the volume of titrant added. In some cases, the redox reaction is self-indicating, and no external indicator is required because the analyte (titrant) changes color intensely at the equivalence point. An example is the permanganate ion, MnO_4^-, whose color changes from purple to pink as it is reduced to Mn^{2+}. If the titrant is not self-indicating, then a redox indicator is added.

Many inorganic and organic substances can be determined via redox titration, including H_2O_2, I_2, Fe^{+3}, MnO_4^-, Sn^{+2}, phenol, aniline, carboxylic acid, aldehydes, and ketones.

22.9.1 Calculations involving redox titration

Here we only deal with basic calculations involving standardization experiments and the determination of an unknown.

Example 22.12

Standardized solutions of potassium permanganate ($KMnO_4$) are frequently used in redox titration. A solution of $KMnO_4$ is standardized by titration with oxalic acid ($H_2C_2O_4$) solution prepared by dissolving 0.260 g of it in 250 mL of 0.5 M H_2SO_4. The balanced equation for the reaction is:

$$5\,H_2C_2O_4\,(aq) + 2\,MnO_4^-\,(aq) + 6\,H^+\,(aq) \longrightarrow 10\,CO_2\,(g) + 2\,Mn^{2+}\,(aq) + 8\,H_2O\,(l)$$

What is the molar concentration of the permanganate solution if 22.40 mL of it were required to react completely with the oxalic acid solution?

Solution

1. First, we need to calculate the number of moles of oxalic acid that has reacted with MnO_4^- We can use the gram-to-mole conversion factor to carry out this calculation.

$$\text{Moles of } H_2C_2O_4 = (0.260 \text{ g}) \left(\frac{1 \text{ mol } H_2C_2O_4}{90 \text{ g } H_2C_2O_4} \right) = 0.00289 \text{ mol}$$

2. Calculate the number of moles of $KMnO_4$ present in 22.40 mL of solution used to reach the end point.

From the balanced equation, 5 mol $H_2C_2O_4$ = 2 mol $KMnO_4$. The conversion factor is :

$$\frac{2 \text{ mol } KMnO_4}{5 \text{ mol } H_2C_2O_4}$$

$$\text{moles KMnO}_4 = (\text{mol H}_2\text{C}_2\text{O}_4) \left(\frac{2 \text{ mol KMnO}_4}{5 \text{ mol H}_2\text{C}_2\text{O}_4} \right)$$

$$= (0.00289 \text{ mol H}_2\text{C}_2\text{O}_4) \times \left(\frac{2 \text{ mol KMnO}_4}{5 \text{ mol H}_2\text{C}_2\text{O}_4} \right) = 0.00116 \text{ mol}$$

3. Calculate the molarity of the permanganate solution from the end-point volume and the number of moles.

$$\text{Molarity} = \frac{\text{number of moles}}{\text{liters of solution}} = \frac{0.0116 \text{ mol}}{0.0224 \text{ L}} = 0.52 \text{ M}$$

Example 22.13

The reagent $KMnO_4$ acts as its own indicator when used in titration. A standard 0.05 M solution of $KMnO_4$ was titrated with 50 mL of 0.20 M acidified $FeSO_4$ solution. Calculate the volume of $KMnO_4$ solution required to completely oxidize the $FeSO_4$ solution.

Solution

We can solve this problem two ways: the conversion factor method and the formula method.

22.9.1.1 The conversion factor method

1. Write a balanced equation for the reaction.

$$MnO_4^- (aq) + 8 H^+ (aq) + 5 Fe^{2+}(aq) \longrightarrow 5 Fe^{3+}(aq) + Mn^{2+}(aq) + 4 H_2O (l)$$

2. Calculate the number of moles of Fe^{2+} titrated from its molarity and volume used.

$$\text{mol Fe}^{2+} = (50 \text{ mL}) \left(\frac{0.20 \text{ mol}}{1000 \text{ mL}} \right) = 0.010 \text{ mol}$$

3. Using the balanced equation, calculate the moles of MnO_4^- required.

$$1 \text{ mol MnO}_4^- = 5 \text{ mol Fe}^{2+}$$

$$\text{Conversion factor} = \frac{1 \text{ mol MnO}_4^-}{5 \text{ mol Fe}^{2+}}$$

$$\text{mol MnO}_4^- = \left(0.010 \text{ mol Fe}^{2+} \right) \left(\frac{1 \text{ mol MnO}_4^-}{5 \text{ mol Fe}^{2+}} \right) = 0.0020 \text{ mol}$$

4. Now, calculate the volume of 0.05 M MnO_4^- which contains 0.0020 mol MnO_4^-.

$$\text{mL of MnO}_4^- (aq) = (0.0020 \text{ mol MnO}_4^-) \left(\frac{1000 \text{ mL}}{0.05 \text{ mol}} \right) = 40 \text{ mL}$$

22.9.1.2 The formula method

This problem can be solved using the formula method.

$$MnO_4^- (aq) + 8 H^+ (aq) + 5 Fe^{2+}(aq) \longrightarrow 5 Fe^{3+}(aq) + Mn^{2+}(aq) + 4 H_2O (l)$$

From the balanced equation,

$$\frac{M_{MnO_4^-} V_{MnO_4^-}}{M_{Fe^{2+}} V_{Fe^{2+}}} = \frac{n_{MnO_4^-}}{n_{Fe^{2+}}} = \frac{1}{5}$$

$$V_{MnO_4^-} = \frac{M_{Fe^{2+}} V_{Fe^{2+}} n_{MnO_4^-}}{M_{MnO_4^-} n_{Fe^{2+}}} = \frac{50 \text{ mL} \times 0.20 \text{ M} \times 1}{0.05 \text{ M} \times 5} = 40 \text{ mL}$$

Example 22.14

Exactly 2.05 g of an iron ore was dissolved in an acid solution and all the Fe^{3+} was converted to Fe^{2+}. The resulting solution required 45 mL of 0.05 M $K_2Cr_2O_8$ solution for oxidation.

(a) Write a balanced equation for the redox reaction.
(b) Calculate the mass of Fe in the ore.
(c) What is the percentage of Fe in the original sample?

Solution

(a) The balanced equation is

$$6 \text{ Fe}^{2+}(aq) + Cr_2O_7^{2-}(aq) + 14 \text{ H}^+(aq) \longrightarrow 6 \text{ Fe}^{3+}(aq) + 2 \text{ Cr}^{3+}(aq) + 7 \text{ H}_2O(l)$$

(b) To calculate the mass of Fe, we need to calculate the moles of Fe reacted. This can readily be obtained from the moles of $Cr_2O_7^{2-}$ used for the titration.

$$\text{mol. } Cr_2O_7^{2-} = (45 \text{ mL}) \left(\frac{0.05 \text{ mol}}{1000 \text{ mL}} \right) = 0.0023 \text{ mol}.$$

From the balanced equation and from the moles of $Cr_2O_7^{2-}$, we can calculate moles of Fe:

$$2 \text{ mol } Cr_2O_6^{2-} = 6 \text{ mol Fe}^{2+}$$

$$\text{Conversion factor} = \frac{6 \text{ mol Fe}^{2+}}{2 \text{ mol } Cr_2O_6^{2-}}$$

$$\text{mol Fe}^{2+} = \left(0.00230 \text{ mol } Cr_2O_6^{2-} \right) \left(\frac{6 \text{ mol Fe}^{2+}}{2 \text{ mol } Cr_2O_6^{2-}} \right) = 0.0068 \text{ mol}$$

$$\text{Mass of Fe} = \left(55.85 \frac{g}{mol} \right) (0.0068 \text{ mol}) \, 0.3770 \text{ g}$$

(c) The percentage of Fe in the original sample is obtained as follows:

$$\% \text{ Fe} = \frac{\text{mass of Fe}}{\text{mass of sample}} \times 100\%$$

$$= \frac{0.3770}{2.05} \times 100\% = 18.4\%$$

Example 22.15

1000 mg of a new calcium supplement tablet for patients suffering calcium deficiency was dissolved in acid and treated with excess sodium oxalate. The resulting solution was made basic, causing the precipitation of the Ca as the oxalate, CaC_2O_4. The precipitate was filtered, washed, and redissolved in dilute acid. The resulting solution was titrated by 28.50 mL of 0.05 M $KMnO_4$ solution. Determine the calcium content in the tablet.
The equations for the reactions are:

$$CaC_2O_4 \text{ (s)} + 2H^+ \text{ (aq)} \longrightarrow Ca^{2+} \text{ (aq)} + H_2C_2O_4 \text{ (aq)}$$

$$2\,MnO_4^- \text{ (aq)} + 5\,H_2C_2O_4 \text{ (aq)} + 6H^+ \text{ (aq)} \longrightarrow 2\,Mn^{2+} \text{ (aq)} + 10CO_2 \text{ (g)} + 8H_2O \text{ (l)}$$

Solution

This is similar to calculations involving back titration (Section 14.5). Using the equation showing the reaction between the permanganate ion and oxalic acid, we can calculate the moles of oxalic acid that reacted. From the equation showing the production of oxalic acid from the dissolution of calcium oxalate, we can determine the moles of calcium oxalate. Since one mole of calcium oxalate contains one mole of calcium, the moles, and hence the mass of calcium, can be calculated.

1. First, calculate the moles of oxalic acid.

$$2\,MnO_4^- \text{ (aq)} + 5\,H_2C_2O_4 \text{ (aq)} + 6\,H^+ \text{ (aq)} \longrightarrow 2\,Mn^{2+} \text{ (aq)} + 10\,CO_2 \text{ (g)} + 8\,H_2O(1)$$

$$\text{Conversion factor}: \frac{5 \text{ mol } H_2C_2O_4}{2 \text{ mol } MnO_4^-}$$

$$\text{mol of } H_2C_2O_4 = \left(28.50 \text{ mL } MnO_4^-\right) \left(0.05\frac{\text{mol } MnO_4^-}{1000 \text{ mL } MnO_4^-}\right) \left(\frac{5 \text{ mol } H_2C_2O_4}{2 \text{ mol } MnO_4^-}\right)$$

$$= 0.0036 \text{ mol}$$

2. Then, calculate the moles of calcium oxalate, and then moles of Ca.

$$CaC_2O_4 \text{ (s)} + 2\,H + \text{(aq)} \longrightarrow Ca^{2+} \text{ (aq)} + H_2C_2O_4 \text{ (aq)}$$

$$\text{Conversion factor}: \frac{1 \text{ mol } CaC_2O_4}{1 \text{ mol } H_2C_2O_4}$$

$$\text{mol of } CaC_2O_4 = (0.0036 \text{ mol } H_2C_2O_4)\left(\frac{1 \text{mol } CaC_2O_4}{1 \text{mol } H_2C_2O_4}\right) = 0.0036 \text{ mol}$$

$$\text{mol of } Ca = (0.0036 \text{ mol } CaC_2O_4)\left(\frac{1 \text{ mol } Ca}{1 \text{ mol } CaC_2O_4}\right) = 0.0036 \text{ mol}$$

3. Last, calculate the mass of Ca and the percentage of Ca in the sample.

$$\text{Mass of Ca} = (0.0036 \text{ mol } Ca)\left(\frac{40.08 \text{ g Ca}}{1 \text{ mol } Ca}\right) = 0.1428 \text{ g or } 142.8 \text{ mg Ca}$$

$$\% \text{ Ca} = \frac{\text{mass of Ca}}{\text{mass of sample}} \times 100 = \frac{142.8}{1000 \text{ mg}} \times 100 = 14.28$$

22.10 Problems

1. Calculate the ON for the element indicated in each of the following compounds or ions.

   (a) Si in $CaMg(SiO_3)_2$
   (b) Xe in Na_4XeO_6
   (c) Mn in Mn_2O_7
   (d) Fe in $FePO_4$
   (e) Cl in $Ca(ClO_2)_2$
   (f) O in $SOCl_2$

2. Calculate the ON for the element indicated in each of the following compounds or ions.

   (a) Cr in $Cr_2O_7^{2-}$
   (b) S in $S_4O_6^{4-}$
   (c) V in $HV_6O_{17}^{3-}$
   (d) As in $H_2AsO_4^-$
   (e) Te in $HTeO_3^-$
   (f) Sn in $Sn(OH)_6^{2-}$

3. Consider the following reactions and determine which substance is oxidized and which substance is reduced.

   (a) $4Fe + 3O_2 \longrightarrow 2Fe_2O_3$
   (b) $Cl_2 + 2NaI \longrightarrow NaCl + I_2$
   (c) $Mg + H_2SO_4 \longrightarrow MgSO_4 + H_2$
   (d) $CuO + H_2 \longrightarrow Cu + H_2O$

4. For each of the following reactions, determine which substance is oxidized and which is reduced.

   (a) $3MnO_2 + 4Al \longrightarrow 2Al_2O_3 + 3Mn$
   (b) $4PH_3 + 8O_2 \longrightarrow P_4O_{10} + 6H_2O$
   (c) $Ca + 2H_2O \longrightarrow Ca(OH)_2 + H_2$
   (d) $2SO_2 + O_2 \longrightarrow 2SO_3$

5. Consider the following reactions and determine which substance is oxidized and which substance is reduced.

   (a) $2FeCl_2(s) + Cl_2(g) \longrightarrow 2FeCl_3(g)$
   (b) $Cl_2(g) + H_2S(g) \longrightarrow 2HCl(g) + S(s)$
   (c) $C(s) + MgO(s) \longrightarrow CO_2(g) + Mg(s)$
   (d) $2Zn(s) + O_2 \longrightarrow 2ZnO(s)$

6. For each of the following reactions, identify the substance that is oxidized, the substance that is reduced, the oxidizing agent, and the reducing agent.

   (a) $3Cu(s) + 8HNO_3(aq) \longrightarrow 3Cu(NO_3)_2(aq) + 4H_2O(l) + 2NO(g)$
   (b) $Zn(s) + CuSO_4(aq) \longrightarrow ZnSO_4(aq) + Cu(s)$
   (c) $2H_2O(l) + 2F_2(g) \longrightarrow 4HF(aq) + O_2(g)$
   (d) $Fe_2O_3(s) + 3CO(g) \longrightarrow 2Fe(s) + 3CO_2(g)$

7. Which of the following are redox reactions? Explain your answer in each case.
  (a) $Ca(OH)_2(s) + HNO_3(aq) \longrightarrow Ca(NO_3)_2 + H_2O(l)$
  (b) $Si(s) + 2Cl_2(g) \longrightarrow SiCl_4(l)$
  (c) $NiCl_2(aq) + Li_2S(aq) \longrightarrow NiS(s) + 2LiCl(aq)$
  (d) $HClO_3(aq) + NH_3(aq) \longrightarrow NH_4ClO_4(aq)$

8. Write a balanced half-reaction for each of the following redox couples in acidic solution.
  (a) $IO_3^- \longrightarrow I_2$
  (b) $O_2 \longrightarrow H_2O_2$
  (c) $S \longrightarrow S_2O_3^{2-}$
  (d) $S_4O_6^{2-} \longrightarrow S_2O_3^{2-}$
  (e) $VO_2^+ \longrightarrow VO^{2+}$
  (f) $CO_2 \longrightarrow H_2C_2O_4$

9. Write a balanced half-reaction for each of the following redox couples in acidic solution.
  (a) $NO_3^- \longrightarrow N_2O_4$
  (b) $Sb_2O_3 \longrightarrow Sb$
  (c) $MnO_2 \longrightarrow Mn^{2+}$
  (d) $HClO \longrightarrow Cl_2$
  (e) $H_2O_2 \longrightarrow H_2O$
  (f) $NO_3^- \longrightarrow HNO_2$

10. Write a balanced half-reaction for each of the following redox couples in alkaline solution.
  (a) $Cu_2O \longrightarrow Cu$
  (b) $MnO_2 \longrightarrow Mn(OH)_2$
  (c) $PbO_2 \longrightarrow PbO$
  (d) $ClO^- \longrightarrow Cl^-$
  (e) $MnO_4^- \longrightarrow MnO_2$
  (f) $Al(OH)_3 \longrightarrow Al$
  (g) $O_3 \longrightarrow O_2$
  (h) $BrO_3^- \longrightarrow Br^-$

11. Balance the following equations in acidic solution using the half-reaction method.
  (a) $Mg + NO_3^- \longrightarrow Mg^{2+} + NO_2$
  (b) $NO_2^- + Cr_2O_7^{2-} \longrightarrow NO_3^- + Cr^{3+}$
  (c) $S^{2-} + NO_3^- \longrightarrow NO + S$

12. Balance the following equations in acidic solution using the half-reaction method.
  (a) $H_2C_2O_4 + MnO_4^- \longrightarrow Mn^{2+} + CO_2$
  (b) $CH_3CH_2OH + Cr_2O_7^{2-} \longrightarrow CH_3COOH + Cr^{3+}$
  (c) $Ca + VO^{2+} \longrightarrow Ca^{2+} + V^{3+}$

13. Balance the following equations in basic solution using the half-reaction method.
  (a) $Zn + MnO_4^- \longrightarrow Zn(OH)_2 + MnO_2$
  (b) $CrO_2^- + IO_3^- \longrightarrow I^- + CrO_4^-$
  (c) $Bi(OH)_3 + Sn(OH)_3^- \longrightarrow Bi + Sn(OH)_6^{2-}$
  (d) $I_2 + Cl_2 \longrightarrow H_3IO_6^{2-} + Cl^-$

14. Balance the following equations in basic solution using the half-reaction method.

(a) $MnO_4^- + C_3H_8O_3 \longrightarrow MnO_4^{2-} + CO_3^{2-}$
(b) $MnO_4^- + SO_3^{2-} \longrightarrow MnO_4^{2-} + SO_4^{2-}$
(c) $MnO_4^- + Br^- \longrightarrow MnO_2 + BrO_3^-$

15. Using the oxidation-number method, balance the following equations.

(a) $Cr(OH)_4^- + H_2O_2 \longrightarrow CrO_4^{2-} + H_2O$ (in basic solution)
(b) $H_2S + Na_2Cr_2O_7 + HCl \longrightarrow S + CrCl_3 + H_2O$
(c) $Sb_2S_5 + NO_3^- + H^+ \longrightarrow HSbO_3 + S + NO + H_2O$

16. Balance the following equations using any method.

(a) $Zn + NO_3^- + H^+ \longrightarrow Zn^{2+} + NH_4^+ + H_2O$
(b) $MnO_4^- + Cl^- + H^+ \longrightarrow Mn^{2+} + Cl_2 + H_2O$
(c) $PtCl_6^{2-} + Sb + H_2O \longrightarrow PtCl_4^{2-} + Cl^- + Sb_2O_3 + H^+$
(d) $Hg_2Cl_2 + I^- + OH^- \longrightarrow Hg + Cl + IO_3^- + H_2O$
(e) $K_2S_2O_8 + Cr_2(SO_4)_3 + H_2O \longrightarrow K_2SO_4 + K_2Cr_2O_7 + H_2SO_4$
(f) $KMnO_4 + Zn + H_2SO_4 \longrightarrow MnSO_4 + ZnSO_4 + H_2O + K_2SO_4$

17. A solution of acidified $KMnO_4$ is standardized with Fe^{2+} solution as the primary standard. To prepare the Fe^{2+} solution, 0.3554 g of pure iron metal was dissolved in acid and then reduced to Fe^{2+}. The iron solution was titrated with permanganate and required 23.50 cm^3 of $KMnO_4$ solution to reach the end point.

(a) Write a balanced ionic equation for the reaction.
(b) What is the molar concentration of the permanganate solution?

18. 25.00 cm^3 of oxalic acid solution was titrated with acidified $KMnO_4$ solution. If 21.75 cm^3 of the 0.05 M $KMnO_4$ solution was required to reach the first end point, what was the molar concentration of the oxalic acid solution?

$$2KMnO_4 + 5H_2C_2O_4 + 3H_2SO_4 \longrightarrow 2MnSO_4 + K_2SO_4 + 10CO_2 + 8H_2O$$

19. A ceric ion (Ce^{+4}) solution was standardized with a solution of H_3AsO_3 as the primary standard. The H_3AsO_3 solution was prepared from 480.5 mg of primary standard As_2O_3 and required 24.95 cm^3 of the cerium solution to reach end point.

(a) Write a balanced equation for the reaction.
(b) What is the molar concentration of the cerium solution?

20. 9.55 g of fresh beef was digested in excess concentrated sulfuric acid, and the solution made up to 250 cm^3 in a standard flask. 25.00 cm^3 of this solution required 22.50 cm^3 of 0.025 M $KMnO_4$ to reach the endpoint. Calculate the percentage of iron in the beef sample (note that the chemistry here is similar to that in Problem 17).

21. An industrial chemistry student doing his summer internship job at the Delta Steel Company was given a sample of iron ore (0.500g of Fe_2O_3) to analyze for the percentage of Fe_2O_3 in the ore. He dissolved the ore sample in boiling dilute nitric acid and evaporated to dryness. He collected the residue and redissolved it in dilute sulfuric acid, filtered off the insoluble residue, and passed it through a Walden silver reductor. The

determination is completed by titrating a sample of the resulting solution with aqueous $KMnO_4$. The solution required 17.5 cm^3 of 0.025 M $KMnO_4$ to reach the end point. Estimate the percentage of Fe_2O_3 in the ore. You may use the following reaction.

$$5Fe^{2+} + MnO_4^- + 8H^+ \longrightarrow 5Fe^{3+} + Mn^{2+} + 4H_2O$$

22. Vitamin C, or ascorbic acid ($C_6H_8O_6$), can be found in citrus fruits, fresh vegetables, tomatoes, and potatoes. It was first isolated in its pure form in 1926 by Azent-Gyorgi and King. The oxidation of ascorbic acid by iodine or iodate salts is rapid and quantitative. Therefore, the exact amount of vitamin C in an unknown sample can be determined by monitoring the amount of iodate used to oxidize a sample containing vitamin C, or by oxidizing the sample with excess iodine followed by titrating the unreacted iodine with sodium thiosulfate. In pill form, vitamin C is often compounded with inactive ingredients. A student was asked to determine the ascorbic acid content of a vitamin C sample by using the iodometric titration method. The equations occurring during the oxidation of the acid and titration of the excess iodine are:

(a) $C_6H_8O_6 + I_2 \longrightarrow C_6H_6O_6 + 2H^+ + 2I^-$
(b) $I_2 + 2 S_2O_3^{2-} \longrightarrow 2I^- + S_4O_6^{2-}$

25.00 cm^3 of 0.125 M I_2 was used to oxidize a tablet of vitamin C weighing 0.2728 g. At the end of the reaction, the excess iodine required 23.55 cm^3 of 0.200 M Na_2 S_2O_3 for the titration to reach the blue starch-iodine equivalence point. Calculate the percent of ascorbic acid in the tablet sample.

23

Fundamentals of Electrochemistry

. . . .

Electrochemistry is the branch of chemistry that deals with the interconversion of chemical and electrical energy.

23.1 Galvanic Cells

A *galvanic* (or voltaic) *cell* is a chemical system that uses an oxidation-reduction reaction to convert chemical energy into electrical energy (hence it is also known as an electrochemical cell). This process is the opposite of electrolysis (explained in Section 23.10), wherein electrical energy is used to bring about chemical changes. The two systems are similar in that both are redox processes; in both, the oxidation takes place at an electrode called the *anode,* while reduction occurs at the *cathode*. Figure 23.1 represents a galvanic cell, showing the half-reactions at the two electrodes. Electrons flow through the external circuit from the anode (Zn) to the cathode (Cu).

The overall reaction, which is obtained by adding the anodic and cathodic half-cell reactions, is

$$Zn\,(s) + Cu^{2+}\,(aq) \longrightarrow Zn^{2+}\,(aq) + Cu\,(s)$$

This cell has a voltage of 1.10 V (see Section 23.2).

23.2 The Cell Potential

The potential energy of electrons at the anode is higher than at the cathode. This difference in potential is the driving force that propels electrons through the external circuit. The cell potential (E_{cell}) is a measure of the potential difference between the two half-cells. It is also known as the electromotive force (emf) of the cell, or, since it is measured in volts, as the cell voltage.

By definition, one volt (V) is the potential difference required to give one joule of energy (1 J) to a charge of one coulomb (C):

$$1\,V = \frac{1J}{1C}, \quad \text{or} \quad 1\,C = \frac{1J}{1V}$$

Chemistry in Quantitative Language: Fundamentals of General Chemistry Calculations. Second Edition.
Christopher O. Oriakhi, Oxford University Press. © Christopher O. Oriakhi 2021.
DOI: 10.1093/oso/9780198867784.003.0023

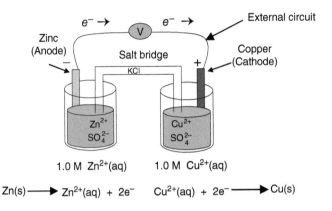

$$Zn(s) \longrightarrow Zn^{2+}(aq) + 2e^- \qquad Cu^{2+}(aq) + 2e^- \longrightarrow Cu(s)$$

Figure 23.1 A galvanic cell showing zinc and copper half-reactions at the electrodes.

The following relationships may be useful in solving electrochemistry problems.

$$1 \text{ faraday (F)} = 96500 \text{ C} = \text{charge on 1 mol of electrons}$$

$$1 \text{ faraday} = 96.5 \frac{kJ}{V \text{ mol } e^-}$$

Hence,

$$96500 \frac{C}{\text{mol } e^-} = 96.5 \frac{kJ}{V \text{ mol } e^-}$$

23.3 Standard Electrode Potential

An electrochemical cell consists of two half-reactions at different potentials, which are known as *electrode potentials*. The electrode potential for the oxidation half-reaction is called the oxidation potential. Similarly, for the reduction half-reaction, we have the reduction potential. The potential of a galvanic cell is determined by the concentrations of the species in solution, the partial pressure of any gaseous reactants or products, and the reaction temperature. When the electrochemical measurement is carried out under standard state conditions, the cell potential is called the *standard electrode potential* and is given the symbol E^o. The standard conditions include a concentration of 1 M, gaseous partial pressure of 1 atm, and a temperature of 25°C.

It is impossible to measure the absolute potential value of a single electrode since every oxidation is accompanied by a reduction. Therefore any measurement is carried out against a reference electrode. The reference electrode that has been selected is the standard hydrogen electrode (SHE). It consists of platinum metal immersed in a 1.0 M H^+ solution. Hydrogen gas at a pressure of 1 atm is bubbled over the platinized electrode. By convention, the SHE is assigned a potential of 0.00 V. The cell reaction is shown below:

$$H_2 \longrightarrow 2H^+ + 2e^- \qquad E^0 = 0.0000 \text{ V (SHE anode)}$$

$$2H^+ + 2e^- \longrightarrow H_2 \qquad E^0 = 0.0000 \text{ V (SHE cathode)}$$

$$\overline{2H^+ + H_2 \longrightarrow 2H^+ + H_2 \qquad E^0 = 0.0000 \text{ V}}$$

Note that when a cell is made up of an SHE and some other standard electrode, the standard cell potential, E^0_{cell}, is simply the standard potential of the other electrode. For example, consider the copper-SHE cell. The SHE half-cell forms the anode with a potential of 0.000 V. The copper half-cell, with a potential of 0.337 V, serves as the cathode. The reaction that occurs in the cell is

$$H_2 \longrightarrow 2\,H^+ + 2e^- \text{(Anode, oxidation)} \qquad E^0 = 0.000 \text{ V}$$

$$Cu^{2+} + 2e^- \longrightarrow Cu \text{ (Cathode, reduction)} \qquad E^0 = 0.337 \text{ V}$$

$$H_2 + Cu^{2+} \longrightarrow 2H^+ + Cu \text{ (Net Cell reaction)} \quad E^0 = 0.337 \text{ V}$$

23.3.1 Standard reduction potential

The *standard reduction potential* of any electrode is defined as the electromotive force of a cell at 25°C wherein the SHE is the anode, and the given electrode is the cathode. The symbol E^0_{red} represents the standard reduction potential, while E^0_{ox} represent the standard reduction potential. For any given half-cell,

$$E^0_{red} = -E^0_{ox}$$

For example, E^0_{red} for the half-cell $Fe^{2+}\,(aq) + 2e^- \longrightarrow Fe\,(s)$ is -0.44 V. Therefore, the E^0_{ox} value will be $+0.44$ V. Values of many standard reduction potentials have been measured. These results, sometimes called the electrochemical series, can be found in most general chemistry textbooks. Representative examples are given in Table 23.1.

23.4 The Electrochemical Series (ECS)

The *electrochemical series* is the arrangement of the standard electrode potentials (for half-cell reactions) in order from the most negative to the most positive or vice-versa. It may also be defined as the arrangement of elements in order of increasing oxidizing or reducing power. Table 23.1 shows the standard electrode potentials of some elements at 25°C. This table may also be called a standard reduction potential table because it is arranged in order of decreasing reducing properties. The higher an element is on this table, the more readily it loses electrons and the stronger a reducing agent it is.

23.5 Applications of Electrode Potential

The electrochemical series has several uses. The most important ones are:

- *Predicting the feasibility of reactions*: Two half-equations can be combined to give a desired net reaction. The general rule is to write the half-reaction lower in the ECS as given in the standard reduction potential table. The one higher in the series will only go in the reverse direction. Therefore, write the reverse form (oxidation) of that half-reaction and then add the two equations to obtain the overall reaction and its potential.

Table 23.1   Some selected standard electrode potentials at 25°C

Reduction half-reaction	E^0, V
Acidic Solution	
$Li^+(aq) + e^- \rightarrow Li(s)$	−3.04
$K^+(aq) + e^- \rightarrow K(s)$	−2.924
$Ca^{2+}(aq) + 2e^- \rightarrow Ca(s)$	−2.84
$Na^+(aq) + e^- \rightarrow Na(s)$	−2.713
$Mg^{2+}(aq) + 2e^- \rightarrow Mg(s)$	−2.356
$Al^{3+}(aq) + 3e^- \rightarrow Al(s)$	−1.676
$V^{2+}(aq) + 2e^- \rightarrow V(s)$	−1.18
$Zn^{2+}(aq) + 2e^- \rightarrow Zn(s)$	−0.763
$Fe^{2+}(aq) + 2e^- \rightarrow Fe(s)$	−0.44
$Cd^{2+}(aq) + 2e^- \rightarrow Cd(s)$	−0.4
$Ni^{2+}(aq) + 2e^- \rightarrow Ni(s)$	−0.25
$Co^{2+}(aq) + 2e^- \rightarrow Co(s)$	−0.28
$Sn^{2+}(aq) + 2e^- \rightarrow Sn(s)$	−0.137
$Pb^{2+}(aq) + 2e^- \rightarrow Pb(s)$	−0.125
$2H^+(aq) + 2e^- \rightarrow H_2(g)$	0
$Sn^{4+}(aq) + 2e^- \rightarrow Sn^{2+}(aq)$	0.13
$S(s) + 2H^+(aq) + 2e^- \rightarrow H_2S(g)$	0.14
$Cu^{2+}(aq) + 2e^- \rightarrow Cu(s)$	0.337
$I_2(s) + 2e^- \rightarrow 2\,I^-(aq)$	0.535
$O_2(g) + 2H^+(aq) + 2e^- \rightarrow H_2O_2(aq)$	0.695
$Fe^{3+}(aq) + e^- \rightarrow Fe^{2+}(aq)$	0.771
$Ag^+(aq) + e^- \rightarrow Ag(s)$	0.8
$Br_2(l) + 2e^- \rightarrow 2\,Br^-(aq)$	1.065
$O_2(g) + 4H^+(aq) + 4e^- \rightarrow 2H_2O(aq)$	1.229
$Cl_2(g) + 2e^- \rightarrow 2\,Cl^-(aq)$	1.358
$MnO_4^-(aq) + 8H^+(aq) + 5e^- \longrightarrow Mn^{2+}(aq) + 4H_2O(l)$	1.52
$F_2(g) + 2e^- \rightarrow 2\,F^-(aq)$	2.866

- *Predicting displacement reactions*: The series can be used to determine which elements will displace each other from solution. Metals higher in the series will displace those lower down since they can readily lose or donate electrons. For example, all metals above hydrogen in the series will displace it from dilute acid.

$$Zn(s) + 2H^+(aq) \longrightarrow Zn^{2+}(aq) + H_2(g) \quad E^0_{cell} = 0.763 \text{ V}$$

Generally the reduced form of an element will reduce the oxidized form of any element below it in the series. Thus calcium can reduce copper ion to copper according to the equation:

$$Ca(s) + Cu^{2+}(aq) \longrightarrow Ca^{2+}(aq) + Cu(s) \quad E^0_{cell} = 3.177 \text{ V}$$

- *Predicting oxidizing and reducing strength*: Those elements with the largest negative E^0_{red} value are stronger reducing agents and most readily lose electrons. On the other hand, elements with largest E^0_{red} values, such as many nonmetals (at the bottom of the series), are stronger oxidizing agents.

- *Predicting the discharge of ions in electrolysis*: During electrolysis of a given electrolyte, more than one ion can migrate to the anode or cathode. The position of an ion in the ECS often helps in determining which ion will be preferentially discharged. Cations lower in the series are generally discharged or deposited in preference to those higher in the series.

23.6 Cell Diagrams

A galvanic cell can be represented schematically as:

$$\text{anode} \,|\, \text{anode electrolyte} \,\|\, \text{cathode electrolyte} \,|\, \text{cathode}$$

1. The anode (oxidation) is placed on the left and the cathode (reduction) is placed on the right side of the diagram.
2. The single vertical line ($|$) indicates a boundary between the different phases in a half-cell.
3. The double vertical line ($\|$) represents a salt bridge separating the two half-cells. It permits the flow of ions between the two cells while preventing them from mixing.
4. Within a half-cell the reactants are listed before the products.
5. The concentrations of ions in solution and the pressure of gases are written in parentheses following the symbol of the ion or molecule. For example, the cell notation for the standard Zn-Cu cell would be $\text{Zn} \,|\, \text{Zn}^{2+} (1\text{M}) \,\|\, \text{Cu}^{2+} (1\text{ M}) \,|\, \text{Cu}$
6. An inert electrode such as platinum or graphite (carbon) is commonly used for an electrode involving a gas like hydrogen or chlorine. The cell notation uses an additional vertical line to denote the extra phase (the inert electrode):

$$\text{Zn(s)} \,|\, \text{Zn}^{2+} \text{(aq)} \,\|\, \text{H}^+ \text{(aq)} \,|\, \text{H}_2 \text{(g)} \,|\, \text{Pt}$$

7. When two ionic species are in the same phase in a cell, commas instead of a vertical line separate them. Neither one can be used as an electrode; hence an inert electrode is employed. This is illustrated by the reaction:

$$\text{Fe(s)} + 2\,\text{Fe}^{3+} \text{(aq)} \longrightarrow 3\,\text{Fe}^{2+} \text{(aq)}$$

The cell notation is $\text{Fe(s)} \,|\, \text{Fe}^{2+} \text{(aq)} \,\|\, \text{Fe}^{3+} \text{(aq)}, \text{Fe}^{2+} \text{(aq)} \,|\, \text{Pt}$

Example 23.1

Write the cell diagram and the cell reaction for the redox couples Ag^+/Ag and Cu^{2+}/Cu.

Solution

1. First, write the half-cell reactions and then determine the net (cell) reaction:

$$\text{Anode:} \quad \text{Cu(s)} \longrightarrow \text{Cu}^{2+} \text{(aq)} + 2\text{e}^-$$

$$\text{Cathode:} \quad 2\,\text{Ag}^+ \text{(aq)} + 2\text{e}^- \longrightarrow 2\,\text{Ag(s)}$$

$$\text{Cell reaction:} \quad 2\,\text{Ag}^+ \text{(aq)} + \text{Cu(s)} \longrightarrow 2\,\text{Ag(s)} + \text{Cu}^{2+} \text{(aq)}$$

2. Next, write the cell diagram following the standard notation.

$$\text{Cu(s)} \left| \text{Cu}^{2+}(\text{aq})(1\text{M}) \right\| \text{Ag}^{+}(\text{aq})(1\text{ M}) \left| \text{Ag(s)} \right.$$

Example 23.2

A galvanic cell was constructed by placing a tin electrode into a tin (II) chloride solution in one chamber, and a platinum electrode into an iron (II) chloride solution in another chamber. The two half-cells were separated by a salt bridge. Write the cell equation and then draw a cell diagram for the cell.

Solution

First write the half-cell reactions and then determine the net (cell) reaction:

$$\text{Anode:} \quad \text{Sn (s)} \longrightarrow \text{Sn}^{2+}(\text{aq}) + 2\text{e}^{-}$$

$$\text{Cathode:} \quad 2\,\text{Fe}^{3+}(\text{aq}) + 2\text{e}^{-} \longrightarrow 2\,\text{Fe}^{2+}(\text{aq})$$

$$\overline{\hspace{8cm}}$$

$$\text{Cell reaction:} \quad 2\,\text{Fe}^{3+}(\text{aq}) + \text{Sn (s)} \longrightarrow 2\,\text{Fe}^{2+}(\text{aq}) + \text{Sn}^{2+}(\text{aq})$$

1. Next write the cell diagram following the standard notation.

$$\text{Sn(s)} \left| \text{Sn}^{2+}(\text{aq}) \right\| \text{Fe}^{2+}(\text{aq}), \text{Fe}^{3+}(\text{aq}) \left| \text{Pt} \right.$$

23.7 Calculating E^0_{cell} from Electrode Potential

The value of E^0_{cell} can be obtained by combining the oxidation half-cell with the reduction half-cell, followed by adding the voltages of the two half-reactions. Generally,

$$E^0_{cell} = E^0(\text{anode}) + E^0(\text{cathode}) \quad \text{or} \quad E^0_{cell} = E^0_{red} + E^0_{ox}$$

The following rules are helpful:

1. Write the oxidation and reduction half-cell reactions.
2. Obtain the standard electrode potential for each half reaction (i.e. E^0_{red} and E^0_{ox}) from the table of potentials. Recall that $E^0_{ox} = -E^0_{red}$.
3. Add the two half-reactions to generate a net cell reaction, and also add the electrode potentials using the equation:

$$E^0_{cell} = E^0_{red} + E^0_{ox}$$

4. Note that changing the coefficients of the half-equations does not affect the values of E^0 and E^0_{cell}.

Example 23.3

Calculate the E^0_{cell} for a proposed cell that uses the following reaction:

$$Mg(s) + Cu^{2+}(aq) \longrightarrow Mg^{2+}(aq) + Cu(s)$$

Solution

1. First, write the half-cell equations for the oxidation and reduction, and their standard electrode potentials, recalling that $E^0_{ox} = -E^0_{red}$.

   Anode: $Mg(s) \longrightarrow Mg^{2+}(aq) + 2e^-$ $E^0_{ox} = +2.363$ V

   Cathode: $Cu^{2+}(aq) + 2e^- \longrightarrow Cu(s)$ $E^0_{red} = +0.337$ V

2. Then, obtain the cell potential using the following equation:

$$E^0_{cell} = E^0_{red} + E^0_{ox}$$
$$E^0_{cell} = +2.363 + 0.337$$
$$= +2.700 \text{ V}$$

Example 23.4

A galvanic cell has an aluminum electrode in 1.0 M aluminum nitrate $(Al(NO_3)_3)$ solution and lead electrode in lead (II) nitrate $(Pb(NO_3)_2)$ solution. Calculate E^0_{cell} for the cell.

Solution

1. First write down the half-cell equations and their standard electrode potentials, recalling that $E^0_{ox} = -E^0_{red}$.

   Anode: $2 Al(s) \longrightarrow 2 Al^{3+}(aq) + 6e^-$ $E^0_{ox} = +1.676$ V

   Cathode: $3 Pb^{2+}(aq) + 6e^- \longrightarrow 3 Pb(s)$ $E^0_{red} = -0.125$ V

2. Obtain the cell potential using the equation

$$E^0_{cell} = E^0_{red} + E^0_{ox}$$
$$E^0_{cell} = +1.676 - 0.125$$
$$= +1.551 \text{ V}$$

23.8 Relationship of the Standard Electrode Potential, the Gibbs Free Energy, and the Equilibrium Constant

The standard Gibbs free energy change, ΔG^0, is related to the thermodynamic equilibrium constant, K, and the standard electrode potential, E^0_{cell}, by the following equations:

$$\Delta G^0 = -2.303RT \log K \quad \text{or} \quad RT \ln K,$$

and

$$\Delta G^0 = -nFE^0_{cell}$$

R = the universal gas constant, T = the Kelvin temperature, n = the moles of electrons involved in the half-reaction, and F = the Faraday constant, which is equal to 96500 C per mole, i.e.,

$$1\,F = 96500\frac{C}{mol} \quad \text{or} \quad 96.5\frac{kJ}{V - mol\ e^{-1}}$$

We can relate E^0_{cell} to the equilibrium constant by combining the two equations for ΔG^0:

$$-nFE^0_{cell} = -RT \ln K \quad \text{or} \quad 2.303 \log K$$

$$\ln K = \frac{nFE^0_{cell}}{RT} \quad \text{or} \quad \log K = \frac{nFE^0_{cell}}{2.303RT}$$

$$E^0_{cell} = \frac{RT}{nF} \ln K \quad \text{or} \quad \frac{2.303RT}{nF} \log K$$

If we substitute $R = 8.314$ J/mol.K, $T = 298$ K, and $F = 96500$C (or J/V.mol e$^-$) into the expression

$$E^0_{cell} = \frac{2.303RT}{nF} \log K,$$

we will obtain a simplified equation for solving problems involving E^0_{cell} and K:

$$E^0_{cell} = \frac{0.0592}{n} \log K$$

23.8.1 Conditions for spontaneous change in redox reactions

Table 23.2 shows the criteria that must hold for a reaction to proceed spontaneously.

Table 23.2   Conditions for spontaneous change in redox reactions

E^0_{cell}	ΔG^0	K	Reaction direction
Positive (+)	Negative (−)	>1	Forward (spontaneous as written)
Zero (0)	Zero(0)	~1	At equilibrium
Negative (−)	Positive (+)	<1	Backward (nonspontaneous as written)

Example 23.5

Calculate ΔG^0 and the equilibrium constant, K, at 25°C for the reaction

$$Ca(s) + Cu^{2+}(aq) \longrightarrow Ca^{2+}(aq) + Cu(s)$$

Solution

1. First, write the half-cell equations and their standard electrode potentials, recalling that $E_{ox}^0 = -E_{red}^0$.

Anode :	$Ca(s) \longrightarrow Ca^{2+}(aq) + 2e^-$		$E_{ox}^0 = +2.87$ V
Cathode :	$Cu^{2+}(aq) + 2e^- \longrightarrow Cu(s)$		$E_{red}^0 = -0.337$ V

2. Next, obtain the cell potential using the following equation:

$$E_{cell}^0 = E_{red}^0 + E_{ox}^0$$
$$= +2.87 + 0.337$$
$$= +3.207 \text{ V}$$

This reaction is spontaneous since E_{cell}^0 is positive.

3. Then, calculate ΔG^0 using the equation $\Delta G^0 = -nFE_{cell}^0$ and the E_{cell}^0 calculated above:

$$\Delta G^0 = -nFE_{cell}^0$$

From the half-reaction, $n = 2$.

$$\Delta G^0 = -nFE_{cell}^0 = -2 \times 96500 \frac{J}{V-mol} \times 3.207 \text{ V}$$
$$= -618,951 \text{ J/mol or} - 618 \text{ kJ/mol}$$

4. Finally, calculate K from the expression $E_{cell}^0 = \dfrac{0.0592}{n} \log K$:

$$E_{cell}^0 = \frac{0.0592}{n} \log K$$

or

$$\log K = \frac{nE_{cell}^0}{0.0592} = \frac{2 \times 3.207}{0.0592} = 108.34$$
$$K = \text{antilog } 108.34 = 2.21 \times 10^{108}$$

Example 23.6

In acidic solution, iron (II) ions are easily oxidized by dissolved oxygen gas:

$$O_2(g) + 4H^+(aq) + 4Fe^{2+} \longrightarrow 4Fe^{3+}(aq) + 2H_2O(l)$$

Calculate E^0_{cell}, ΔG^0, and K for the reaction.

Solution

1. First, write the half-cell equations and their standard electrode potentials, recalling that $E^0_{ox} = -E^0_{red}$.

   Anode : $4Fe^{2+}(aq) \longrightarrow 4Fe^{3+}(aq) + 4e^-$ $E^0_{ox} = -0.77$ V

   Cathode : $O_2(g) + 4H^+(aq) + 4e^- \longrightarrow 2H_2O(l)$ $E^0_{red} = +1.23$ V

2. Next, obtain the cell potential using the following equation:

$$E^0_{cell} = E^0(\text{anode}) + E^0(\text{cathode})$$

$$E^0_{cell} = -0.77 + 1.23$$

$$= +0.46 \text{ V}$$

   This reaction is spontaneous since E^0_{cell} is positive.
3. Then, calculate ΔG^0 using the equation $\Delta G^0 = -nFE^0_{cell}$ and the E^0_{cell} calculated above.

$$\Delta G^0 = -nFE^0_{cell}$$

From the half-reaction, $n = 4$.

$$\Delta G^0 = -nFE^0_{cell} = -4 \times 96500 \frac{J}{V.mol} \times 0.46 \text{ V}$$

$$= -177,560 \text{ J/mol or} - 177.56 \text{ kJ/mol}$$

4. Finally, calculate K from $E^0_{cell} = \frac{0.0592}{n} \log K$:

$$E^0_{cell} = \frac{0.0592}{n} \log K$$

   or

$$\log K = \frac{nE^0_{cell}}{0.0592} = \frac{4 \times 0.46}{0.0592} = 31.1$$

$$K = \text{antilog } 31.1 = 1.2 \times 10^{31}$$

23.9 Dependence of Cell Potential on Concentration (the Nernst Equation)

Potentials for cells not under standard conditions (e.g. concentration other than 1M) can be derived by considering the dependence of the standard free energy change on concentration, given by the following equation:

$$\Delta G = \Delta G^0 + RT \ln Q$$

Here, Q is the reaction quotient, which has the form of an equilibrium constant, except that the concentrations and gas pressures are those that exist in the reaction mixture at any given moment. If we substitute $\Delta G = -nFE$ and $\Delta G^0 = -nFE_{cell}^0$ into the above expression, we have

$$-nFE = -nFE_{cell}^0 + RT \ln Q$$

If we solve this equation for E, we obtain the so-called Nernst equation:

$$E = E^0 - \frac{RT}{nF} \ln Q \text{ or } E = E^0 - \frac{2.303\, RT}{nF} \log Q$$

If we express the equation in the common logarithm form and substitute $T = 298$ K as well as the values of R and F, the Nernst equation becomes:

$$E = E^0 - \frac{0.0592}{n} \log Q$$

This equation can be used to calculate the cell potential under nonstandard conditions.

Example 23.7

Calculate the potential of the cell $Zn(s)\,|\,Zn^{2+}\,(10^{-5}M)\,\|\,Cu^{2+}\,(0.01M)\,|\,Cu(s)$ at 298 K, if its standard potential (with all concentrations $= 1M$) is 1.10 V.

Solution

1. First, from the cell diagram, write down the net cell reaction and determine the number of electrons transferred.

Anode: $\quad\quad\quad\quad$ $Zn(s) \longrightarrow Zn^{2+}(aq) + 2e^-$ $\quad$ $E_{ox}^0 = +0.76$ V

Cathode: $\quad Cu^{2+}(aq) + 2e^- \longrightarrow Cu(s)$ $\quad\quad\quad$ $E_{red}^0 = +0.337$ V

$\quad\quad\quad$ $Zn(s) + Cu^{2+}(aq) \longrightarrow Zn^{2+}(aq) + Cu(s)$ $\quad\quad$ $E_{cell}^0 = 1.10$ V

For this reaction, $n = 2$

2. Then, calculate the reaction quotient, Q

$$Q = \frac{[Zn^{2+}]}{[Cu^{2+}]} = \frac{10^{-5}M}{10^{-2}M} = 10^{-3}$$

3. Finally, calculate E_{cell} using the Nernst equation.

$$E_{cell} = E_{cell}^0 - \frac{0.0592}{n} \log Q$$

$$E_{cell} = 1.10 - \frac{0.0592}{2} \log 10^{-3}$$

$$E_{cell} = 1.19 \text{ V}$$

So, the potential of this cell at 298 K is 1.19 V.

Example 23.8

Using the Nernst equation for the cell

$$\text{Mg(s)} \left| \text{Mg}^{2+}(0.010\text{M}) \right\| \text{Sn}^{2+}(0.10\text{M}) \left| \text{Sn(s)} \right.$$

calculate (a) the actual cell potential, (b) the ratio of concentrations, $\dfrac{[\text{Mg}^{2+}]}{[\text{Sn}^{2+}]}$, which would cause the cell potential to be 2.15 V.

Solution

(a) To calculate the actual cell potential:

1. From the cell diagram, write down the half-cells and the net cell reaction, calculate the standard electrode potential, and determine the number of electrons transferred.

Anode: $\qquad\qquad$ $\text{Mg(s)} \longrightarrow \text{Mg}^{2+}(\text{aq}) + 2e^- \quad E_{ox}^0 = +2.37 \text{ V}$

Cathode: $\quad \text{Sn}^{2+}(\text{aq}) + 2e^- \longrightarrow \text{Sn(s)} \qquad\qquad E_{red}^0 = -0.14 \text{ V}$

$$\overline{\text{Mg(s)} + \text{Sn}^{2+}(\text{aq}) \longrightarrow \text{Mg}^{2+}(\text{aq}) + \text{Sn(s)} \quad E_{cell}^0 = +2.23 \text{ V}}$$

For this reaction $n = 2$.

2. To obtain the nonstandard cell potential we need to calculate the reaction quotient, Q, using the concentrations given.

$$Q = \frac{[\text{Mg}^{2+}]}{[\text{Sn}^{2+}]} = \frac{10^{-2}\text{M}}{10^{-1}\text{M}} = 0.10$$

3. Calculate E_{cell} using the value of Q in the Nernst equation.

$$E_{cell} = E_{cell}^0 - \frac{0.0592}{n} \log Q$$

$$E_{cell} = 2.23 - \frac{0.0592}{2} \log 0.10$$

$$E_{cell} = 2.23 - (-0.03 \text{ V}) = 2.26 \text{ V}$$

(b) Make $E_{cell} = 2.15V$ and solve for Q which is essentially the desired ratio.

1. From part (a) we know that $E^0_{cell} = +2.23$ V and $n = 2$.
2. Calculate E_{cell} using the Nernst equation.

$$E_{cell} = 2.15 \text{ V} = E^0_{cell} - \frac{0.0592}{n} \log \left(\frac{[Mg^{2+}]}{[Sn^{2+}]} \right)$$

$$2.15 \text{ V} = 2.23 - \frac{0.0592}{2} \log \left(\frac{[Mg^{2+}]}{[Sn^{2+}]} \right)$$

$$0.08 = \frac{0.0592}{2} \log \left(\frac{[Mg^{2+}]}{[Sn^{2+}]} \right)$$

$$\log \left(\frac{[Mg^{2+}]}{[Sn^{2+}]} \right) = 2.71$$

$$\left(\frac{[Mg^{2+}]}{[Sn^{2+}]} \right) = 515$$

Example 23.9

The following cell reaction has a standard state potential of $+0.58$ V:

$$Pt \left| Fe^{2+}(0.25M), Fe^{3+}(0.85M) \right| \left| Cr_2O_7^{2-}(2.50M), \right.$$

$$Cr^{3+}(3.50M), H^+(1.0M) \right| Pt$$

Is the reaction spontaneous under these concentration conditions?

Solution

All we need to answer this is the cell potential under the actual conditions. So follow the usual procedure, as in previous examples.

1. From the cell diagram, write down the half-cells and the net cell reaction; calculate the standard electrode potential, and determine the number of electrons transferred.

$$6 \text{ Fe}^{2+}(aq) \longrightarrow 6 \text{ Fe}^{3+}(aq) + 6e^- \quad E^0_{ox} = -0.77 \text{ V}$$

$$14 \text{ H}^+(aq) + 6e^- + Cr_2O_7^{2-}(aq) \longrightarrow 2 \text{ Cr}^{3+}(aq) + 7 \text{ H}_2O(l) \quad E^0_{red} = +1.33 \text{ V}$$

$$14 \text{ H}^+(aq) + 6 \text{ Fe}^{2+}(aq) + Cr_2O_7^{2-}(aq) \longrightarrow 6 \text{ Fe}^{3+}(aq) + 2 \text{ Cr}^{3+}(aq)$$

$$+ 7 \text{ H}_2O(l) \quad E^0_{cell} = +0.58 \text{ V}$$

For this reaction $n = 6$.

2. To obtain the nonstandard cell potential we need to calculate the reaction quotient Q using the concentrations given:

$$Q = \frac{\left[Fe^{3+}\right]^6\left[Cr^{3+}\right]^2}{\left[Fe^{2+}\right]^6\left[Cr_2O_7^{2-}\right]\left[H^+\right]^{14}} = \frac{(0.85)^6(3.50)^2}{(0.25)^6(2.5)(1.0)^{14}} = 7,700$$

3. Calculate E_{cell} using the value of Q in the Nernst equation.

$$E_{cell} = E_{cell}^0 - \frac{0.0592}{n}\log Q$$

$$E_{cell} = 0.58 - \frac{0.0592}{6}\log 7,700$$

$$E_{cell} = 0.58 - 0.04 = 0.54 \text{ V}$$

Since the cell potential is positive, the reaction will be spontaneous under these conditions.

23.10 Electrolysis

Electrolysis: The chemical decomposition of an electrolyte in the molten state or in solution by passing an electric current through it.

Electrolyte: A substance which, when molten or in aqueous solution, dissociates into ions and conducts an electric current.

Non-electrolyte: A substance which, when molten or in solution, neither allows the passage of an electric current nor is decomposed by it.

Anode: The negative electrode through which electrons leave an electrolyte and at which oxidation takes place.

Cathode: The positive electrode through which electrons enter an electrolyte and at which reduction takes place.

Electric current: The flow of electrons through a conductor or electrolyte.

Coulomb: The unit of electric charge. One coulomb is the quantity of charge or electricity that passes through an electrolytic cell when a current of one ampere (A) flows for one second (s).

23.11 Faraday's Laws of Electrolysis

23.11.1 First law of electrolysis

Faraday's first law of electrolysis states that *the mass (m) of a substance produced at the electrodes during electrolysis is directly proportional to the quantity of charge or electricity (Q) that passed through the electrolytic cell.*

Mathematically,

$$m \propto Q$$

This implies that the moles of product formed during electrolysis depend on the number of moles of electrons that pass through the electrolyte. The unit of the quantity of electricity is the Coulomb, which is the quantity of electricity flowing through a conductor when a current of 1 ampere (A) flows for 1 second.

$$Q = I \times t$$

$$(C) = (A) \times (s)$$

(Note: 1 coulomb = 1 ampere - second.)
Therefore, Faraday's first law states that

$$m \propto I \times t$$

$$m = E \times I \times t$$

where E is a constant called the electrochemical equivalent of the substance reduced or oxidized.

23.11.2 The Faraday constant

The charge on a single electron is 1.602×10^{-19} C. One mole of electrons contains Avogadro's number of electrons, i.e., 6.023×10^{23} electrons. This corresponds to a total charge of

$$1.602 \times 10^{-19}C \times 6.023 \times 10^{23} = 96500\,C$$

This quantity of electricity (96500 C), the charge carried by one mole of electrons, is known as the Faraday constant (F).

During electrolysis, one Faraday of electricity (96500 C) is needed to reduce and oxidize one equivalent weight (relative atomic mass divided by charge) of the oxidizing and reducing agents.

For example, consider the generation of one mole of sodium atoms (i.e. 23 g) as represented by the following half-reaction:

$$Na^+ + e^- \longrightarrow Na$$

1 Faraday $\Rightarrow$ 1 mol of $e^- \Rightarrow$ one atom of Na

1 Faraday $\Rightarrow$ 23 g Na

Similarly, 3 Faradays will be required to liberate 1 mole of a trivalent element such as Al or Cr, because one atom requires three electrons:

$$Al^{3+} + 3e^- \longrightarrow Al$$

3 Faradays $\Rightarrow$ 3 mol of $e^- \Rightarrow$ 1 mol of Al

3 Faradays $\Rightarrow$ 27 g of Al

Example 23.10

Calculate the mass of aluminum that would be deposited on the cathode when a current of 3.5 A is passed through a molten aluminum salt for 120 minutes. (Al = 27 g/mol; 1 Faraday = 96500 C).

Solution

Quantity of electricity, $Q = I \times t = 3.5$ A $\times$ 120 min $\times$ 60 s/min $= 25200$ C.
   Cathodic half-reaction:

$$Al^{3+}(1) + 3e^- \longrightarrow Al(s)$$

3 moles of $e^- = 3$ Faradays $= 3 \times 96500$ C $= 1$ mole of Al

We can use the following conversions:

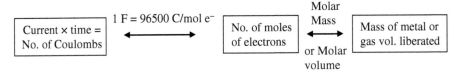

Cathode half-reaction:

$$Al^{3+}(1) + 3e^- \longrightarrow Al(s)$$

$$Q = I \times t = 3.5 \text{ A} \times 120 \text{ min} \times 60 \text{ s/min} = 25,200 \text{ C}.$$

$$\text{Moles of electrons passed} = \frac{\text{quantity of electricity (C)}}{1 \text{ Faraday}}$$

$$= \frac{25,200 \text{ C}}{96500 \text{ C/mole}} = 0.2611 \text{ mol of } e^-$$

3 moles of e^- deposit 1 mol of Al

$$0.2611 \text{ mol of } e^- \text{ deposit} \frac{1 \times 0.2611}{3} \text{ moles of Al} = 0.087 \text{ mol of Al}$$

Mass of Al deposited $=$ mol of Al $\times$ molar mass of Al

$$\text{Mass of Al deposited} = 0.087 \text{ mol} \times 27 \frac{g}{mol} = 2.35 \text{ g Al.}$$

Example 23.11

A current of 3.0 A is passed through a solution of copper (II) sulfate for 1 hour.

(a) Calculate the mass of copper deposited.
(b) What is the volume of oxygen gas produced at STP by the oxidation of water?

Solution

$$Q = I \times t = 3.0 \text{ A} \times 1\text{h} \times 60\frac{min}{h} \times 60\frac{sec}{min} = 10800 \text{ C}$$

$$\text{Moles of electrons passed} = \frac{Q}{1 \text{ faraday}} = \frac{10,800}{1 \times 96500} = 0.112 \text{ mol}$$

(a) Cathode half-cell reaction:

$$Cu^{2+}(aq) + 2e^- \longrightarrow Cu(s)$$

2 mol of electrons = 1 mol of Cu

$$0.112 \text{ mol of electrons} = 0.112 \text{ mol of } e^- \times \frac{1 \text{ mol Cu}}{2 \text{ mol } e^-} = 0.056 \text{ mol of Cu}$$

Mass of Cu deposited = mol of Cu × molar mass of Cu

$$= 0.056 \text{ mole} \times 63.5\frac{g}{\text{mole}} = 3.556 \text{ g}$$

(b) Anode half-cell reaction

$$4\,OH^-(aq) \longrightarrow O_2(g) + 2H_2O(1) + 4e^-$$

4 mol of electrons = 1 mol of O_2 = 22.4 liters O_2

$$0.112 \text{ mol of electrons} = 0.112 \text{ mol of } e^- \times \frac{22.4 \text{ LO}_2}{4 \text{ mol } e^-} = 5.61 \text{ of } O_2$$

23.11.3 Second law of electrolysis

Faraday's second law states that *when the same quantity of electricity is passed through different electrolytes, the number of moles or masses of the different ions produced is inversely proportional to their ionic charges.*

If the same quantity of electricity is passed through two electrolytic cells, say, one containing Cu^{2+} ions and the other Ag^+ ions, for the same period of time, then by Faraday's second law

$$\frac{\text{Number of moles of } Cu^{2+} \text{ liberated}}{\text{Number of moles of } Ag^+ \text{ liberated}} = \frac{\text{Charge on } Ag^+}{\text{Charge on } Cu^{2+}}$$

Example 23.12

A current is passed through two electrolytic cells containing solutions of copper (II) sulfate and silver nitrate, both connected in series. If 1.08 g of silver is deposited on the cathode of the cell containing silver nitrate solution, calculate the mass of copper deposited on the cathode of the second cell containing copper (II) sulfate solution.

Solution

Cathode reaction in cell 1, containing $AgNO_3$ solution:

$$Ag^+(aq) + 1e^- \longrightarrow Ag(s)$$

$$1 \text{ mol of } e^- = 1 \text{ mol of } Ag(s)$$

Cathode reaction in cell 2, containing $CuSO_4$ solution:

$$Cu^{2+}(aq) + 2e^- \longrightarrow Cu(s)$$

$$2 \text{ mol of } e^- = 1 \text{ mol of } Cu(s)$$

Since the same quantity of electricity is flowing through the cells, $Q_{cell1} = Q_{cell2}$

$$\text{mol of Ag deposited} = \frac{\text{mass of Ag in grams}}{\text{molar mass of Ag}} = \frac{5.4 \text{ g}}{108.0 \text{ g}} = 0.050 \text{ mol}$$

From the above half-reaction,

$$\text{mol of Ag}^+ \text{ liberated} = 0.050 \text{ mol}$$

From Faraday's second law,

$$\frac{\text{Number of moles of } Cu^{2+} \text{ liberated}}{\text{Number of moles of } Ag^+ \text{ liberated}} = \frac{\text{Charge on } Ag^+}{\text{Charge on } Cu^{2+}} = \frac{1}{2}$$

$$\frac{\text{Number of moles of } Cu^{2+} \text{ liberated}}{0.050 \text{ mole } Ag^+} = \frac{1}{2}$$

$$\text{mol of } Cu^{2+} \text{ liberated} = 0.025 \text{ mol}$$

$$\text{mol of Cu deposited} = 0.025 \text{ mol}$$

$$\text{Mass of Cu deposited} = 0.025 \text{ mol} \times 63.5 \frac{\text{g}}{\text{mol}} = 1.588 \text{ g}$$

Example 23.13

A steady current is passed through a series of solutions of $AgNO_3$, $CrCl_3$, $ZnSO_4$, and $NiSO_4$ for three hours. If 1.5 g of Ag is deposited from the first solution, calculate:

(a) The masses of metals (Cr, Zn, and Ni) deposited simultaneously at the cathodes in the remaining solutions.
(b) The current flowing through the solutions.

Solution

(a) The half-cell reactions at the cathodes are:

$$Ag^+(aq) + e^- \longrightarrow Ag(s); 1F \Rightarrow 1 \text{ mol of } e^- \Rightarrow 1 \text{ mol Ag (s)}$$

$$Cr^{3+}(aq) + 3e^- \longrightarrow Cr(s); 3F \Rightarrow 3 \text{ mol of } e^- \Rightarrow 1 \text{ mol Cr(s)}$$

$$Zn^{2+}(aq) + 2e^- \longrightarrow Zn(s); 2F \Rightarrow 2 \text{ mol of } e^- \Rightarrow 1 \text{ mol Zn(s)}$$

$$Ni^{2+}(aq) + 2e^- \longrightarrow Ni(s); 2F \Rightarrow 2 \text{ mol of } e^- \Rightarrow 1 \text{ mol Ni(s)}$$

The same quantity of electricity is passed through all the cells. Therefore,

$$q_{cell\ Ag^+} = q_{cell\ Cr^{3+}} = q_{cell\ Zn^{2+}} = q_{cell\ Ni^{2+}}$$

$$\text{mol of Ag deposited} = \frac{\text{mass of Ag in grams}}{\text{molar mass of Ag}} = \frac{1.5\ g}{108.0\ g} = 0.0139$$

mol of Ag deposited $= 0.0139$
From Faraday's second law,

$$\frac{\text{Number of moles of Cr}^{3+}\text{ deposited}}{\text{Number of moles of Ag deposited}} = \frac{\text{Charge on Ag}^+}{\text{Charge on Cr}^{3+}} = \frac{1}{3}$$

$$\frac{\text{Number of moles of Cr}^{3+}\text{ deposited}}{0.0139\ \text{mole Ag}} = \frac{1}{3}$$

Number of moles of Cr deposited $= 0.0046$ mole

$$\text{and mass of Cr deposited} = 0.0046\ \text{moles} \times 52.0\frac{g}{\text{mole}} = 0.241\ g$$

Mass of Zn deposited is obtained similarly:

$$\frac{\text{Number of moles of Zn deposited}}{0.0139\ \text{mole Ag}} = \frac{1}{2}$$

Number of moles of Zn deposited $= 0.007$ mol

$$\text{Mass of Zn deposited} = 0.007\ \text{mol} \times 65.4\frac{g}{\text{mol}} = 0.455\ g$$

Mass of Ni deposited:

Moles of Ni deposited $=$ moles of Zn (from half-cell equation) $= 0.007$

$$\text{Mass of Ni deposited} = 0.007\ \text{mol} \times 58.7\frac{g}{\text{mol}} = 0.4110\ g$$

So the masses of Cr, Zn, and Ni deposited at the various cathodes are 0.241 g, 0.455 g, and 0.411 g, respectively.
(b) Current flowing through the solution:
Using the Ag^+ half-cell we have:

1 mol of Ag $= 1$ mol of $e^- = 1 \times 96500$ C

0.0139 mol of Ag $= 1341.35$ C

Quantity of electricity $Q = I \times t$

$Q = 1341.35$ C; $t = 3$ hours $= 3 \times 60 \times 60 = 10800$ s

$$I = \frac{Q}{t} = \frac{1341.35\ C}{10800\ s} = 0.124\ A$$

23.12 Problems

1. Write the cell diagram and the cell reaction for the following redox couples.

   (a) Cu^{2+}/Cu and Sn^{2+}/Sn
   (b) Ag^+/Ag and Zn^{2+}/Zn
   (c) Fe^{2+}/Fe and Mg^{2+}/Mg
   (d) Pb^{2+}/Pb and Sn^{2+}/Sn

2. Write the cell diagrams corresponding to the following reactions.

   (a) $Mn(s) + Ti^{2+}(aq) \longrightarrow Mn^{2+}(aq) + Ti(s)$
   (b) $Mg(s) + Zn^{2+}(aq) \longrightarrow Mg^{2+}(aq) + Zn(s)$
   (c) $Sn^{2+}(aq) + Pb^{4+}(aq) \longrightarrow Sn^{4+}(aq) + Pb^{2+}(aq)$
   (d) $Ca(s) + Cu^{2+}(aq) \longrightarrow Ca^{2+}(aq) + Cu(s)$

3. Write the cell notation for the following overall reactions.

   (a) $Sn(s) + Cu^{2+}(aq) \longrightarrow Sn^{2+}(aq) + Cu(s)$
   (b) $Co^{2+}(s) + Ag^+(aq) \longrightarrow Co^{3+}(aq) + Ag(s)$
   (c) $Sn^{2+}(aq) + Fe(s) \longrightarrow Sn(s) + Fe^{2+}(aq)$
   (d) $Zn(s) + 2H^+(aq) \longrightarrow Zn^{2+}(aq) + H_2(g)$

4. The action of zinc metal on a solution of copper (II) sulfate is represented by the redox equation

$$Zn + Cu^{2+}(aq) \longrightarrow Zn^{2+}(aq) + Cu$$

   (a) Write the half-cell reactions for both the oxidation and the reduction.
   (b) Is zinc oxidized or reduced?
   (c) Write the cell notation for the reaction.

5. Write balanced equations for the cells represented by the following shorthand notation.

   (a) $Pb(s)\,|\,Pb^{2+}(1M)\,\|\,Br_2(1\ atm)\,|\,Br^-(aq)\,|\,Pt(s)$
   (b) $Zn(s)\,|\,Zn^{2+}(1M)\,\|\,Eu^{3+}(aq),Eu^{2+}(aq)\,|\,Pt(s)$
   (c) $Co(s)\,|\,Co^{2+}(1M)\,\|\,Cu^{2+}(aq)\,|\,Cu(s)$
   (d) $Cu(s)\,|\,Cu^{2+}(1M)\,\|\,Cl^-(aq)\,|\,Cl_2(1\ atm)\,|\,C(s)$

6. Using the table of standard electrode potentials, calculate the emf (E^0_{cell}) for the following cells:

   (a) $Zn(s)\,|\,Zn^{2+}(1M)\,\|\,Sn^{2+}(aq)\,|\,Sn(s)$
   (b) $Co(s)\,|\,Co^{2+}(1M)\,\|\,Cu^{2+}(aq)\,|\,Cu(s)$

7. The standard electrode potential for the galvanic cell $Cd(s)\,|\,Cd^{2+}(aq)\,\|\,Pb^{2+}(aq)\,|\,Pb(s)$ is $+0.27$ V. Given that the standard reduction potential for the Cd^{2+}/Cd half-cell is -0.40 V, calculate the standard reduction potential for the Pb^{2+}/Pb half-cell.

8. Using the table of standard electrode potentials, determine if copper metal will spontaneously react with a solution of tin (II) ions.

9. The standard reduction potential is -1.80 V for the U^{3+}/U couple and -0.61 V for the U^{4+}/U^{3+} couple. What is the potential (emf) for the U^{4+}/U couple?

10. The table of standard electrode potentials can be used to predict chemical reactivity. Will Cu metal reduce Sn(IV) to Sn(II)?

11. Will Fe(II) ion reduce MnO_4^- in acidic solution? Hint: Use the table of standard electrode potential and calculate the cell voltage.

12. Calculate the potential of the half-cell reaction:

$$Mg(s) \rightleftharpoons Mg^{2+}(aq) + 2e^- \quad E^0_{cell} = -2.37 \text{ V}$$

when the concentration of Mg^{2+} ion is 0.01 M.

13. Given that $[Fe^{2+}] = 0.05M$ and $[Fe^{3+}] = 1.25$ M, calculate the potential of the half-cell reaction: $Fe^{2+}(aq) \rightleftharpoons Fe^{3+}(aq) + e^- \quad E^0_{cell} = -0.77$ V

14. Using the table of standard electrode potential calculate the (emf) for the following cells:

    (a) $Zn(s)|Zn^{2+}(0.200M)||Sn^{2+}(0.100 \text{ M})|Sn(s)$

    (b) $Mg(s)|Mg^{2+}(0.010M)||Cu^{2+}(0.100M)|Cu(s)$

    (c) $Zn(s)|Zn^{2+}(0.002M)||H^+(1M), H_2(0.1 \text{ atm})|Pt(s)$

15. Will the reactions for the following cells take place as written?

    (a) $Mg(s)|Mg^{2+}(0.020M)||Sn^{2+}(0.2500M)|Sn(s)$

    (b) $Cd(s)|Cd^{2+}(0.010M)||Cu^{2+}(0.100M)|Cu(s)$

    (c) $Ni(s)|Ni^{2+}(0.200M)||Hg^{2+}(0.200M)|Hg(s)$

16. Calculate the standard Gibbs free energy change and the equilibrium constant at 25°C for the reaction $\left(\text{note : } 1 \text{ V} = 1\frac{J}{C}\right)$

$$Cu(s) + 2Ag^+(aq) \longrightarrow Cu^{2+}(aq) + 2Ag(s)$$

17. Calculate the standard Gibbs free energy change and the equilibrium constant at 25°C for the reaction

$$2Br^-(aq) + Cl_2(g) \longrightarrow Br_2(l) + 2Cl^-(aq)$$

18. Calculate the standard cell potential, E^0, and the equilibrium constant, K, for the following reaction at 25°C.

$$Fe(s) + Cd^{2+}(aq) \longrightarrow Fe^{2+}(aq) + Cd(s)$$

19. Design a battery from aluminum metal and chlorine gas using the couples $Al^{3+}(aq)/Al(s)$ and $Cl_2(g)/Cl^-(aq)$.

    (a) Write a balanced equation for the electrode reaction in the battery.
    (b) Calculate the standard electrode potential for the battery.
    (c) What is the voltage of the battery if the pressure of chlorine gas is increased from 1 to 5 atm?

20. The net reaction occurring in a lead-acid battery is

$$Pb(s) + PbO_2(s) + H_2SO_4(aq) \longrightarrow 2\,PbSO_4(s) + 2\,H_2O(l)$$

Given that the standard potential E^0 for the cell is 2.04V, determine the potential of the battery at 25°C when sulfuric acid with a concentration of 10 M is used in the cell.

21. Calculate the value of E^0 for the following reactions and determine if each reaction is spontaneous in the direction written.

(a) $Zn(s) + Sn^{4+} \longrightarrow Zn^{2+}(aq) + Sn^{2+}(aq)$
(b) $2\,I^-(aq) + Zn^{2+}(aq) \longrightarrow I_2(s) + Zn(s)$
(c) $Cl_2(g) + V(s) \longrightarrow 2Cl^-(aq) + V^{2+}(aq)$
(d) $I_2(s) + 2Br^-(aq) \longrightarrow 2I^-(aq) + Br_2(l)$

22. How many Faradays are required to reduce 1 mole of the following?

(a) $Ag^+(aq) + e^- \longrightarrow Ag(s)$
(b) $Ca^{2+}(aq) + 2e^- \longrightarrow Ca(s)$
(c) $Al^{3+}(aq) + 3e^- \longrightarrow Al(s)$
(d) $Fe^{3+}(aq) + 3e^- \longrightarrow Fe(s)$

23. How many Faradays are required to oxidize or reduce 1 mole of the following?

(a) $Mg(OH)_2(s) + 2e^- \longrightarrow Mg(s) + OH^-(aq)$
(b) $PbSO_4(s) + 2e^- \longrightarrow Pb(s) + SO_4^{2-}(aq)$
(c) $Sn^{4+}(aq) + 2e^- \longrightarrow Sn^{2+}(aq)$
(d) $2Br^-(aq) \longrightarrow Br_2(l) + 2e^-$

24. Balance the following redox reactions in acidic solution and determine the number of coulombs of electricity required to reduce one mole of the oxide or anion.

(a) $Sb_2O_3(s) \longrightarrow 2Sb(s)$
(b) $SO_4^{2-}(aq) \longrightarrow H_2SO_3(aq)$
(c) $IO_3^-(aq) \longrightarrow I_2(s)$
(d) $H_2O_2 \longrightarrow H_2O$

25. Given that 1 F is equivalent to 96500 coulombs per mole of electron, what is the charge on an individual electron? (Note : 1 mole of electrons contains 6.022×10^{23} electrons.)

26. A solution of copper (II) sulfate is electrolyzed between copper electrodes for two hours using a steady direct current of 50 A. Calculate the mass of elemental copper deposited.

27. Aluminum is extracted from its ore by the electrolysis of a molten Al salt. Calculate the mass of Al produced by a current of 5 A flowing for twelve hours.

28. When a current of 0.25 A was passed through a divalent metal salt solution for 3.83 hours, 2.00g of the metal was deposited. Determine the atomic mass of the metal. Identify the metal M using the periodic table.

29. A steady current of 0.5 A flows through aqueous solutions of zinc nitrate and silver nitrate connected in series for ten hours. In a particular experiment, 10.5 g of zinc was deposited from the $Zn(NO_3)_2$ electrolytic cell. Calculate:

(a) the mass of Ag metal deposited at the same time
(b) the quantity of electricity (in coulombs) that passed through the electrolytes
(c) the time it took to run the experiment.

30. A direct current is passed through three electrolytic cells of copper (II) sulfate, gold (III) nitrate and dilute sulfuric acid joined together in series such that the same quantity of electricity flows through each cell. If 6.355 g of copper are deposited in the first cell, calculate:

(a) the mass of Au deposited in the second cell, and
(b) the volume of hydrogen gas liberated from the third cell, measured at STP.

24

Radioactivity and Nuclear Reactions

. . . .

24.1 Definitions

Nuclide: an atom containing a specified number of protons and neutrons in its nucleus — in other words, any particular atom under discussion.

Unstable nuclide: one that will spontaneously disintegrate or emit radiation, thus giving off energy and altering to some new form (often another element). The new form may also be unstable; often it will be *stable*, that is, with no tendency to disintegrate. Unstable nuclides are often referred to as *radioactive*.

Radioactivity: the spontaneous emission of radiation by elements with unstable nuclei.

Radionuclide: a radioactive (that is, unstable) nuclide.

Radioisotope: another more commonly seen term for a radionuclide.

Radioactive decay: the process whereby a radionuclide is converted to another form (usually another element) by emission of radiation.

Parent nuclide: a nuclide undergoing radioactive decay.

Daughter nuclide: the nuclide produced when a parent nuclide decays.

Activity: the rate at which a radionuclide decays, usually expressed as the number of disintegrations per unit time.

24.2 Radioactive Decay and Nuclear Equations

Naturally radioactive elements decay spontaneously by emitting *alpha particles*, *beta particles*, or *gamma radiation*. Other elements can be induced to decay by bombarding them with high-energy particles; this is known as artificial radioactivity.

Like chemical reactions, equations representing nuclear reactions must be balanced. However, the method for balancing nuclear equations differs from that used for chemical equations. To balance a nuclear equation, the sums of the atomic numbers or particle charges (subscripts) and the sums of the mass numbers (superscripts) on both sides of the equations must be equal.

24.2.1 Alpha emission $\left(^4_2\alpha\right)$

A nucleus undergoing alpha decay emits a particle identical to a helium nucleus, with an atomic number of 2 and a mass number of 4.

Chemistry in Quantitative Language: Fundamentals of General Chemistry Calculations. Second Edition.
Christopher O. Oriakhi, Oxford University Press. © Christopher O. Oriakhi 2021.
DOI: 10.1093/oso/9780198867784.003.0024

Since the emission of an alpha particle from the nucleus results in a loss of two protons and two neutrons, when writing a nuclear reaction involving an alpha decay, subtract 4 from the mass number, and 2 from the atomic number. Alpha-particle decay can be represented by the general equation:

$$^A_Z C \longrightarrow\ ^4_2\alpha\ +\ ^{(A-4)}_{(Z-2)}D$$

$$\text{Parent} \hspace{3cm} \text{Daughter}$$
$$\text{nuclide} \hspace{3cm} \text{nuclide}$$

For example, uranium-238 emits an alpha particle and forms thorium-234.

$$^{238}_{92}U \longrightarrow\ ^{234}_{90}Th + ^4_2\alpha$$

Remember: To balance a nuclear equation, the sums of the mass numbers and atomic numbers must be the same on both sides of the equation. This is true of all types of nuclear equations.

Example 24.1

Write balanced equations for the following radioactive nuclides undergoing alpha decay:

1. Bismuth-214
2. Polonium-212
3. Platinum-190

Solution

1. $^{214}_{83}Bi \longrightarrow\ ^{210}_{81}Tl + ^4_2\alpha$

2. $^{212}_{84}Po \longrightarrow\ ^{210}_{82}Pb + ^4_2\alpha$

3. $^{190}_{78}Pt \longrightarrow\ ^{186}_{76}Os + ^4_2\alpha$

24.2.2 Beta emission ($^0_{-1}\beta$)

Beta radiation consists of particles with a charge of −1 and mass very close to zero: in other words, electrons. Beta particles are written as $^0_{-1}e$ and $^0_{-1}\beta$. Nuclear disintegration by beta emission always results in the formation of a nuclide of a new element with the same mass number as the parent nuclide and its atomic number increased by one unit. Beta-particle decay can be represented by the general equation

$$^A_Z C \longrightarrow\ ^0_{-1}\beta\ +\ ^A_{(Z+1)}D$$

For example, thorium-234 emits a beta particle and forms protactinium-234:

$$^{234}_{90}Th \longrightarrow\ ^{234}_{91}Pa + ^0_{-1}\beta$$

Example 24.2

What products are formed when (a) Radium-226, (b) lead-208, and (c) iodine-131 undergo beta decay?

Solution

1. $^{226}_{88}\text{Ra} \longrightarrow {}^{0}_{-1}\beta + {}^{226}_{89}\text{Ac}$

2. $^{207}_{82}\text{Pa} \longrightarrow {}^{0}_{-1}\beta + {}^{207}_{83}\text{Bi}$

3. $^{131}_{53}\text{I} \longrightarrow {}^{0}_{-1}\beta + {}^{131}_{54}\text{Xe}$

24.2.3 Gamma radiation (γ)

Gamma rays are high-energy electromagnetic radiation of very short wavelength, with no associated mass or charge. Gamma radiation always accompanies other radioactive emissions like alpha and beta radiations. It may also occur on its own as a way for a high-energy nuclide to get rid of excess energy. Decay by gamma radiation leaves both the atomic and mass numbers unchanged and is represented as $^{0}_{0}\gamma$ or simply γ.

24.2.4 Positron emission ($^{0}_{1}e$)

A positron is a particle with the same mass as an electron (or beta particle) but the opposite charge (positive). A positron is represented as $^{0}_{1}e$ or $^{0}_{1}\beta$. It is produced when a proton is converted to a neutron within the nucleus:

$$^{1}_{1}p \longrightarrow {}^{1}_{0}n + {}^{0}_{1}\beta$$

When a nucleus decays by positron emission, the mass number remains the same, but the atomic number decreases by one unit. Positron decay can be represented by the general equation:

$$^{A}_{Z}C \longrightarrow {}^{0}_{1}\beta + {}^{A}_{(Z-1)}D$$

For example, carbon-11 decays by emitting a positron:

$$^{11}_{6}C \longrightarrow {}^{0}_{1}\beta + {}^{11}_{5}B$$

24.2.5 Electron capture ($_{-1}^{0}e$)

In electron capture, the nucleus pulls in an electron from a low-energy orbital such as the $1s$ orbital; the electron then combines with a proton to form a neutron.

$$_{-1}^{0}e + {}^{1}_{1}p \longrightarrow {}^{1}_{0}n$$

When a nucleus decays by electron capture, the mass number remains the same, but the atomic number decreases by one unit. Electron capture decay can be represented by the general equation:

$$^{A}_{Z}C + {}_{-1}^{0}e \longrightarrow {}^{A}_{(Z-1)}D$$

An example of such a process is the reaction

$$^{18}_{9}F + {}_{-1}^{0}e \longrightarrow {}^{18}_{8}O$$

Example 24.3

Write balanced nuclear equations for the following radioactive decay processes:

1. $^{21}_{11}$Na (positron emission)
2. $^{62}_{29}$Cu (electron capture)
3. $^{82}_{37}$Rb (gamma radiation)

Solution

1. $^{21}_{11}$Na $\longrightarrow$ $^{0}_{1}\beta + ^{21}_{10}$Ne
2. $^{62}_{29}$Cu $+ _{-1}^{0}$e $\longrightarrow$ $^{62}_{28}$Ni
3. $^{82}_{37}$Rb $\longrightarrow$ $^{0}_{0}\gamma + ^{82}_{37}$Rb (no change)

24.3 Nuclear Transmutations

Naturally occurring isotopes of some elements like uranium and polonium spontaneously and uncontrollably disintegrate by emitting radiation. These are called radioactive isotopes. On the other hand, stable isotopes can be made artificially radioactive by bombarding them with subatomic particles. The result is the production of other nuclides or isotopes. This type of nuclear reaction is known as a *nuclear transmutation*. A nuclear transmutation is brought about by bombarding the nucleus of an atom with high-energy particles, such as protons (p), neutrons (n), or alpha particles (α).

An example of a nuclear transmutation is

$$^{14}_{7}N + ^{4}_{2}He \longrightarrow ^{1}_{1}H + ^{17}_{8}O$$

Nuclear equations like the preceding one can be written in a shorthand form, showing (in order), the target nucleus, the bombarding particle, the particle ejected, and the nuclide produced. Thus, the above example becomes: $^{14}_{7}N$ (α, p) $^{17}_{8}O$.

Example 24.4

Write balanced equations for the following nuclear reactions:

1. $^{239}_{94}$Pu$+? \longrightarrow ^{1}_{0}n + ^{242}_{96}$Cm
2. $^{32}_{16}$S $+ ^{1}_{0}n \longrightarrow ^{1}_{1}H+?$
3. $^{235}_{92}$U $+ ^{1}_{0}n \longrightarrow 3^{1}_{0}n+?+ ^{92}_{36}$Kr
4. $^{241}_{95}$Am $+ ^{4}_{2}He \longrightarrow 2^{1}_{0}n+?$

Solution

1. $^{239}_{94}$Pu $+ ^{4}_{2}He \longrightarrow ^{1}_{0}n + ^{242}_{96}$Cm
2. $^{32}_{16}$S $+ ^{1}_{0}n \longrightarrow ^{1}_{1}H + ^{32}_{15}$P
3. $^{235}_{92}$U $+ ^{1}_{0}n \longrightarrow 3^{1}_{0}n + ^{141}_{56}$Ba $+ ^{92}_{36}$Kr
4. $^{241}_{95}$Am $+ ^{4}_{2}He \longrightarrow 2^{1}_{0}n + ^{243}_{97}$Bk

Example 24.5

Write balanced reactions for the following nuclear processes, and identify the product nuclides (E):

1. $^{138}_{92}U \left(^{1}_{0}n, \; ^{0}_{-1}\beta \right) \; ^{A}_{Z}E$

2. $^{24}_{12}Mg \left(^{4}_{2}He, \; ^{1}_{0}n \right) \; ^{A}_{Z}E$

3. $^{55}_{25}Mn \left(^{2}_{1}D, \; 2^{1}_{0}n \right) \; ^{A}_{Z}E$

Note: $^{2}_{1}D$ is the nucleus of deuterium, a heavy isotope of hydrogen.

Solution

First, determine A and Z by balancing the mass number and the atomic number. Then check the periodic table and use the value of Z to identify the element E.

1. $^{138}_{92}U \left(^{1}_{0}n, \; ^{0}_{-1}\beta \right) \; ^{A}_{Z}E$

   $^{138}_{92}U + ^{1}_{0}n \longrightarrow \; ^{0}_{-1}\beta + ^{A}_{Z}E$

   Mass number balance: $138 + 1 = 0 + A; A = 139$
   Atomic number balance: $92 + 0 = -1 + Z; Z = 93$
   Check the periodic table for element with $Z = 93$. The element is neptunium. The balanced equation is:

$$^{138}_{92}U + ^{1}_{0}n \longrightarrow \; ^{0}_{-1}\beta + ^{139}_{93}Np$$

2. $^{24}_{12}Mg \left(^{4}_{2}He, \; ^{1}_{0}n \right)^{A}_{Z}E$

$$^{24}_{12}Mg + ^{4}_{2}He \longrightarrow \; ^{1}_{0}n + ^{A}_{Z}E$$

$$A = 27; Z = 14; E = Si$$

   The balanced equation is

$$^{24}_{12}Mg + ^{4}_{2}He \longrightarrow \; ^{1}_{0}n + ^{27}_{14}Si$$

3. $^{55}_{25}Mn \left(^{2}_{1}D, \; 2^{1}_{0}n \right)^{A}_{Z}E$

$$^{55}_{25}Mn + ^{2}_{1}D \longrightarrow \; ^{1}_{0}n + ^{A}_{Z}E$$

   $A = 56; Z = 26; E = Fe.$
   The balanced equation is

$$^{55}_{25}Mn + ^{2}_{1}D \longrightarrow \; ^{1}_{0}n + ^{56}_{26}Fe$$

24.4 Rates of Radioactive Decay and Half-Life

Radioactive nuclides decay at different rates. Some disintegrate within seconds, while others take millions of years. Radioactive decay follows first-order kinetics, and the rate of decay is directly proportional to the number of radioactive nuclei N in the sample:

$$\text{Decay rate} = kN$$

where k is the first-order rate constant, called the decay constant. Like any first-order rate constant, it has the units of 1/time.

By using calculus, a first-order rate law can be transformed into an integrated rate equation that is very convenient:

$$\ln\left(\frac{N_t}{N_0}\right) = -kt \text{ or } \log\left(\frac{N_t}{N_0}\right) = \frac{-kt}{2.303}$$

In this equation, N_0 is the initial number of radioactive nuclei in the sample, N_t is the number remaining after time t, k is the decay constant, and t is the time interval of decay. Note that N can be any measurement of amount, activity, or molar concentration as long as both N and N_0 are in the same units.

24.4.1 Half-life

The half-life of a radioactive nuclide is defined as the time required for half of any given sample to decay; this number is a constant for any particular radionuclide.

Substituting $N_t = \frac{1}{2}N_0$ at time $t = t_{1/2}$ in the integrated rate equation gives the expression:

$$t_{1/2} = \frac{0.693}{k} \text{ or } k = \frac{0.693}{t_{1/2}}$$

Example 24.6

The half-life of vanadium-50 is 6×10^{15} years. It disintegrates by β emission to produce $^{50}_{24}\text{Cr}$. What is the rate constant for this disintegration process?

Solution

Use the equation:

$$k = \frac{0.693}{t_{1/2}}$$

$$t_{1/2} = 6 \times 10^{15}\text{yr}$$

$$k = \frac{0.693}{t_{1/2}} = \frac{0.693}{6 \times 10^{15}\text{yr}} = 1.16 \times 10^{-16} \text{ yr}^{-1} \text{ or } 3.73 \times 10^{-24} \text{ s}^{-1}$$

Example 24.7

Radioactive iodine finds application in cancer therapy. In particular, large doses of $^{131}_{53}\text{I}$ are used in the treatment of thyroid cancer. If the half-life is 8.06 days, how long will it take a sample of $^{131}_{53}\text{I}$ to decay to 5% of the original concentration?

Solution

First, calculate the decay constant:

$$k = \frac{0.693}{t_{1/2}} = \frac{0.693}{8.06 \text{ day}} = 0.086 \text{ day}^{-1}$$

Use the equation:

$$\log\left(\frac{N_t}{N_0}\right) = \frac{-kt}{2.303}$$

$$\log\left(\frac{5.0}{100}\right) = \frac{-\left(0.086 \text{ day}^{-1}\right)t}{2.303}$$

$$-1.3010 = \frac{-\left(0.086 \text{ day}^{-1}\right)t}{2.303}$$

$$t = 34.8 \text{ days}$$

Example 24.8

The half-life of ^{32}P, a radioisotope used medically in leukemia therapy, is 14.28 days. It decays by emitting beta particles and gamma rays. How much of a 20-μg ^{32}P sample remains in a patient's blood after sixty days?

Solution

First, determine the decay constant:

$$k = \frac{0.693}{t_{1/2}} = \frac{0.693}{14.28 \text{ days}} = 0.049 \text{ day}^{-1}$$

Next we determine the ratio of N_t to N_0:

$$\log\left(\frac{N_t}{N_0}\right) = \frac{-kt}{2.303}$$

$$\log\left(\frac{N_t}{N_0}\right) = \frac{-\left(0.0485 \text{ day}^{-1}\right)(60 \text{ days})}{2.303} = -1.2643$$

$$\frac{N_t}{N_0} = 0.0544$$

But $N_0 = 20 \ \mu$g, so $N_t = N_0 \times 0.0544 = 20 \ \mu$g $\times 0.0544 = 1.088 \ \mu$g

The amount of the radioisotope remaining after 60 days is 1.09μg.

24.5 Energy of Nuclear Reactions

Like chemical reactions, nuclear reactions involve energy changes. However, the energy changes in nuclear reactions are several orders of magnitude larger than those in chemical reactions. The large amount of energy liberated in a nuclear reaction is due to a loss in mass during the process. The energy equivalent of mass is given by Einstein's equation

$$E = mc^2$$

E is energy released, m is mass decrease, and c is the speed of light, 3.0×10^8 m/s.

24.5.1 Mass defect

The nucleus of an atom is made up of protons and neutrons (collectively known as nucleons). When a stable nucleus is formed, the mass of the nucleus is always less than the mass of its constituent protons and neutrons. This difference in mass is known as the *mass defect* and is due to the packing of the nucleons into small elements of volume. The mass defect, Δm, is calculated from the following expression:

$$\Delta m = \left(\text{total masses of all particles, e}^-, \text{p}^+, \text{and n}^0 \right) - (\text{actual mass of atom})$$

or more simply,

$$\Delta m = \text{mass reactants—mass of products}$$

When calculating mass defects, the masses of the nucleons must be expressed in atomic mass units (amu). Recall that

$$1 \text{ amu} = 1.660 \times 10^{-24} \text{ g or } 1.660 \times 10^{-27} \text{ kg and } 1 \text{ g} = 6.023 \times 10^{23} \text{amu.}$$

Example 24.9

Calculate the mass of the proton and the neutron in amu given that the mass of the proton is 1.6724×10^{-24} g, and the mass of the neutron is 1.6747×10^{-24} g.

Solution

Use the conversion factor:

$$6.023 \times 10^{23} \text{amu} = 1 \text{ g}$$

$$\text{mass of proton} = \left(\frac{6.023 \times 10^{23} \text{amu}}{1 \text{ g}} \right) \left(1.6724 \times 10^{-24} \text{ g} \right) = 1.0073 \text{ amu}$$

$$\text{mass of neutron} = \left(\frac{6.023 \times 10^{23} \text{amu}}{1 \text{ g}} \right) \left(1.6747 \times 10^{-24} \text{ g} \right) = 1.0087 \text{ amu}$$

Example 24.10

Calculate the mass defect for the aluminum-27 nucleus if the isotopic mass of $^{27}_{13}\text{Al}$ is 26.9815 amu. The mass of a proton is 1.0073 amu and the mass of a neutron is 1.0087 amu.

Solution

Use the expression:

$$\Delta m = (\text{mass of separated neutrons} + \text{protons}) - (\text{mass of nucleus})$$

Aluminum has an atomic number of 13; therefore it contains thirteen protons. Since the mass number of Al is 27, the number of neutrons will be 14 ($27 - 13 = 14$). Hence,

$$\Delta m = \left[(14 \times \text{mass of neutron}) + (13 \times \text{mass of proton})\right] - \left(\text{mass of } ^{27}\text{Al nucleus}\right)$$

$$\Delta m = [(14 \times 1.0087 \text{ amu}) + (13 \times 1.0073 \text{ amu})] - (26.9815 \text{ amu})$$

$$\Delta m = (27.2167 \text{ amu}) - (26.9815 \text{ amu}) = 0.2352 \text{ amu}$$

24.5.2 Binding energy

The *binding energy* of the nucleus is the energy holding the nucleon together and is defined as the energy released when the neutrons and protons in the nucleus combine to form the nucleus. Binding energy gives a direct measure of the stability of the nucleus. The larger the binding energy, the more stable the nucleus. The binding energy can be calculated by using Einstein's equation:

$$B.E. = \Delta mc^2$$

The symbol *B.E.* is the binding energy, and Δm is the change in mass, or mass defect. The nuclear binding energy is often expressed as kJ/mol of nuclei or as MeV/nucleon. Note that 1 MeV = 1.602×10^{-13} J.

Example 24.11

What is the binding energy (a) in kJ/mol, and (b) kJ/nucleon, of iron-56, which has an atomic mass of 55.9349 amu? The mass of a neutron is 1.0087 amu, and that of a proton is 1.0073 amu.

Solution

Use the expression:

$$\Delta m = (\text{mass of separated neutrons} + \text{protons}) - (\text{mass of nucleus})$$

The atomic number of iron is 26, so it has twenty-six protons. Since the mass number of Fe is 56, it also has thirty neutrons. Therefore,

$$\Delta m = [(30 \times \text{mass of neutron}) + (26 \times \text{mass of proton})] - (\text{mass of } {}^{56}\text{Fe nucleus})$$

$$\Delta m = [(30 \times 1.0087 \text{ amu}) + (26 \times 1.0073 \text{ amu})] - (55.9349 \text{ amu})$$

$$\Delta m = (56.4508 \text{ amu}) - (55.9349 \text{ amu}) = 0.5159 \text{ amu}$$

$$\Delta m \text{ in kg} = (0.5159 \text{ amu}) \left(\frac{1.660 \times 10^{-27} \text{ kg}}{1 \text{ amu}} \right) = 8.56 \times 10^{-28} \text{ kg}$$

(a) *B. E.* in kJ/mol

$$B.E. = \Delta mc^2 = \left(8.56 \times 10^{-28} \text{ kg} \right) \left(3.0 \times 10^8 \ \frac{\text{m}}{\text{s}} \right)^2 = 7.71 \times 10^{-11} \text{ J or } 7.71 \times 10^{-14} \text{ kJ}$$

Now convert this to kJ per mol by using Avogadro's number.

$$B.E. = \left(7.71 \times 10^{-14} \ \frac{\text{kJ}}{\text{atom}} \right) \left(6.023 \times 10^{23} \ \frac{\text{atom}}{\text{mol}} \right) = 4.64 \times 10^{10} \ \frac{\text{kJ}}{\text{mol}}$$

(b) *B.E.* in kJ/nucleon

One iron atom has fifty-six nucleons. Therefore the binding energy per nucleon is obtained as follows:

$$B.E. = \left(7.71 \times 10^{-14} \ \frac{\text{kJ}}{\text{atom}} \right) \left(\frac{1 \text{ atom}}{56 \text{ nucleon}} \right) = 1.38 \times 10^{-15} \ \frac{\text{kJ}}{\text{nucleon}}$$

Example 24.12

When a uranium-235 nuclide is bombarded with a fast-moving neutron in a fission reaction, one possible reaction pathway is the production of zirconium-97 and tellurium-137:

$$ {}^{235}_{92}\text{U} + {}^{1}_{0}\text{n} \longrightarrow {}^{97}_{40}\text{Zr} + {}^{137}_{52}\text{Te} + 2{}^{1}_{0}\text{n}$$

Calculate the mass defect and the binding energy (in MeV/nucleon) for the daughter nuclides. Which of the two has more binding energy?

(a) ${}^{97}\text{Zr}$ (atomic mass = 96.9110 amu)
(b) ${}^{137}\text{Te}$ (atomic mass = 136.9254 amu)

Solution

(a) In ${}^{97}_{40}\text{Zr}$, there are forty protons and fifty-seven neutrons. The mass defect is obtained as

$$\Delta m = [(57 \times \text{mass of neutron}) + (40 \times \text{mass of proton})] - (\text{mass of } {}^{97}\text{Zr nucleus})$$

$$\Delta m = [(57 \times 1.0087 \text{ amu}) + (40 \times 1.0073 \text{ amu})] - (96.9110 \text{ amu}) = 0.8769 \text{ amu}$$

$$\Delta m \text{ in kg} = (0.8769 \text{ amu}) \left(\frac{1.660 \times 10^{-27} \text{ kg}}{1 \text{ amu}} \right) = 1.456 \times 10^{-27} \text{ kg}$$

$$B.E. = \Delta mc^2 = \left(1.456 \times 10^{-27} \text{ kg}\right)\left(3.0 \times 10^8 \, \frac{m}{s}\right)^2 = 1.31 \times 10^{-10} \text{ J or } 1.31 \times 10^{-13} \text{ kJ}$$

B.E. in MeV:
Use the conversion factor $1 \text{ MeV} = 1.602 \times 10^{-13}$ J. Therefore, B.E. per nucleon is

$$B.E. = \left(\frac{1.31 \times 10^{-10} \text{ J}}{\text{nucleus}}\right)\left(\frac{1 \text{ MeV}}{1.602 \times 10^{-13}}\right)\left(\frac{1 \text{ nucleus}}{97 \text{ nucleons}}\right) = 8.43 \, \frac{\text{MeV}}{\text{nucleon}}$$

(b) In $^{137}_{52}$Te, there are fifty-two protons and eighty-five neutrons. The mass defect is
     therefore

$$\Delta m = \left[(85 \times \text{mass of neutron}) + (52 \times \text{mass of protons})\right] - (\text{mass of } ^{137}\text{Te nucleus})$$

$$\Delta m = [(85 \times 1.0087 \text{ amu}) + (52 \times 1.0073 \text{ amu})] - (136.9254 \text{ amu}) = 1.1937 \text{ amu}$$

$$\Delta m \text{ in kg} = (1.1977 \text{ amu})\left(\frac{1.660 \times 10^{-27} \text{ kg}}{1 \text{ amu}}\right) = 1.982 \times 10^{-27} \text{ kg}$$

$$B.E. = \Delta mc^2 = \left(1.982 \times 10^{-27} \text{ kg}\right)\left(3.0 \times 10^8 \, \frac{m}{s}\right)^2 = 1.78 \times 10^{-10} \text{ J}$$

$$\text{or } 1.78 \times 10^{-13} \text{ kJ}$$

B.E. in MeV:
Use the conversion factor $1 \text{ MeV} = 1.602 \times 10^{-13}$ J. Therefore B.E. per nucleon is

$$B.E. = \left(\frac{1.78 \times 10^{-10} \text{ J}}{\text{nucleus}}\right)\left(\frac{1 \text{ MeV}}{1.602 \times 10^{-13}}\right)\left(\frac{1 \text{ nucleus}}{137 \text{ nucleons}}\right) = 8.13 \, \frac{\text{MeV}}{\text{nucleon}}$$

The Zr nuclide is more stable than the Te nuclide by about 0.304 MeV/nucleon.

Example 24.13

Calculate the amount of energy released when 50 g of deuterium (^{2}H) undergoes fusion to
form ^{3}He:

$$^2_1\text{H} + ^2_1\text{H} \longrightarrow ^3_2\text{He} + ^1_0\text{n}$$

The atomic masses of the species involved are:

$$^2_1\text{H} (2.0140 \text{ amu}); \, ^3_2\text{He} (3.01605 \text{ amu}); \text{ and } ^1_0\text{n} (1.0087 \text{ amu})$$

Solution

The energy released is obtained using Einstein's equation

$$\Delta E = \Delta mc^2$$

The mass defect can be calculated using the following expression:

$$\Delta m = (\text{mass of reactants}) - (\text{mass of products})$$

$$\Delta m = \left(2 \times \text{mass of } {}^2_1\text{H}\right) - \left(\text{mass of } {}^3_2\text{He} + \text{mass of } {}^1_0\text{n}\right)$$

$$\Delta m = (2 \times 2.0140 \text{ amu}) - (3.01605 \text{ amu} + 1.0087 \text{ amu}) = 0.0033 \text{ amu}$$

$$\Delta m \text{ in kg} = (0.0033 \text{ amu}) \left(\frac{1.660 \times 10^{-27} \text{ kg}}{1 \text{ amu}}\right) = 5.48 \times 10^{-30} \text{ kg}$$

$$\Delta E = \Delta mc^2$$

$$\Delta E = \left(5.48 \times 10^{-30} \text{ kg}\right) \left(3.0 \times 10^8 \text{ m/s}\right)^2 = 4.93 \times 10^{-13} \text{ J per 2 atoms of } {}^2\text{H}$$

or

$$\Delta E = \left(4.93 \times 10^{-13} \text{ J}\right) \left(\frac{1}{2 \text{ atoms } {}^2\text{H}}\right) = 2.46 \times 10^{-13} \text{ J/atoms of } {}^2\text{H}$$

Calculate the number of atoms in 50 g of ^{2}H:

$$\left(50.0 \text{ g } {}^2\text{H}\right) \times \left(\frac{1 \text{ mol } {}^2\text{H}}{2.01 \text{ g}}\right) \times \left(6.023 \times 10^{23} \frac{\text{atom}}{\text{mol}}\right) = 1.50 \times 10^{25} \text{ atom}$$

1.5×10^{25} atoms of ^{2}H undergo the fusion reaction. The energy released by this number of atoms is:

$$\Delta E = \left(2.46 \times 10^{-13} \frac{\text{J}}{\text{atom}}\right) \left(1.50 \times 10^{25} \text{ atom}\right) = 3.69 \times 10^{12} \text{ J or } 3.69 \times 10^9 \text{ kJ}$$

24.6 Problems

1. Write balanced equations for the following radioactive nuclides undergoing alpha decay.

   (a) Uranium-238
   (b) Actinium-228
   (c) Thorium-232
   (d) Radium-226

2. Write balanced equations for the following radioactive nuclides undergoing beta decay.

   (a) Barium-140
   (b) Iodine-131
   (c) Thallium-232
   (d) Sulfur-36

3. Write a balanced nuclear equation for each of the following.

   (a) Alpha emission of bismuth-213
   (b) Electron capture of lead-208
   (c) Beta emission of tungsten-188
   (d) Positron emission of potassium-40

4. Complete the following nuclear reactions.

   (a) $^{14}_{7}N + ^{4}_{2}He \longrightarrow ? + ^{1}_{1}H$

   (b) $^{55}_{25}Mn + ^{2}_{1}H \longrightarrow ^{56}_{26}Fe + ?$

   (c) $^{235}_{92}U + ^{1}_{0}n \longrightarrow ^{94}_{38}Sr + ? + 3^{1}_{0}n$

   (d) $^{27}_{13}Al + ^{4}_{2}He \longrightarrow ? + ^{1}_{0}n$

   (e) $? + ^{4}_{2}He \longrightarrow ^{14}_{7}N + ^{1}_{0}n$

5. Complete the following nuclear reactions.

   (a) $^{18}_{8}O + ^{1}_{1}H \longrightarrow ? + ^{1}_{0}n$

   (b) $? + ^{1}_{0}n \longrightarrow ^{85}_{36}Kr + ^{1}_{1}H$

   (c) $^{14}_{7}N + ^{4}_{2}He \longrightarrow ? + ^{1}_{1}H$

   (d) $^{11}_{5}B + ? \longrightarrow ^{11}_{6}C + ^{1}_{0}n$

   (e) $^{23}_{12}Mg \longrightarrow ? + ^{0}_{+1}e$

6. Write the symbols for the daughter nuclei in the following reactions.

   (a) $^{14}_{7}N\left(^{4}_{2}\alpha, ^{1}_{1}p\right)$

   (b) $^{68}_{30}Zn\left(^{4}_{2}\alpha, ^{1}_{0}n\right)$

   (c) $^{31}_{15}P\left(^{1}_{0}n, \gamma\right)$

   (d) $^{191}_{77}Ir\left(^{4}_{2}\alpha, ^{1}_{0}n\right)$

7. Complete each of the following and write it as a nuclear reaction.

   (a) $^{18}_{8}O\left(^{1}_{0}n, ^{1}_{1}p\right)?$

   (b) $^{85}_{37}Rb\left(?, ^{1}_{0}n\right)^{85}_{36}Kr$

   (c) $^{58}_{26}Fe\left(2^{1}_{0}n, ?\right)^{60}_{27}Co$

   (d) $?\left(^{4}_{2}\alpha, ^{0}_{-1}\beta\right)^{243}_{97}Bk$

8. Using the shorthand notation, write an equation for each of the following nuclear processes.

   (a) $^{2}_{1}H + ^{3}_{1}H \longrightarrow ^{4}_{2}He + ^{1}_{0}n$

   (b) $^{27}_{13}Al + ^{4}_{2}He \longrightarrow ^{30}_{14}Si + ^{1}_{1}H$

   (c) $^{6}_{3}Li + ^{1}_{0}n \longrightarrow ^{4}_{2}He + ^{3}_{1}H$

   (d) $^{63}_{29}Cu + ^{1}_{1}H \longrightarrow ^{62}_{29}Cu + ^{2}_{1}H$

   (e) $^{40}_{18}Ar + ^{4}_{2}He \longrightarrow ^{43}_{19}K + ^{1}_{1}H$

9. The decay constant for Cobalt-60, used in cancer therapy, is 0.131/year. If a 1.0-g sample of cobalt is stored for twenty-five years, what mass of the isotope will remain?

10. Strotium-90 is one of the nuclides that occur in the fallout from nuclear explosions, and its decay constant is 0.0247 per year. What mass of the isotope will remain after a 0.0258-g sample of strontium-90 is kept for five years?

11. Phosphorus-32 is a radioisotope used in the treatment of blood cancer. A leukemia patient is subjected to a dose of phosphorus-32 radiation therapy. After thirty days, about 25% of the isotope remains in the patient's body. Calculate the rate of disintegration and the half-life of this isotope.

12. The half-life of $^{24}_{11}Na$ is 15.0 h. Calculate:

   (a) the decay constant
   (b) the fraction of the isotope originally present that is left after 6 h
   (c) the mass of the radioactive isotope left when a 10.0mg sample decays for 12 h (use the decay constant from (a)).

13. At 10:00 am on a given day, a nuclear chemist observed that a certain transuranium element stored in one of the national defense labs decayed at a rate of 55,000 counts per minute. At 3:00 pm of the same day the rate of decay had reduced to 15,000 counts per minute. Calculate the half-life of the radioisotope if the rate of decay is proportional to the number of radioatoms in the sample.

14. The production of ammonia by the Haber process is represented by the equation:

$$N_2(g) + 3 H_2(g) \longrightarrow 2 NH_3(g) \quad \Delta H^0 = -92.2 \text{ kJ}$$

Calculate the mass defect (in grams) accompanying the formation of 1 mol of NH_3 from N_2 and H_2 gases.

15. Calculate the mass defects (in g/mol) for the following nuclides.

   (a) ^{56}Fe (atomic mass = 55.9207 amu)
   (b) ^{239}Pu (atomic mass = 239.0006 amu)
   (c) ^{210}Po (atomic mass = 209.9368 amu)
   (d) ^4He (atomic mass = 4.0015 amu)

16. Determine the mass defect for the following reaction.

$$^{235}_{92}U + {}^1_0n \longrightarrow {}^{142}_{56}Ba + {}^{91}_{39}Kr + 3{}^1_0n$$

from the following known masses : ^{235}U (235.0439 amu);
^{142}Ba (141.9164 amu); ^{91}Kr (90.9234 amu); and ${}^1_0n$ (1.00866 amu)

17. Calculate the energy released (in MeV) during each of the following fusion reactions.

   (a) ${}^1_1H + {}^2_1H \longrightarrow {}^3_2He$
   (b) ${}^1_1H + {}^3_2He \longrightarrow {}^4_2He + {}^0_1e$
   (c) ${}^3_1H + {}^2_1H \longrightarrow {}^4_2He + {}^1_0n$
   (d) ${}^1_1H + {}^7_3Li \longrightarrow 2{}^4_2He$

   The atomic masses are ${}^1_1H$ (1.00783 amu); ${}^2_1H$ (2.01410 amu); ${}^3_1H$ (3.01605 amu); ${}^3_2He$ (3.01603 amu); ${}^7_3Li$ (7.01600 amu); ${}^4_2He$ (4.00260 amu); and ${}^0_1e$ (0.0005486 amu); ${}^1_0n$ (1.00866 amu).

18. (a) Write a balanced equation for the nuclear reaction represented by the notation

$$^{56}_{26}Fe \left(^2_1H, \, ^4_2\alpha \right) ^{54}_{25}Mn.$$

   (b) Calculate the energy change in Joules per mole of ^2_1H.

19. Given that $1\,kg = 6.022 \times 10^{26}\,amu$ and that $1\,J = 1\,kgm^2s^{-2}$, show that 1 amu is equivalent to 931.25 MeV of energy. (Hint : $1\,MeV = 1.6 \times 10^{-13}\,J$.)

20. $^{127}_{53}I$ is one of the most stable nuclei known and has an atomic mass of 126.9004 amu. Calculate the total binding energy and the average binding energy per nucleon for this radioisotope (1 amu = 931.25 MeV).

21. The half-life of $^{201}_{81}Tl$, a radioisotope used for parathyroid imaging, is 3.05 days. Calculate the total binding energy per nucleon for the formation of a $^{201}_{81}Tl$ nucleus.

22. A skeleton suspected to be that of the first Onojie (King) of Opoji was recovered by a local archeologist from an ancient palace site. Radiocarbon measurements made of the bones showed a decay rate of 18.5 disintegration of ^{14}C per minute per gram of carbon. Determine the age of the suspected Onojie if the carbon from living material taken from the same area decays at the rate of 23.1 disintegrations per minute per gram of carbon. The half-life of ^{14}C is 5715 years.

Appendix A

....

A.1 Essential mathematics

A.1.1 Significant figures

Significant figures are the number of digits in a measured or calculated value that are statistically significant or offer reasonable and reliable information. Significant figures in any measurement usually contain digits that are known with certainty and one digit that is uncertain, starting with the first non-zero digit. To determine the number of significant figures, certain rules must be followed.

1. All non-zero digits (1–9) are significant. For example, 125 has three significant figures and 14.44 has four significant figures.

    (a) Leading zeros to the left of the first non-zero digit in the number are not significant. They are only used to fix the position of the decimal. For example:

      - 0.007 has one significant figure
      - 0.000105 has three significant figures
      - 0.000000000015 has two significant figures

2. Zeros between non-zero digits are significant. For example:

    - 5.005 has four significant figures
    - 50.05 has four significant figures
    - 500.0000075 has ten significant figures

3. All zeros to the right of the decimal point are significant. For example:

    - 25.00 has four significant figures
    - 0.1250 has four significant figures
    - 0.2000 has four significant figures

4. Zeros at the end of a number may or may not be significant. They are significant only if there is a decimal point in the number. For example:

    - 1500 has two significant figures
    - 602,000,000,000,000,000,000 has three significant figures
    - 404,570,000 has five significant figures
    - 0.00580 has three significant figures

A.1.2 Rounding-off

Rounding off is a process of eliminating non-significant digits from a calculated number. There are three rules that govern rounding off:

1. If the non-significant digit is less than 5, remove it and all digits to its right. For example, 100.5129 is equal to 100.5 if rounded off to four significant figures.
2. If the non-significant digit is more than or equal to 5, remove it and increase the last reported digit by 1. For example 15.58729 is equal to 15.59 to four significant figures.
3. If the non-significant digit to the right of the last digit to be retained is a 5 followed by zeros, the last digit is increased by one if it is odd and is left unchanged if it is even. For example, 64.750 is equal to 64.8 to three significant digits. Also, 25.850 is equal to 25.8 to three significant figures.

A.2 Significant figures and mathematical operations

Two rules guide the operations of multiplication, division, addition, and subtraction such that the precision of the measurement may be maintained throughout a calculation.

A.2.1 Multiplication and division

For mathematical operations involving multiplication and division, the number of significant figures in the product or quotient is the same as the number involved in the operation with the least number of significant digits. For example: in the operation 5.0085 g $\times$ 3.55, the number 3.55 limits our answer to three significant figures. Although the calculator gives the product as 17.7802 g, the answer should be expressed as 17.8 g, i.e. (i.e., $5.0085 \times 3.55 = 17.7802 = 17.8$).

Also, in the operation 2.1259/5.5, the number 5.5 limits the final answer to two significant figures (sf):

$$\frac{2.1259}{5.5} = 0.3865 = 0.39 \,(2\,\text{s.f.})$$

Example A.1

Multiply 20.25 by 125.125 and give the answer with the proper significant figures.

Solution

$20.25 \times 125.125 = 253378.125$

Final answer is 253400 (4 sf)

Example A.2

Divide 1960.2008 by 2005.252 and give the answer with the proper significant figures.

Solution

$$\frac{1960.2008}{2005.252} = 0.9775333972 = 0.9775334\,(7\,\text{s.f.})$$

A.2.2 Addition and subtraction

For mathematical operations involving addition and subtraction, the last digit retained in the sum or difference is determined by the position of the first doubtful digit. This means that the answer cannot have more digits to the right of the decimal point than either of the original numbers.

Example A.3

Add the following and give your answer in the proper number of significant figures.

$$12.08 + 1.2575$$

Solution

$$
\begin{array}{r}
12.08 \\
+\,1.2575 \\
\hline
13.3375
\end{array}
\Rightarrow 13.34 \ (\text{final answer})
$$

Example A.4

Subtract 34 .2345 g from 39.251 and express your answer in the proper significant figures.

Solution

$$
\begin{array}{r}
39.251 \\
34.2346 \\
\hline
5.0164
\end{array}
\Rightarrow 5.016 \ (\text{final answer})
$$

A.3 Scientific notation and exponents

In chemistry one frequently encounters numbers that are either very large or very small. For example, Avogadro's number is 602,200,000,000,000,000,000,000 per mole, and the mass of an electron is 0.00000000000000000000000000000091091 kg. These types of numbers are difficult to handle in routine calculations, and chances of errors are high. To reduce the chances of typographical error, and convey explicit information about the precision of measurements, scientists usually write such very large and small numbers in a more convenient way known as scientific notation. A number is expressed in *scientific notation* or *standard form* if it is written as a power of 10, i.e. A $\times$ 10^n, where A is A is a number between 1 and 10, and the exponent n is a positive or negative integer. To write a number in scientific notation, simply move the decimal in the original number to get A, the first nonzero digit. A is greater than or equal to 1 but less than 10. Then multiply A by 10^n, where n is the number of decimal places that the decimal point has been moved. If the decimal point has been moved to the left, the exponent n will have a positive value. If the decimal point has been moved to the right, n will have a negative value and A will be multiplied by 10^{-n}.

Example A.5

Express 1,250,000 in scientific notation.

Solution

Place the decimal between 1 and 2 to get the number A $=$ 1.25. Since the decimal point was moved 6 places to the left, the power n will be equal to $+6$. Therefore, we can write the number in scientific notation as: $1{,}250{,}000 = 1.25 \times 10^6$.

Example A.6

Write 0.0000000575 in scientific notation.

Solution

Place the decimal point between the first 5 and the 7 to get the number A $=$ 5.75. Since the decimal was moved 8 places to the right, the exponent n will be -8. The number in exponential form is $0.0000000575 = 5.75 \times 10^{-8}$.

A.3.1 Addition and subtraction

To add or subtract numbers in scientific notation, it is necessary to first express each quantity to the same power of ten. The digit terms are added or subtracted in the usual manner.

Example A.7

Carry out the following arithmetic operation and give your answer in scientific notation.

(a) $\left(2.35 \times 10^{-5}\right) + \left(1.8 \times 10^{-6}\right)$
(b) $\left(3.33 \times 10^{4}\right) - \left(1.21 \times 10^{3}\right)$

Solution

(a) $\left(2.35 \times 10^{-5}\right) + \left(1.8 \times 10^{-6}\right) = \left(2.35 \times 10^{-5}\right) + \left(0.18 \times 10^{-5}\right) = 2.53 \times 10^{-5}$
(b) $\left(3.33 \times 10^{4}\right) - \left(1.21 \times 10^{3}\right) = \left(3.33 \times 10^{4}\right) - \left(0.121 \times 10^{4}\right) = 3.21 \times 10^{4}$

A.3.2 Multiplication and division

To multiply numbers in scientific notation, first multiply the decimals in the usual way, and then add their exponents. In division, the digit term in the numerator is divided by the digit term in the denominator in the usual manner. Then the denominator's exponent is subtracted from the numerator's exponent to obtain the power of 10. In either case, the decimal point may have to be adjusted so that the result has one nonzero digit to the left of the decimal.

Example A.8

Perform the following arithmetic operations:

(a) $(5.5 \times 10^3) \times (2.8 \times 10^5)$

(b) $\left(\dfrac{2.1 \times 10^4}{9.7 \times 10^7}\right)$

(c) $\left(\dfrac{4.3 \times 10^{-4}}{9.7 \times 10^7}\right) \times \left(\dfrac{5.1 \times 10^{-4}}{3.7 \times 10^{-7}}\right)$

Solution

(a) $(5.5 \times 10^3) \times (2.8 \times 10^5) = (5.5 \times 2.8) \times 10^{(5+3)}$

$$(15.4 \times 10^8) = 1.54 \times 10^9$$

(b) $\left(\dfrac{2.1 \times 10^4}{9.7 \times 10^7}\right) = \left(\dfrac{2.1}{9.7}\right) \times 10^{(4-7)} = 0.217 \times 10^{-3} = 2.17 \times 10^{-4}$

(c) $\left(\dfrac{4.3 \times 10^{-4}}{9.7 \times 10^7}\right) \times \left(\dfrac{5.1 \times 10^{-4}}{3.7 \times 10^{-7}}\right)$

$$= \left(\dfrac{4.3}{9.7}\right) \times 10^{-11} \times \left(\dfrac{5.1}{3.7}\right) \times 10^3$$

$$= \left(0.443 \times 1.378\right) \times 10^{(-11+3)}$$

$$= 0.6106 \times 10^{-8} = 6.11 \times 10^{-9}$$

A.3.3 Powers and roots

Generally a number in scientific notation $A \times 10^n$ raised to the power b can be evaluated as follows:

$$(A \times 10^n)^b = A^b \times 10^{n \times b}$$

For example, $(3.05 \times 10^3)^2 = (3.05)^2 \times 10^{3 \times 2} = 9.30 \times 10^6$.

The root of an exponential number of the general form $A \times 10^n$ can be extracted by raising the number to fractional power such as $1/2$, for square root, and $1/3$ for cube root. To evaluate the root of an exponential number, move the decimal point in A such that the exponent in the power of base 10 is exactly divisible by the root. Generally:

$$\sqrt[p]{(A \times 10^n)} = \left(\sqrt[p]{A} \times 10^{\frac{n}{p}}\right)$$

Recall the following laws of exponents: $\sqrt[n]{A} = A^{\frac{1}{n}}$ and $\sqrt[n]{A^m} = A^{\frac{m}{n}}$.

Example A.9

Evaluate the following expressions:

(a) $\sqrt[3]{0.64 \times 10^{-4}}$

(b) $\sqrt{8.1 \times 10^8}$

Solution

(a) $\sqrt[3]{0.64 \times 10^{-4}} = \sqrt[3]{64 \times 10^{-6}} = \sqrt[3]{64} \times 10^{\frac{-6}{3}}$

$\quad = 4.0 \times 10^{-2}$

(b) $\sqrt{8.1 \times 10^8} = \sqrt{8.1} \times 10^{8/2} = 2.8 \times 10^4$

A.4 Logarithms

Many problems such as those involving pH, chemical kinetics, or nuclear chemistry require the use of logarithms and antilogarithms. There are two types of logarithm: common and natural.

A.4.1 Common logarithm

The common logarithm of a number N (abbreviated as log) is the exponent to which 10 must be raised to obtain that number N. Mathematically, this is expressed as:

$$N = 10^x \,(10 \text{ is the base}; x \text{ is the exponent}), \text{ and}$$
$$\log_{10}N = x \,(10 \text{ is the base}; x \text{ is the logarithm})$$

For example:

$$
\begin{aligned}
10{,}000 &= 10^4 & \log 10{,}000 &= 4 \\
1{,}000 &= 10^3 & \log 1{,}000 &= 3 \\
10 &= 10^1 & \log 10 &= 1 \\
1 &= 10^0 & \log 1 &= 0 \\
0.1 &= 10^{-1} & \log 0.1 &= -1 \\
0.001 &= 10^{-3} & \log 0.001 &= -3 \\
0.0001 &= 10^{-4} & \log 0.0001 &= -4
\end{aligned}
$$

The common logarithm of numbers between 1 and 10 or between 10 and 100 can be read from a logarithm table or from a calculator. For example, the logarithm of 45, which lies between 10 and 100, is 1.6532. Hence, 45 may be expressed as $45 = 10^{1.6532}$. The logarithm of a number greater than 1 is positive, and that of a number lesser than 1 is negative.

A.4.2 Antilogarithms

The common antilogarithm (or inverse logarithm) of a number (abbreviated as antilog) is the value obtained by raising the base to the power of the logarithm. Thus finding the antilogarithm of a number is the reverse of evaluating the logarithm of the number. For example, if log $_{10}N = 3$, the antilog of $N = 10^3$, or 1000.

A.4.3 Natural logarithms

The natural logarithm of a number N (abbreviated as ln N) is the power to which the number e (where $e = 2.718$) must be raised to give N. The base e is a mathematical constant encountered in many science and engineering problems. The natural antilogarithm of x is also referred to as the power of e, represented as e^x.

$$N = e^x = 2.718^x, \text{ and}$$
$$\ln e^x = x$$
$$\Rightarrow \ln e^5 = 5$$

Common and natural logarithms are related as follows:

$$\ln x = \ln 10 \cdot \log x$$
$$= 2.303 \log x$$

Thus it is possible to express the natural logarithm of a number in terms of the common logarithm (to the base 10).

A.4.4 Calculations involving logarithms

The following relationships hold true for all logarithms:

1. $\log_a (AB) = \log_a A + \log_a B$, and ln (AB) = ln A + ln B

2. $\log_a \left(\dfrac{A}{B} \right) = \log_a A - \log_b B$, and $\ln \left(\dfrac{A}{B} \right) = \ln A - \ln B$

3. $\log_a A^x = x \log_a A$ and ln $A^x = x$ ln A

4. $\log_a \sqrt[x]{A} = \log_a A^{1/x} = \dfrac{1}{x} \log_a A$ and ln $\sqrt[x]{A} = \ln A^{1/x} = \dfrac{1}{x}$ ln A

Example A.10

Find the log of the following:

1. $\log \left(2.51 \times 10^5 \right)$

2. $\log \left(8.6 \times 10^{-4} \right)$

Solution

1. $\log(2.51 \times 10^5) = \log 2.51 + \log 10^5$

$$= 0.3997 + 5.0000$$

$$= 5.3997$$

2. $\log(8.6 \times 10^{-4}) = \log 8.6 + \log 10^{-4}$

$$= 0.9345 + (-4.0000)$$

$$= 3.0655$$

Example A-11

Find the antilog of the following:

(a) 23.7796

(b) −15.5515

Solution

(a) antilog of $23.7796 = 10^{23.7796} = 6.02 \times 10^{23}$
    The antilog of 23.7796 is 6.02×10^{23}

(b) antilog of -15.5515

The antilog of a negative log such as -15.5515 cannot be found directly.
The following steps are helpful in finding the antilogarithm of a negative number:

1. Rewrite the log as the sum of a negative whole number and a positive decimal number
   $-15.5515 = (-16 + 0.4485)$
2. Then find the antilog of the decimal portion
   antilog of $0.4485 = 10^{0.4485} = 2.8087$
3. Multiply the antilog by 10 raised to the power of the negative whole number,
   i.e. 2.8087×10^{-16}
4. Thus the antilog of -15.7796 is equal to 2.8×10^{-16}.

A.5 Algebraic equations

Chemistry students will invariably encounter mathematical problems that require basic knowledge of algebraic operations. Two common cases are simple linear equations and quadratic equations.

A.5.1 Linear equations

A linear equation is an equation that contains one unknown variable whose largest exponent is 1. It can be expressed in the following general form:

$$bx^1 = c \ (b \neq 0, c = \text{real number})$$

Basic rules of algebraic transformation are needed to solve for the unknown in a linear equation. The necessary steps are:

- Clear any parentheses
- Isolate the unknown or collect like terms
- Solve for the unknown.

Recall the following:

- Multiplication and division must be completed before addition and subtraction
- Adding or subtracting the same number to both sides of the equation does not alter the equation
- Multiplying or dividing both sides of the equation by the same number does not alter the equation.

Example A.12

Solve for x in each of the following equations.

1. $4x = 12$
2. $\dfrac{3x}{5} = \dfrac{2}{7}$
3. $106 = \dfrac{9}{5}x + 32$
4. $\dfrac{5}{3x} = \dfrac{2}{7}$

Solution

1. $4x = 12$

   Divide both sides by 4
   $$\frac{\cancel{4}x}{\cancel{4}} = \frac{\cancel{12}}{\cancel{4}} = 3$$
   $$x = 3$$

2. $\dfrac{3x}{5} = \dfrac{2}{7}$

   Multiply both sides by $\dfrac{5}{3}$
   $$\frac{\cancel{5} \times \cancel{3}x}{\cancel{3} \times \cancel{5}} = \frac{5 \times 2}{3 \times 7}$$
   $$x = \frac{10}{21}$$

3. $106 = \dfrac{9}{5}x + 32$

First substract 32 from both sides of the equation and rearrange

$$106 - 32 = \dfrac{9}{5}x + 32 - 32$$

$$\dfrac{9}{5}x = 74$$

Multiply both sides by $\dfrac{5}{9}$

$$\dfrac{\cancel{5}}{\cancel{9}} \times \dfrac{\cancel{9}}{\cancel{5}}x = \dfrac{5}{9} \times 74$$

$$x = \dfrac{370}{9} \text{ or } 41.1$$

4. $\dfrac{5}{3x} = \dfrac{2}{7}$

First cross-multiply the expression

$$3x \times 2 = 5 \times 7 \Rightarrow 6x = 35$$

Now divide both sides of the equation by 6

$$\dfrac{\cancel{6}x}{\cancel{6}} = \dfrac{35}{6}$$

$$x = 5.83$$

A.5.2 Straight-line graphs and linear equations

The equation of a straight-line graph can generally be written as $y = mx + b$. In graphing this linear equation, y is treated as the dependent variable, while x is the independent variable. The constant m is called the slope (or gradient) of the line, while b is the point at which the line intercepts the y-axis.

The slope of a line is obtained as the ratio of the rate of change in y to that in x:

$$m = \frac{\Delta y}{\Delta x} = \frac{y_2 - y_1}{x_2 - x_1}$$

Let us consider the linear equation $y = 4x + 25$. The slope is 4, and the intercept is 25, occurring when $x = 0$. The plot is illustrated by Figure A.1.

The slope can be verified directly from the graph by considering any two points, say $(x_1, y_1 = 5, 45)$ and $(x_2, y_2 = 15, 85)$.

$$m = \frac{\Delta y}{\Delta x} = \frac{y_2 - y_1}{x_2 - x_1} = \frac{85 - 45}{15 - 5} = \frac{40}{10} = 4$$

Examples of some linear equations you will encounter in chemistry include:

1. First-order rate law
$$\underbrace{\ln [A]_t}_{y} = \underbrace{(-k)t}_{mx} + \underbrace{\ln [A]_0}_{b}$$

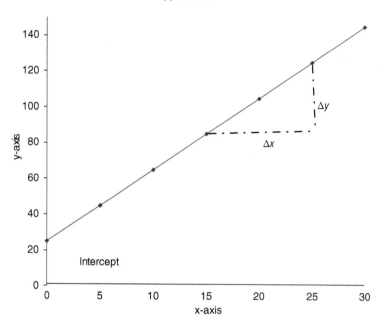

Figure A.1   The graph of $y = 4x + 25$.

2. Second-order rate law
$$\underbrace{\frac{1}{[A]_t}}_{y} = \underbrace{kt}_{mx} + \underbrace{\frac{1}{[A]_0}}_{b}$$

3. The Arrhenius equation
$$\underbrace{\ln k}_{y} = \underbrace{-\left(\frac{Ea}{R}\right)}_{m}\underbrace{\left(\frac{1}{T}\right)}_{x} + \underbrace{\ln A}_{b}$$

4. The van't Hoff equation
$$\underbrace{\ln K_{eq}}_{y} = \underbrace{\left(-\frac{\Delta H_{vap}}{R}\right)}_{m}\underbrace{\left(\frac{1}{T}\right)}_{x} + \underbrace{C}_{b}$$

A.5.3 Quadratic equations

A quadratic equation is a polynomial equation in which the largest exponent of the unknown variable is 2. It can be expressed in the following general form:

$$ax^2 + bx + c = 0, \text{ (where } a, b, \text{ and } c \text{ have real values, and } a \neq 0)$$

The second degree term, ax^2, is called the quadratic term. The first degree term bx is called the linear term, while the numerical term, c, is the constant term.

Many general chemistry problems involving chemical equilibria require knowledge of quadratic equation. The goal here is to summarize some of the methods that can be used to solve quadratic equations.

A.5.3.1 Solution of quadratic equations

Three solutions of quadratic equations are described below. They include:

1. Extraction of the square root
2. Factorization
3. The quadratic formula.

A.5.3.2 Extraction of the square root

If $b = 0$, the general equation above becomes $ax^2 + c = 0$. The roots or solutions of this type of quadratic equation are numerically equal but opposite in sign.

Example A.13

Solve the following quadratic equations

(a) $x^2 - 25 = 0$
(b) $3x^2 - 6.02 \times 10^{23} = 0$

Solution

(a) $x^2 - 25 = 0$
$$\sqrt{x^2} = \sqrt{25}$$
$$x = \pm 5$$

(b) $3x^2 - 6.02 \times 10^{23} = 0$
$$3x^2 = 6.02 \times 10^{23}$$
$$x^2 = \frac{6.02 \times 10^{23}}{3} = 2.01 \times 10^{23}$$
$$x = \pm\sqrt{\left(2.01 \times 10^{23}\right)} = \pm 4.48 \times 10^{11}$$

A.5.3.3 The factor method

Every reducible quadratic trinomial of the form $ax^2 + bx + c = 0$, can be solved by the factor method. This involves binomial factors represented as $Ax + B$ and $Cx + D$. The product of the two binomial terms gives:

$$(Ax + B)(Cx + D) = ACx^2 + (AD + BC)x + BD$$

Therefore, solving a quadratic equation by the factor method involves finding the values of the coefficients A, B, C, and D that will satisfy the equation, i.e. make $AC = a$, $AD + BC = b$, and $BD = c$.

Example A.14

Solve $x^2 + 7x + 12$ by the factor method.

Solution

1. Identify the possible combination of factors which when multiplied gives the original equation or makes $AD + BC = b$ or 7

$AC = 1, AD + BC = 7, BD = 12$

$(x+2)(x+6) = x^2 + 8x + 12$

$(x+3)(x+4) = x^2 + 7x + 12$

$(x+1)(x+12) = x^2 + 13x + 12$

The factors of 12 which make $AD + BC = 7$ and $BD = 12$ are 3 and 4.

2. Set each factor equal to zero and solve the resulting linear equations.

$(x+3)(x+4) = 0$

$x + 3 = 0$ or $x + 4 = 0$

$x = -3$ or $x = -4$

A.5.3.4 Quadratic formula

Many quadratic equations cannot be solved by the factor method. In such cases the quadratic formula may be used. The solution of a quadratic equation of the general form $ax^2 + bx + c = 0$ is given by the quadratic formula:

$$x = \frac{-b \pm \sqrt{b^2 - 4ac}}{2a}$$

Example A.15

Solve $2x^2 + x - 16 = 0$ using the formula method.

Solution

1. Comparing this to the general equation, $a = 2$, $b = 1$, and $c = -16$.
2. Solve for x by substituting the values of a, b, and c in the quadratic formula.

$$x = \frac{-b \pm \sqrt{b^2 - 4ac}}{2a}$$

$$= \frac{-1 \pm \sqrt{(-1)^2 - 4(2)(-16)}}{2(2)}$$

$$= \frac{-1 \pm \sqrt{129}}{4}$$

$$= \frac{-1 \pm 11.36}{4}$$

$$x = 2.56 \text{ or } -3.09$$

Thus the solution set is $\{2.56, -3.09\}$

Appendix A: Sample Problems

1. Determine the number of significant figures in each of the following:

   (a) 0.0038   (b) 2.0038   (c) 125   (d) 50,008   (e) 0.00022500   (f) 40.000

2. Correctly round off the final answer in each of the following:

   (a) $(2.254)(5.55) = 12.5097$

   (b) $\dfrac{18.322}{191750} = 9.55515$

   (c) $\dfrac{(4.3)(8.51)(20.5360)}{6.750} = 11.32946$

3. Carry out the following mathematical operations and express the answer the proper number of significant figures.

   (a) $55.58\ \text{g} + 20.0015\ \text{g} + 33.2\ \text{g}$

   (b) $18.25\ \text{mL} - 1.55\ \text{mL}$

   (c) $\dfrac{(98,365)(0.02365)}{(44.0556)} - 2.3$

   (d) $(3.1258 - 2.303)(0.00022)$

4. Express the following numbers in standard scientific notation.

   (a) 525000   (b) 0.00000911   (c) 602,000,000,000,000,000,000,000

   (d) -0.00010

5. Express the following in decimal form:

   (a) 2.345×10^0 (b) 7.85×10^5 (c) -2.345×10^{-7} (d) 0.0345×10^4 (e) 25.0×10^{-3}

6. Carry out the indicated mathematical operations and leave your final answer in exponential form:

   (a) $(1.25 \times 10^2) + (7.25 \times 10^3) + (2.5 \times 10^4)$
   (b) $(3.15 \times 10^{-2}) - (7.25 \times 10^{-3})$
   (c) $(5.1 \times 10^{-2}) + (6.23 \times 10^3) - (1.15 \times 10^2)$
   (d) $(0.81 \times 10^6) - (2.25 \times 10^4) - (1.15 \times 10^0)$

7. Perform the indicated mathematical operations, expressing the result in exponential form:

   (a) $(2.55 \times 10^4)(1.85 \times 10^{-6})$

   (b) $(4.45 \times 10^{-12})(9.55 \times 10^{-4})(2.55 \times 10^9)$

   (c) $\sqrt{(8.2 \times 10^7)}$

   (d) $(3.2 \times 10^{-5})^2$

   (e) $\left(\sqrt[3]{6.4 \times 10^{10}}\right)^2$

8. Carry out the following calculations, expressing your result in exponential form:

(a) $\dfrac{\left(2.3 \times 10^{-3}\right)}{\left(5.3 \times 10^{-6}\right)}$

(b) $\dfrac{\left(9.3 \times 10^{-3}\right)}{\left(0.35 \times 10^{6}\right)}$

(c) $\dfrac{\left(8.3 \times 10^{3}\right)\left(7.4 \times 10^{-9}\right)}{\left(0.35 \times 10^{6}\right)\left(1.3 \times 10^{-13}\right)}$

(d) $\dfrac{\left(5.3 \times 10^{-5}\right)}{\left(7.5 \times 10^{6}\right)}\left(6.3 \times 10^{12}\right) - \left(9.3 \times 10^{-3}\right)$

9. Find the approximate value for

$$\frac{(987{,}000{,}234)(0.000, 000, 000, 2156)}{(675.87)(0.000, 000, 321456)}$$

10. If $a = 3{,}000$, $b = 0.0009825$, $c = 3.142$, determine the approximate value for $\dfrac{a^2 b}{c^3}$

11. Solve for x in the following

(a) $\log_{10} x = 6$
(b) $\log_3 81 = x$
(c) $\log_x 25 = 2$
(d) $\log_{\frac{1}{3}} 9 = x$

12. Evaluate X in the following expression

(a) $\log_{10} X = \log_{10} 5 + \log_{10} 3$
(b) $\log_{10} X = \log_{10} 8 - \log_{10} 12$
(c) $\log_{10} X = 1 + \log_{10} 6$
(d) $X = -\log_{10} 10^{-12.5}$
(e) $\log_{10} (x+1) - \log_{10} (x-1) = 1$

13. Solve for x in the following equations:

(a) $12x - 12 = 36$
(b) $\dfrac{x}{5} = \dfrac{5}{75}$
(c) $6x + 12 = 3x - 6$
(d) $5(2x+1) - 3(x-2) = 4$

14. Solve for E_a (in kJ) in the following expression:

$$\ln\frac{k_2}{k_1} = \frac{-E_a}{R}\left(\frac{1}{T_2} - \frac{1}{T_1}\right)$$

given that $k_1 = 7.5 \times 10^{-8} \text{s}^{-1}$ at 298 K, $k_2 = 4.7 \times 10^{-6} \text{s}^{-1}$ at 318 K, and $R = 8.314$ J/K/mol

15. Determine k at $T = 1500$ K from the following expression:

$$\ln k = A + \frac{\ln T}{2} - \frac{22000}{T}$$

where $A = 21.3$ L.mol$^{-1}\cdot$s^{-1}

16. Solve for x in the following equations

    (a) $x^2 + 7x + 12 = 0$

    (b) $(x+3)^2 = 16$

    (c) $10x^2 - 9x + 2 = 0$

    (d) $\dfrac{2}{x} - \dfrac{3}{x+1} - 1 = 0$

    (e) $\log_{10}(x+2)(x-1) = 1$

Appendix A: Solutions to Problems

1. (a) Two   (b) Five   (c) Three   (d) Five   (e) Five   (f) Five

2. (a) The least certain number here is 5.55. Therefore, round the final answer to three significant figures, i.e. $(2.254)(5.55) = 12.5$.

   (b) The least certain number here is 18.322. Round the final answer to five significant figures, i.e. $\dfrac{18.322}{1.91750} = 9.5552$.

   (c) The least certain number here is 4.3. Round the final answer to two significant figures, i.e. $\dfrac{(4.3)(8.51)(20.5360)}{6.750} = 110$.

3. (a) 108.8

   (b) 16.7

   (c) 50.5

   (d) 0.00018

4. (a) 5.25×10^5

   (b) 9.11×10^{-6}

   (c) 6.02×10^{23}

   (d) -1.0×10^{-4}

5. (a) 2.345

   (b) 785,000

   (c) -0.0000002345

   (d) 345

   (e) 0.025

6. (a) (3.2×10^4)

   (b) (2.43×10^{-2})

   (c) (6.12×10^3)

   (d) $7.9 \times 10^5)$

7. (a) (4.72×10^{-2})
   (b) (1.08×10^{-5})
   (c) $9.1 \times 10^{3})$
   (d) $1.0 \times 10^{-9})$
   (e) (1.6×10^{7})

8. (a) (4.3×10^{2})
   (b) (2.7×10^{-8})
   (c) (1.3×10^{3})
   (d) (4.5×10^{1})

9. 979.4

10. 285.1

11. (a) $x = 1{,}000{,}000$
    (b) $x = 4$
    (c) $x = 5$
    (d) $x = -2$

12. (a) 15
    (b) 0.67
    (c) 60.0
    (d) 12.5
    (e) 1.2

13. (a) 4
    (b) 0.33
    (c) −6
    (d) −1

14. 163 kJ

15. 2.94×10^{4} L.mol^{-1}s^{-1}

16. (a) $x = -4$ or $x = -3$
    (b) $x = 1$ or $x = -7$
    (c) $x = 0.4$ or $x = 0.5$
    (d) $x = 0.732$ or $x = -2.732$
    (e) $x = -4$ or $x = 3$

Appendix B

. . . .

B.1 Systems of measurement

B.1.1 Measurements in chemistry

Chemistry is an experimental science that involves measuring the magnitudes of various properties of substances. Measurements generate numbers, and we need units attached to these numbers so we can tell what exactly is being measured. Some examples of quantities measured include amount, mass, pressure, size, temperature, time and volume. There are several systems of units, including the English system and the metric system of units. The metric system is the most commonly used system of measurement in chemistry. In 1960, a modernized version of the metric system known as the Systeme Internationale (or SI system) was recommended for worldwide adoption. The SI, based on the metric system, consists of seven fundamental units and several units derived from them. These units serve all scientific measurements. The seven fundamental units are listed in Table B.1.

The fundamental units do have some shortcomings. For example, in some cases the base unit is inconveniently large or small. To overcome this, the SI units can be modified through the use of prefixes. The prefixes define multiples or fractions of the base or fundamental units. Some examples are listed in Table B.2.

Table B.1   The Seven Fundamental Units of Measurement (SI)

Physical quantity	Name of unit	Symbol
Length	meter	m
Mass	kilogram	kg
Time	seconds	s
Amount of substance	mole	mol
Electric current	ampere	A
Temperature	kelvin	K
Luminous intensity	candela	cd

Table B.2   SI prefixes

Prefix	Symbol	Numerical value	Factor
exa	E	1,000,000,000,000,000,000	10^{18}
peta	P	1,000,000,000,000,000	10^{15}
tera	T	1,000,000,000,000	10^{12}
giga	G	1,000,000,000	10^{9}
mega	M	1,000,000	10^{6}
kilo	k	1,000	10^{3}
hecto	h	100	10^{2}
deka	da	10	10^{1}
deci	d	0.1	10^{-1}
centi	c	0.01	10^{-2}
milli	m	0.001	10^{-3}
micro	μ	0.000001	10^{-6}
nano	n	0.000000001	10^{-9}
pico	p	0.00000000000 1	10^{-12}
femto	f	0.00000000000000 1	10^{-15}
atto	a	0.00000000000000000 1	10^{-18}

Notes:
Commonly used prefixes are highlighted in bold

B.2 Measurement of mass, length, and time

Mass, length, and time are fundamental quantities commonly measured in science and are assigned fundamental units.

B.2.1 Measurement of mass

The standard unit of mass in the SI unit is the kilogram (kg), which is defined as the mass of a certain block of platinum-iridium alloy, also known as the prototype kilogram, kept in a vault at the International Bureau of Weights and Measures, in Sevres, France. Other mass units that are related to the kilogram include gram (g), milligram (mg), and microgram (µg). Here is how they are related:

$$1000 \text{ g} = 1 \text{kg}$$
$$1000 \text{ mg} = 1 \text{g}$$
$$1,000,000 \text{ } \mu g = 1 \text{g}$$

Note that the symbol for grams is 'g', not gm, gms, gr, or anything else.

B.2.2 Measuring length

The SI unit of length is the meter, which is defined as the distance light travels in approximately $\frac{1}{299,792,458}$ second. Another convenient unit of length is the centimeter (cm) which is related to the meter as follows:

$$100 \text{ cm} = 1 \text{ m or } 1 \text{ cm} = \frac{1}{100} \text{ (or } 0.01) \text{m}$$

B.2.3 Measuring time

The SI unit of time is the second (s) and is defined as 9,192,631,770 oscillations of the light emitted by cesium atoms.

B.3 Temperature

Temperature is a measure of the intensity of heat, or hotness, of a body. Three temperature scales are commonly encountered in chemistry:

1. Fahrenheit scale (°F)
2. Celsius (centigrade) scale (°C)
3. Kelvin (absolute) scale (K).

Note that no degree symbol is used with Kelvin temperatures.

On the Fahrenheit scale, water freezes at 32°F and boils at 212°F. On the Celsius scale, water freezes at 0°C and boils at 100°C. On the Kelvin scale, water freezes at 273 K and boils at 373 K. There are 100° between the freezing and boiling points of water on both the Celsius and Kelvin scales and 180° difference on the Fahrenheit scale. Thus:

$$1 \text{ Fahrenheit degree} = \frac{100}{180} \left(\text{or } \frac{5}{9} \right) \times \text{Celsius degree or}$$

$$1 \text{ Celsius degree} = \frac{180}{100} \left(\text{or } \frac{9}{5} \right) \times \text{Fahrenheit degree}$$

We can convert a temperature from one scale to another by using the following equations:

1. Converting from Celsius to Fahrenheit

$$°F = \left(\frac{9}{5} \times °C \right) + 32$$

2. Converting from Fahrenheit to Celsius

$$°C = \frac{5}{9} \times \left(°F - 32 \right)$$

3. Converting from Celsius to Kelvin

$$K = °C + 273$$

4. Converting from Celsius to Kelvin

$$°C = K - 273$$

Example B.1

Express 37°C on the Fahrenheit scale

Solution

$$^\circ F = \left(\frac{9}{5} \times {}^\circ C\right) + 32$$

$$^\circ F = \left(\frac{9}{5} \times 37\right) + 32 = 98.6^\circ F$$

Example B.2

Helium boils at −452°F. Convert this temperature to °C and K.

Solution

1. Converting −452°F to °C.

$$^\circ C = \frac{5}{9} \times \left({}^\circ F - 32\right)$$

$$^\circ C = \frac{5}{9} \times (-452 - 32)$$

$$= -269^\circ C$$

2. Converting −452°F to K.

From part 1 above, −452°F = −269°C

$$K = {}^\circ C + 273$$
$$= -269 + 273$$
$$= 4 \text{ K}$$

Thus −452°F corresponds to −269°C and 4 K.

B.4 Derived units

Derived units are units obtained by applying algebraic operations (multiplication or division) to the fundamental units. For example, the unit for volume is the cubic meter, which is derived as follows:

$$\text{Volume of a solid} = \text{length} \times \text{width} \times \text{height} = m \times m \times m = m^3$$

Examples of some derived units are shown in Table B.3.

Table B.3   Examples of some derived units

Quantity	Derived unit	Symbol for unit
Area	square meter	m^2
Volume	cubic meter	m^3
Density	kilogram per cubic meter	kgm^{-3}
Force	Newton	N (i.e. $kgms^{-1}$)
Energy	joule	J (i.e. $kgm2s^{-2}$)
Heat capacity	joule per degree kelvin	JK^{-1}
Electric charge	coulomb (amper.second)	Q (A.s)
Gas constant	joule per kevin per mole	$JK^{-1}mol^{-1}$

B.5 Density and specific gravity

B.5.1 Density

Density is defined as the mass of a substance per unit volume. In mathematical terms:

$$\text{Density } (\rho) = \frac{\text{Mass}}{\text{Volume}}$$

Any units of mass and volume may be used but g/mL (or cm^3) and kg/m^3 are the most common. Like the volume of a substance, density varies with temperature. For example, water has a density of 1.0000 g/mL at 4°C and 0.9718 g/mL at 80°C.

Example B.3

The density of carbon tetrachloride is 1.595 g/mL at 20°C. Calculate the volume occupied by 770 g of carbon tetrachloride at this temperature.

Solution

The unknown in this case is volume. Use the density formula and solve for the unknown.

$$\text{Density} = \frac{\text{Mass}}{\text{Volume}} \quad \text{or} \quad \text{Volume} = \frac{\text{Mass}}{\text{Density}}$$

$$\text{Volume} = \frac{770 \text{ g}}{1.595 \text{ g/mL}}$$

$$= 484.8 \text{ mL} = 485 \text{ mL}$$

B.5.2 Specific gravity

The specific gravity of a substance is the ratio of the density of that substance to the density of some substance taken as a standard. The standards commonly used are water in the case of liquids or solids, and air in the case of gases.

$$\text{Specific gravity} = \frac{\text{Density of a liquid or solid}}{\text{Density of water}} \text{ or } \frac{\text{Density of a gas}}{\text{Density of air}}$$

In terms of masses, specific gravity may also be expressed as:

$$\text{Specific gravity} = \frac{\text{Mass of a substance}}{\text{Mass of an equal volume of water at } 4°C}$$

A careful look at the information given will allow one to decide on the appropriate formula to use. For calculation purpose densities must be expressed in the same units, hence specific gravity does not carry any unit.

Example B.4

The specific gravity of concentrated acetic acid, CH_3COOH, is 1.05. Calculate the mass of the acid that would have a volume of 500 mL (NB: the density of water is 1.00 g/mL.)

Solution

First calculate the density of acetic acid. Then use the result to calculate the mass of the acid, which correspond to 500 mL.

$$\text{Specific gravity} = \frac{\text{Density of acetic acid}}{\text{Density of water}}, \text{ and}$$

$$\text{Density of acetic acid} = \text{Specific gravity} \times \text{Density of water}$$

$$= 1.05 \times 1.00 \text{ g/mL}$$

$$= 1.05 \text{ g/mL}$$

$$\text{Density} = \frac{\text{Mass}}{\text{Volume}}$$

$$\text{Mass} = \text{Density} \times \text{Volume} = 1.05 \frac{g}{mL} \times 500 \text{ mL} = 525 \text{ g}$$

Example B.5

A block of aluminum (Al) measuring 5.0 cm × 10.0 cm × 15.0 cm weighs 2027 g. Calculate the specific gravity of aluminum (NB: the density of water is 1.00 g/mL.)

Solution

First calculate the volume of aluminum. Use the result to calculate the density of Al. Then calculate the specific gravity.

Volume of Al $= 5.0 \text{ cm} \times 10.0 \text{ cm} \times 15.0 \text{ cm} = 750 \text{ cm}^3$

$$\text{Density} = \frac{\text{Mass}}{\text{Volume}} = \frac{2027 \text{ g}}{750 \text{ cm}^3} = 2.70 \text{ g/cm}^3$$

$$\text{Specific gravity} = \frac{\text{Density of Al}}{\text{Density of H}_2\text{O}} = \frac{2.70 \text{ g/cm}^3}{1.00 \text{ g/cm}^3} = 2.70$$

B.6 Dimensional analysis and conversion factors

Most numerical calculations in science require conversion from one system of units to another. The unit conversion may be from metric to metric (e.g. grams to kilograms) or English to metric (e.g. inches to centimeters) equivalents. Therefore, chemistry students and scientists should be able to correctly carry out the needed transformation. The best and simplest method is called the *factor-unit method* or *dimensional analysis*. Table B.4 shows representative examples of conversion factors.

The dimensional analysis method of solving problems makes use of *conversion factors* to convert one unit to another unit. Table B.4 shows representative examples of conversion factors. The general setup is as follows:

$$\text{Unit A} \times \text{conversion factor} = \text{Unit B}$$

For example, if you want to find out how many pounds (lb) are in 10 kg, you will need to determine the conversion factor. Once you figure out the conversion factor, use it to multiply the known (10 kg) so as to obtain the unknown:

$$\text{quantity in kg} \times \text{conversion factor} = \text{quantity in lb}$$

A conversion factor is a fraction in which the numerator and the denominator are the same type of measurement but expressed in different units. For example, we can convert kilograms to pounds by using the following equivalence statement:

$$1 \text{ kg} = 2.205 \text{ lb}$$

If we divide both sides of this expression by 1 kg or 2.205 lb, we get the conversion factors

$$\frac{1 \text{ kg}}{2.205 \text{ lb}} = 1 \text{ or } \frac{2.205 \text{ lb}}{1 \text{ kg}} = 1$$

A conversion factor is always equal to 1, hence it is sometimes called a factor-unit. If it is not $= 1$, it is not a valid conversion factor. To convert from pounds to kilograms we simply multiply the given quantity in pounds by the first conversion factor. This ensures that the pound (lb) in the numerator and denominator cancel out, yielding the desired unit (kg). For example, 155.0 lb may be converted to kilograms:

$$155.0 \text{ lb} \times \frac{1 \text{ kg}}{2.205 \text{ lb}} = 70.3 \text{ kg}$$

Table B.4   Common conversion factors

Physical quantity	Symbol	Conversion factor
Length		
Meter	m	$1 \text{ m} = 1.0936 \text{ yd} = 39.37 \text{ in} = 100 \text{ cm}$
Centimeter	cm	$1 \text{ cm} = 0.3937 \text{ in}$
Inch	in	$1 \text{ in} = 2.54 \text{ cm}$
Feet	ft	$1 \text{ ft} = 12 \text{ in}$
Yard	yd	$1 \text{ yd} = 0.9144 \text{ m}$
Mile	mi	$1 \text{ mi} = 1.609 \text{ km} = 5280 \text{ ft}$
Kilometer	km	$1 \text{ km} = 1000 \text{ m} = 0.6215 \text{ mi}$
Angstrom	Å	$1 \text{ Å} = 10^{-10} \text{ m}$
Mass		
Kilogram	kg	$1 \text{ kg} = 1000 \text{ g} = 2.205 \text{ lb}$
Gram	g	$1 \text{ g} = 1000 \text{ mg} = 0.03527 \text{ oz}$
Pound	lb	$1 \text{ lb} = 453.59 \text{ g}$
Ton	ton	$1 \text{ ton} = 907.2 \text{ kg} = 2000 \text{ lb}$
Ounce	oz	$1 \text{ oz} = 28.35 \text{ g}$
Atomic mass unit	amu	$1 \text{ amu} = 1.66056 \times 10^{27} \text{ kg}$
Volume		
Liter	L	$1 \text{ L} = 1000 \text{ cm}^3 = 0.001 \text{ m}^3 = 1.056 \text{ qt}$
Cubic decimeter	dm^3	$1 \text{ dm}^3 = 1 \text{ L}$
Cubic centimeter	cm^3	$1 \text{ cm}^3 = 10^{-6} \text{ m}^3$
Milliliter	mL	$1 \text{ cm}^3 = 0.001 \text{ L} = 10^{-6} \text{ m}^3$
Gallon	gal	$1 \text{ gal} = 4 \text{ qt} = 3.7854 \text{ L}$
Quart	qt	$1 \text{ qt} = 0.9463 \text{ L}$
Cubic foot	ft^3	$1 \text{ ft}^3 = 7.474 \text{ gal} = 28.316 \text{ L}$
Time		
Minute	min	$1 \text{ min} = 60 \text{ s}$
Hour	h	$1 \text{ h} = 3,600 \text{ s}$
Day		$1 \text{ day} = 86,400 \text{ s}$
Energy		
Calorie	cal	$1 \text{ cal} = 4.184 \text{ J}$
Joule	J	$1 \text{ J} = 1 \text{ kg-m}^2/\text{s}^2 = 0.2390 \text{ cal}$
Electron volt	eV	$1 \text{ eV} = 1.602 \times 10^{-19} \text{ J}$

Example B.6

How many kilometers are there in a marathon of 26 miles (NB: 1 mile = 1.6093 km)?

Solution

1. Find the relationship between the units involved in the problem and give the equivalence statement. The relationship between mile and kilometer is 1 mile = 1.6093 km.
2. Next generate the conversion factors.

The conversion factors are

$$\frac{1 \text{ mile}}{1.6093 \text{ km}} \text{ and } \frac{1.6093 \text{ km}}{1 \text{ mile}}$$

3. Multiply the quantity to be converted (26 miles) by the conversion factor that gives the desired unit.

$$26.0 \text{ miles} \times \frac{1.6093 \text{ km}}{1.0 \text{ mile}} = 41.8 \text{ km}$$

In many cases, problems may require a combination of two or more conversion factors in order to obtain the desired units. For example, we may want to express one week in seconds. This would require the following sequence:

$$\text{Weeks} \longrightarrow \text{days} \longrightarrow \text{hours} \longrightarrow \text{minutes} \longrightarrow \text{seconds}$$

There are seven days in one week, twenty-four hours in one day, sixty minutes in one hour, and sixty seconds in one minute. From these relationships, we can generate conversion factors, which we can multiply to get the desired result.

$$1 \text{ wk} \times \frac{7 \text{ day}}{1 \text{ wk}} \times \frac{24 \text{ hr}}{1 \text{ day}} \times \frac{60 \text{ min}}{1 \text{ hr}} \times \frac{60 \text{ s}}{1 \text{ min}} = 604,800 \text{ s}$$

Example B.7

Express 4.5 gallons of vegetable oil in milliliters.

Solution

1. The quantity given is 4.5 gallons, and the result is to be expressed in milliliters. Use the sequence:

$$\text{gallons} \longrightarrow \text{quarts} \longrightarrow \text{liters} \longrightarrow \text{milliliters}$$

2. State the relationship and the conversion factors:

$$1 \text{ gal} = 4 \text{ qt}; \quad \frac{1 \text{ gal}}{4 \text{ qt}} \quad \text{or} \quad \frac{4 \text{ qt}}{1 \text{ gal}}$$

$$1 \text{ qt} = 0.9463 \text{ L}; \quad \frac{0.9463 \text{ L}}{1 \text{ qt}} \quad \text{or} \quad \frac{1 \text{ qt}}{0.9463 \text{ L}}$$

$$1 \text{ L} = 1000 \text{ mL}; \quad \frac{1000 \text{ mL}}{1 \text{ L}} \quad \text{or} \quad \frac{1 \text{ L}}{1000 \text{ mL}}$$

3. Set up and multiply the appropriate conversion factors:

$$4.5 \text{ gal} \times \frac{4 \text{ qt}}{1 \text{ gal}} \times \frac{0.9463 \text{ L}}{1 \text{ qt}} \times \frac{1000 \text{ mL}}{1 \text{ L}} = 17.033 \text{ mL} \simeq 20,000 \text{ mL}$$

B.7 Problems

1. Convert the following temperatures:
   (a) 30°C to °F
   (b) −196°C to °F

(c) 0.0500°C to °F

(d) 1200°F to °C

(e) −68°F to °C

(f) 173°F to °C

2. Convert the following temperatures:

(a) 300°C to K

(b) −196°C to K

(c) 105°F to K

(d) 1200 K to °C

(e) 350 K to °F

(f) 173°F to K

3. Propane gas (a) boils at −42°C and (b) melts at −188°C. Express these temperatures in °F and K.

4. What is the temperature at which the Celsius and Fahrenheit thermometers have the same numerical value?

5. A steel tank has the dimensions 200 cm × 110 cm × 45 cm. Calculate the volume of the tank in liters.

6. The volume of a cylindrical container of radius r and height h is given by $\pi r^2 h$. (a) What is the volume of a cylinder with a radius of 8.15 cm and a height of 106 cm? (b) Calculate the mass in kg of a volume of carbon tetrachloride equal to the volume of the cylinder obtained in part (a) given that the density of carbon tetrachloride is 1.60 g/cm^3.

7. The density of gold is 19.3 g/cm^3. Calculate the mass of 500 cm^3 of gold.

8. A block of a certain precious stone 8.00 cm long, 5.50 cm wide, and 3.8 cm thick has a mass of 8.00 kg. Calculate its density in g/cm^3.

9. A steel ball weighing 3.250 g has a density of 8.00×10^3 kg/m^3. Calculate the volume (in mm^3) and the radius (in mm) of the steel ball (NB: the volume of a sphere with radius r is $V = \frac{4}{3}\pi r^3$).

10. The density of a sphere of edible clay is 1.4 g/cm^3. Calculate the mass of the clay if the radius of the sphere is 10 cm. (Hint: volume of a sphere with radius r is $V = \frac{4}{3}\pi r^3$)

11. Concentrated sulfuric acid assays 98 % by weight of sulfuric acid (the remainder is water) and has a density of 1.84 g/cm^3. Calculate the mass of pure sulfuric acid in 80.00 cm^3 of concentrated acid.

12. Express 0.55 cm in nanometers (nm).

13. The density of blood plasma is 1.027 g/cm^3. Express the density in kg/m^3.

14. Which has the greater mass, 125 cm^3 of olive oil or 130 cm^3 of ethyl alcohol? The density of olive oil is 0.92 g/cm^3, and the density of ethyl alcohol is 0.79 g/cm^3.

15. The specific gravity of concentrated hydrochloric acid, HCl, is 1.18. Calculate the volume of the acid that would have a mass of 750 g. (The density of water is 1.00 g/cm^3.)

16. Calculate the length, in millimeters, of a 250-ft steel rod given that 12 in = 1 ft, 1 m = 39.37 in, and 1000 mm = 1m.

17. How many seconds are there in 365 days?

18. What is the mass in pounds of 25.5 kg of lead?

19. Express 350 lb in kg.

20. A common unit of energy is the electron volt (eV) (see Table B.4). Convert 12 eV to the following units of energy: (a) Joule, (b) Kilojoule, (c) Calorie, (d) ergs ($1 J = 1 \times 10^7$ ergs).

Appendix B: Solutions to Problems

1. (a) 86°F
   (b) −321°F
   (c) 32.1°F
   (d) 648.9°C
   (e) −56°C
   (f) 78.3°C

2. (a) 573K
   (b) 77K
   (c) 314K
   (d) 927°C
   (e) 171°F
   (f) 351K

3. (a) −43.6°F, 231K
   (b) −306°F, 85K

4. $T = \dfrac{5}{9}(T - 32)$ or $T = -40°C = -40°F$

5. 990 L

6. (a) $2.21 \times 10^4 \text{ cm}^3$
   (b) 35.4 kg

7. 9.65×10^3 g

8. 12 g/cm^3

9. $V = 406 \text{ mm}^3$; $r = 4.59$ mm

10. Mass $= 5{,}865$ g $\simeq 5.9 \times 10^3$ g

11. 144g

12. 5.5×10^6 nm

13. 1.027×10^3 kg/m^3

14. 125 cm^3 olive oil has a greater mass (115 g) than 130 cm^3 ethyl alcohol (103 g).

15. 636 cm^3

16. 7.62×10^4 mm

17. 31,536,00 s

18. 56.1 lb

19. 159 kg

20. (a) 1.92×10^{-18} J
    (b) 1.92×10^{-21} kJ
    (c) 4.59×10^{-19} cal
    (d) 1.92×10^{-11} ergs

Solutions

· ·

Chapter 1

1.

	Symbol	Protons	Neutrons	Electrons
(a)	$^{197}_{79}Au$	79	118	79
(b)	$^{10}_{5}B$	5	5	5
(c)	$^{40}_{20}Ca$	20	20	20
(d)	$^{163}_{66}Dy$	66	97	66

2.

	Symbol	Protons	Neutrons	Electrons
(a)	$^{84}_{26}Kr$	36	48	36
(b)	$^{24}_{12}Mg$	12	12	12
(c)	$^{69}_{31}Ga$	31	38	31
(d)	$^{75}_{33}As$	33	42	33

3.

	Species	Proton	Neutron	Electron
(a)	$^{59}_{27}Co^{2+}$	27	32	25
(b)	$^{24}_{12}Mg^{2+}$	12	12	10
(c)	$^{69}_{31}Ga^{3+}$	31	38	28
(d)	$^{118}_{50}Sn^{2+}$	50	68	48

4. (a) Br^- (b) Cs^+ (c) Mn^{2+}

5.

Symbol	$^{40}_{20}Ca^{2+}$	$^{31}_{15}P^{3-}$	$^{79}_{34}Se^{2-}$	$^{238}_{92}U^{4+}$	$^{15}_{7}N^{3-}$	$^{137}_{56}Ba^{2+}$
Mass number	40	31	79	238	15	137
Protons	20	15	34	92	7	56
Neutrons	20	16	45	146	8	81
Electrons	18	18	36	88	10	54
Net Charge	2	-3	-2	4	-3	2

6. (a) $^{56}_{26}Fe$ (b) $^{131}_{53}I$ (c) $^{202}_{80}Hg$ (d) $^{19}_{9}F$

7.

Particle	Atomic number	Mass number	Protons	Number of Neutrons	Electrons	Net charge	Symbol
A	11	23	11	12	10	1	Na^+
B	13	27	13	14	10	3	Al^{3+}
C	58	140	58	82	58	0	Ce
D	24	52	24	28	24	0	Cr
E	3	7	3	4	2	1	Li^+
F	54	131	54	77	54	0	Xe
G	81	204	81	123	80	1	TI^+
H	40	91	40	51	36	4	Zr^{4+}

8.

Species	Protons	Neutrons	Electrons
(a) $^{32}_{16}S^{2-}$	16	16	18
(b) $^{80}_{35}Br^-$	35	45	36
(c) $^{52}_{128}Te^{2-}$	52	76	54
(d) $^{7}_{14}N^{3-}$	7	7	10

9. 35.45
10. 11.01
11. 16.0
12. 60.3% ^{69}Ga and 39.7% ^{71}Ga

Chapter 2

1. (a) 84 g/mol (b) 100 g/mol (c) 120 g/mol (d) 328 g/mol (e) 342 g/mol
2. (a) 342 g/mol (b) 34 g/mol (c) 101 g/mol (d) 191 g/mol (e) 142 g/mol
3. 339.5 amu
4. 1006 g/mol
5. 180 g/mol
6. 162 g/mol
7. 227 g/mol
8. 154 g/mol
9. 192.2 amu, M = Ir
10. 31 amu, X = P

Chapter 3

1. (a) 1.00 mol $CaCo_3$ (b) 5.56 mol H_2O (c) 2.50 mol NaOH (d) 1.02 mol H_2SO_4
   (e) 3.57 mol N_2
2. (a) 0.0919 mol $CaSO_4$ (b) 7.35 × 10^{-3} mol H_2O_2 (c) 4.76 × 10^{-4} mol $NaHCO_3$
   (d) 0.100 mol H_2SO_4

3. (a) 1.01g MgO (b) 1.55g Na_2O (c) 0.425g NH_3 (d) 0.70g C_2H_4 (e) 4.45g $H_4P_2O_7$

4. (a) 28g CaO (b) 0.15g Li_2O (c) 710g P_4O_{10} (d) 42.0g C_2H_4 (e) 0.80g $B_3N_3H_6$

5. (a) 8×10^{-4} mol (b) 415.3 mol (c) 0.19 mol (d) 1.0×10^{-23} mol
   (e) 5.0×10^{-47} mol

6. (a) 1.20×10^{24} Al atoms & 1.81×10^{24} O atoms;
   (b) 1.20×10^{24} H atoms & 6.02×10^{23} S atoms.
   (c) 6.02×10^{23} C atoms & 1.20×10^{24} O atoms;
   (d) 6.02×10^{23} H atoms & 6.02×10^{23} Cl atoms.

7. (a) 3.32×10^{-2}g Ca (b) 2.33×10^4g Fe (c) 20 g Ca (d) 6.48×10^{-22}g K
   (e) 1.0×10^{-46}g H

8. 6327 kg

9. (a) 1.77×10^{25} O (b) 9.68×10^{24} O (c) 9.03×10^{24} O (d) 7.76×10^{24} O

10. 2.4×10^{24} Fe^{3+} ions and 3.6×10^{24} SO_4^{2-} ions

11. (a) 0.75 mol Y; 1.50 mol Ba; 2.25 mol Cu (b) 4.52×10^{23} Y atoms; 9.03×10^{23} Ba atoms; 1.36×10^{24} Cu atoms

12. 9.03×10^{23} Si atoms and 1.2×10^{24} N atoms

Chapter 4

1. (a) H_2S: S = 94.1% (b) SO_3: S = 40% (c) H_2SO_4: S = 32.65% (d) $Na_2S_2O_3$: S = 40.5%

2. (a) K_2CO_3: K = 56.6%, C = 8.7%, O = 34.7%
   (b) $Ca_3(PO_4)_2$: Ca = 38.7%, P = 20.0%, O = 41.3%
   (c) $Al_2(SO_4)_3$: Al = 15.8%, S = 28.1%, O = 56.1%
   (d) C_6H_5OH: C = 76.6%, H = 6.4%, O = 17.0%

3. (a) $C_{20}H_{25}N_3O$: C = 74.3%, H = 7.74%, N = 13.0%, O = 4.95%
   (b) $C_{10}H_{16}N_5P_3O_{13}$: C = 23.7%, H = 3.2%, N = 13.8%, P = 18.3%, O = 41.0%
   (c) $C_6H_4N_2O_4$: C = 42.9%, H = 2.4%, N = 16.7%, O = 38.1%
   (d) $C_6H_5CONH_2$: C = 69.4%, H = 5.8%, N = 11.6%, O = 13.2%

4. Mg = 12.0%, Cl = 34.9%, H = 5.9%, O = 47.2%

5. 6.83 g

6. 55.28 g of C and 3.80 g of N

7. 70 g of Fe

8. 0.6657 g

9. (a) SO_3 (b) $CaWO_3$ (c) CH_2 (d) CCl_2F_2 (e) $Cr_2S_3O_{12}$ or $Cr_2(SO_4)_3$

10. (a) $Mg_2P_2O_7$ (b) $C_2H_3O_5N$ (c) $C_{14}H_9Cl_5$ (d) $Co_3Mo_2Cl_{11}$ (e) $C_3H_2O_3$

11. (a) CH_2 (b) Fe_2S_3 (c) $C_2H_3O_2$ (d) $C_4H_5N_2O$ (e) $Al_2(SO_4)_3$

12. P_4O_{10}

13. $C_{27}H_{46}O$, $C_{27}H_{46}O$

14. $C_{18}H_{24}O_2$

15. $Na_2CO_3 \cdot 10H_2O$

16. $C_3H_8O_2$, $C_9H_{24}O_6$

17. $FeCl_3 \cdot 6H_2O$

18. $C_6H_{10}O_5$

19. $C_3H_6O_5$, $C_6H_{12}O_{10}$

20. (a) C = 57.14%, H = 6.12%, N = 9.53%, O = 27.23% (b) $C_{14}H_{18}N_2O_5$

Chapter 5

1. (a) +4 for C, −2 for O (b) +6 for S, −2 for O (c) +6 for S, −1 for F
   (d) −2 for N, +1 for H (e) +2 for Pb, −2 for O

2. (a) +2 (b) +4 (c) +1 (d) 0 (e) −3

3. (a) −3 (b) −2 (c) P = +3, O = −2 (d) S = +6, O = −2 (e) Cl = +5, O = −2

4. (a) Mn = +7, O = −2 (b) Cr = +6, O = −2 (c) U = +6, O = −2
   (d) S = +2, O = −2 (e) S = $+\frac{5}{2}$, O = −2

5. (a) +3 (b) +1 (c) +2 (d) +3 (e) $+\frac{8}{3}$

6. (a) +3 (b) +4 (c) +5 (d) +6 (e) +5

7. (a) +6 (b) +3 (c) +7 (d) +5 (e) +3

8. (a) Mg_3N_2 (b) SnF_2 or SnF_4 (c) H_2S (d) InI_3 (e) $AlBr_3$

9. (a) Li_3N (b) B_2O_3 (c) CaO (d) RbCl (e) Cs_2S

10. (a) $Fe_2(CO_3)_3$ (b) $(NH_4)_3PO_4$ (c) $Ca(NO_3)_2$ (d) $LiClO_4$ (e) $K_2Cr_2O_7$

11. (a) $MnCO_3$ (b) $Sn_3(AsO_4)_2$ (c) $Ca(C_2H_3O_2)_2$ (d) $Fe(ClO_4)_3$ (e) Na_3BO_3

12. (a) SO_2 (b) CO_2 (c) NO_2 (d) N_2O_5 (e) CCl_4 (f) ClO_2 (g) LiI (h) SeO_2 (i) $FeCl_2$
    (j) Ba_3P_2

13. (a) V_2O_5 (b) CuS (c) Fe_2S_3 (d) GaN (e) $HgCl_2$

14. (a) Na_2CO_3 (b) NH_4Cl (c) K_3PO_4 (d) $Ca(HSO_4)_2$ (e) $Fe_2(CrO_4)_3$ (f) $Pd_3(PO_4)_2$
    (g) $Al(HCO_3)_3$ (h) $K_2Cr_2O_7$ (i) $Fe(OH)_3$ (j) $Mg_3(BO_3)_2$

15. (a) Phosphorus trichloride   (b) Phosphorus pentachloride   (c) Carbon monoxide
    (d) Carbon dioxide   (e) Sulfur dioxide   (f) Sulfur trioxide   (g) Silicon dioxide
    (h) Diphosphorus pentasulfide   (i) Dinitrogen pentoxide

16. (a) Nickel (II) nitride   (b) Iron (III) Chloride   (c) Aluminum sulfide   (d) Ger-
    manium disulfide   (e) Titanium disulfide   (d) Calcium hydride   (g) Mercury (II)
    chloride   (h) Mercury (I) phosphide   (i) Copper (I) nitride

17. (a) Hydrofluoric acid (b) Hydrobromic acid (c) Hydroselenic acid (d) Hydrosul-
    furic acid (e) Hydrocyanic acid

18. (a) Nitric acid   (b) Bromic acid   (c) Hypochloric acid   (d) Oxalic acid   (e) Phos-
    phorous acid   (f) Iodic acid

19. (a) Gallium nitrate, Nitric acid   (b) Cobalt (II) sulfate, sulfuric acid   (c) Iron (II)
    acetate, acetic acid   (d) Calcium borate, boric acid   (e) Lead (II) oxalate, oxalic acid
    (f) Chromium (III) phosphate, phosphoric acid   (g) Nickel carbonate, carbonic acid
    (h) Iron (III) cyanide, hydrocyanic acid   (i) Aluminum iodide, hydroiodic acid
    (j) Rubidium bromate, bromic acid.

20. (a) Potassium hydroxide   (b) Ammonium hydroxide   (c) Cobalt (II) hydroxide
    (d) Barium hydroxide   (e) Chromium (III) hydroxide

Chapter 6

1. (a) $SO_3 + H_2O \rightarrow H_2SO_4$
   (b) $2 H_2O \rightarrow 2 H_2 + O_2$
   (c) $PCl_3 + Cl_2 \rightarrow PCl_5$
   (d) $2 C + O_2 \rightarrow 2 CO$
   (e) $2 Mg + O_2 \rightarrow 2 MgO$

2. (a) $4 NH_3 + 3 O_2 \rightarrow 2 N_2 + 6 H_2O$
   (b) $I_4O_9 \xrightarrow{\Delta} I_2O_5 + I_2 + 2 O_2$
   (c) $K_2S_2O_3 + 4 Cl_2 + 5 H_2O \xrightarrow{\Delta} 2 KHSO_4 + 8 HCl$
   (d) $CS_2 + 3 O_2 \rightarrow CO_2 + 2 SO_2$

3. (a) $CaCO_3 \xrightarrow{\Delta} CaO + CO_2$
   (b) $2 KClO_3 \xrightarrow{\Delta} 2 KCl + 3 O_2$
   (c) $2 Li_3N \xrightarrow{\Delta} 6 Li + N_2$
   (d) $2 H_2O_2 \xrightarrow{\Delta} 2 H_2O + O_2$

4. (a) $TiCl_4 + 2 H_2S \rightarrow TiS_2 + 4 HCl$
   (b) $LaCl_3 \cdot 7H_2O \xrightarrow{\Delta} LaOCl + 2 HCl + 6 H_2O$
   (c) $3 CsCl + 2 ScCl_3 \rightarrow Cs_3Sc_2Cl_9$
   (d) $Li_2CO_3 + 5 Fe_2O_3 \rightarrow 2 LiFe_5O_8 + CO_2$
   (e) $2 La_2O_3 + 12 B_2O_3 \rightarrow 4 LaB_6 + 21 O_2$
   (f) $4 ScCl_3 + 7 SiO_2 \rightleftharpoons 2 Sc_2Si_2O_7 + 3 SiCl_4$

5. (a) $2 NaN_3 \rightarrow 2 Na + 3 N_2$
   (b) $(NH_4)_2 Cr_2O_7 \rightarrow Cr_2O_3 + N_2 + 4 H_2O$
   (c) $2 Ag_2CO_3 \rightarrow 4 Ag + 2 CO_2 + O_2$
   (d) $2 NaHCO_3 \rightarrow Na_2CO_3 + CO_2 + H_2O$
   (e) $Al_2(CO_3)_3 \rightarrow Al_2O_3 + 3 CO_2$

6. (a) $H_2SO_4 + 2 KOH \rightarrow K_2SO_4 + 2 H_2O$
   (b) $2 H_3PO_4 + 3 Ba(OH)_2 \rightarrow Ba_3(PO_4)_2 + 6 H_2O$
   (c) $4 HBr + Sn(OH)_4 \rightarrow SnBr_4 + 4 H_2O$
   (d) $HBr + NaOH \rightarrow NaBr + H_2O$
   (e) $H_4P_2O_7 + 4 NaOH \rightarrow Na_4P_2O_7 + 4 H_2O$

7. (a) $Ca(NO_3)_2 + Na_2SO_4 \rightarrow 2 NaNO_3 + CaSO_4$
   (b) $Mg(NO_3)_2 + Li_2S \rightarrow MgS + 2 LiNO_3$
   (c) $Pb(NO_3)_2 + Cs_2CrO_4 \rightarrow PbCrO_4 + 2 CsNO_3$
   (d) $2 FeCl_3 + 3 CaS \rightarrow Fe_2S_3 + 3 CaCl_2$
   (e) $2 Na_3PO_4 + 3 Zn(OH)_2 \rightarrow Zn_3(PO_4)_2 + 6 NaOH$

8. $2 Ca_3(PO_4)_2 + 6 SiO_2 + 5 C \rightarrow P_4 + 6 CaSiO_3 + 5 CO_2$

9. (a) $C_4H_8 + 6 O_2 \rightarrow 4 CO_2 + 4 H_2O$
   (b) $2 C_6H_6 + 15 O_2 \rightarrow 12 CO_2 + 6 H_2O$
   (c) $2 C_6H_{12}O + 17 O_2 \rightarrow 12 CO_2 + 12 H_2O$
   (d) $C_{12}H_{22}O_{11} + 12 O_2 \rightarrow 12 CO_2 + 11 H_2O$
   (e) $2 C_4H_{10} + 13 O_2 \rightarrow 8 CO_2 + 10 H_2O$

10. (a) $2 Fe + 3 S \xrightarrow{\Delta} Fe_2S_3$
    (b) $2 C_8H_{18} + 25 O_2 \rightarrow 16 CO_2 + 18 H_2O$
    (c) $2 SO_2 + O_2 \rightarrow 2 SO_3$
    (d) $3 Ba(OH)_2 + 2 H_3PO_4 \rightarrow Ba_3(PO_4)_2 + 6 H_2O$
    (e) $2 Al(s) + 6 HCl(aq) \rightarrow 2 AlCl_3(s) + 3 H_2(g)$

11. (a) $2\,Al + 3\,Cl_2 \rightarrow 2\,AlCl_3$
     (b) $2\,K + 2\,H_2O \rightarrow 2\,KOH + H_2$
     (c) $BCl_3 + 3\,H_2O \rightarrow B(OH)_3 + 3\,HCl$
     (d) $AgNO_3 + NaCl \rightarrow AgCl + NaNO_3$
     (e) $SiF_4 + 8\,NaOH \rightarrow Na_4SiO_4 + 4\,NaF + 4\,H_2O$
     (f) $Ca(OH)_2 + 2\,NH_4Cl \rightarrow 2\,NH_3 + CaCl_2 + 2H_2O$

12. (a) $2\,Si_4H_{10} + 13\,O_2 \rightarrow 8\,SiO_2 + 10\,H_2O$
     (b) $CH_3NO_2 + 3\,Cl_2 \rightarrow CCl_3NO_2 + 3\,HCl$
     (c) $2\,NaF + CaO + H_2O \rightarrow CaF_2 + 2\,NaOH$
     (d) $Al_4C_3 + 12\,H_2O \rightarrow 4\,Al(OH)_3 + 3\,CH_4$
     (e) $2\,TiO_2 + B_4C + 3\,C \rightarrow 2\,TiB_2 + 4\,CO$
     (f) $C_7H_{16}O_4S_2 + 11\,O_2 \rightarrow 7\,CO_2 + 8\,H_2O + 2\,SO_2$

Chapter 7

1. 128g C_2H_5OH and 122g CO_2

2. 80.00g of O_2

3. (a) 511g $KClO_3$ (b) 0.0418 mol KCl (c) 0.0627 mol O_2 and 2.01g O_2

4. 6.10g MnO_2

5. 3.51 g Si

6. 87.92 g W

7. 7.47 liters

8. 4.96 liters

9. 44.07 liters

10. 4.48 liters

11. (a) $2\,C_2H_2 + 5\,O_2 \rightarrow 2\,H_2O + 4\,CO_2$
    (b) 0.505 mol
    (c) 9.05 liters

12. (a) $2\,Al(OH)_3 + 3\,H_2SO_4 \rightarrow Al_2(SO_4)_3 + 6\,H_2O$
    (b) $Al(OH)_3$ is the limiting reagent
    (c) 96.15 g $Al_2(SO_4)_3$
    (d) 87.61 g H_2SO_4

13. (a) $2\,Na_2S_2O_3 + AgBr \rightarrow Na_3Ag(S_2O_3)_2 + NaBr$
    (b) $Na_2S_2O_3$ is the limiting reagent
    (c) 0.0484 mol

14. (a) $C_7H_8 + 3\,HNO_3 \rightarrow C_7H_5N_3O_6 + 3\,H_2O$
    (b) The limiting reagent is HNO_3
    (c) Theoretical yield = 60.04 g
    (d) 91.6%

15. (a) $Ca_3P_2 + 6\,H_2O \rightarrow 3\,Ca(OH)_2 + 2\,PH_3$
    (b) 8.262 g of PH_3

16. 14.24 g of aspirin

17. 296 g of acetic acid

18. 56% Fe, and 44% Cu

19. (a) $4\,C_{21}H_{23}O_5N + 101\,O_2 \rightarrow 84\,CO_2 + 4\,NO_2 + 46\,H_2O$
    (b) O_2 is the limiting reagent and heroin is the excess
    (c) 8.9 g of heroin unreacted
    (d) 64.4 %
    (e) $40.31\text{ g }CO_2 + 2.01\text{ g }NO_2 + 9.03\text{ g }H_2O = 51.35$ g of products

20. 17.49 kg of Fe

Chapter 8

1. (a) 2.34×10^{-5} nm
   (b) 0.173 nm
   (c) 5.44×10^{12} nm
   (d) 5.75×10^8 nm

2. (a) 5.34×10^{-2} cm
   (b) 2.65×10^{-5} cm
   (c) 6.64×10^4 cm
   (d) 8.79×10^{-6} cm

3. $5.09 \times 10^{14}\text{s}^{-1}$

4. 3.23×10^{-7} nm

5. $1.70 \times 10^4 \text{ cm}^{-1}$

6. (a) 432 nm;
   (b) $2.31 \times 10^4 \text{ cm}^{-1}$

7. 2.45×10^{-24} kJ

8. 4.88×10^{-19} J

9. (a) $5.71 \times 10^{14} \text{ s}^{-1}$
   (b) $1.91 \times 10^4 \text{ cm}^{-1}$
   (c) 3.79×10^{-19} J

10.

Frequency (MHz)	Energy (kJ)
60	3.98×10^{-29}
200	1.33×10^{-28}
400	2.65×10^{-28}
600	3.98×10^{-28}
700	4.64×10^{-28}

11. 2.42×10^{-19} J

12. (a) $1.10 \times 10^5 \text{ cm}^{-1}$
    (b) 2.09×10^{-18} J

13. 2.18×10^{-18} J

14. (a) 2.04×10^{-18} J, 97.3 nm
    (b) 6.54×10^{-19} J, 304 nm

15. 8.71×10^{-3} nm

16. $2.91 \times 10^4 \text{ ms}^{-1}$

17. 0.0549 Å

18. (a) 2 (b) 8 (c) 18 (d) 32

19. $l = 0, 1, 2$ corresponding to $s, p,$ and d sublevels

20. (a) $3p$ (b) $3d$ (c) $4f$

21. (a) $n = 2; l = 1; m_l = 1, 0, -1$
    (b) $n = 3; l = 2; m_l = 2, 1, 0, -1, -2$
    (c) $n = 4; l = 1; m_l = 1, 0, -1$
    (d) $n = 4; l = 3; m_l = 3, 2, 1, 0, -1, -2, -3$

22. (a) Na $1s^2\, 2s^2\, 2p^6\, 3s^1$
    (b) P $1s^2\, 2s^2\, 2p^6\, 3s^2\, 3p^3$
    (c) Ca $1s^2\, 2s^2\, 2p^6\, 3s^2\, 3p^6\, 4s^2$
    (d) Rb $1s^2\, 2s^2\, 2p^6\, 3s^2\, 3p^6\, 4s^2\, 3d^{10}\, 4p^6\, 5s^1$
    (e) Fe $1s^2\, 2s^2\, 2p^6\, 3s^2\, 3p^6\, 4s^2\, 3d^6$

23. (a) S^{2-} $1s^2\, 2s^2\, 2p^6\, 3s^2\, 3p^6$
    (b) Ni^{2+} $1s^2\, 2s^2\, 2p^6\, 3s^2\, 3p^6\, 4s^2\, 3d^6$
    (c) Cu^{2+} $1s^2\, 2s^2\, 2p^6\, 3s^2\, 3p^6\, 4s^2\, 3d^7$
    (d) Ti^{4+} $1s^2\, 2s^2\, 2p^6\, 3s^2\, 3p^6$
    (e) Br^- $1s^2\, 2s^2\, 2p^6\, 3s^2\, 3p^6\, 4s^2\, 3d^{10}\, 4p^6$

24. (a) O $[He]\, 2s^2\, 2p^4$
    (b) Zn^{2+} $[Ar]\, 4s^2\, 3d^8$
    (c) Mg $[Ne]\, 3s^2$
    (d) Pb^{2+} $[Xe]\, 6s^2\, 5d^{10}\, 4f^{14}$
    (e) V $[Ar]\, 4s^2\, 3d^3$
    (f) Te^{2-} $[Kr]\, 4d^{10}\, 5s^2\, 5p^2$ *or* [Xe]
    (g) Mn^{7+} [Ar]

25. (a) B (b) Ge (c) I (d) Bi (e) Mn (f) P

26. (a) F (b) P (c) Sc (d) Ni (e) C (f) Ti

27. (a) Oxygen

(b) Sodium

(c) Titanium

(d) Zinc

28. (a) A = Oxygen (^{18}O), B = Magnesium (^{26}Mg)
    (b) A has twenty-six subatomic particles; B has thirty-eight subatomic particles.

Chapter 9

1. a. Rb b. Mg• c. •Si• d. •As• e. •Se•

2. a. $\left[\text{B}\right]^{3+}$ b. $\left[\text{Mg}\right]^{2+}$ c. $\left[\ddot{\overset{..}{\text{N}}}\ddot{}\right]^{3-}$ d. $\left[\ddot{\overset{..}{\text{Cl}}}\ddot{}\right]^{-}$

3. The Lewis symbols for Be, Mg, Ca, and Sr are

   Be• Mg• Ca• Sr•

   They all have the same Lewis symbol since each of the elements in the group has two valence electrons.

4. (a)

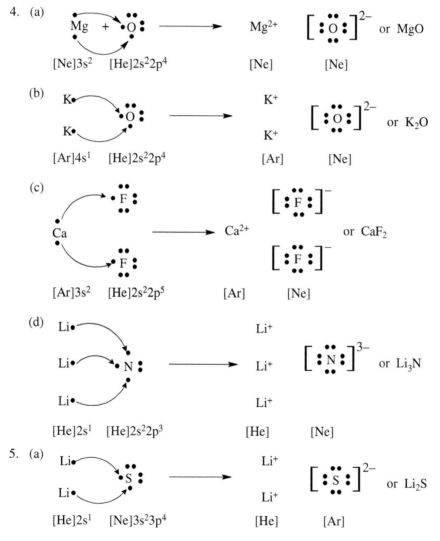

   Mg + •Ö: ⟶ Mg^{2+} $\left[\ddot{\overset{..}{\text{O}}}\ddot{}\right]^{2-}$ or MgO

   $[Ne]3s^2$ $[He]2s^22p^4$ [Ne] [Ne]

   (b)

   K• + Ö: ⟶ K^+

   K• K^+ $\left[\ddot{\overset{..}{\text{O}}}\ddot{}\right]^{2-}$ or K_2O

   $[Ar]4s^1$ $[He]2s^22p^4$ [Ar] [Ne]

   (c)

   Ca + •F: ⟶ Ca^{2+} $\left[\ddot{\overset{..}{\text{F}}}\ddot{}\right]^{-}$ or CaF_2

   •F: $\left[\ddot{\overset{..}{\text{F}}}\ddot{}\right]^{-}$

   $[Ar]3s^2$ $[He]2s^22p^5$ [Ar] [Ne]

   (d)

   Li• Li^+

   Li• + •N: ⟶ Li^+ $\left[\ddot{\overset{..}{\text{N}}}\ddot{}\right]^{3-}$ or Li_3N

   Li• Li^+

   $[He]2s^1$ $[He]2s^22p^3$ [He] [Ne]

5. (a)

   Li• + •S: ⟶ Li^+

   Li• Li^+ $\left[\ddot{\overset{..}{\text{S}}}\ddot{}\right]^{2-}$ or Li_2S

   $[He]2s^1$ $[Ne]3s^23p^4$ [He] [Ar]

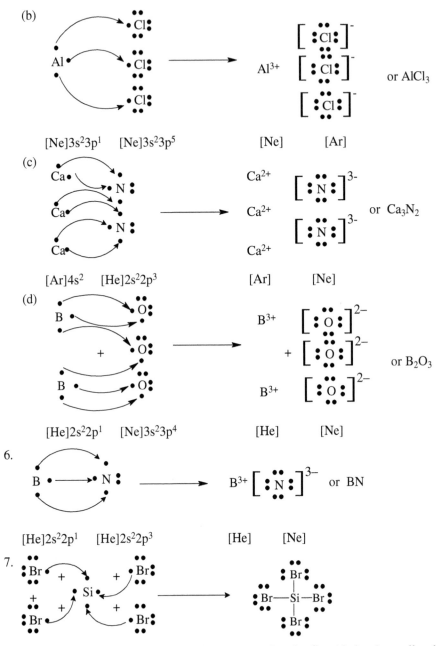

(b)

$[Ne]3s^23p^1$ $[Ne]3s^23p^5$ [Ne]   [Ar]

or $AlCl_3$

(c)

$[Ar]4s^2$ $[He]2s^22p^3$ [Ar]   [Ne]

or Ca_3N_2

(d)

$[He]2s^22p^1$ $[Ne]3s^23p^4$ [He]   [Ne]

or B_2O_3

6.

$[He]2s^22p^1$ $[He]2s^22p^3$ [He]   [Ne]

or BN

7.

8. (a) MgF_2. They have the Mg ion in common, but the fluoride ion is smaller than the
         iodide ion. Therefore, the Mg^{2+} is closer to the F^- than the I^- and so can interact
         more strongly, and hence will have greater lattice energy.
    (b) CaO. The higher charge on CaO $(Ca^{2+}$ and $O^{2-})$ and smaller interionic distance
         compared to CsI allows it to have higher lattice energy relative to CsI.
    (c) CaS. Sulfur has a small ionic radius and hence a smaller interionic distance than the
         tellurium ion, and thus a larger lattice energy than CaTe.

(d) SrO. The higher charges on Sr and O, and the relatively similar radii of the ions in RbF and SrO, leads to a larger lattice energy for SrO.

9. Considering LiF and KF, the lattice energy of LiF is greater because KF has a much larger interionic distance. Considering BeO, CaO and BaO,the ionic radii and hence the interionic distance increase going from BeO to BaO. Therefore, one would expect BaO to have the lowest lattice energy and BeO the highest. With respect to LiF, BaO should have roughly four times the lattice energy of LiF given the charges on the ions, that is $(+2), (-2)$ for BaO vs$(1+), (1-)$ for LiF. The order is therefore KF $<$ LiF $<$ BaO $<$ CaO $<$ BeO.

10. All are hydrides of group 2A of the periodic table and so ionic charge is the same for all the metals. Also, all have hydrogen as their anion counterpart. The ionic radius and hence the interionic distance for these hydrides will increase from Be through Ba. Different interionic distances results in different lattice energies for similar charge combinations. Therefore, the lattice energy will be largest for BeH_2 and lowest for BaH_2.

11. -2318 kJ/mol

12. -2637 kJ/mol

13. Cs (1), Mg (2), Ga (3), Cl (7)

14. (a) HCN : $1+4+5 = 10$
(b) $ClO_4^- : 7+6\times4+1 = 32$
(c) $IF_6^+ : 7+7\times6-1 = 48$
(d) $XeF_6 : 8+7\times6 = 50$

15. (a) $CCl_4 : 4+7\times4 = 32$
(b) $PH_4^+ : 5+4\times1-1 = 8$
(c) $H_3PO_4 : 1\times3+5+6\times4 = 32$
(d) $SO_4^{2-} : 6+6\times4+2 = 32$

16. (a) CaSe (b) $BaBr_2$ (c) Mg_3N_2 (d) B_2O_3

17. (a)

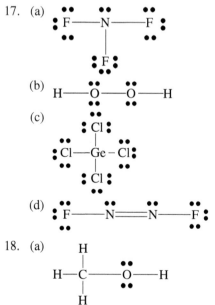

(b)

(c)

(d)

18. (a)

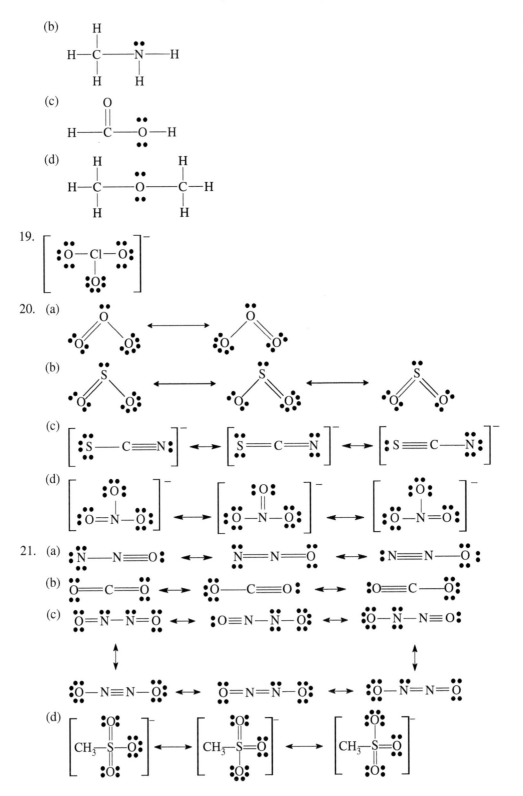

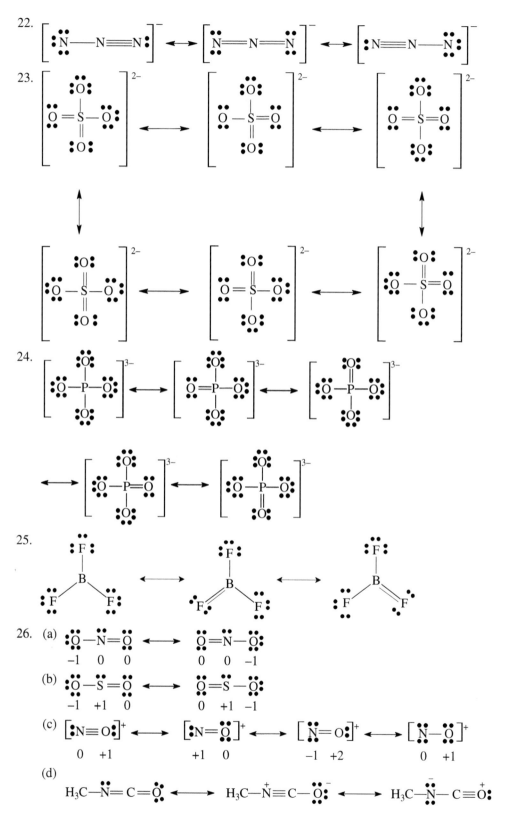

27. (a) All the atoms have formal charges of zero, consistent with a net zero charge on the molecule.

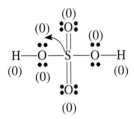

(b) All the atoms have formal charges of zero, consistent with a net zero charge on the molecule.

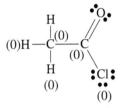

(c) All the atoms have formal charges of zero, consistent with a net zero charge on the molecule.

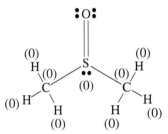

(d) The formal charges add to +1, the charge for the entire ion.

28.

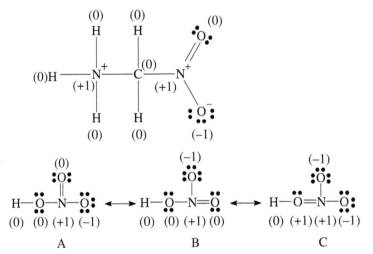

Formal charges are minimized in structures A and B, so both are equally important, and more important than C.

29.

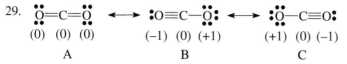

(0) (0) (0)   (−1) (0) (+1)   (+1) (0) (−1)

A   B   C

Comparing the formal charges, structure A is more stable and contributes the most.

30. (a)

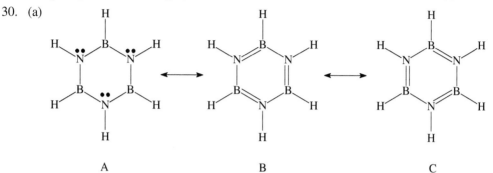

A   B   C

(b) All hydrogen atoms are identical, and each has a formal charge of zero.

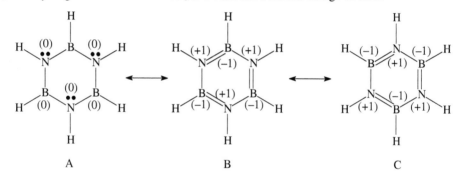

A   B   C

31. Structure A has minimal formal charge and so is preferred.

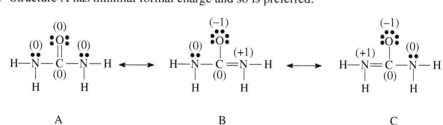

A   B   C

32. (a)

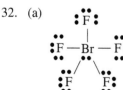

(b)

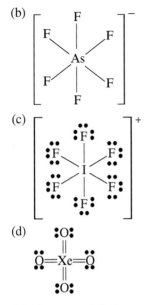

(c)

(d)

33. (a) $\Delta\chi_{H-F} = 1.9$; $\Delta\chi_{H-Cl} = 0.9$. HF is more polar.
    (b) $\Delta\chi_{H-H} = 0$; $\Delta\chi_{C-O} = 1.0$, C – O is more polar.
    (c) $\Delta\chi_{B-N} = 1.0$; $\Delta\chi_{C-F} = 1.5$, C – F is more polar.
    (d) $\Delta\chi_{H-N} = 0.9$; $\Delta\chi_{H-F} = 1.9$, H – F is more polar.
    (e) $\Delta\chi_{O-C} = 1.0$; $\Delta\chi_{O-N} = 0.5$, O – C is more polar.

34. (a) Li – F (b) K – O (c) Na – S (d) Li – N (e) K – Cl

35. (a) B – N is polar
    (b) Rb – F is ionic
    (c) S – S is non–polar
    (d) H – I is polar

36. (a) HF and (b) SO_2

37. (b) H_2S and (d) NH_3

38. 2.03 %

39. (a) 13.5 D (b) 77.1 % (c) $\overset{+0.77}{K} — \overset{-0.77}{Br}$

40. (a) 8.5 D (b) 15.2 % (c) $\overset{+0.15}{Br} — \overset{-0.15}{F}$

Chapter 10

1. (a) Tetrahedral; T – shaped
   (b) Trigonal planar; bent
   (c) Trigonal bipyramidal; linear
   (d) Octahedral; square planar

2. (a) Tetrahedral; trigonal pyramidal
   (b) Tetrahedral; bent
   (c) Tetrahedral; bent
   (d) Tetrahedral; tetrahedral

3. (a) CHF$_3$

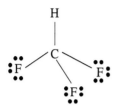

The electron geometry is tetrahedral; the molecular geometry is tetrahedral.

(b) XeO$_3$

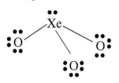

The electron geometry is tetrahedral; the molecular geometry is trigonal pyramidal.

(c) IF$_4^-$

The electron geometry is octahedral; the molecular geometry is square planar.

(d) S$_3^{2-}$

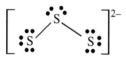

The electron geometry is tetrahedral; the molecular geometry is bent.

4. (a) PH$_4^+$

The electron geometry is tetrahedral; the molecular geometry is tetrahedral.

(b) Si(OH)$_4$

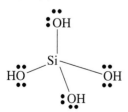

The electron geometry is tetrahedral; the molecular geometry is tetrahedral.

(c) SF_5^+

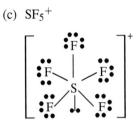

The electron geometry is octahedral; the molecular geometry is square pyramidal.

(d) OSF_4

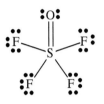

The electron geometry is trigonal bipyramidal; the molecular geometry is trigonal bipyramidal.

5. The Lewis structures for H_2NOH and H_3NOH^+ are

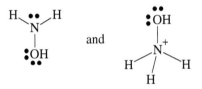

The molecular geometry for H_2NOH is trigonal pyramidal and the molecular geometry for H_3NOH^+ is tetrahedral.

6. The Lewis structures are

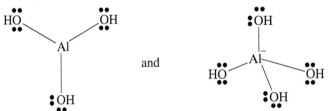

The molecular geometry for $Al(OH)_3$ is trigonal planar, and for its complex ion $Al(OH)_4^-$, tetrahedral.

7. The molecular geometry for PBr_3 is trigonal pyramidal. PBr_5 is trigonal bipyramidal.

8. (a) BF_3 trigonal planar, nonpolar
   (b) PBr_3. trigonal pyramid, polar
   (c) ClF_3 T-shaped, polar
   (d) BrF_5 square pyramid, polar

9. (a) CS_2 linear, nonpolar
   (b) NO_2 bent, polar
   (c) CH_3OH tetrahedral, polar
   (d) BHF_2 trigonal planar, polar

10. (a) BeF_2, linear, nonpolar
    (b) H_2S, bent, polar
    (c) $SeBr_2$, bent, polar
    (d) CHI_3, tetrahedral, polar

11. (a) $COBr_2$, trigonal planar, polar
    (b) HCOOH, trigonal planar, polar
    (c) SO_3, trigonal planar, nonpolar
    (d) CBr_4, tetrahedral, nonpolar

12. (a) CF_4, tetrahedral, nonpolar
    (b) SeF_4, see-saw, polar
    (c) KrF_4, square planar, nonpolar
    (d) SiF_4, tetrahedral, nonpolar

13. (a) BH_4^- has tetrahedral electron geometry and so uses sp^3 hybrid orbitals.
    (b) ICl_4^- has octahedral electron geometry and so uses $sp^3 d^2$ hybrid orbitals.
    (c) F_4ClO^- has octahedral electron geometry and so uses $sp^3 d^2$ hybrid orbitals.
    (d) $F_2BrO_2^+$ has tetrahedral electron geometry and so uses sp^3 hybrid orbitals.

14. (a) BeF_2 has linear electron geometry and uses sp hybrid orbitals.
    (b) NH_3 has tetrahedral electron geometry and so uses sp^3 hybrid orbitals.
    (c) SO_3 has trigonal planar electron geometry and so uses sp^2 hybrid orbitals.
    (d) SiH_4 has tetrahedral electron geometry and so uses sp^3 hybrid orbitals.

15. (a) NH_2^- has tetrahedral electron geometry and is sp^3 hybridized.
    (b) BrF_3 has trigonal bipyramidal geometry and so uses $sp^3 d$ hybrid orbitals.
    (c) SO_3^{2-} has tetrahedral electron geometry and is sp^3 hybridized.
    (d) IF_5 has octahedral electron geometry and so uses $sp^3 d^2$ hybrid orbitals.

16. The Al atom in $Al(OH)_3$ is sp^2 hybridized. Upon reaction to form $Al(OH)_4^-$ the hybridization of Al changes to sp^3.

17. The Br atom is $sp^3 d$ hybridized in BrF_3, sp^3 in BrF_2^+, and $sp^3 d^2$ in BrF_4^-.

18. C(2) is sp^3 hybridized; C(3) is sp^2 hybridized; C(5) is sp^2 hybridized; C(11) is sp^3 hybridized; O(1) is sp^3 hybridized; O(7) is sp^3 hybridized; O(10) is sp^2 hybridized.

19. (a) There are four π bonds and seventeen σ bonds.
    (b) The N atom is sp^3 hybridized; C(1), C(3), and C(7) are each sp^2 hybridized.

20. The ground state electronic configuration of Se is $[Ar]4s^2 3d^{10} 4p^4$. Selenium would promote one s and one p electron to the electron to the $4d$ orbitals, which upon hybridization gives six $sp^3 d^2$ orbitals in an octahedral arrangement (90°) available for bonding to six fluorine atoms, $[Ar] 3d^{10} 4s^1 4p^1 4p^1 4p^1 4d^1 4d^1$ Therefore, the hybridization state of Se in SeF_6 is $sp^3 d^2$.

21. MO theory predicts a molecule does not exist if the bond order (BO) is zero.
    (a) $H_2^+ : (\sigma_{1s})^1, BO = 1/2 \, (1-0) = 1/2$, stable
    (b) $H_2^- : (\sigma_{1s})^2 (\sigma_{1s}*)^1, BO = 1/2(2-1) = 1/2$, stable
    (c) $H_2^{2-} : (\sigma_{1s})^2 (\sigma_{1s}*)^2, BO = 1/2(2-2) = 0$, not stable
    (d) $He_2^{2+} : (\sigma_{1s})^2, BO = 1/2(2-0) = 1$, stable

22. MO theory predicts a molecule does not exist if the BO (BO) is zero.
    (a) $F_2^- : (\sigma_{2s})^2 (\sigma_{2s}*)^2 (\sigma_{2p})^2 (\pi_{2p})^4 (\pi_{2p}*)^4 (\sigma_{2p}*)^1, BO = 1/2(8-7) = 1/2$, stable
    (b) $O_2^{4-} : (\sigma_{2s})^2 (\sigma_{2s}*)^2 (\sigma_{2p})^2 (\pi_{2p})^4 (\pi_{2p}*)^4 (\sigma_{2p}*)^2, BO = 1/2(8-8) = 0$, unstable

(c) $N_2^{2-} : (\sigma_{2s})^2 (\sigma_{2s}{}^*)^2 (\pi_{2p})^4 (\sigma_{2p})^2 (\pi_{2p}{}^*)^2, BO = 1/2(8-4) = 2, stable$

(d) $C_2^{2+} : (\sigma_{2s})^2 (\sigma_{2s}{}^*)^2 (\pi_{2p})^2, BO = 1/2(4-2) = 1, stable$

23. Two, paramagnetic

24. $B_2^{2-} : (\sigma_{2s})^2 (\sigma_{2s}{}^*)^2 (\pi_{2p})^4, BO = 1/2(6-2) = 3$

25. $N_2^{2+} : (\sigma_{2s})^2 (\sigma_{2s}{}^*)^2 (\pi_{2p})^4, BO = 1/2(6-2) = 2, diamagnetic$

26. $CO^+ : (\sigma_{2s})^2 (\sigma_{2s}{}^*)^2 (\sigma_{2p})^2 (\pi_{2p})^3, BO = 1/2(7-2) = 2.5, paramagnetic$ (one unpaired electron)

27. (a) F_2^{2-} (has electron in π^* – orbital)
    (b) O_2^{2-} (has electron in π^* – orbital)
    (c) Does not
    (d) Does not

28. $NO : (\sigma_{2s})^2 (\sigma_{2s}{}^*)^2 (\sigma_{2p})^2 (\pi_{2p})^4 (\pi_{2p}{}^*)^1, BO = 1/2(8-3) = 2.5, paramagnetic$ (one unpaired electron)
    $NO^+ : (\sigma_{2s})^2 (\sigma_{2s}{}^*)^2 (\sigma_{2p})^2 (\pi_{2p})^4, BO = 1/2(8-2) = 3, diamagnetic$ (zero unpaired electrons)
    $NO^- : (\sigma_{2s})^2 (\sigma_{2s}{}^*)^2 (\sigma_{2p})^2 (\pi_{2p})^4 (\pi_{2p}{}^*)^2, BO = 1/2(8-4) = 2, paramagnetic$ (two unpaired electrons)

29. $CN^- : (\sigma_{2s})^2 (\sigma_{2s}{}^*)^2 (\sigma_{2p})^2 (\pi_{2p})^4, BO = 1/2(8-2) = 3, diamagnetic$ (zero unpaired electrons)
    $CO : (\sigma_{2s})^2 (\sigma_{2s}{}^*)^2 (\sigma_{2p})^2 (\pi_{2p})^4, BO = 1/2(8-2) = 3, diamagnetic$ (zero unpaired electrons)

30. Based on the energies of the AO, the $1s$ orbital of hydrogen interacts mostly with a $2p$ orbital of fluorine. The F $2s$ is nonbonding.

Chapter 11

1. 233.3 mL
2. 0.215 L
3. $3.33 \times 10^4 \ Nm^{-2}$
4. $279.8 \ cm^3$
5. $619 \ cm^3$
6. $-121.5°C$
7. 1.26 L
8. $-67.4°C$
9. $566.3°C$
10. 9.31 atm
11. $240.3 \ cm^3$
12. $9.10 \ cm^3$; $2 \ C_4H_{10} + 13 \ O_2 \rightarrow 8 \ CO_2 + 10 \ H_2O$
13. 13.0 L
14. 6.67 L ClF_3
15. 2.50 L

16. $0.0820 \ L - atm/mol - K$

17. $5.22 \ L$

18. $5.43 \ l$

19. $7.15 \ g/mol$

20. $44.82 \ g/mol$

21. $22.24 \ g/mol$

22. $2.99 \ g/L$

23. $3.60 \ g/L$

24. (a) $1500 \ L$ of O_2
    (b) $1000 \ L$ of SO_2

25. $24.37 \ \%$

26. C_3H_8

27. C_3H_6O (b) $C_3H_6O(g) + 4 \ O_2(g) \rightarrow 3 \ CO_2(g) + 3 \ H_2O(g)$

28. (a) $2 \ KClO_3(s) \rightarrow 2 \ KCl(s) + 3 \ O_2(g)$
    (b) $38.41 \ g \ KClO_3$

29. $57.5 \ L$

30. $P_{N_2} = 1483.5 \ mm \ Hg; P_{O_2} = 398.1 \ mm \ Hg; P_{Ar} = 17.7 \ mm \ Hg; P_{CO_2} = 0.7 \ mm \ Hg$

31. (a) $\chi_{He} = 0.250; \chi_{Ne} = 0.333; \chi_{Ar} = 0.417$
    (b) $P_{He} = 0.625 \ atm; \ P_{Ne} = 0.833 \ atm; P_{Ar} = 1.042 \ atm$

32. (a) $n_{O_2} = 24.3\%; \ n_{N_2} = 42.0\%; n_{CO_2} = 33.7\%$
    (b) Volume$\%N_2 = 40.0\%$

33. (a) N_2 is 1.25 times faster than CO_2
    (b) H_2 is 4 times faster than O_2
    (c) Ne is 1.41 times faster than Ar

34. O_2 is 2.82 times faster than He

35. (a) Rate of diffusion of $SO_2 = 12 \ cm^3/s$ and rate of diffusion of $Y = 27.43 \ cm^3/s$
    (b) Molecular mass of $Y = 28.0 \ g/mol$

36. (a) $15.3 \ atm$ (b) $14.18 \ atm$

37. (a) $66.6 \ atm$ (b) $70.0 \ atm$

38. (a) $P_{N_2} = 25.03 \ atm; P_{O_2} = 6.79 \ atm; P_{Ar} = 0.303 \ atm; \ P_{CO_2} = 0.012 \ atm$
    (b) $P_{Total} = 32.14 \ atm$

Chapter 12

1. $0.058 \ e$

2. $41 \ \%$

3. $15.3 \ D$

4. $41.2 \ kJ/mol$

5. $375 \ mmHg$

6. $331.5 \ K$

7. 31.3 kJ/mol

8. 1001.6 mmHg

9. (a) 35.1 kJ/mol (b) 23 mmHg (c) 353.2 K

10. (a) 29.7 kJ/mol (b) 4381 mmHg (c) 268.3 K

11. Four Al atoms

12. Two V atoms per unit cell

13. 11.37 g/cm^3

14. (a) 1 Po atom per unit cell
    (b) 3.72×10^{-23} cm^3
    (c) Density $= 9.33$ g/cm^3

15. (a) 4 Na$^+$ and 4 Cl$^-$
    (b) 5.65×10^{-8} cm or 5.65 Å

16. 197.3 g/mol

17. 6.05×10^{23} atoms/mol

18. 1.44 Å

19. 1.99 Å

20. Four Al atoms per unit cell

21. $\frac{r_{Zn^{2+}}}{r_{O^{2-}}} = 0.529$. ZnO should have a structure in which Zn^{2+} cations fit into the octahedral holes in a closest-packed array of O^{2-} anions. Since the stoichiometry is 1 : 1, all the holes should be occupied. ZnO should have FCC NaCl structure.

22. $\frac{r_{In^{3+}}}{r_{p^{3-}}} = 0.382$. The predicted structure for InP is zincblende.

23. $d = 3.50$ Å

24. $d = 0.61$ Å

25. For n $= 1, 2, 3$, the grazing angles are $16°, 33.4°$, and $55.6°$, respectively.

26. $d_{100} = 4.08$ Å, $d_{111} = 2.35$ Å, and $d_{311} = 1.23$ Å.

27. (a) 4.30 Å
    (b) $d_{002} = 2.15$ Å, $d_{211} = 1.76$ Å, and $d_{222} = 1.24$ Å.

28. (a) 3.57 Å
    (b) density $= 3.51$ g/cm^3

Chapter 13

1. 12 %

2. 30.625 g solution

3. 10 g NaCl

4. 1.10 g anhydrous HCl

5. (a) 1323.5 ppm (b) 1.323 ppt

6. 750 ppm

7. (a) 10 ppb
   (b) 0.01 ppm

8. 4.72×10^{-3} mol/L

9. (a) 0.165 M   (b) 0.225 M   (c) 1.00 M   (d) 0.25 M

10. 0.291 M

11. 11.50 cm^3

12. 1120 cm^3

13. 0.40 N

14. 130.72 cm^3

15. 65.32 g H_3PO_4

16. $\chi_{NaNO_3} = 0.014; \chi_{H_2O} = 0.986$

17. $\chi_{CH_3OH} = 0.110; \chi_{CH_3COCH_3} = 0.375; \chi_{CCl_4} = 0.515$

18. $\chi_{NaCl} = 0.110, \quad \chi_{H_2O} = 0.89$

19. 12.12 m

20. 2.28 g $FeSO_4$

21. 1.90 m

22. 4.89 M

23. 2.40 M; 2.55 m

24. (a) 0.0067 M   (b) 0.0067 m   (c) 1.19×10^{-4}

25. 0.52 M

26. 0.24 M

27. 1.33 M

28. Measure 0.965 L (965 mL) of stock concentrated HCl into a 5.00 L volumetric flask. Dilute with distilled water to the 5.0 L mark.

29. (a) 0.0094 M   (b) 1.00 g/L

30. (a) $Li^+ = 2.25$ M; $PO_4^{3-} = 0.75$ M
    (b) $Li_3PO_4 = 0.075$ M, $Li^+ = 0.225$ M; $PO_4^{3-} = 0.075$ M

31. (a) 106 g per 100 g H_2O
    (b) 6.67 moles/L

32. 23.4 g of solute

33. 92.1 g $AgNO_3$

34. 12.07 g of salt

35. 31.03 g of salt

36. (a) 121 g of KNO_3
    (b) 57.5 g KNO_3
    (c) 560 b $NaNO_3$
    (d) Dissolve 26.25 g of $NaNO_3$ in 25 g of water and mix thoroughly.
    (e) 60 g NH_3 per 100 g of H_2O
    (f) 40 g of NH_3 liberated.

37. (a1) 35°C; (a2) 10°C
    (b) 70°C
    (c) 51 g of KNO_3
    (d) 23.2 g of NH_3

38. 5.3×10^{-4} M
39. 5.3×10^{-6} M
40. (a) 1.3×10^{-5} M   (b) 0.144 M

Chapter 14

1. 0.195 M
2. 0.20 M
3. 7.0 g KOH
4. 0.176 g/L
5. (a) 0.165 M   (b) 6.609 g/L
6. (a) 0.0836 M   (b) atomic mass of M = 85.47   (c) M = Rb
7. 87.4 %
8. 3.72 %
9. (a) 63 %
   (b) 10
10. 133 g/mol
11. 11.52 %
12. 16.5 mg natural product/mL
13. 23.5 %
14. 81.3 %

Chapter 15

1. 23.7 mmHg
2. 23.6 mmHg
3. 264 g sucrose
4. 0.533 m
5. 22.03 mmHg
6. 23.4 mmHg
7. 55.5 mmHg
8. 0.085
9. 168.7°C
10. –0.170°C
11. 130 g/mol
12. 238 g/mol
13. 215.0°C
14. (a) 182 g/mol   (b) $C_3H_7O_3$, $C_6H_{14}O_6$
15. (a) –0.55°C   (b) –1.40°C   (c) –0.47°C

16. 443 g/mol

17. $(C_6H_{10}O_5)_{65}$ or $C_{390}H_{650}O_{325}$

18. 1.23 M

19. 3731 g/mol

20. 0.345 M

21. 29.35 mmHg

22. (a) 46.2 atm   (b) 6.90 atm   (c) 14 atm

Chapter 16

1. (a) Rate $= -\dfrac{1}{2}\dfrac{\Delta[H_2O_2]}{\Delta t} = \dfrac{1}{2}\dfrac{\Delta[H_2O]}{\Delta t} = \dfrac{\Delta[O_2]}{\Delta t}$

   (b) Rate $= -\dfrac{1}{2}\dfrac{\Delta[N_2O_5]}{\Delta t} = \dfrac{1}{4}\dfrac{\Delta[NO_2]}{\Delta t} = \dfrac{\Delta[O_2]}{\Delta t}$

   (c) Rate $= -\dfrac{\Delta[SO_2Cl_2]}{\Delta t} = \dfrac{\Delta[SO_2]}{\Delta t} = \dfrac{\Delta[Cl_2]}{\Delta t}$

   (d) Rate $= -\dfrac{1}{4}\dfrac{\Delta[PH_3]}{\Delta t} = \dfrac{\Delta[P_4]}{\Delta t} = \dfrac{1}{6}\dfrac{\Delta[H_2]}{\Delta t}$

2. (a) Rate $= -\dfrac{1}{2}\dfrac{\Delta[NH_3]}{\Delta t} = -\dfrac{1}{5}\dfrac{\Delta[O_2]}{\Delta t} = \dfrac{1}{4}\dfrac{\Delta[NO]}{\Delta t} = \dfrac{\Delta[H_2O]}{\Delta t}$

   (b) Rate $= -\dfrac{1}{2}\dfrac{\Delta[NO]}{\Delta t} = -\dfrac{1}{2}\dfrac{\Delta[H_2]}{\Delta t} = \dfrac{\Delta[N_2]}{\Delta t} = \dfrac{\Delta[H_2O]}{\Delta t}$

   (c) Rate $= -\dfrac{\Delta[CH_3CHO]}{\Delta t} = \dfrac{\Delta[CH_4]}{\Delta t} = \dfrac{\Delta[CO]}{\Delta t}$

   (d) Rate $= -\dfrac{1}{2}\dfrac{\Delta[SO_2]}{\Delta t} = -\dfrac{\Delta[O_2]}{\Delta t} = \dfrac{1}{2}\dfrac{\Delta[SO_3]}{\Delta t}$

3. 0.0125 M/min

4. 0.024 M/s

5. 0.04 M/hr

6. (a) Rate $= k[X]^1[Y]^2$

   (b) $288\,M^{-1}min^{-1}$

7. (a) Rate $= k[NH_4^+]^2[CNO]^1$

   (b) $510\,M^{-1}s^{-1}$

8. (a) Rate $= -\dfrac{k[I^-][OCl^-]}{[OH^-]}$

   (b) First−order

9. (a) Zero−order in $S_2O_3^{2-}$ and first−order in I_2

   (b) Rate $= k[I_2]$

   (c) $0.04\,s^{-1}$

10. (a) First-order in H_2O_2, first-order in I^-, and second-order in H^+

    (b) Rate $= k[H_2O_2][I^-][H^+]^2$

    (c) $2.13\,M^{-3}\,s^{-1}$

    (d) $8.3 \times 10^{-7}\,Ms^{-1}$

11. (a) 0.085 M   (b) 602 s

12. (a) $4.0 \times 10^{-7}\,Ms^{-1}$ (b) 0.243 M

13. (a) $2.67 \times 10^3\,s$ (b) 3.125×10^{-4} M

14. (a) $0.053\,h^{-1}$

    (b) 0.014 M

    (c) Fraction remining $= 0.28$

    (d) 30.2 h

15. $5.8 \times 10^{-4}\ s^{-1}$

16. (a) First$-$order (b) $9.0 \times 10^{-4}s^{-1}$

17. (a) $1.01 \times 10^{-4}M$ (b) $t = 327\ s$ (c) $t_{1/2} = 36\ s$

18. (a) Reaction is second$-$order
    (b) $k = 0.079\ M^{-1}min^{-1}$
    (c) $t_{1/2} = 84\ min$

19. $3.68\ s^{-1}$

20. $T = 731\ K$ or $458°C$

21. (a) $E_a = 51.7\ kJ/mol$
    (b) $k_{355\ K} = 6.61 \times 10^{-6}s^{-1}$

22. (a) $E_a = 223.9kJ/mol$
    (b) $k_{25°C} = 3.71 \times 10^{-22}s^{-1}$

23. $E_a = 90.4\ kJ/mol; A = 4.2 \times 10^{12}s^{-1}$

24. $E_a = 46.6\ kJ/mol; A = 2.9 \times 10^{7}s^{-1}$

Chapter 17

1. (a) $K_c = \dfrac{[SO_3]^2}{[SO_2]^2[O_2]}$

   (b) $K_c = \dfrac{[CO_2][H_2]}{[CO][H_2O]}$

   (c) $K_c = \dfrac{[CS_2][H_2]^4}{[CH_4][H_2S]^2}$

   (d) $K_c = \dfrac{[NO_2]^2}{[N_2O_4]}$

2. (a) $K_P = \dfrac{P_{SO_3}^2}{P_{SO_2}^2 P_{O_2}}$

   (b) $K_P = \dfrac{P_{CO_2} P_{H_2}}{P_{CO} P_{H_2O}}$

   (c) $K_P = \dfrac{P_{CS_2} P_{H_2}^4}{P_{CH_4} P_{H_2S}^2}$

   (d) $K_P = \dfrac{P_{NO_2}^2}{P_{N_2O_4}}$

3. (a) $K_c = \dfrac{[CH_3OH]}{[CO][H_2]^2}$

   (b) $K_c = \dfrac{[CH_3OH][Cl^-]}{[CH_3Cl][OH^-]}$

   (c) $K_c = \dfrac{[Ag(NH_3)_2^+]}{[Ag^+][NH_3]^2}$

   (d) $K_c = \dfrac{[CH_3COOC_2H_5]}{[CH_3COOH][C_2H_5OH]}$

4. (a) $K_P = \dfrac{P_{PCl_5}}{P_{PCl_3} P_{PCl_2}}$

   (b) $K_P = \dfrac{P_{POCl_3}^2}{P_{PCl_3}^2 P_{O_2}}$

   (c) $K_P = \dfrac{P_{NO_2}^2 P_{H_2O}^8}{P_{N_2H_4} P_{H_2O_2}^6}$

   (d) $K_P = \dfrac{P_{NO} P_{SO_3}}{P_{NO_2} P_{SO_2}}$

5. (a) $2\,HF \rightleftharpoons H_2 + F_2$
   (b) $2\,C_2H_2 + 5\,O_2 \rightleftharpoons 4\,CO_2 + 2H_2O$
   (c) $CO + H_2O \rightleftharpoons CO_2 + H_2$
   (d) $2\,Cl_2 + 2\,H_2O \rightleftharpoons 4\,HCl + O_2$

6. (a) $2\,NH_3 \rightleftharpoons N_2 + 3\,H_2$
   (b) $2\,CH_4 + 2\,H_2S \rightleftharpoons CS_2 + 4\,H_2$
   (c) $CH_2O \rightleftharpoons CO + H_2$
   (d) $N_2 + O_2 \rightleftharpoons 2\,NO$

7. (a) $K_P = \dfrac{P_{N_2} P_{H_2}^3}{P_{NH_3}^2}$

   (b) $K_P = \dfrac{P_{CS_2} P_{H_2}^4}{P_{CH_4} P_{H_2S}^2}$

   (c) $K_P = \dfrac{P_{H_2} P_{CO}}{P_{CH_2O}}$

   (d) $K_P = \dfrac{P_{NO}^2}{P_{N_2} P_{O_2}}$

8. (a) $K_P = \dfrac{P_{O_2} P_{HCl}^4}{P_{Cl_2}^2 P_{H_2O}^2}$, $K_P = 3.2 \times 10^{-14}$; $K_P \ll 1$, hence reactant is favored.

   (b) $K_P = \dfrac{P_{HI}^2}{P_{H_2} P_{I_2}}$, $K_P = 51$; $K_P \approx 1$, neither is favored.

9. (a) $K_c = 3.5 \times 10^6$; $K_c \gg 1$, hence SO_3 formation is favored.
   (b) $K_p = 6.33 \times 10^4$
   (c) $K_c = 2.9 \times 10^{-7}$

10. $K_c = 0.23$

11. $K_c = 1.17 \times 10^7$

12. Equilibrium composition : $[CH_3CO_2H] = 0.33$ mol; $[C_2H_2OH] = 0.33$ mol, $[CH_3CO_2C_2H_5] = 0.67$ mol; $[H_2O] = 0.67$ mol.

13. Equilibrium composition : $[H_2] = 2.11$ mol; $[I_2] = 0.11$ mol, $[HI] = 3.78$ mol

14. $[COCl_2] = 0.999$ mol; $[CO] = [Cl_2] = 1.48 \times 10^{-5}$ mol

15. (a) $K_p = P_{H_2O}^2$ (b) $K_p = P_{H_2O}^6$ (c) $K_p = P_{H_2O}$ (d) $K_p = P_{H_2O}^{10}$

16. Vapor pressure $P = 6.1 \times 10^{-8}$ atm and 0.025 atm.

17. $P_{NH_3} = P_{H_2S} = 0.26$ atm

18. (a) $K_c = 0.045$,
    (b) $P = 2.32$ atm,
    (c) $P_{NH_4HS} = 0.58$ atm, $P_{NH_3} = P_{H_2S} = 0.87$ atm,
    (d) $K_p = 1.5 \times 10^{-5}$.

19. (a) More products formed
    (b) More reactants formed
    (c) Equilibrium shift from left to right favoring product formation
    (d) No effect on position of equilibrium

20. (a) Less NO is formed
    (b) No effect
    (c) No effect
    (d) Less NO is formed

Chapter 18

1. (a) 5.0 (b) 1.3 (c) 10.7 (d) 6.0

2. (a) 9.0 (b) 12.7 (c) 3.3 (d) 8.0

3. (a) 1.58×10^{-12} M (b) 1.0×10^{-2} M (c) 1.25×10^{-8} M (d) 3.0×10^{-7} M
   (e) 0.10 M

4. (a) 6.33×10^{-3} M (b) 1.0×10^{-12} M (c) 8.0×10^{-7} M (d) 3.33×10^{-8} M
   (e) 1.0×10^{-13} M

5. (a) 1.0 (b) 1.7 (c) 0.7 (d) 1.3

6. (a) $[H^+] = 0.025\,M - 0.016\,M$
   (b) $[H^+] = 1.58 \times 10^{-5}\,M - 1.58 \times 10^{-9}\,M$
   (c) $[H^+] = 5.00 \times 10^{-7}\,M - 2.51 \times 10^{-7}\,M$
   (d) $[H^+] = 1.60 \times 10^{-2}\,M - 7.90 \times 10^{-3}\,M$
   (e) $[H^+] = 4.47 \times 10^{-8}\,M - 3.55 \times 10^{-8}\,M$
   (f) $[H^+] = 1.26 \times 10^{-10}\,M - 7.94 \times 10^{-11}\,M$
   (g) $[H^+] = 1.00 \times 10^{-4}\,M - 3.98 \times 10^{-5}\,M$

7. 8.30×10^{-10} M

8. (a) 18.4 M; (b) 1.56

9. At 100°C pH = 6.0 and at 25°C pH = 7.0
   Conclusion:
   (i) pH is high at 100°C (1 unit more) than at 25°C
   (ii) Water at 25°C has a neutral pH
   (iii) Water at elevated temperatures has more ions in it and so is more reactive.

10. pH = pOH = 6.51

11. (a) 0.1 (b) 2.4 (c) 13.4 (d) 11.5

12. $[C_6H_5CO_2H] = 0.019\,M;\ [C_6H_5CO_2^-] = [H_3O^+] = 1.12 \times 10^{-3}M;\ pH = 3.0$

13. $K_a = 1.71 \times 10^{-4};\ [HCO_2H] = 5.8 \times 10^{-4}$ M

14. 2.7

15. pH = 2.2; pOH = 11.8

16. pH = 12

17. pH = 10.8; pOH = 3.2

18. 4.1×10^{-6} M

19. 0.036 %

20. 0.025 M

21. 64.9 %

22. 8.16×10^{-5}

23. (a) 1.30×10^{-3} M (b) 87% (c) 11.9

24. 9.0 and 5.3×10^{-3}%

25. 3.2

26. 3.3×10^{-6}

27. 1.84×10^{-4} M

28. 4.7

29. 8.9 and $6.0 \times 10^{-3}\%$

30. $pH = 10.5, [CH_3NH_2] = 5.47 \times 10^{-2}$ M; $[CH_3NH_3^+] = 6.53 \times 10^{-2}$ M; $[OH^-] = 3.1 \times 10^{-4}$M; $[H^+] = 3.3 \times 10^{-11}$M.

31. 4.8

32. The ratio is $\frac{[CH_3CO_2H]}{[CH_3CO_2^-]} = 5.5 \times 10^{-3}$

33. The required ratio is $\frac{[H_2PO_4^-]}{[HPO_4^{2-}]} = 0.51$

   To prepare the phosphate buffer with a pH of 7.5, dissolve 0.51 mole of NaH_2PO_4 and 1.0 mole of Na_2HPO_4 in enough water to make up a 1.0 L solution in a volumetric flask.

34. (a) 2.9×10^{-6} (b) The pK_a is fairly close to the pH of interest. So it will be a good buffer.

35. 0.02 mole

36. 8.6

37. $[H_3O^+] = [HCO_3^-] = 3.16 \times 10^{-3}$M; $[CO_3^{2-}] = 5.6 \times 10^{-11}$M; $[CO_2] = 0.047$; pH = 2.5.

38. $[H^+] = [HS^-] = 1.41 \times 10^{-4}$M; $[S^{2-}] = 1.0 \times 10^{-19}$M; pH = 3.9

39. (a) $[H_3O^+] = [HTeO_3^-] = 0.019$ M (b) $[TeO_3^{2-}] = 1.0 \times 10^{-8}$ M

40. (a) Phenolphthalein, litmus, methyl orange
    (b) Phenolphthalein
    (c) Bromophenol blue, methyl orange, methyl red

41. (a) $pH = 8.7$; $pOH = 5.3$ (b) Phenolphthalein or Thymol Blue

42. 10.2

43. (a) 1.0 (b) 1.2 (c) 1.6 (d) 7.0 (e) 12.2

44. (a) 13.0 (b) 12.0

45. (a) 2.9 (b) 4.6 (c) 8.7 (d) 12.2

46. (a) 11.1 (b) 9.4 (c) 8.7 (d) 1.7

Chapter 19

1. (a) $K_{sp} = [Ag^+]^2 [CrO_4^{2-}]$
   (b) $K_{sp} = [Ca^{2+}][F^-]^2$
   (c) $K_{sp} = [Ca^{2+}][CO_3^{2-}]$
   (d) $K_{sp} = [Ca^{2+}]^3 [PO_4^{3-}]^2$
   (e) $K_{sp} = [Pb^{2+}]^3 [AsO_4^{3-}]^2$

2. (a) $K_{sp} = [Ag^+][I^-]$
   (b) $K_{sp} = [Ca^{2+}][C_2O_4^{2-}]$
   (c) $K_{sp} = [Hg_2^{2+}][Cl^-]^2$
   (d) $K_{sp} = [Pb^{2+}][I^-]^2$

    (e) $K_{sp} = \left[Fe^{3+}\right]\left[OH^-\right]^3$

3. 8.5×10^{-5}

4. 6.63×10^{-11}

5. 2.9×10^{-9} M

6.   (a) 2.5×10^{-6} M   (b) 7.69×10^{-4} g/L

7. 4.3×10^{-8} M

8. $\left[NO_3^-\right] = 0.0833$ M; $\left[K^+\right] = 0.1167$ M; $\left[Ag^+\right] = 8.48 \times 10^{-6}$ M, and $\left[CrO_4^{2-}\right] = $ 0.0167M.

9. 3.0×10^{-6} M

10. 4.4×10^{-12} M

11. Hg_2I_2 will precipitate first, since the $\left[I^-\right]$ required is less.

12.   AgI will precipitate first, since the $\left[Ag^+\right]$ required to cause its precipitation is less.

13.   0.006 M

14.   Since $Q > K_{sp}$, $PbCO_3$ will precipitate from the solution.

15. CaF_2 will precipitate first, since the $\left[F^-\right]$ required is less; $\left[F^-\right]$ needed to precipitate BaF_2 is 0.15 M, for CaF_2 is 2.0×10^{-5} M.

16.   1.109 mol

17.   0.3570 M

18.   0.0505 M

19. 6.2×10^{-7} M

20. 1.0×10^{-7} M

Chapter 20

1.   38.85 kJ

2. -8678 kJ/mol

3.   3.36 J/g.-C

4.   29.6°C

5.   23.3 kJ

6.   (a) Exothermic   (b) 690 kJ   (c) 85.2 g

7. $-26,000$ kJ

8.   (a) The reaction is exothermic
     (b) $2CH_4 + 2NH_3 + 3O_2 \rightarrow 2HCN + 6H_2O + 930$ kJ
     (c) -465 kJ/mol CH_4
     (d) -1395 kJ

9.   (a) Exothermic
     (b) Endothermic
     (c) Exothermic
     (d) Exothermic
     (e) Endothermic

10. −177.4 kJ

11. −227 kJ/mol

12. 9.65 kJ/mol

13. −2535.5 kJ

14. −89.3 kJ/mol

15. 34 kJ/mol

16. −565.6 kJ

17. −1819.1 kJ

18. −1153 kJ

19. −202.3 kJ

20. (a) −713 kJ   (b) Exothermic

21. 2406 kJ/mol

22. −3887.7 kJ/mol

23. (a) −543.1 kJ   (b) −537.2 kJ/mol

24. (a) 132.6 kJ/mol   (b) −2694 kJ/mol   (c) −489.2 kJ/mol

25. 0 kJ/mol

26. 557 kJ/mol

Chapter 21

1. −100 J

2. 172.7 kJ

3. −7.6 kJ

4. −1932 kJ

5. −87.2 kJ

6. −6.81 kJ

7. 176.6 J/K

8. −145.3 J/K or −72.7 J/K per mol of NO_2

9. (a) Increase
   (b) Increase
   (c) No change
   (d) Decrease
   (e) Increase
   (f) Decrease

10. (a) Negative
    (b) Positive
    (c) Positive
    (d) Positive
    (e) Positive
    (f) Negative

11. (a) $6.4\,J/K - mol$ (b) $1.2\,J/K - mol$

12. (a) $-186.7\,J/K.mol$ (b) $-306.5\,J/K - mol$ (c) $169.8\,J/K - mol$

13. $112\,J/K$

14. $129.4\,J/K$

15. $98.2\,J/K - mol$

16. (a) $-44.03\,J/K.mol$
    (b) The reaction is a spontaneous process since ΔG is negative
    (c) $\Delta G = -102.7\,kJ$. Yes, the reaction is spontaneous at this temperature.

17. $\Delta G = -190.6\,kJ$. Yes, the reaction is spontaneous at this temperature.

18. $\Delta G = 841.8\,kJ$. Therefore the engineer will not make NH_3 under these conditions.

19. (a) $T = 617.5\,K$. At all temperatures above 617.5 K, $T\Delta S$ will be larger than ΔG which would make ΔG negative.
    (b) The reaction will be spontaneous at all temperatures since ΔH°_{rxn} is negative and ΔS°_{rxn} is positive.

20. (a) $-394.9\,kJ/mol$ and $-393.5\,kJ/mol$
    (b) $15.4\,Jmol^{-1}K^{-1}$

21. $-53.9\,kJ/mol$; the reaction will occur as written.

22. 2.8×10^9

23. 1.96×10^{-34}

24. $-976\,kJ/mol$

25. $0.08\,atm$ or $60.8\,mmHg$

26. $31.9\,kJ/mol$

27. $8.8\,kJ/mol$

28. 0.51

Chapter 22

1. (a) $+4$ (b) $+8$ (c) $+7$ (d) $+3$ (e) $+3$ (f) -2

2. (a) $+6$ (b) $+2$ (c) $+5$ (d) $+5$ (e) $+4$ (f) $+4$

3. (a) Fe is oxidized, and O_2 is reduced
   (b) Cl_2 is reduced, and NaI is oxidized
   (c) Mg is oxidized and H_2SO_4 is reduced
   (d) H_2 is oxidized and CuO is reduced

4. (a) Al is oxidized and MnO_2 is reduced
   (b) PH_3 is oxidized and O_2 is reduced
   (c) Ca is oxidized and H_2O is reduced
   (d) SO_2 is oxidized and O_2 is reduced

5. (a) $2\,FeCl_2\,(aq) + Cl_2\,(l) \rightarrow 2\,FeCl_3\,(aq)$
       R.A.   O.A.
   (b) $Cl_2\,(g) + H_2\,S\,(g) \rightarrow 2HCl\,(g) + S\,(s)$
       O.A.   R.A.

(c) $C(s) + MgO(s) \rightarrow CO_2(g) + Mg(s)$
R.A.   O.A.

(d) $2 Zn(s) + O_2(g) \rightarrow 2 ZnO(s)$
R.A.   O.A.

6. (a) $3 Cu(s) + 8 HNO_3(aq) \rightarrow 3 Cu(NO_3)_2(aq) + 4 H_2O(l) + 2 NO(g)$
Oxidized Reduced
   R.A.   O.A.

(b) $Zn(s) + CuSO_4(aq) \rightarrow ZnSO_4(aq) + Cu(s)$
Ox   Red
R.A. O.A

(c) $2 H_2O(l) + 2 F_2(g) \rightarrow 4 HF(aq) + O_2$
Ox   Red
R.A. O.A

(d) $Fe_2O_3(s) + 3 CO(g) \rightarrow 2 Fe(s) + 3 CO_2(g)$
Red Ox
O.A R.A

7. (a) Acid-base reaction: No net change in oxidation number
   (b) Redox reaction : There is a net change in oxidation state of reactants & products
   (c) Precipitation reaction : NiS precipitated; No change in oxidation state
   (d) Acid – base reaction : No net change in oxidation state of reactants and products

8. (a) $2 IO_3^- + 12 H^+ + 10e^- \rightarrow I_2 + 6 H_2O$
   (b) $O_2 + 2 H^+ + 2e^- \rightarrow H_2O_2$
   (c) $2 S + 3 H_2O \rightarrow S_2O_3^{2-} + 6 H^+ + 6e^-$
   (d) $S_4O_6^{2-} + 2e^- \rightarrow 2 S_2O_3^{2-}$
   (e) $VO_2^+ + 2 H^+ + e^- \rightarrow VO^{2+} + H_2O$
   (f) $2 CO_2 + 2 H^+ + 2e^- \rightarrow H_2C_2O_4$

9. (a) $2 NO_3^- + 4 H^+ + 2e^- \rightarrow N_2O_4 + 2 H_2O$
   (b) $Sb_2O_3 + 6 H^+ + 6e^- \rightarrow 2 Sb + 3 H_2O$
   (c) $MnO_2 + 4 H^+ + 2e^- \rightarrow Mn^{2+} + 2 H_2O$
   (d) $2 HClO + 2H^+ + 2e^- \rightarrow Cl_2 + 2 H_2O$
   (e) $H_2O_2 + 2 H^+ + 2e^- \rightarrow 2 H_2O$
   (f) $NO_3^- + 3 H^+ + 2e^- \rightarrow HNO_2 + H_2O$

10. (a) $Cu_2O + H_2O + 2e^- \rightarrow 2 Cu + 2 OH^-$
    (b) $MnO_2 + 2 H_2O + 2e^- \rightarrow Mn(OH)_2 + 2 OH^-$
    (c) $PbO_2 + H_2O + 2e^- \rightarrow PbO + 2 OH^-$
    (d) $ClO^- + H_2O + 2e^- \rightarrow Cl^- + 2 OH^-$
    (e) $MnO_4^- + 2 H_2O + 3e^- \rightarrow MnO_2 + 4 OH^-$
    (f) $Al(OH)_3 + 3e^- \rightarrow Al + 3 OH^-$
    (g) $O_3 + H_2O + 2e^- \rightarrow O_2 + 2 OH^-$
    (h) $BrO_3^- + 3 H_2O + 6e^- \rightarrow Br^- + 6 OH^-$

11. (a) $Mg + 2 NO_3^- + 4 H^+ \rightarrow Mg^{2+} + 2 NO_2 + 2 H_2O$
    (b) $3 NO_2^- + CrO_7^{2-} + 8 H^+ \rightarrow 3 NO_3^- + 2 Cr^{3+} + 4 H_2O$
    (c) $3 S^{2-} + 8 H^+ + 3 NO_3^- \rightarrow 3 S + 2 NO + 4 H_2O$

12. (a) $5 H_2C_2O_4 + 2 MnO_4^- + 6H^+ \rightarrow 2 Mn^{2+} + 10 CO_2 + 8 H_2O$
    (b) $3 CH_3CH_2OH + 2 Cr_2O_7^{2-} + 16 H^+ \rightarrow 3 CH_3CO_2H + 4 Cr^{3+} + 11 H_2O$
    (c) $Ca + 2 VO^{2+} + 4H^+ \rightarrow Ca^{2+} + 2 V^{3+} + 2 H_2O$

13. (a) $3\,Zn + 2\,MnO_4^- + 4\,H_2O \rightarrow 3\,Zn(OH)_2 + 2\,MnO_2 + 2\,OH^-$

     (b) $2\,CrO_2^- + IO_3^- + 2\,OH^- \rightarrow 2\,CrO_4^{2-} + I^- + H_2O$

     (c) $2\,Bi(OH)_3 + 3\,Sn(OH)_3^- + 3\,OH^- \rightarrow 2\,Bi + 3\,Sn(OH)_6^{2-}$

     (d) $I_2 + 7\,Cl_2 + 10\,OH^- \rightarrow 3\,H_3IO_6^{2-} + 14\,Cl^- + 6\,H_2O$

14. (a) $14\,MnO_4^- + 20\,OH^- + C_3H_8O_3 \rightarrow 14\,MnO_4^{2-} + 3\,CO_3^{2-} + 14\,H_2O$

     (b) $2\,MnO_4^- + SO_3^{2-} + 2\,OH^- \rightarrow 2\,MnO_4^{2-} + 3\,CO_3^{2-} + 14\,H_2O$

     (c) $2\,MnO_4^- + H_2O + Br^- \rightarrow 2\,MnO_2 + BrO_3^- + 2\,OH^-$

15. (a) $2\,Cr(OH)_4^- + 3\,H_2O_2 + 2\,OH^- \rightarrow 2\,CrO_2^{2-} + 8\,H_2O$

     (b) $3\,H_2S + Na_2Cr_2O_7 + 8\,HCl \rightarrow 3\,S + 2\,CrCl_3 + 2\,NaCl + 7\,H_2O$

     (c) $3\,Sb_2S_5 + 10\,NO_3^- + 10\,H^+ \rightarrow 6\,HSbO_3 + 15\,S + 10\,NO + 2\,H_2O$

16. (a) $4\,Zn + NO_3^- + 10\,H^+ \rightarrow 4\,Zn^{2+} + NH_4^+ + 3\,H_2O$

     (b) $2\,MnO_4^- + 10\,Cl^- + 16\,H^+ \rightarrow 2\,Mn^{2+} + 5\,Cl_2 + 8\,H_2O$

     (c) $3\,PtCl_6^{2-} + Sb + 3\,H_2O \rightarrow 3\,PtCl_4^{2-} + 6\,Cl^- + Sb_2O_3 + 6\,H^+$

     (d) $3\,Hg_2Cl_2 + I^- + 6\,OH^- \rightarrow 6\,Hg + 6\,Cl^- + IO_3^- + 3\,H_2O$

     (e) $3\,K_2S_2O_8 + Cr_2(SO_4)_3 + 7\,H_2O \rightarrow 2\,K_2SO_4 + K_2Cr_2O_7 + 7\,H_2SO_4$

     (f) $2\,KMnO_4 + 5\,Zn + 8\,H_2SO_4 \rightarrow 2\,MnSO_4 + 5\,ZnSO_4 + 8\,H_2O + K_2SO_4$

17. (a) $MnO_4^- + 5\,Fe^{2+} + 8\,H^+ \rightarrow Mn^{2+} + 5\,Fe^{3+} + 4\,H_2O$

     (b) 0.05 M

18. 0.109 M

19. 0.388 M

20. 16.5 % Fe

21. 34.9 % Fe_2O_3

22. 50 %

Chapter 23

1. (a) $Sn(s)\,|\,Sn^{2+}\,(1\,M)\,||\,Cu^{2+}\,(1\,M)\,|\,Cu(s)$

     (b) $Zn(s)\,|\,Zn^{2+}\,(1\,M)\,||\,Ag^{2+}\,(1\,M)\,|\,Ag(s)$

     (c) $Mg(s)\,|\,Mg^{2+}\,(1\,M)\,||\,Fe^{2+}\,(aq)\,|\,Fe(s)$

     (d) $Sn(s)\,|\,Sn^{2+}\,(1\,M)\,||\,Pb^{2+}\,(1\,M)\,|\,Pb(s)$

2. (a) $Mn(s)\,|\,Mn^{2+}\,(1\,M)\,||\,Ti^{2+}\,(1\,M)\,|\,Ti(s)$

     (b) $Mg(s)\,|\,Mg^{2+}\,(1\,M)\,||\,Zn^{2+}\,(1\,M)\,|\,Zn(s)$

     (c) $Pt(s)\,|\,Sn^{2+}\,(1\,M),\,Sn^{4+}\,(1\,M)\,||\,Pb^{4+}\,(1\,M),\,Pb^{2+}\,(1\,M)\,|\,Pt(s)$

     (d) $Ca(s)\,|\,Ca^{2+}\,(1\,M)\,||\,Cu^{2+}\,(1\,M)\,|\,Cu(s)$

3. (a) $Sn(s)\,|\,Sn^{2+}\,(1\,M)\,||\,Cu^{2+}\,(1\,M)\,|\,Cu(s)$

     (b) $Pt(s)\,|\,Co^{2+}\,(1\,M),\,Co^{3+}\,(1\,M)\,||\,Ag^+\,(1\,M)\,|\,Ag(s)$

     (c) $Fe(s)\,|\,Fe^{2+}\,(1\,M)\,||\,Sn^{2+}\,(1\,M)\,|\,Sn(s)$

     (d) $Zn(s)\,|\,Zn^{2+}\,(1\,M)\,||\,2\,H^+\,(1\,M)\,|\,H_2\,(1\,atm)\,|\,Pt(s)$

4. (a) $Zn \rightarrow Zn^{2+} + 2e^-$

        $Cu^{2+} + 2e^- \rightarrow Cu$

     (b) Zn is oxidized

     (c) $Zn(s)\,|\,Zn^{2+}\,(1\,M)\,||\,Cu^{2+}\,(1\,M)\,|\,Cu(s)$

5. (a) $Pb(s) + Br_2(1) \rightarrow Pb^{2+} + 2\,Br^-$ (aq)
   (b) $Zn(s) + 2\,Eu^{3+}$ (aq) $\rightarrow Zn^{2+}$ (aq) $+ 2\,Eu^{2+}$ (aq)
   (c) $Co(s) + Cu^{2+}$ (aq) $\rightarrow Co^{2+}$ (aq) $+ Cu$ (s)
   (d) $Cu(s) + 2\,Cl^-$ (aq) $\rightarrow Cu^{2+}$ (aq) $+ Cl_2$ (g)

6. (a) $E^o_{Cell} = 0.62$ V (b) $E^o_{Cell} = 1.57$ V

7. $E^o_{Cell} = -0.13$ V

8. Yes. $E^o_{Cell} = 0.62$ V

9. $E^o_{U^{4+}/U} = -1.5$ V

10. $E^o_{Cell} = -0.187$ V. Therefore the reaction will not occur.

11. $E^o_{Cell} = 0.75$ V. Therefore in acidic solution Fe^{2+} will reduce MnO_4^-.

12. $E_{Cell} = -2.43$ V

13. $E_{Cell} = -0.793$ V

14. (a) $E_{Cell} = 0.67$ V (b) $E_{Cell} = 2.74$ V (c) $E_{Cell} = 0.81$ V

15. (a) $E_{Cell} = 2.26$ V (b) $E_{Cell} = 0.65$ V (c) $E_{Cell} = 3.10$ V

16. $\Delta G = -88.8$ kJ; $K = 3.92 \times 10^{15}$

17. $\Delta G = -57.9$ kJ; $K = 1.48 \times 10^{10}$

18. $E^o = 0.04$ V; $K = 22$

19. (a) $2\,Al(s) + 3\,Cl_2$ (g) $\rightarrow 2\,Al^{3+}$ (aq) $+ 6\,Cl^-$ (aq)
    (b) $E^o_{Cell} = 3.02$ V
    (c) $E^o_{Cell} = 3.04$ V

20. $E_{Cell} = 2.10$ V

21. (a) $E^o_{Cell} = 0.91$ V; Reaction is spontaneous in the direction written.
    (b) $E^o_{Cell} = -1.30$ V; Reaction is not spontaneous in the direction written.
    (c) $E^o_{Cell} = 2.54$ V; Reaction is spontaneous in the direction written.
    (d) $E^o_{Cell} = -0.53$ V; Reaction is not spontaneous in the direction written.

22. (a) 1 F (b) 2 F (c) 3 F (d) 3 F

23. (a) 2 F (b) 2 F (c) 2 F (d) 1 F

24. (a) 5.79×10^5 C (b) 5.79×10^5 C (c) 9.65×10^5 C

25. 1.61×10^{-19} C/e^-

26. 118.5 g Cu

27. 20.15 g Al

28. 112 g/mol

29. (a) 34.9 g (b) $31,177$ C (c) 17.3 hours

30. (a) 13.13 g (b) 2.24 dm^3

Chapter 24

1. (a) $^{238}_{92}U \rightarrow {}^{238}_{90}Th + {}^{4}_{2}He$
   (b) $^{227}_{89}Ac \rightarrow {}^{223}_{87}Fr + {}^{4}_{2}He$

(c) $^{234}_{90}\text{Th} \rightarrow {}^{230}_{88}\text{Ra} + {}^{4}_{2}\text{He}$

(d) $^{226}_{88}\text{Ra} \rightarrow {}^{230}_{88}\text{Ra} + {}^{4}_{2}\text{He}$

2. (a) $^{140}_{56}\text{Ba} \rightarrow {}^{140}_{57}\text{La} + {}^{0}_{-1}\beta$

   (b) $^{131}_{53}\text{I} \rightarrow {}^{131}_{54}\text{Xe} + {}^{0}_{-1}\beta$

   (c) $^{209}_{81}\text{Tl} \rightarrow {}^{209}_{82}\text{Pb} + {}^{0}_{-1}\beta$

   (d) $^{36}_{16}\text{S} \rightarrow {}^{36}_{17}\text{Cl} + {}^{0}_{-1}\beta$

3. (a) $^{213}_{83}\text{Bi} \rightarrow {}^{209}_{81}\text{Tl} + {}^{4}_{2}\text{He}$

   (b) $^{208}_{82}\text{Pb} + {}^{0}_{-1}e \rightarrow {}^{208}_{81}\text{Tl}$

   (c) $^{188}_{74}\text{W} \rightarrow {}^{188}_{75}\text{Re} + {}^{0}_{-1}\beta$

   (d) $^{40}_{19}\text{K} \rightarrow {}^{40}_{18}\text{Ar} + {}^{0}_{+1}e$

4. (a) $^{14}_{7}\text{N} + {}^{4}_{2}\text{He} \rightarrow {}^{17}_{8}\text{O} + {}^{1}_{1}\text{H}$

   (b) $^{55}_{25}\text{Mn} + {}^{2}_{1}\text{H} \rightarrow {}^{56}_{38}\text{Fe} + {}^{1}_{0}\text{n}$

   (c) $^{235}_{92}\text{U} + {}^{1}_{0}\text{n} \rightarrow {}^{94}_{38}\text{Sr} + {}^{139}_{54}\text{Xe} + 3{}^{1}_{0}\text{n}$

   (d) $^{27}_{13}\text{Al} + {}^{4}_{2}\text{He} \rightarrow {}^{30}_{15}\text{P} + {}^{0}_{1}\text{n}$

   (e) $^{11}_{5}\text{B} + {}^{4}_{2}\text{He} \rightarrow {}^{14}_{7}\text{N} + {}^{1}_{0}\text{n}$

5. (a) $^{18}_{8}\text{O} + {}^{1}_{1}\text{H} \rightarrow {}^{18}_{9}\text{F} + {}^{1}_{0}\text{n}$

   (a) $^{85}_{37}\text{Rb} + {}^{1}_{0}\text{n} \rightarrow {}^{85}_{36}\text{Kr} + {}^{1}_{1}\text{H}$

   (b) $^{14}_{7}\text{N} + {}^{4}_{2}\text{He} \rightarrow {}^{17}_{8}\text{O} + {}^{1}_{1}\text{H}$

   (c) $^{11}_{5}\text{B} + {}^{1}_{1}\text{H} \rightarrow {}^{11}_{6}\text{C} + {}^{1}_{0}\text{n}$

   (d) $^{23}_{12}\text{Mg} \rightarrow {}^{23}_{11}\text{Na} + {}^{0}_{+1}e$

6. (a) $^{14}_{7}\text{N} \left({}^{4}_{2}\alpha, {}^{1}_{1}\text{p}\right) {}^{17}_{8}\text{O}$

   (b) $^{68}_{30}\text{Zn} \left({}^{4}_{2}\alpha, {}^{1}_{0}\text{n}\right) {}^{71}_{31}\text{Ge}$

   (c) $^{31}_{15}\text{P} \left({}^{1}_{0}\text{n}, \gamma\right) {}^{32}_{15}\text{P}$

   (d) $^{191}_{77}\text{Ir} \left({}^{4}_{2}\alpha, {}^{1}_{0}\text{n}\right) {}^{194}_{79}\text{Au}$

7. (a) $^{18}_{8}\text{O} \left({}^{1}_{0}\text{n}, {}^{1}_{1}\text{p}\right) {}^{18}_{7}\text{N}; \quad {}^{18}_{8}\text{O} + {}^{1}_{0}\text{n} \rightarrow {}^{18}_{7}\text{N} + {}^{1}_{1}\text{P}$

   (b) $^{85}_{37}\text{Rb} \left({}^{1}_{1}\text{H}, {}^{0}_{1}\text{n}\right) {}^{85}_{36}\text{Kr}; \quad {}^{85}_{37}\text{Rb} + {}^{1}_{1}\text{H} \rightarrow {}^{85}_{36}\text{Kr} + {}^{1}_{0}\text{n}$

   (c) $^{58}_{26}\text{Fe} \left(2{}^{1}_{0}\text{n}, \beta\right) {}^{60}_{27}\text{Co}; \quad {}^{58}_{26}\text{Fe} + 2{}^{1}_{0}\text{n} \rightarrow {}^{60}_{27}\text{Co} + {}^{0}_{-1}\beta$

   (d) $^{243}_{94}\text{Po} \left({}^{4}_{2}\alpha, {}^{0}_{-1}\beta\right) {}^{243}_{97}\text{BK}; \quad {}^{243}_{94}\text{Po} + {}^{4}_{2}\alpha \rightarrow {}^{243}_{97}\text{BK} + {}^{0}_{-1}\beta$

8. (a) $^{1}_{2}\text{H} \left({}^{3}_{1}\text{H}, {}^{0}_{1}n\right) {}^{2}_{4}\text{He}$

   (b) $^{27}_{13}\text{Al} \left({}^{4}_{2}\alpha, {}^{1}_{1}\text{p}\right) {}^{30}_{14}\text{Si}$

   (c) $^{6}_{3}\text{Li} \left({}^{1}_{0}n, {}^{3}_{1}\text{H}\right) {}^{4}_{2}\text{He}$

(d) $^6_3Cu \left(^1_1H, ^2_1H\right) ^{62}_{29}Cu$

(e) $^{40}_{18}Ar \left(^4_2\alpha, ^1_1H\right) ^{43}_{19}K$

9. 0.03765 g

10. 0.0228 g

11. 0.0462/day $\left(\text{or } k = 0.0462 \text{ day}^{-1}\right)$ and $t_{1/2} = 15$ days

12. (a) 0.0462 h^{-1} (b) 75.8% (c) 5.744 mg

13. $t_{1/2} = 160$ min

14. 1.02×10^{-9} g/mol

15. (a) 0.53 g/mol (b) 1.9417 g/mol (c) 1.7726 g/mol (d) 0.0305 g/mol

16. 0.1868 amu

17. (a) 5.5 MeV (b) 19.3 MeV (c) 17.59 MeV (d) 17.34 MeV.

18. (a) $^{56}_{26}Fe + ^2_1H \rightarrow ^4_2\alpha + ^{54}_{25}Mn$

(b) 1.5×10^{-13} kJ/mol ^2_1H

19. $1 \text{ amu} = \dfrac{1 \text{ Kg}}{6.022 \times 10^{26}}$

$E = mc^2 = \left(\dfrac{1 \text{ kg}}{6.022 \times 10^{26}}\right) \left(3.00 \times 10^8 \text{ m/s}\right)^2$

$= 1.49 \times 10^{-10}$ J

But 1 MeV $= 1.6 \times 10^{-13}$ J

$E = \left(1.49 \times 10^{-10} \text{ J}\right) \left(\dfrac{1 \text{ MeV}}{1.6 \times 10^{-13} \text{ J}}\right)$

$E = 934.07$ MeV

20. (a) -1056.3 MeV

(b) $8.288 \text{ MeV}/\text{Nucleon}$

21. $7.59 \text{ MeV}/\text{Nucleon}$

22. 1832 years

Index